Fluid Mechanics

Fluid Mechanics

Edited by **Fay McGuire**

NY RESEARCH
P R E S S

New York

Published by NY Research Press,
23 West, 55th Street, Suite 816,
New York, NY 10019, USA
www.nyresearchpress.com

Fluid Mechanics
Edited by Fay McGuire

International Standard Book Number: 978-1-63238-495-9 (Hardback)

Contents

Preface

Application of the laws of force and motion to fluids is what fluid mechanics deals with. It has two main branches fluid statics and fluid dynamics. Fluid statics studies fluids at rest while fluid dynamics focuses on fluids in motion. It is applied to the areas such as industrial gas supply, designing of a hydro power plant, even water supply of a household. It is an active field of research, because of many problems that are fully or partially unresolved. This book is a compilation of chapters that discuss the most vital concepts and emerging trends in this area. It aims to shed light on some of the unexplored aspects of fluid mechanics and the recent researches in this field. It will help readers in keeping pace with the rapid changes in this discipline.

The researches compiled throughout the book are authentic and of high quality, combining several disciplines and from very diverse regions from around the world. Drawing on the contributions of many researchers from diverse countries, the book's objective is to provide the readers with the latest achievements in the area of research. This book will surely be a source of knowledge to all interested and researching the field.

In the end, I would like to express my deep sense of gratitude to all the authors for meeting the set deadlines in completing and submitting their research chapters. I would also like to thank the publisher for the support offered to us throughout the course of the book. Finally, I extend my sincere thanks to my family for being a constant source of inspiration and encouragement.

Editor

Effects of Heat and Mass Transfer on the Peristaltic Transport of MHD Couple Stress Fluid through Porous Medium in a Vertical Asymmetric Channel

K. Ramesh and M. Devakar

Department of Mathematics, Visvesvaraya National Institute of Technology, Nagpur 440010, India

Correspondence should be addressed to M. Devakar; m_devakar@yahoo.co.in

Academic Editor: Jamshid M. Nouri

The intrauterine fluid flow due to myometrial contractions is peristaltic type motion and the myometrial contractions may occur in both symmetric and asymmetric directions. The channel asymmetry is produced by choosing the peristaltic wave train on the walls to have different amplitude, and phase due to the variation of channel width, wave amplitudes and phase differences. In this paper, we study the effects of heat and mass transfer on the peristaltic transport of magnetohydrodynamic couple stress fluid through homogeneous porous medium in a vertical asymmetric channel. The flow is investigated in the wave frame of reference moving with constant velocity with the wave. The governing equations of couple stress fluid have been simplified under the long wave length approximation. The exact solutions of the resultant governing equations have been derived for the stream function, temperature, concentration, pressure gradient, and heat transfer coefficients. The pressure difference and frictional forces at both the walls are calculated using numerical integration. The influence of diverse flow parameters on the fluid velocity, pressure gradient, temperature, concentration, pressure difference, frictional forces, heat transfer coefficients, and trapping has been discussed. The graphical results are also discussed for four different wave shapes. It is noticed that increasing of couple stresses and heat generation parameter increases the size of the trapped bolus. The heat generation parameter increases the peristaltic pumping and temperature.

1. Introduction

In recent years, the flow of non-Newtonian fluids has received much attention due to the increasing industrial, medical, and technological applications. Various researchers have attempted diverse flow problems related to several non-Newtonian fluids and couple stress fluid is one of them. The theory of couple stress fluids originated by Stokes [1] has many biomedical, industrial, and scientific applications and was used to model synthetic fluids, polymer thickened oils, liquid crystals, animal blood, and synovial fluid. Some earlier developments in couple stress fluid theory with some basic flows can be found in the book by Stokes [2]. Recently, few researchers have studied some couple stress fluid flows for different flow geometries [3–8].

Nowadays, peristaltic flows have gained much attention because of their applications in physiology and industry.

Peristaltic transport is a form of fluid transport induced by a progressive wave of area contraction or expansion along the length of a distensible tube/channel and transporting the fluid in the direction of the wave propagation. This phenomenon is known as peristalsis. In physiology this plays an important role in various situations such as the food movement in the digestive tract, urine transport from kidney to bladder through ureter, movement of lymphatic fluids in lymphatic vessels, bile flow from the gall bladder into the duodenum, spermatozoa in the ductus efferentes of the male reproductive tract, ovum movement in the fallopian tube, blood circulation in the small blood vessels, the movement of the chyme in the gastrointestinal tract, intrauterine fluid motion, swallowing food bolus through esophagus, and transport of cilia. Many industrial and biological instruments such as roller pumps, finger pumps, heart-lung machines, blood pump machines, and dialysis machines are engineered based on the peristaltic mechanism [9]. The intrauterine fluid flow due to myometrial contractions is peristaltic in nature and these myometrial contractions occur in both symmetric and asymmetric directions and also when embryo enters the uterus for implantation there start the asymmetric contractions. The contractions inside the nonpregnant uterus are very complicated because they are composed of variable amplitudes and different wave lengths [10]. In view of this, Pandey and Chaube [11] have investigated the peristaltic transport of a couple stress fluid in a symmetrical channel using perturbation method in terms of small amplitude ratio. Ali and Hayat [12] have studied the peristaltic motion of micropolar fluid in an asymmetric channel. Naga Rani and Sarojamma [13] have analyzed the peristaltic transport of a Casson fluid in an asymmetric channel. Hayat et al. [14] have discussed the peristaltic flow of a Johnson-Segalman fluid in an asymmetric channel. Hayat and Javed [15] have studied the peristaltic transport of power-law fluid in asymmetric channel.

The porous medium plays an important role in the study of transport process in biofluid mechanics, industrial mechanics, and engineering fields. The fluid transport through porous medium is widely applicable in the vascular beds, lungs, kidneys, tumorous vessels, bile duct, gall bladder with stones, and small blood vessels. In the pathological situations, the distribution of fatty cholesterol, artery clogging, blood clots in the lumen of coronary artery, transport of drugs and nutrients to brain cells, and functions of organs are modeled as porous medium [16]. Recently, Tripathi [17] studied the peristaltic hemodynamic flow of couple stress fluids through a porous medium. Tripathi and Bég [18] have investigated the peristaltic flow of generalized Maxwell fluid through a porous medium using homotopy perturbation method. Abd elmaboud and Mekheimer [19] have discussed peristaltic transport of a second-order fluid through a porous medium using regular perturbation method. The magneto-hydrodynamic flows also gained much attention due to the widespread applications in biofluid mechanics and industry. It is the fact that many fluids like blood are conductive in nature and gave a new direction for research. The indispensable role of biomagnetic fluid dynamics in medical science has been very helpful with many problems of physiology. It has wide range of applications, such as magnetic wound or cancer tumour treatment, bleeding reduction during surgeries, provocation of occlusion of feeding vessels of cancer tumor, cell separation, transport of drugs, blood pump machines, and magnetic resonance imaging to diagnose the disease and the influence of magnetic field which may be utilized as a blood pump in carrying out cardiac operations for the blood flow in arteries with arterial disease like arterial stenosis or arteriosclerosis. Specifically, the magnetohydro-dynamic flows of non-Newtonian fluids are of great interest in magnetotherapy. The noninvasive radiological tests use the magnetic field to evaluate organs in abdomen [20]. Hayat et al. [21] have studied the peristaltic transport of magnetohydrodynamic Johnson-Segalman fluid for the case of a planar channel. Wang et al. [22] have investigated the peristaltic motion of a magnetohydrodynamic generalized second-order fluid in an asymmetric channel. Nadeem and

Akram [23, 24] have discussed the peristaltic transport of a couple stress fluid and Williamson fluid in an asymmetric channel with the effect of the magnetic field.

Heat transfer plays a significant role in the cooling processes of industrial and medical applications. Such consideration is very important since heat transfer in the human body is currently considered as an important area of research. In view of the thermotherapy and the human thermoregulation system, the model of bioheat transfer in tissues has been attracted by the biomedical engineers. In fact the heat transfer in human tissues involves complicated processes such as heat conduction in tissues, heat transfer due to perfusion of the arterial-venous blood through the pores of the tissue, metabolic heat generation, and external interactions such as electromagnetic radiation emitted from cell phones [25]. Heat transfer also involves many complicated processes such as evaluating skin burns, destruction of undesirable cancer tissues, dilution technique in examining blood flow, paper making, food processing, vasodilation, and radiation between surface and its environment [26]. Mustafa et al. [27] have studied the peristaltic transport of nanofluid in a channel. The heat transfer characteristics of a couple stress fluid in an asymmetric channel have been analyzed by Abd elmaboud et al. [28]. Nadeem and Akbar [29] have discussed the influence of heat transfer and magnetic field on peristaltic flow of a Johnson-Segalman fluid in a vertical symmetric channel. Some more works regarding peristaltic flows with the effect of heat transfer and magnetic field can be seen in [30–33]. Srinivas et al. [34] have studied the effects of both wall slip conditions and heat transfer on peristaltic flow of MHD Newtonian fluid in a porous channel with elastic wall properties. Mass transfer is another important phenomenon in physiology and industry. This phenomenon has great applications such as nutrients' diffusion out from the blood to neighboring tissues, membrane separation process, reverse osmosis, distillation process, combustion process, and diffusion of chemical impurities [35]. Recently, Noreen [36] studied the problem of mixed convection peristaltic flow of third-order nanofluid with an induced magnetic field. Saleem and Haider [37] have discussed the peristaltic transport of Maxwell fluid with heat and mass transfer in an asymmetric channel. Some more relevant works on the peristaltic transport with heat and mass transfer can be seen in [38–42].

The aim of the present study is to investigate the influence of heat and mass transfer on the peristaltic flow of magnetohydrodynamic couple stress fluid through homogeneous porous medium in a vertical asymmetric channel. This paper is arranged as follows. Section 2 presents the mathematical formulation for the problem. The solution of the problem is obtained in Section 3. The four different wave forms are presented in Section 4 while the computational results are discussed in Section 5. The last section, Section 6, presents the conclusions of the present study.

2. Formulation of the Problem

Let us consider magnetohydrodynamic couple stress fluid in a vertical asymmetric channel through the porous medium

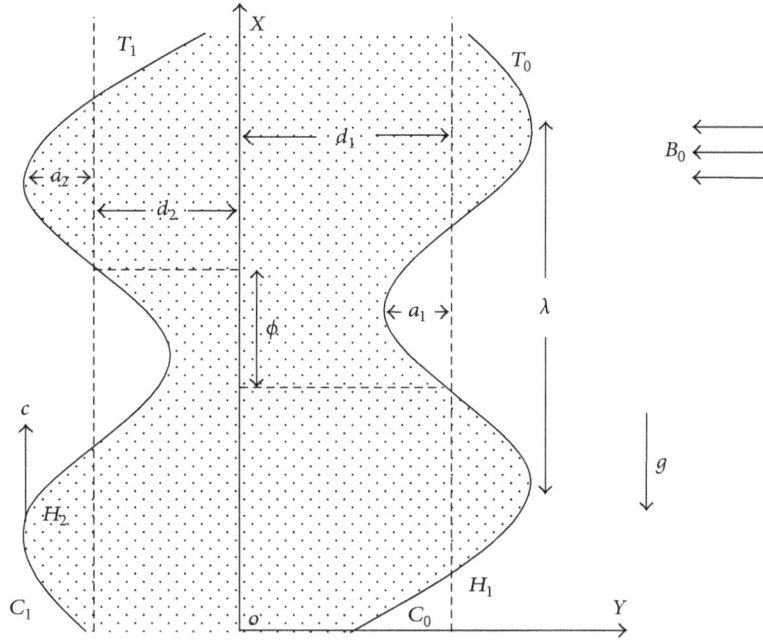

FIGURE 1: Physical model of the problem.

with the width of $d_1 + d_2$. The surfaces H_1 and H_2 of the asymmetric channel are maintained at constant temperatures T_0 and T_1 and the constant concentrations C_0 and C_1, respectively (see Figure 1). The porous medium is assumed to be homogeneous. The motion is induced by sinusoidal wave trains propagating with constant speed c along the channel walls as defined by the following:

$$H_1 = d_1 + a_1 \cos\left(\frac{2\pi}{\lambda}(X - ct)\right) \quad \text{(right side wall)},$$

$$H_2 = -d_2 - a_2 \cos\left(\frac{2\pi}{\lambda}(X - ct) + \phi\right) \quad \text{(left side wall)},$$

$$(1)$$

where a_1 and a_2 are the wave amplitudes, λ is the wave length, $d_1 + d_2$ is the channel width, c is the velocity of propagation, t is the time, and X is the direction of wave propagation. The phase difference ϕ varies in the range $0 \leq \phi \leq \pi$, in which $\phi = 0$ corresponds to symmetric channel with waves out of phase and $\phi = \pi$ corresponds to waves in phase, and further a_1, a_2, d_1, d_2, and ϕ meet the following relation $a_1^2 + a_2^2 + 2a_1 a_2 \cos\phi \leq (d_1 + d_2)^2$.

The continuity, momentum, energy, and concentration equations for an MHD incompressible couple stress fluid, in the absence of body couples, are [8, 16]

$$\nabla \cdot \overline{q} = 0,$$

$$\rho \frac{D\overline{q}}{Dt} = -\nabla p - \mu \nabla^2 \overline{q} - \eta \nabla^4 \overline{q} + J \times B + R$$

$$+ \rho g \beta_T (T - T_0) + \rho g \beta_C (C - C_0),$$

$$\rho c_p \frac{DT}{Dt} = k^* \nabla^2 T + Q_0,$$

$$\frac{DC}{Dt} = D\nabla^2 C + \frac{DK_T}{T_m} \nabla^2 T,$$

$$(2)$$

in which D/Dt represents the material derivative and R is Darcy's resistance in the porous medium which are given by

$$\frac{D}{Dt} = \frac{\partial}{\partial t} + u\frac{\partial}{\partial x} + v\frac{\partial}{\partial y}, \qquad R = -\frac{\mu}{k_0}\overline{q},$$

$$(3)$$

where $\overline{q}$ is the velocity vector, ρ is the density, p is the pressure, μ is the viscosity, η is material constant associated with couple stress, J is the electric current density, B is the total magnetic field, g is the acceleration due to the gravity, β_T is the coefficient of thermal expansion, β_C is the coefficient of expansion with concentration, c_p is the specific heat at constant pressure, T is the temperature, C is the mass concentration, k^* is the thermal conductivity, Q_0 is the constant heat generation parameter, D is the coefficient of mass diffusivity, K_T is the thermal diffusion ratio, T_m is the mean temperature, and k_0 is the permeability parameter. The viscous dissipation is neglected in the energy equation.

In the fixed frame, governing equations for the peristaltic motion of an incompressible magnetohydrodynamic couple stress fluid through homogeneous porous medium in the two-dimensional vertical channel are

$$\frac{\partial U}{\partial X} + \frac{\partial V}{\partial Y} = 0,$$

$$\rho\left(\frac{\partial U}{\partial t} + U\frac{\partial U}{\partial X} + V\frac{\partial U}{\partial Y}\right)$$

$$= -\frac{\partial P}{\partial X} + \mu\left(\frac{\partial^2 U}{\partial X^2} + \frac{\partial^2 U}{\partial Y^2}\right)$$

$$-\eta\left(\frac{\partial^4 U}{\partial X^4} + 2\frac{\partial^4 U}{\partial X^2 \partial Y^2} + \frac{\partial^4 U}{\partial Y^4}\right)$$

$$-\sigma B_0^2 U - \frac{\mu}{k_0}U + \rho g\beta_T(T - T_0) + \rho g\beta_C(C - C_0),$$

$$\rho\left(\frac{\partial V}{\partial t} + U\frac{\partial V}{\partial X} + V\frac{\partial V}{\partial Y}\right)$$

$$= -\frac{\partial P}{\partial Y} + \mu\left(\frac{\partial^2 V}{\partial X^2} + \frac{\partial^2 V}{\partial Y^2}\right)$$

$$-\eta\left(\frac{\partial^4 V}{\partial X^4} + 2\frac{\partial^4 V}{\partial X^2 \partial Y^2} + \frac{\partial^4 V}{\partial Y^4}\right) - \frac{\mu}{k_0}V,$$

$$\rho c_p\left(\frac{\partial T}{\partial t} + U\frac{\partial T}{\partial X} + V\frac{\partial T}{\partial Y}\right) = k^*\left(\frac{\partial^2 T}{\partial X^2} + \frac{\partial^2 T}{\partial Y^2}\right) + Q_0,$$

$$\left(\frac{\partial C}{\partial t} + U\frac{\partial C}{\partial X} + V\frac{\partial C}{\partial Y}\right)$$

$$= D\left(\frac{\partial^2 C}{\partial X^2} + \frac{\partial^2 C}{\partial Y^2}\right) + \frac{DK_T}{T_m}\left(\frac{\partial^2 T}{\partial X^2} + \frac{\partial^2 T}{\partial Y^2}\right), \tag{4}$$

in which U and V are the respective velocity components, P is the pressure, T is the temperature and C is the concentration in the reference to fixed frame system, σ is the electrical conductivity of the fluid, and B_0 is the applied magnetic field. In the above, the induced magnetic field is neglected since the magnetic Reynolds number is assumed to be small.

The coordinates, velocities, pressure, temperature, and concentration in the fixed frame (X, Y) and wave frame (x, y) are related by the following expressions:

$$x = X - ct, \qquad y = Y, \qquad u = U - c, \qquad v = V,$$

$$p(x, y) = P(X, Y, t), \qquad \overline{T}(x, y) = T(X, Y, t), \tag{5}$$

$$\overline{C}(x, y) = C(X, Y, t),$$

in which u, v, p, $\overline{T}$, and $\overline{C}$ are velocity components, pressure, temperature, and concentration in the wave frame, respectively.

Using (5), the governing equations in the wave frame are given as follows:

$$\frac{\partial u}{\partial x} + \frac{\partial v}{\partial y} = 0,$$

$$\rho\left(u\frac{\partial u}{\partial x} + v\frac{\partial u}{\partial y}\right)$$

$$= -\frac{\partial p}{\partial x} + \mu\left(\frac{\partial^2 u}{\partial x^2} + \frac{\partial^2 u}{\partial y^2}\right)$$

$$-\eta\left(\frac{\partial^4 u}{\partial x^4} + 2\frac{\partial^4 u}{\partial x^2 \partial y^2} + \frac{\partial^4 u}{\partial y^4}\right)$$

$$-\sigma B_0^2(u + c) - \frac{\mu}{k_0}(u + c) + \rho g\beta_T(\overline{T} - \overline{T}_0)$$

$$+ \rho g\beta_C(\overline{C} - \overline{C}_0),$$

$$\rho\left(u\frac{\partial v}{\partial x} + v\frac{\partial v}{\partial y}\right)$$

$$= -\frac{\partial p}{\partial y} + \mu\left(\frac{\partial^2 v}{\partial x^2} + \frac{\partial^2 v}{\partial y^2}\right)$$

$$-\eta\left(\frac{\partial^4 v}{\partial x^4} + 2\frac{\partial^4 v}{\partial x^2 \partial y^2} + \frac{\partial^4 v}{\partial y^4}\right) - \frac{\mu}{k_0}v,$$

$$\rho c_p\left(u\frac{\partial \overline{T}}{\partial x} + v\frac{\partial \overline{T}}{\partial y}\right) = k^*\left(\frac{\partial^2 \overline{T}}{\partial x^2} + \frac{\partial^2 \overline{T}}{\partial y^2}\right) + Q_0,$$

$$\left(u\frac{\partial \overline{C}}{\partial x} + v\frac{\partial \overline{C}}{\partial y}\right)$$

$$= D\left(\frac{\partial^2 \overline{C}}{\partial x^2} + \frac{\partial^2 \overline{C}}{\partial y^2}\right) + \frac{DK_T}{T_m}\left(\frac{\partial^2 \overline{T}}{\partial x^2} + \frac{\partial^2 \overline{T}}{\partial y^2}\right). \tag{6}$$

We introduce the following dimensionless parameters:

$$\overline{x} = \frac{x}{\lambda}, \qquad \overline{y} = \frac{y}{d_1}, \qquad \overline{u} = \frac{u}{c}, \qquad \overline{v} = \frac{v}{c},$$

$$h_1 = \frac{H_1}{a_1}, \qquad h_2 = \frac{H_2}{d_1}, \qquad \overline{t} = \frac{ct}{\lambda},$$

$$\overline{p} = \frac{d_1^2}{\lambda\mu c}p,$$

$$\delta = \frac{d_1}{\lambda}, \qquad d = \frac{d_2}{d_1}, \qquad a = \frac{a_1}{d_1}, \qquad b = \frac{a_2}{d_1},$$

$$\text{Re} = \frac{\rho c d_1}{\mu}, \qquad M = \sqrt{\frac{\sigma}{\mu}}B_0 d_1, \qquad D_a = \frac{k_0}{d_1^2},$$

$$\gamma = \sqrt{\frac{\mu}{\eta}}d_1, \qquad \text{Gr} = \frac{\rho g d_1^2 \beta_T(\overline{T}_1 - \overline{T}_0)}{\mu c},$$

$$\text{Gc} = \frac{\rho g d_1^2 \beta_C(\overline{C}_1 - \overline{C}_0)}{\mu c}, \qquad \overline{\psi} = \frac{\psi}{c d_1}, \qquad \text{Pr} = \frac{\mu c_p}{k^*},$$

$$\theta = \frac{\overline{T} - \overline{T}_0}{\overline{T}_1 - \overline{T}_0}, \qquad \Phi = \frac{\overline{C} - \overline{C}_0}{\overline{C}_1 - \overline{C}_0}, \qquad \beta = \frac{Q_0 d_1^2}{k^*(\overline{T}_1 - \overline{T}_0)},$$

$$\text{Sc} = \frac{\mu}{\rho D}, \qquad \text{Sr} = \frac{\rho DK_T(\overline{T}_1 - \overline{T}_0)}{\mu T_m(\overline{C}_1 - \overline{C}_0)}, \tag{7}$$

where δ is the dimensionless wave number, Re is the Reynolds number, M is the Hartmann number, D_a is the Darcy number, γ is the couple stress parameter, Gr is the local temperature Grashof number, Gc is the local concentration

Grashof number, Pr is the Prandtl number, θ is the dimensionless temperature, Φ is the dimensionless concentration, β is the heat generation parameter, Sc is the Schmidt number, and Sr is the Soret number.

In terms of these nondimensional variables, the governing equations (6), after dropping the bars, become

$$\delta \frac{\partial u}{\partial x} + \frac{\partial v}{\partial y} = 0, \tag{8}$$

$$\operatorname{Re} \delta \left(u \frac{\partial u}{\partial x} + \frac{1}{\delta} v \frac{\partial u}{\partial y} \right)$$

$$= -\frac{\partial p}{\partial x} + \left(\delta^2 \frac{\partial^2 u}{\partial x^2} + \frac{\partial^2 u}{\partial y^2} \right)$$

$$- \frac{1}{\gamma^2} \left(\delta^4 \frac{\partial^4 u}{\partial x^4} + 2\delta^2 \frac{\partial^4 u}{\partial x^2 \partial y^2} + \frac{\partial^4 u}{\partial y^4} \right)$$

$$- \left(M^2 + \frac{1}{D_a} \right)(u+1) + \operatorname{Gr}\theta + \operatorname{Gc}\Phi, \tag{9}$$

$$\operatorname{Re} \delta^2 \left(u \frac{\partial v}{\partial x} + \frac{1}{\delta} v \frac{\partial v}{\partial y} \right)$$

$$= -\frac{\partial p}{\partial y} + \left(\delta^3 \frac{\partial^2 v}{\partial x^2} + \delta \frac{\partial^2 v}{\partial y^2} \right)$$

$$- \frac{1}{\gamma^2} \left(\delta^5 \frac{\partial^4 v}{\partial x^4} + 2\delta^3 \frac{\partial^4 v}{\partial x^2 \partial y^2} + \delta \frac{\partial^4 v}{\partial y^4} \right) - \frac{1}{D_a} \delta v, \tag{10}$$

$$\operatorname{RePr}\delta \left(u \frac{\partial \theta}{\partial x} + v \frac{\partial \theta}{\partial y} \right) = \left(\delta^2 \frac{\partial^2 \theta}{\partial x^2} + \frac{\partial^2 \theta}{\partial y^2} \right) + \beta, \tag{11}$$

$$\operatorname{Re} \delta \left(u \frac{\partial \Phi}{\partial x} + v \frac{\partial \Phi}{\partial y} \right)$$

$$= \frac{1}{\operatorname{Sc}} \left(\frac{\partial^2 \Phi}{\partial x^2} + \frac{\partial^2 \Phi}{\partial y^2} \right) + \operatorname{Sr} \left(\delta^2 \frac{\partial^2 \theta}{\partial x^2} + \frac{\partial^2 \theta}{\partial y^2} \right). \tag{12}$$

The dimensionless velocity components (u, v) in terms of stream function ψ are related by the following relations:

$$u = \frac{\partial \psi}{\partial y}, \qquad v = -\delta \frac{\partial \psi}{\partial x}. \tag{13}$$

Using (13), the governing equations (9)–(12) reduced to

$$\operatorname{Re} \delta \left[\left(\frac{\partial \psi}{\partial y} \frac{\partial}{\partial x} - \frac{\partial \psi}{\partial x} \frac{\partial}{\partial y} \right) \frac{\partial \psi}{\partial y} \right]$$

$$= -\frac{\partial p}{\partial x} + \left(\delta^2 \frac{\partial^3 \psi}{\partial x^2 \partial y} + \frac{\partial^3 \psi}{\partial y^3} \right)$$

$$- \frac{1}{\gamma^2} \left(\delta^4 \frac{\partial^5 \psi}{\partial x^4 \partial y} + 2\delta^2 \frac{\partial^5 \psi}{\partial x^2 \partial y^3} + \frac{\partial^5 \psi}{\partial y^5} \right)$$

$$- \left(M^2 + \frac{1}{D_a} \right) \left(\frac{\partial \psi}{\partial y} + 1 \right) + \operatorname{Gr}\theta + \operatorname{Gc}\Phi,$$

$$- \operatorname{Re} \delta^3 \left[\left(\frac{\partial \psi}{\partial y} \frac{\partial}{\partial x} - \frac{\partial \psi}{\partial x} \frac{\partial}{\partial y} \right) \frac{\partial \psi}{\partial x} \right]$$

$$= -\frac{\partial p}{\partial y} - \left(\delta^4 \frac{\partial^3 \psi}{\partial x^3} + \delta^2 \frac{\partial^3 \psi}{\partial x \partial y^2} \right)$$

$$+ \frac{1}{\gamma^2} \left(\delta^6 \frac{\partial^5 \psi}{\partial x^5} + 2\delta^4 \frac{\partial^5 \psi}{\partial x^3 \partial y^2} + \delta^2 \frac{\partial^5 \psi}{\partial x \partial y^4} \right)$$

$$+ \frac{1}{D_a} \delta^2 \frac{\partial \psi}{\partial x},$$

$$\operatorname{RePr}\delta \left(\frac{\partial \psi}{\partial y} \frac{\partial \theta}{\partial x} - \frac{\partial \psi}{\partial x} \frac{\partial \theta}{\partial y} \right) = \left(\delta^2 \frac{\partial^2 \theta}{\partial x^2} + \frac{\partial^2 \theta}{\partial y^2} \right) + \beta,$$

$$\operatorname{Re} \delta \left(\frac{\partial \psi}{\partial y} \frac{\partial \Phi}{\partial x} - \frac{\partial \psi}{\partial x} \frac{\partial \Phi}{\partial y} \right)$$

$$= \frac{1}{\operatorname{Sc}} \left(\delta^2 \frac{\partial^2 \Phi}{\partial x^2} + \frac{\partial^2 \Phi}{\partial y^2} \right) + \operatorname{Sr} \left(\delta^2 \frac{\partial^2 \theta}{\partial x^2} + \frac{\partial^2 \theta}{\partial y^2} \right). \tag{14}$$

The dimensionless boundary conditions can be put in the forms

$$\psi = \frac{F}{2}, \qquad \frac{\partial \psi}{\partial y} = -1, \qquad \frac{\partial^3 \psi}{\partial y^3} = 0,$$

$$\theta = 0, \qquad \Phi = 0$$

$$\text{at } y = h_1 = 1 + a \cos(2\pi x),$$

$$\psi = -\frac{F}{2}, \qquad \frac{\partial \psi}{\partial y} = -1, \qquad \frac{\partial^3 \psi}{\partial y^3} = 0,$$

$$\theta = 1, \qquad \Phi = 1$$

$$\text{at } y = h_2 = -d - b \cos(2\pi x + \phi). \tag{15}$$

The dimensionless mean flow rate Θ in fixed frame is related to the nondimensional mean flow rate F in wave frame by

$$\Theta = F + 1 + d, \tag{16}$$

in which

$$F = \int_{h_2}^{h_1} \frac{\partial \psi}{\partial y} dy = \psi(h_1(x)) - \psi(h_2(x)). \tag{17}$$

We note that h_1 and h_2 represent the dimensionless forms of the peristaltic walls

$$h_1(x) = 1 + a \cos(2\pi x),$$

$$h_2(x) = -d - b \cos(2\pi x + \phi), \tag{18}$$

where a, b, d, and ϕ satisfy the relation $a^2 + b^2 + 2ab \cos\phi \leq (1+d)^2$.

3. Solution of the Problem

Assuming that the wave length of the peristaltic wave is very large compared to the width of the channel, the wave number

δ becomes very small. This assumption is known as long wave length approximation. Since δ is very small, all the higher powers of δ are also very small. Therefore, neglecting terms containing δ and its higher powers from (14), we get

$$\frac{\partial p}{\partial x} = \frac{\partial^3 \psi}{\partial y^3} - \frac{1}{\gamma^2} \frac{\partial^5 \psi}{\partial y^5}$$
$$- \left(M^2 + \frac{1}{D_a} \right) \left(\frac{\partial \psi}{\partial y} + 1 \right) + Gr\theta + Gc\Phi, \tag{19}$$

$$\frac{\partial p}{\partial y} = 0, \tag{20}$$

$$\frac{\partial^2 \theta}{\partial y^2} + \beta = 0, \tag{21}$$

$$\frac{1}{Sc} \frac{\partial^2 \Phi}{\partial y^2} + Sr \frac{\partial^2 \theta}{\partial y^2} = 0. \tag{22}$$

Elimination of pressure form from (19) and (20) yields

$$\frac{\partial^4 \psi}{\partial y^4} - \frac{1}{\gamma^2} \frac{\partial^6 \psi}{\partial y^6} - \left(M^2 + \frac{1}{D_a} \right) \frac{\partial^2 \psi}{\partial y^2} + Gr \frac{\partial \theta}{\partial y} + Gc \frac{\partial \Phi}{\partial y} = 0. \tag{23}$$

Solving (21) and (22) with the boundary conditions (15), the temperature and concentration are obtained as

$$\theta = -\beta \frac{y^2}{2} + A_1 y + A_2, \tag{24}$$

$$\Phi = SrSc\beta \frac{y^2}{2} + B_1 y + B_2,$$

where

$$A_1 = \frac{\beta \left(h_1^2 - h_2^2 \right) - 2}{2 \left(h_1 - h_2 \right)}; \qquad A_2 = \frac{\beta h_1^2 - 2 A_1 h_1}{2};$$

$$B_1 = -\frac{ScSr\beta \left(h_1^2 - h_2^2 \right) + 2}{2 \left(h_1 - h_2 \right)}; \qquad B_2 = -\frac{ScSr\beta h_1^2 + 2 B_1 h_1}{2}. \tag{25}$$

Inserting (24) in (23), with the help of boundary conditions (15), we obtain the stream function as

$$\psi = D_1 y^3 + D_2 y^2 + C_3 y + C_4 + C_5 \cosh (n_1 y)$$
$$+ C_6 \sinh (n_1 y) + C_7 \cosh (n_2 y) + C_8 \sinh (n_2 y). \tag{26}$$

in which

$$m_1 = \sqrt{M^2 + \frac{1}{D_a}}; \qquad n_1 = \sqrt{\frac{\gamma^2 + \gamma \sqrt{\gamma^2 - 4 m_1^2}}{2}};$$

$$n_2 = \sqrt{\frac{\gamma^2 - \gamma \sqrt{\gamma^2 - 4 m_1^2}}{2}};$$

$$C_1 = \beta \gamma^2 (GcScSr - Gr); \qquad C_2 = \gamma^2 (GrA_1 + GcB_1);$$

$$C_3 = D_{19} + C_8 D_{20};$$

$$C_4 = \frac{F}{2} - D_1 h_1^3 - D_2 h_1^2 - C_3 h_1 - C_5 \cosh (n_1 h_1)$$
$$- C_6 \sinh (n_1 h_1) - C_7 \cosh (n_2 h_1) - C_8 \sinh (n_2 h_1);$$

$$C_5 = D_{13} + C_8 D_{14}; \qquad C_6 = D_{15} + C_8 D_{16};$$

$$C_7 = D_{17} + C_8 D_{18}; \qquad C_8 = \frac{D_{21}}{D_{22}};$$

$$D_1 = \frac{C_1}{6 m_1^2 \gamma^2}; \qquad D_2 = \frac{C_2}{2 m_1^2 \gamma^2};$$

$$D_3 = \sinh (n_1 h_1) - \sinh (n_1 h_2);$$

$$D_4 = \cosh (n_1 h_1) - \cosh (n_1 h_2);$$

$$D_5 = \sinh (n_2 h_1) - \sinh (n_2 h_2);$$

$$D_6 = \cosh (n_2 h_1) - \cosh (n_2 h_2);$$

$$D_7 = n_1^3 \left(\sinh (n_1 h_1) - \frac{D_3 \sinh (n_2 h_1)}{D_5} \right);$$

$$D_8 = n_1^3 \left(-\cosh (n_1 h_1) + \frac{D_4 \sinh (n_2 h_1)}{D_5} \right);$$

$$D_9 = n_2^3 \left(\cosh (n_2 h_1) + \frac{D_6 \sinh (n_2 h_1)}{D_5} \right);$$

$$D_{10} = 3 D_1 \left(h_2^2 - h_1^2 \right) + 2 D_2 \left(h_2 - h_1 \right)$$
$$- \frac{6 n_1 D_1 D_4}{D_8} + \frac{6 n_1^3 D_1 D_4}{n_2^2 D_8};$$

$$D_{11} = n_1 D_3 + \frac{n_1 D_4 D_7}{D_8} - \frac{n_1^3 D_3}{n_2^2} - \frac{n_1^3 D_4 D_7}{n_2^2 D_8};$$

$$D_{12} = \frac{n_1 D_4 D_9}{D_8} - \frac{n_1^3 D_4 D_9}{n_2^2 D_8};$$

$$D_{13} = \frac{D_{10}}{D_{11}}; \qquad D_{14} = \frac{D_{12}}{D_{11}};$$

$$D_{15} = \frac{D_7 D_{13} + 6 D_1}{D_8}; \qquad D_{16} = \frac{D_7 D_{14} - D_9}{D_8};$$

$$D_{17} = -\frac{n_1^3 \left(D_3 D_{13} + D_4 D_{15} \right)}{n_2^3 D_5};$$

$$D_{18} = -\frac{n_1^3 \left(D_3 D_{14} + D_4 D_{16} \right) + n_2^3 D_6}{n_2^3 D_5};$$

$$D_{19} = -1 - 3 D_1 h_1^2 - 2 D_2 h_1 - n_1 D_{13} \sinh (n_1 h_1)$$
$$- n_1 D_{15} \cosh (n_1 h_1) - n_2 D_{17} \sinh (n_2 h_1);$$

$$D_{20} = -n_1 D_{14} \sinh (n_1 h_1) - n_1 D_{16} \cosh (n_1 h_1)$$
$$- n_2 D_{18} \sinh (n_2 h_1) - n_2 \cosh (n_2 h_1);$$

$$D_{21} = F - D_1 \left(h_1^3 - h_2^3 \right) - D_2 \left(h_1^2 - h_2^2 \right) - D_{19} \left(h_1 - h_2 \right)$$
$$- D_4 D_{13} - D_3 D_{15} - D_6 D_{17};$$

$$D_{22} = D_{20} \left(h_1 - h_2 \right) + D_4 D_{14} + D_3 D_{16} + D_6 D_{18} + D_5. \tag{27}$$

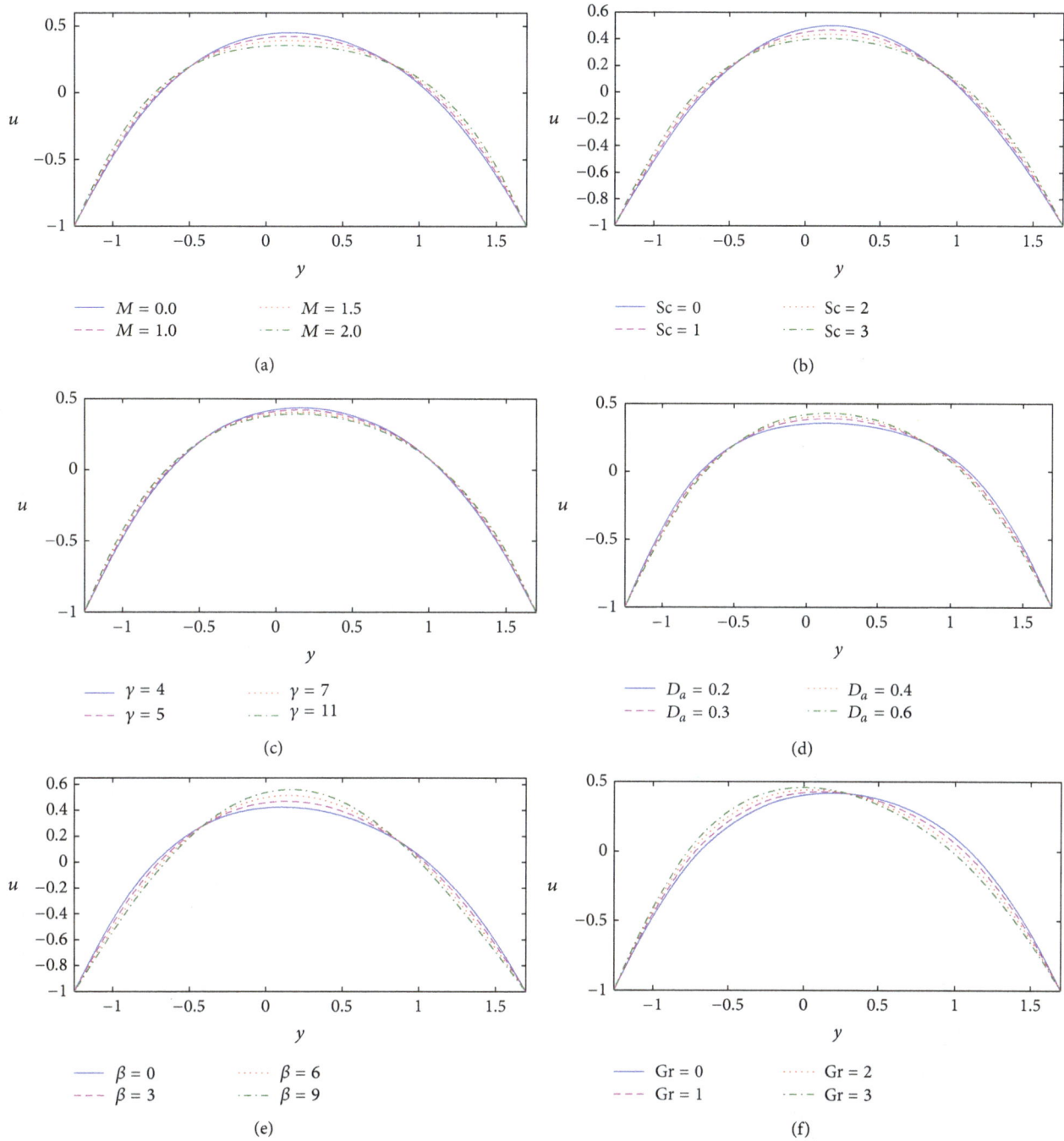

FIGURE 2: Velocity profile for (a) $x = 1$, $a = 0.7$, $b = 0.5$, $d = 1$, $D_a = 0.5$, $\gamma = 5$, $\phi = \pi/3$, Gr = 0.5, Gc = 0.5, Sr = 0.4, Sc = 0.4, $\Theta = 2$, and $\beta = 0.2$; (b) $x = 1$, $a = 0.7$, $b = 0.5$, $d = 1$, $M = 1$, $D_a = 0.5$, $\gamma = 5$, $\phi = \pi/3$, Gr = 0.5, Gc = 0.5, Sr = 0.4, $\Theta = 2$, and $\beta = 7$; (c) $x = 1$, $a = 0.7$, $b = 0.5$, $d = 1$, $M = 1$, $D_a = 0.5$, $\phi = \pi/3$, Gr = 0.5, Gc = 0.5, Sr = 0.4, Sc = 0.4, $\Theta = 2$, and $\beta = 0.2$; (d) $x = 1$, $a = 0.7$, $b = 0.5$, $d = 1$, $M = 1$, $\gamma = 5$, $\phi = \pi/3$, Gr = 0.5, Gc = 0.5, Sr = 0.4, Sc = 0.4, $\Theta = 2$, and $\beta = 0.2$; (e) $x = 1$, $a = 0.7$, $b = 0.5$, $d = 1$, $M = 1$, $D_a = 0.5$, $\gamma = 5$, $\phi = \pi/3$, Gr = 0.5, Gc = 0.5, Sr = 0.8, Sc = 0.8, and $\Theta = 2$; (f) $x = 1$, $a = 0.7$, $b = 0.5$, $d = 1$, $M = 1$, $D_a = 0.5$, $\gamma = 5$, $\phi = \pi/3$, Gc = 0.5, Sr = 0.4, Sc = 0.4, $\Theta = 2$, and $\beta = 0.2$.

Using (24) and (26) in (19), the pressure gradient is given by

$$\frac{\partial p}{\partial x} = E_1 y^2 + E_2 y + E_3 + E_4 \cosh(n_1 y) + E_5 \sinh(n_1 y) + E_6 \cosh(n_2 y) + E_7 \sinh(n_2 y),$$

(28)

where

$$E_1 = \frac{\text{GcScSr}\beta - \text{Gr}\beta - 6m_1^2 D_1}{2};$$

$$E_2 = \text{Gr}A_1 + \text{Gc}B_1 - 2m_1^2 D_2;$$

$$E_3 = \text{Gr}A_2 + \text{Gc}B_2 + 6D_1 - m_1^2(1 + C_3);$$

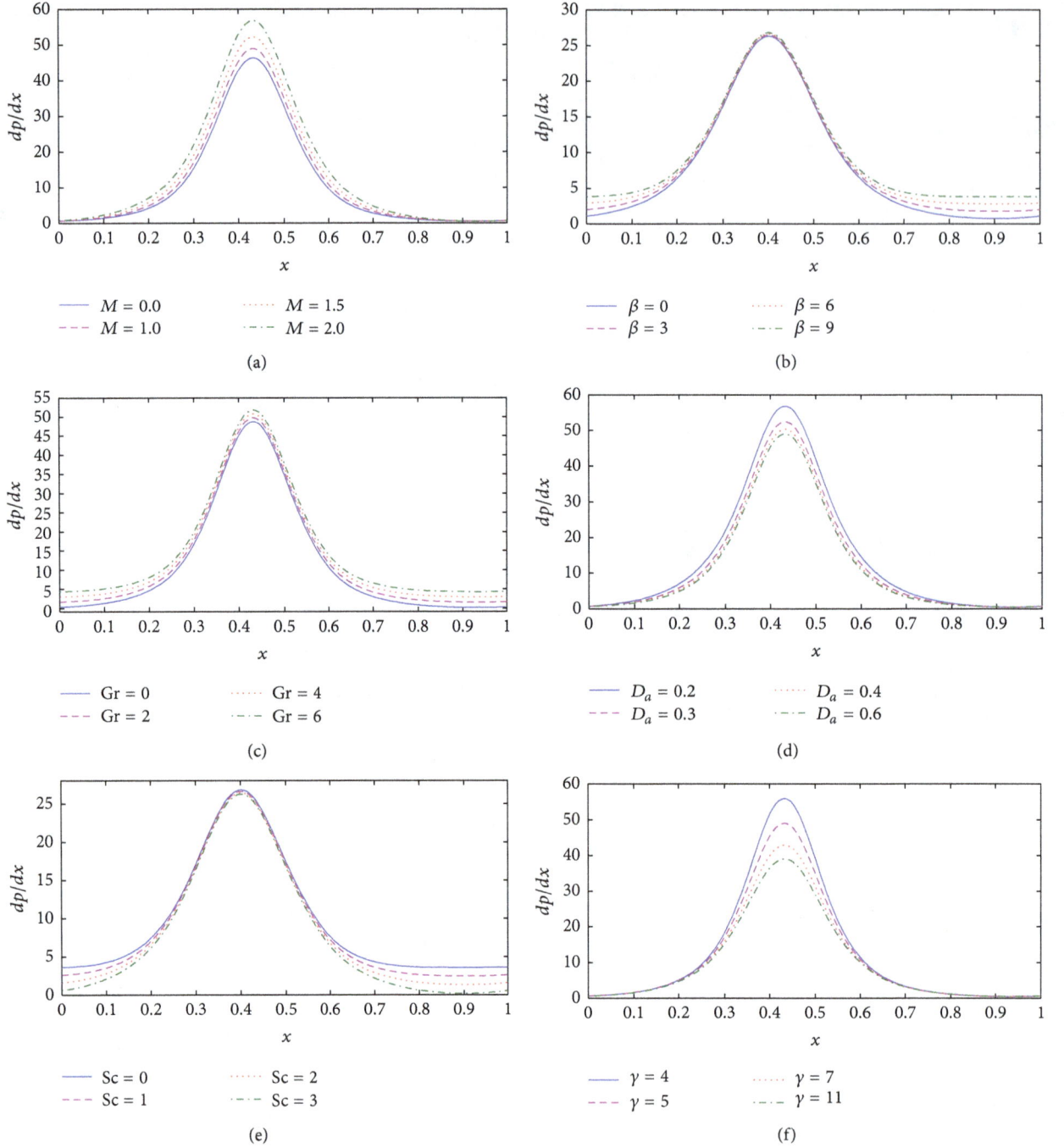

FIGURE 3: Pressure gradient for (a) $y = 0$, $a = 0.7$, $b = 0.5$, $d = 1$, $D_a = 0.5$, $\gamma = 5$, $\phi = \pi/3$, Gr $= 0.5$, Gc $= 0.5$, Sr $= 0.4$, Sc $= 0.4$, $\Theta = -1$, and $\beta = 0.2$; (b) $y = 0$, $a = 0.7$, $b = 0.5$, $d = 1$, $M = 1$, $D_a = 0.5$, $\gamma = 5$, $\phi = \pi/3$, Gr $= 0.5$, Gc $= 0.5$, Sr $= 0.4$, Sc $= 0.4$, and $\Theta = -1$; (c) $y = 0$, $a = 0.7$, $b = 0.5$, $d = 1$, $M = 1$, $D_a = 0.5$, $\gamma = 5$, $\phi = \pi/2$, Gc $= 0.5$, Sr $= 0.4$, Sc $= 0.4$, $\Theta = -1$, and $\beta = 0.2$; (d) $y = 0$, $a = 0.7$, $b = 0.5$, $d = 1$, $M = 1$, $\gamma = 5$, $\phi = \pi/3$, Gr $= 0.5$, Gc $= 0.5$, Sr $= 0.4$, Sc $= 0.4$, $\Theta = -1$, and $\beta = 0.2$; (e) $y = 0$, $a = 0.7$, $b = 0.5$, $d = 1$, $M = 1$, $D_a = 0.5$, $\gamma = 5$, $\phi = \pi/2$, Gr $= 0.5$, Gc $= 0.5$, Sr $= 0.4$, $\Theta = -1$, and $\beta = 0.2$; (f) $y = 0$, $a = 0.7$, $b = 0.5$, $d = 1$, $M = 1$, $D_a = 0.5$, $\phi = \pi/3$, Gr $= 0.5$, Gc $= 0.5$, Sr $= 0.4$, Sc $= 0.4$, $\Theta = -1$, and $\beta = 0.2$.

$$E_4 = C_6\left(n_1^3 - m_1^2 n_1 - \frac{n_1^5}{\gamma^2}\right);$$

$$E_5 = C_5\left(n_1^3 - m_1^2 n_1 - \frac{n_1^5}{\gamma^2}\right);$$

$$E_6 = C_8\left(n_2^3 - m_1^2 n_2 - \frac{n_2^5}{\gamma^2}\right);$$

$$E_7 = C_7\left(n_2^3 - m_1^2 n_2 - \frac{n_2^5}{\gamma^2}\right).$$

$$(29)$$

(a) Sinusoidal wave

(b) Triangular wave

(c) Square wave

(d) Trapezoidal wave

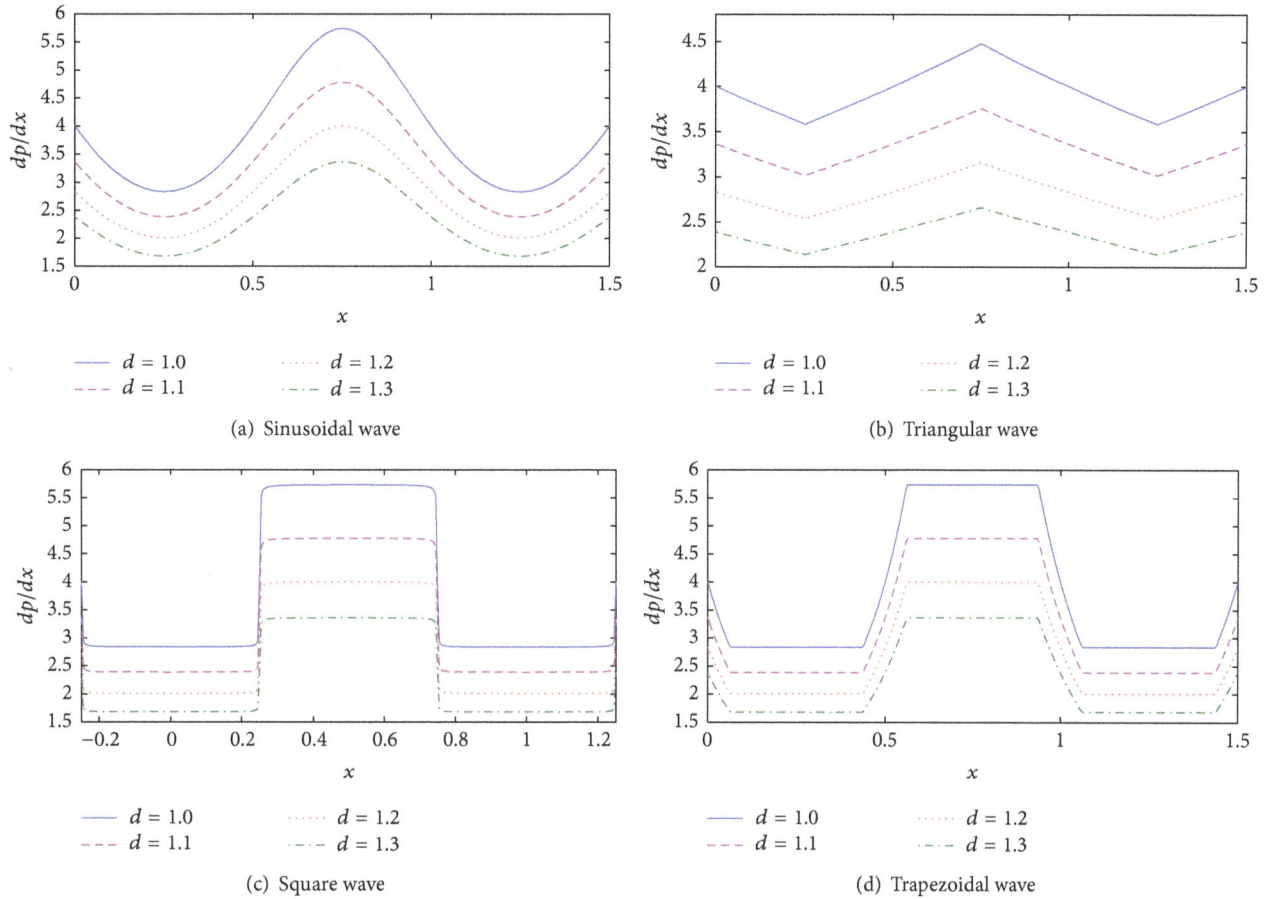

FIGURE 4: Pressure gradient for various wave forms for fixed values of $y = 0$, $a = 0.7$, $b = 0.5$, $D_a = 0.5$, $M = 1$, $\gamma = 5$, $\phi = 0$, $Gr = 0.5$, $Gc = 0.5$, $Sr = 0.4$, $Sc = 0.4$, and $\beta = 0.2$.

The nondimensional expressions for the pressure difference for one wave length Δ_{p_λ}, the frictional forces at both walls F_{λ_1} at $y = h_1$ and F_{λ_2} at $y = h_2$, and the heat transfer coefficients Z_{h_1} and Z_{h_2} at the right and left walls are defined as follows [32]:

$$\Delta p_\lambda = \int_0^1 \left(\frac{dp}{dx} \right) dx, \qquad F_{\lambda_1} = \int_0^1 h_1^2 \left(-\frac{dp}{dx} \right) dx,$$

$$F_{\lambda_2} = \int_0^1 h_2^2 \left(-\frac{dp}{dx} \right) dx, \qquad (30)$$

$$Z_{h_1} = \frac{\partial h_1}{\partial x} \frac{\partial \theta}{\partial y}, \qquad Z_{h_2} = \frac{\partial h_2}{\partial x} \frac{\partial \theta}{\partial y}.$$

Using (30), the heat transfer coefficients at the right and left walls, respectively, are obtained as

$$Z_{h_1} = 2a\pi \left(\beta y - A_1 \right) \sin (2\pi x),$$

$$Z_{h_2} = -2b\pi \left(\beta y - A_1 \right) \sin (2\pi x + \phi). \qquad (31)$$

The expressions for pressure rise Δp_λ and frictional forces at both walls F_{λ_1} at $y = h_1$ and F_{λ_2} at $y = h_2$ involve the integration of dp/dx. Due to the complexity of dp/dx, the analytical integration of integrals of (30) is not possible. In view of this, a numerical integration scheme is used for the evaluation of the integrals.

4. Expressions for Wave Shapes

The nondimensional expressions for the four considered wave forms are given in the following.

(1) Sinusoidal wave:

$$h_1 (x) = 1 + a \sin (2\pi x),$$

$$h_2 (x) = -d - b \sin (2\pi x + \phi). \qquad (32)$$

(2) Triangular wave:

$$h_1 (x) = 1 + a \left(\frac{8}{\pi^3} \sum_{n=1}^{\infty} \frac{(-1)^{n+1}}{(2n-1)^2} \sin (2\pi (2n-1) x) \right),$$

$$h_2 (x) = -d - b \left(\frac{8}{\pi^3} \sum_{n=1}^{\infty} \frac{(-1)^{n+1}}{(2n-1)^2} \sin (2\pi (2n-1) x + \phi) \right). \qquad (33)$$

(3) Square wave:

$$h_1 (x) = 1 + a \left(\frac{4}{\pi} \sum_{n=1}^{\infty} \frac{(-1)^{n+1}}{2n-1} \cos (2\pi (2n-1) x) \right),$$

$$h_2 (x) = -d - b \left(\frac{4}{\pi} \sum_{n=1}^{\infty} \frac{(-1)^{n+1}}{2n-1} \cos (2\pi (2n-1) x + \phi) \right). \qquad (34)$$

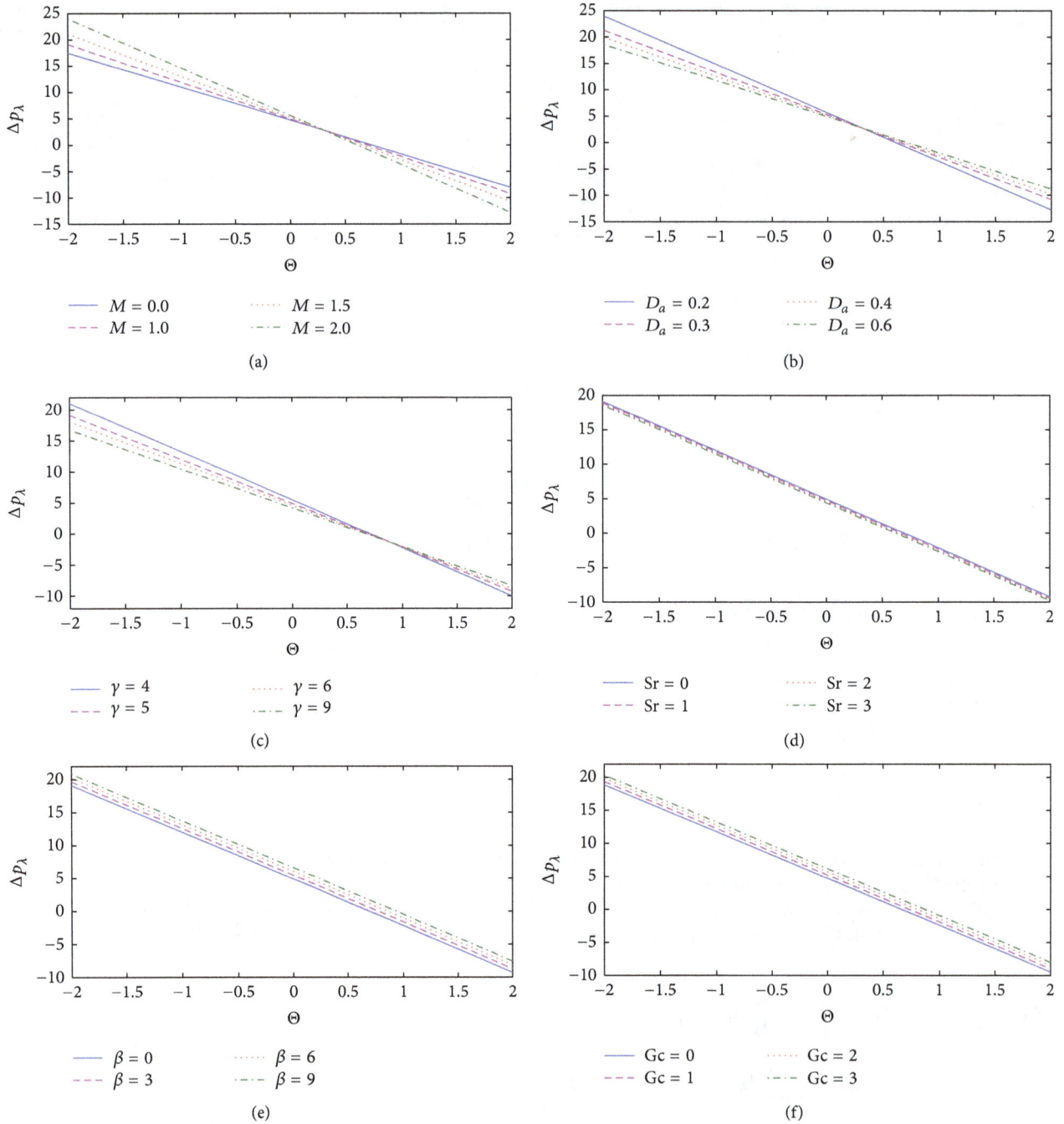

FIGURE 5: Pressure difference for (a) $y = 0$, $a = 0.7$, $b = 0.5$, $d = 1$, $D_a = 0.5$, $\gamma = 5$, $\phi = \pi/3$, $Gr = 0.5$, $Gc = 0.5$, $Sr = 0.4$, $Sc = 0.4$, $\Theta = 2$, and $\beta = 0.2$; (b) $y = 0$, $a = 0.7$, $b = 0.5$, $d = 1$, $M = 1$, $\gamma = 5$, $\phi = \pi/3$, $Gr = 0.5$, $Gc = 0.5$, $Sr = 0.4$, $Sc = 0.4$, $\Theta = 2$, and $\beta = 0.2$; (c) $y = 0$, $a = 0.7$, $b = 0.5$, $d = 1$, $M = 1$, $D_a = 0.5$, $\phi = \pi/3$, $Gr = 0.5$, $Gc = 0.5$, $Sr = 0.4$, $Sc = 0.4$, $\Theta = 2$, and $\beta = 0.2$; (d) $y = 0$, $a = 0.7$, $b = 0.5$, $d = 1$, $M = 1$, $D_a = 0.5$, $\gamma = 5$, $\phi = \pi/2$, $Gr = 0.5$, $Gc = 0.5$, $Sc = 0.4$, $\Theta = 2$, and $\beta = 0.2$; (e) $y = 0$, $a = 0.7$, $b = 0.5$, $d = 1$, $M = 1$, $D_a = 0.5$, $\gamma = 5$, $\phi = \pi/3$, $Gr = 0.5$, $Gc = 0.5$, $Sr = 0.4$, $Sc = 0.4$, and $\Theta = 2$; (f) $y = 0$, $a = 0.7$, $b = 0.5$, $d = 1$, $M = 1$, $D_a = 0.5$, $\gamma = 5$, $\phi = \pi/3$, $Gr = 0.5$, $Sr = 0.4$, $Sc = 0.4$, $\Theta = 2$, and $\beta = 0.2$.

(4) Trapezoidal wave:

$h_1(x)$

$$= 1 + a \left(\frac{32}{\pi^2} \sum_{n=1}^{\infty} \frac{\sin\left((\pi/8)(2n-1)\right)}{(2n-1)^2} \sin\left(2\pi(2n-1)x\right) \right),$$

$h_2(x)$

$$= -d$$

$$- b \left(\frac{32}{\pi^2} \sum_{n=1}^{\infty} \frac{\sin\left((\pi/8)(2n-1)\right)}{(2n-1)^2} \sin\left(2\pi(2n-1)x + \phi\right) \right).$$

$$(35)$$

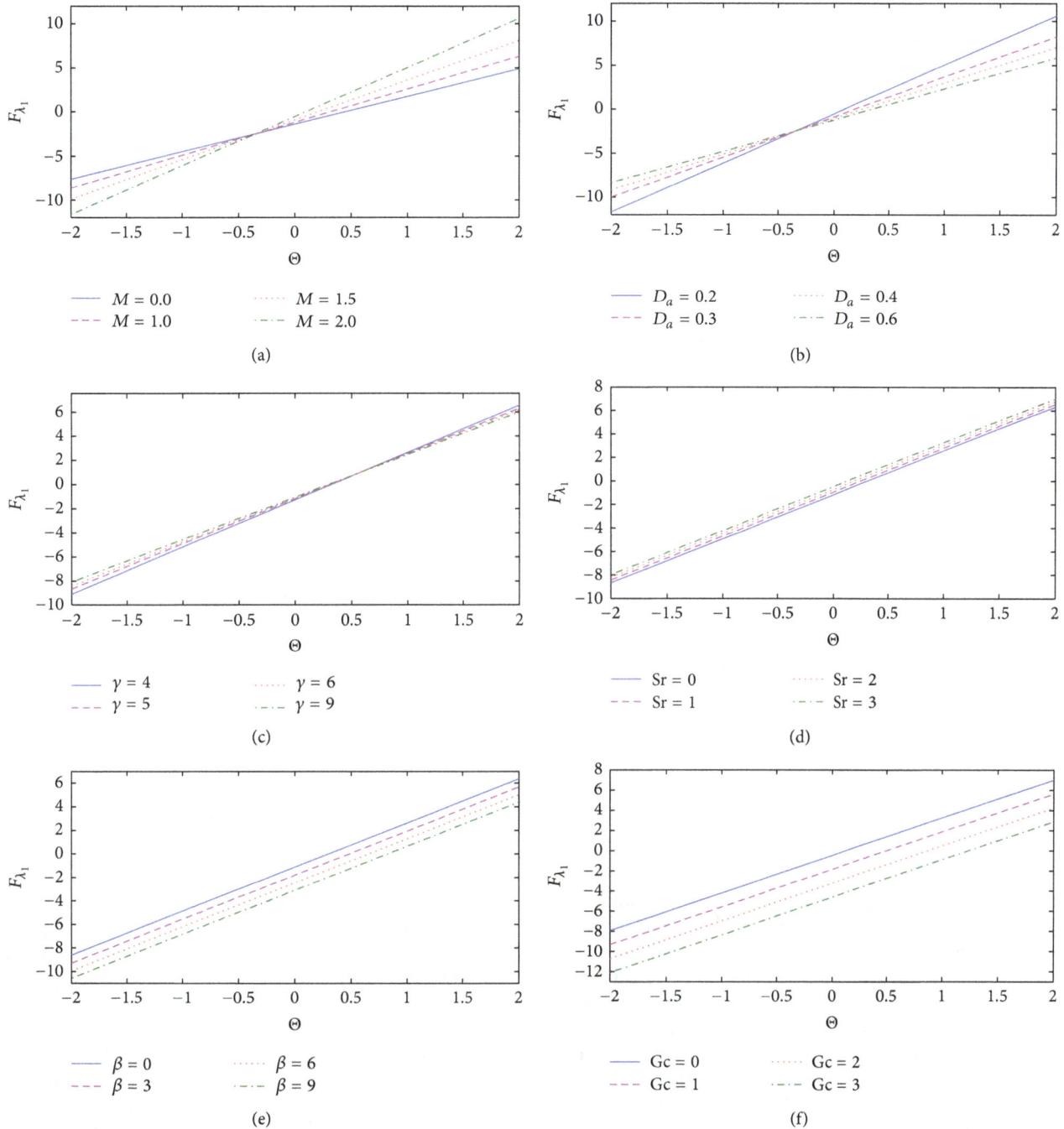

FIGURE 6: Frictional force at the right wall for (a) $y = 0$, $a = 0.7$, $b = 0.5$, $d = 1$, $D_a = 0.5$, $\gamma = 5$, $\phi = \pi/3$, $Gr = 0.5$, $Gc = 0.5$, $Sr = 0.4$, $Sc = 0.4$, $\Theta = 2$, and $\beta = 0.2$; (b) $y = 0$, $a = 0.7$, $b = 0.5$, $d = 1$, $M = 1$, $\gamma = 5$, $\phi = \pi/3$, $Gr = 0.5$, $Gc = 0.5$, $Sr = 0.4$, $Sc = 0.4$, $\Theta = 2$, and $\beta = 0.2$; (c) $y = 0$, $a = 0.7$, $b = 0.5$, $d = 1$, $M = 1$, $D_a = 0.5$, $\phi = \pi/3$, $Gr = 0.5$, $Gc = 0.5$, $Sr = 0.4$, $Sc = 0.4$, $\Theta = 2$, and $\beta = 0.2$; (d) $y = 0$, $a = 0.7$, $b = 0.5$, $d = 1$, $M = 1$, $D_a = 0.5$, $\gamma = 5$, $\phi = \pi/2$, $Gr = 0.5$, $Gc = 0.5$, $Sc = 0.4$, $\Theta = 2$, and $\beta = 0.2$; (e) $y = 0$, $a = 0.7$, $b = 0.5$, $d = 1$, $M = 1$, $D_a = 0.5$, $\gamma = 5$, $\phi = \pi/3$, $Gr = 0.5$, $Gc = 0.5$, $Sr = 0.4$, $Sc = 0.4$, and $\Theta = 2$; (f) $y = 0$, $a = 0.7$, $b = 0.5$, $d = 1$, $M = 1$, $D_a = 0.5$, $\gamma = 5$, $\phi = \pi/3$, $Gr = 0.5$, $Sr = 0.4$, $Sc = 0.4$, $\Theta = 2$, and $\beta = 0.2$.

5. Results and Discussion

This section is dedicated to discussion and analysis of the velocity distribution, pumping characteristics, heat and mass characteristics, and trapping phenomena for different flow parameters.

5.1. Flow Characteristics. Figures 2(a)–2(f) illustrate the influence of Hartmann number M, Schmidt number Sc, couple stress parameter γ, Darcy number D_a, heat generation parameter β, and Grashof number Gr on axial velocity profile across the channel. From these figures, it is observed that the maximum velocities are always located near the centre

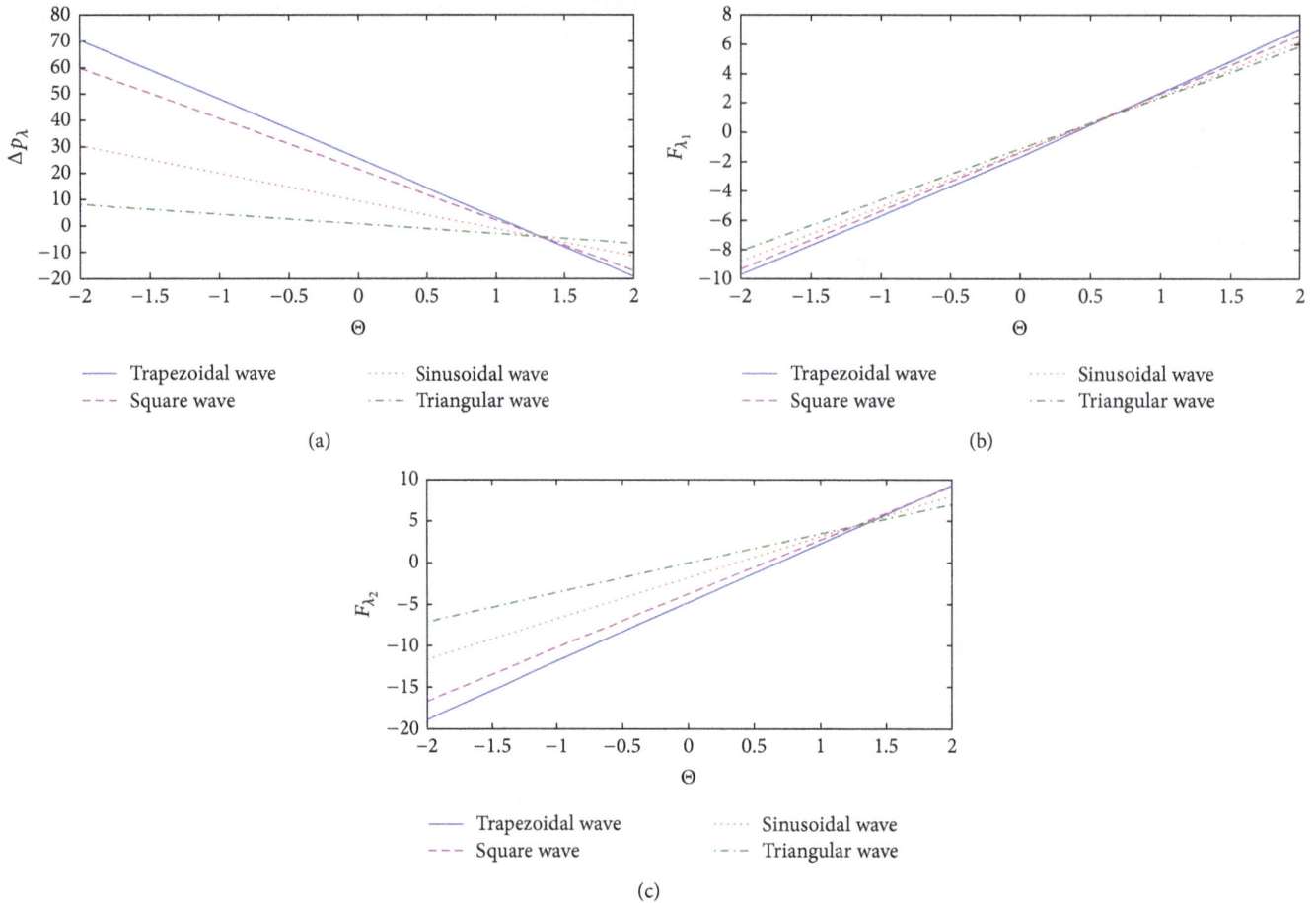

(a)

(b)

(c)

FIGURE 7: (a) Pressure difference, (b) frictional force at right wall, and (c) frictional force at left wall for different wave forms when $y = 0$, $a = 0.7$, $b = 0.5$, $d = 1$, $D_a = 0.5$, $M = 1$, $\gamma = 5$, $\phi = 0$, Gr $= 0.5$, Gc $= 0.5$, Sr $= 0.4$, Sc $= 0.4$, $\Theta = 2$, and $\beta = 0.2$.

of the channel and the velocity profiles are nearly parabolic in all cases. It is noted from Figure 2(a) that as Hartmann number M increases, the velocity decreases near the centre of the channel and it is increased in the neighborhood of the walls. This seems realistic because the magnetic field acts in the transverse direction to the flow and magnetic force resists the flow. The similar behavior is observed in [25]. The same behavior can be seen with increasing of Schmidt number Sc and couple stress parameter γ (see Figures 2(b) and 2(c)). It is observed from Figure 2(d) that increasing of Darcy number D_a increases the velocity near the centre of the channel and decreases the velocity of the fluid near the peristaltic walls. The same trend is followed with the increasing of heat generation parameter β (see Figure 2(e)). It is noticed from Figure 2(f) that, with increasing of Grashof number Gr, the velocity at the left wall increases while a reverse trend is seen at the right wall.

5.2. Pumping Characteristics. Figure 3 illustrates the variation of pressure gradient over one wave length $x \in [0, 1]$. The effects of M, β, and Gr on pressure gradient are displayed in Figures 3(a)–3(c). It can be seen from Figure 3(a) that increasing of Hartmann number M increases the pressure gradient. It shows that when strong magnetic field is applied

to the flow field then higher pressure gradient is needed to pass the flow. This result suggests that the fluid pressure can be controlled by the application of suitable magnetic field strength. This phenomenon is useful during surgery and critical operation to control excessive bleeding. It is also observed that increasing of β and Gr increases the pressure gradient. From Figures 3(d)–3(f), it is noted that with the increasing of D_a, Sc, and γ the pressure gradient decreases. It is noticed that, in the wider part of the channels $x \in [0, 0.2]$ and $x \in [0.7, 1]$, the pressure gradient is small, so the flow can be easily passed without the imposition of large pressure gradient. However, in the narrow part of the channel $x \in [0.2, 0.7]$ the pressure gradient is large; that is, much larger pressure gradient is needed to maintain the same given volume flow rate. Figure 4 is prepared to see the behaviour of pressure gradient for different four wave forms. It is observed from Figures 4(a)–4(d) that, in all the wave forms, increase in d decreases pressure gradient.

The dimensionless pressure difference per unit wave length versus time mean flow rate Θ has been plotted in Figure 5. We split the whole region into four segments as follows: peristaltic pumping region where $\Delta p_\lambda > 0$ and $\Theta > 0$ and augmented pumping region when $\Delta p_\lambda < 0$ and $\Theta > 0$. There is retrograde pumping region when $\Delta p_\lambda > 0$

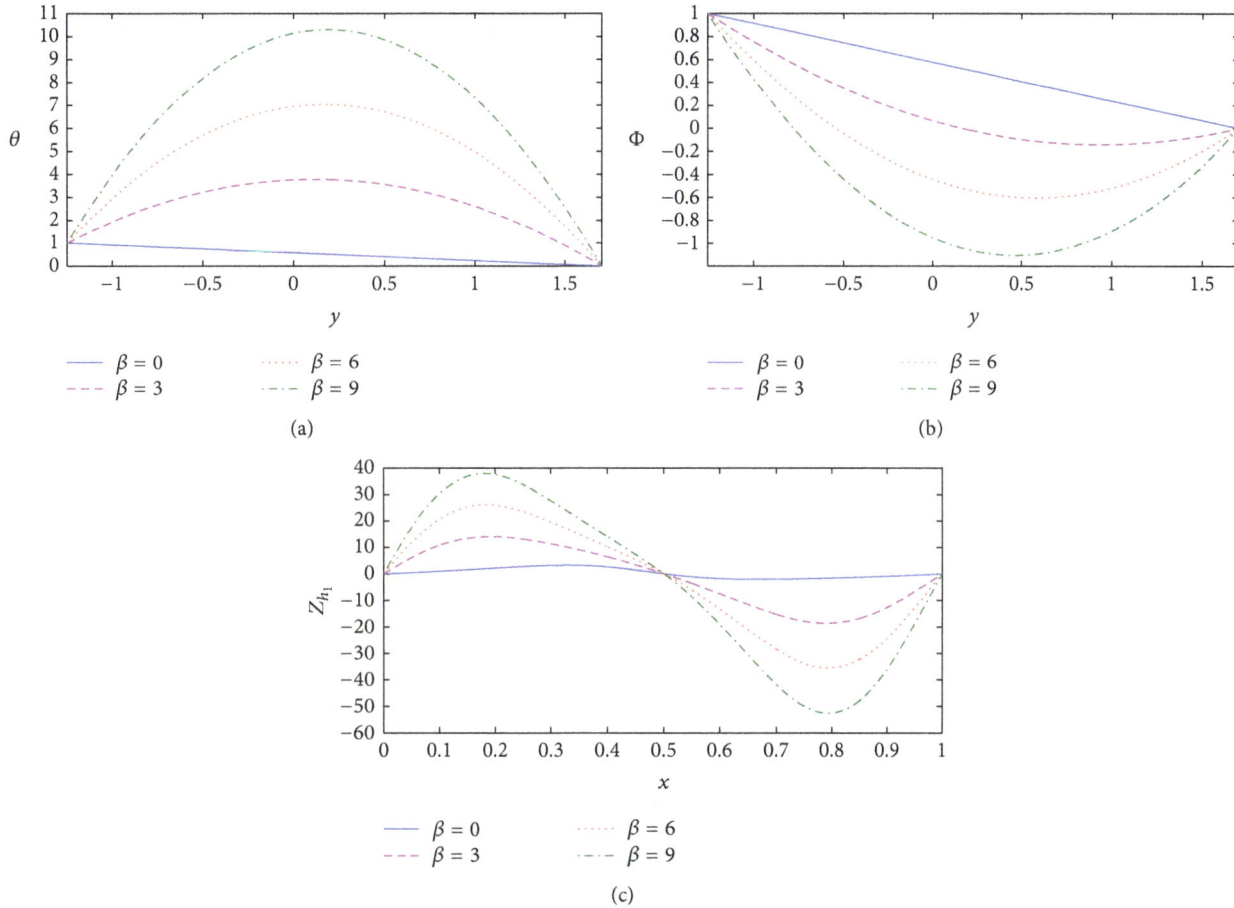

FIGURE 8: (a) Temperature profile, (b) concentration profile, and (c) heat transfer coefficient at the right wall for $a = 0.7$, $b = 0.5$, $d = 1$, $D_a = 0.5$, $M = 1$, $\gamma = 5$, $\phi = \pi/3$, Gr $= 0.5$, Gc $= 0.5$, Sr $= 0.4$, Sc $= 0.4$, and $\Theta = 2$.

and $\Theta < 0$. Free pumping region corresponds to $\Delta p_\lambda = 0$. The region where $\Delta p_\lambda > 0$ and $\Theta > 0$ is known as peristaltic pumping region. In this region, the positive value of Θ is entirely due to the peristalsis after overcoming the pressure difference. The region where $\Delta p_\lambda < 0$ and $\Theta > 0$ is known as copumping or augmented pumping region. In this region, a negative pressure difference assists the flow due to the peristalsis of the walls. The region where $\Delta p_\lambda > 0$ and $\Theta < 0$ is called retrograde pumping region. In this region, the flow is opposite to the direction of the peristaltic motion. In the free pumping region, the flow is caused purely by the peristalsis of the walls. It is evident from Figure 5 that there is an inversely linear relation between Δp_λ and Θ. From Figure 5(a), it is clear that with the increasing of M, in the augmented pumping and free pumping regions, the pumping decreases, in the peristaltic pumping region, the pumping increases up to a critical value of Θ and decreases after the critical value, and in the retrograde pumping region the pumping increases. It is observed from Figure 5(b) that, with the increasing of D_a, the behaviour is quite opposite with M. It is noticed from Figure 5(c) that in the augmented pumping region the pumping increases and in the peristaltic pumping and retrograde pumping regions the pumping decreases. Figure 5(d) depicts that, in all the pumping regions, the

pumping decreases by increasing Sr. It is noted from Figures 5(e)-5(f) that the behaviour is quite opposite with Sr while increasing β and Gc.

Figure 6 describes the variation of frictional forces against flow rate Θ for different values of M, D_a, γ, Sr, β, and Gc. It is observed that there is a direct linear relation between frictional forces and Θ. The frictional forces have exactly opposite behaviour when compared with that of pressure difference. Figure 7(a) indicates the effects of four different wave forms on pressure difference. It is noticed that the trapezoidal wave has best peristaltic pumping characteristics, while the triangular wave has poor peristaltic pumping as compared to the other waves. Figures 7(b)-7(c) show that the frictional forces at the walls have opposite behaviour as compared to the pressure difference.

5.3. Heat and Mass Characteristics. Figure 8 depicts the effects of heat transfer, concentration, and heat transfer coefficient on the peristaltic transport for various values of β. We can observe that the temperature and concentration profiles are almost parabolic except when $\beta = 0$. It is observed from Figure 8(a) that the temperature increases with β. It is clear from Figure 8(b) that the concentration profile has quite opposite behaviour of temperature profile. It is noticed

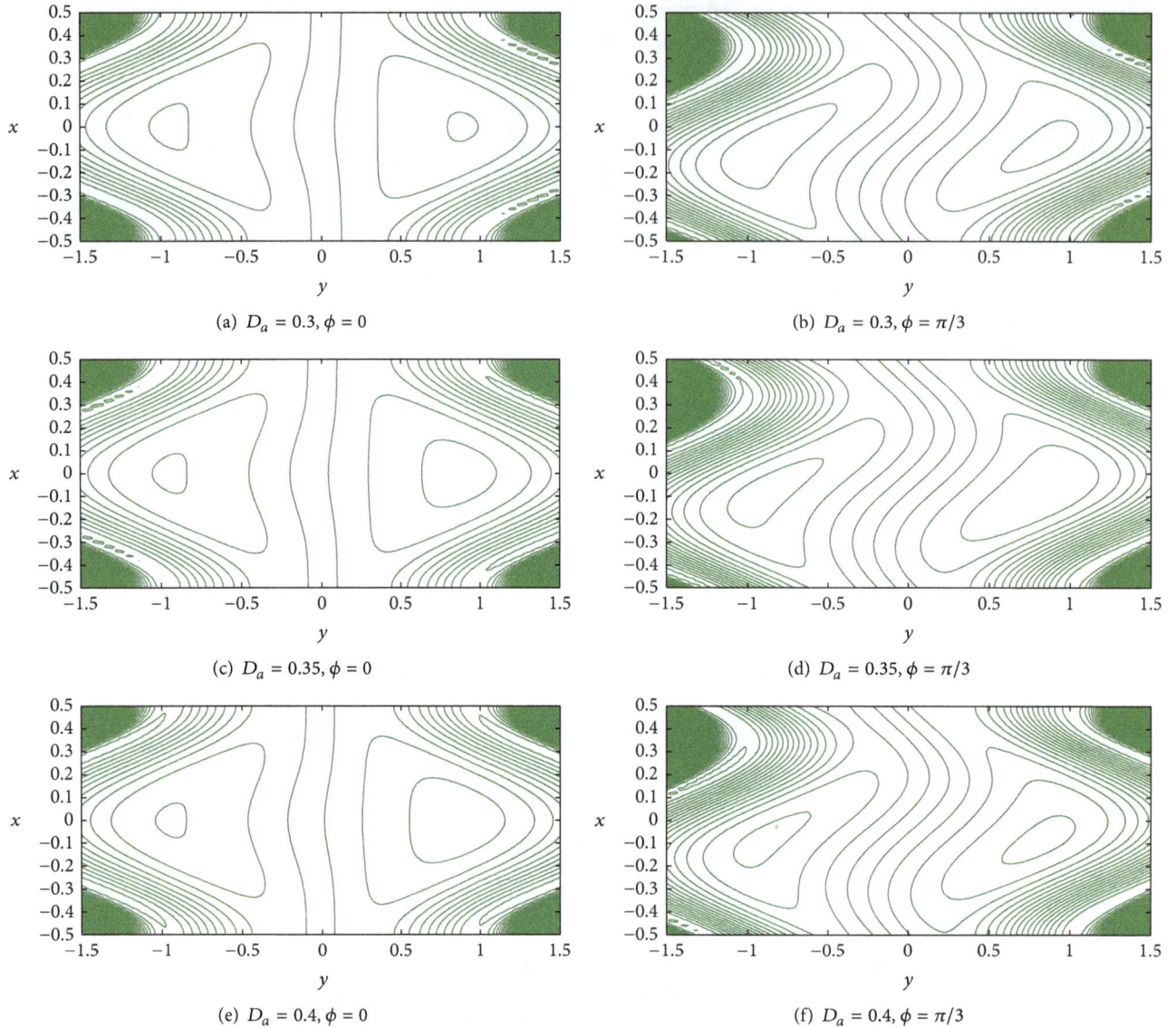

(a) $D_a = 0.3, \phi = 0$

(b) $D_a = 0.3, \phi = \pi/3$

(c) $D_a = 0.35, \phi = 0$

(d) $D_a = 0.35, \phi = \pi/3$

(e) $D_a = 0.4, \phi = 0$

(f) $D_a = 0.4, \phi = \pi/3$

FIGURE 9: Streamlines for $a = 0.5$, $b = 0.5$, $d = 1$, $M = 1$, $\gamma = 5$, Gr = 0.5, Gc = 0.5, Sr = 0.4, Sc = 0.4, $\Theta = 2$, and $\beta = 0.2$.

from Figure 8(c) that, due to the peristalsis, the heat transfer coefficient is in oscillatory behaviour. Moreover, the absolute value of heat transfer coefficient increases with increase of β.

5.4. Trapping Phenomenon. In the wave frame, the streamlines, in general, have a shape similar to the walls as the walls are stationary. However under certain conditions some streamlines can split to enclose a bolus of fluid particles in closed streamlines. Hence some circulating regions occur. In the fixed frame of reference the fluid bolus is trapped with the wave and it moves as a whole with the wave speed. To examine the effects of D_a, γ, and ϕ in the symmetric and asymmetric channels we have plotted Figures 9–11. In Figures 9 and 10, the left panels (a), (c), and (e) related to symmetric channel and the right panels (b), (d), and (f) are corresponding to asymmetric channel. It is observed from Figure 9 that when D_a increases, the size of the trapped bolus decreases near

the left wall and increases near the right wall in both panels. However, the increase of Darcy number D_a increases the trapped bolus near both walls when Gc = 0 and Gr = 0. The effects of γ on the trapping phenomena are displayed in Figure 10. It is shown that with the increasing γ decrease the trapping phenomena for the symmetric and asymmetric channels. Since $\gamma = \sqrt{\mu/\eta}d_1$, γ decreases as η increases and hence increasing of couple stresses increases the size of trapped bolus. Figure 11 gives the trapping behaviour for various values of ϕ. It is evident that as ϕ increases, the trapping bolus decreases and when it reaches to π the trapping disappears. Moreover, with the increase of ϕ the bolus moves upward with decreasing effect. Figures 12 and 13 provide the variations of β on trapping for different wave shapes: (a) sinusoidal, (b) triangular, (c) square, and (d) trapezoidal. From these figures we observe that the size of the trapped bolus increases with increasing β in all the wave forms.

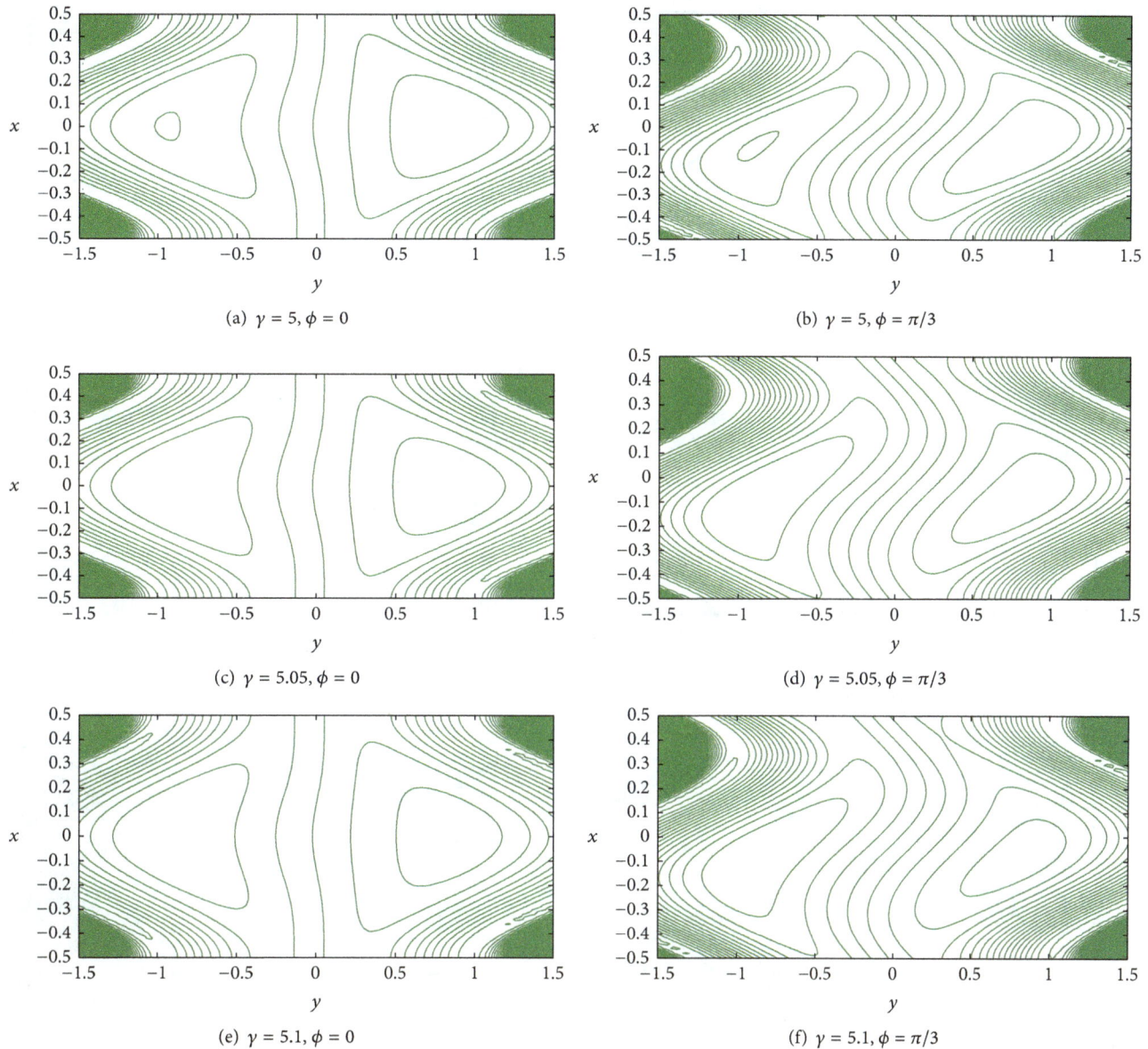

(a) $\gamma = 5, \phi = 0$

(b) $\gamma = 5, \phi = \pi/3$

(c) $\gamma = 5.05, \phi = 0$

(d) $\gamma = 5.05, \phi = \pi/3$

(e) $\gamma = 5.1, \phi = 0$

(f) $\gamma = 5.1, \phi = \pi/3$

FIGURE 10: Streamlines for $a = 0.5, b = 0.5, d = 1, M = 1, D_a = 0.5, Gr = 0.5, Gc = 0.5, Sr = 0.4, Sc = 0.4, \Theta = 2,$ and $\beta = 0.2$.

6. Conclusions

The effects of heat and mass transfer on the peristaltic flow of magnetohydrodynamic couple stress fluid through porous medium in a vertical asymmetric channel have been analyzed. The governing equations are modeled under the assumption of long wave length approximation. The exact solutions for the stream function, pressure gradient, temperature, heat transfer coefficients, and concentration are obtained. The effects of involved parameters on the velocity characteristics, pumping characteristics, heat and mass characteristics, and the trapping due to the peristalsis of the walls are discussed in detail. From the analysis the main findings can be summarized as follows:

(i) Increasing of heat generation increases the peristaltic pumping, size of the trapped bolus, and the magnitude of heat transfer coefficient at the peristaltic walls.

(ii) Increasing of couple stresses increases the size of trapped bolus.

(iii) Increasing of heat generation increases the temperature and decreases the concentration.

(iv) The trapezoidal wave has best peristaltic pumping as compared to the other wave shapes.

(v) The frictional forces have an opposite behaviour as compared to the pressure difference.

Symbols

a_1, a_2: Wave amplitudes
λ: Wave length
c: Propagation velocity
t: Time

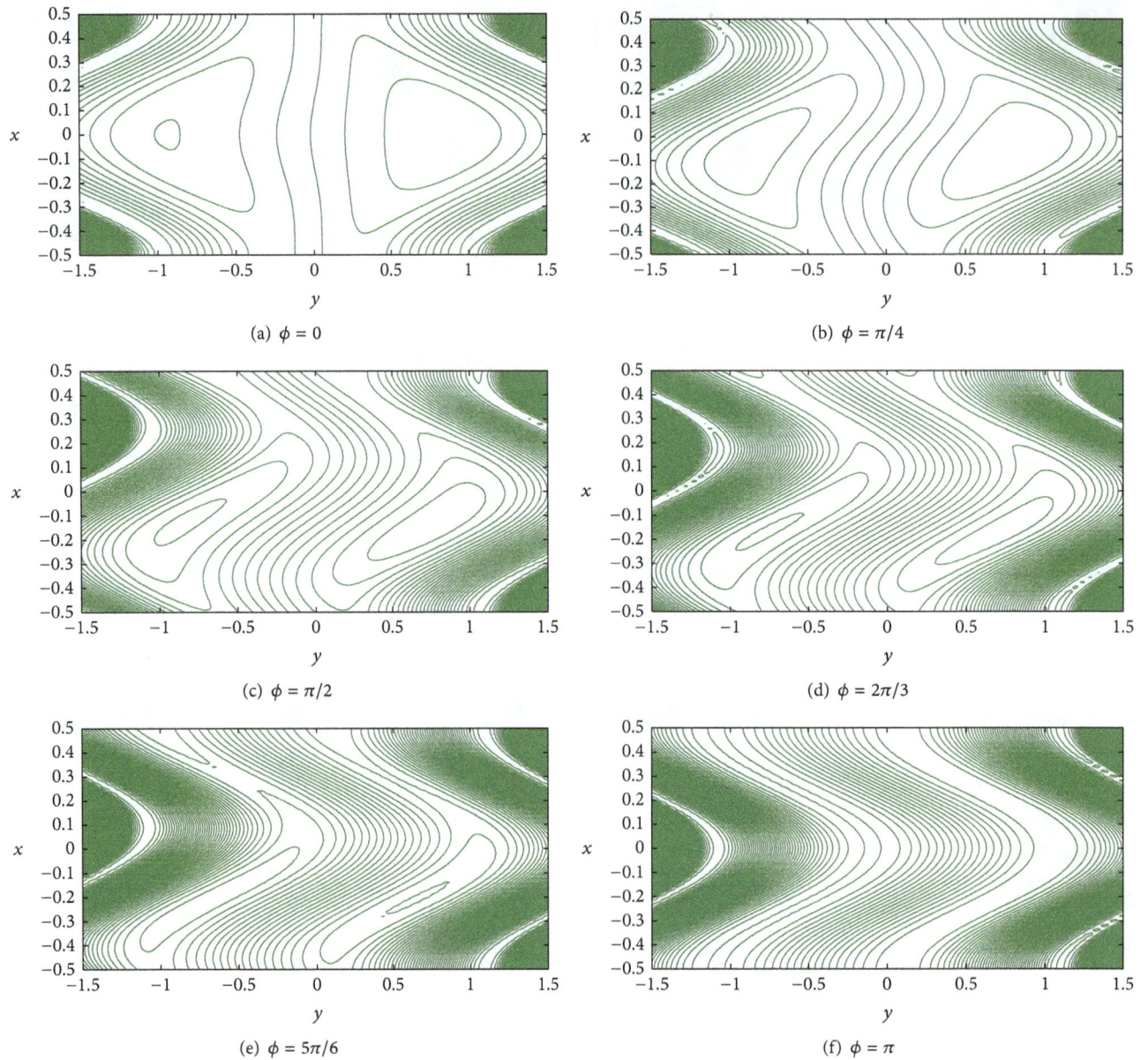

(a) $\phi = 0$

(b) $\phi = \pi/4$

(c) $\phi = \pi/2$

(d) $\phi = 2\pi/3$

(e) $\phi = 5\pi/6$

(f) $\phi = \pi$

FIGURE 11: Streamlines for $a = 0.5$, $b = 0.5$, $d = 1$, $M = 1$, $D_a = 0.5$, Gr $= 0.5$, Gc $= 0.5$, $\gamma = 5$, Sc $= 0.4$, Sr $= 0.4$, $\Theta = 2$, and $\beta = 0.2$.

X,Y:	Coordinates of fixed frame	k^*:	Thermal conductivity
P:	Pressure in the fixed frame	Q_0:	Heat generation parameter
$\overline{q}$:	Velocity vector	D:	Coefficient of mass diffusivity
R:	Darcy's resistance in the porous medium	K_T:	Thermal diffusion ratio
ρ:	Density	T_m:	Mean temperature
μ:	Viscosity	k_0:	Permeability parameter
η:	Material constant associated with couple stress	U,V:	Velocity components in the fixed frame
J:	Electric current density	P:	Pressure in the fixed frame
B:	Total magnetic field	σ:	Electrical conductivity of the fluid
g:	Acceleration due to the gravity	B_0:	Uniform applied magnetic field
β_T:	Coefficient of thermal expansion	δ:	Dimensionless wave length
β_C:	Coefficient of expansion with concentration	x, y:	Coordinates of wave frame
c_p:	Specific heat at constant pressure	p:	Pressure in the wave frame
T:	Temperature	u, v:	Velocity components in the wave frame
C:	Mass concentration	Re:	Reynolds number

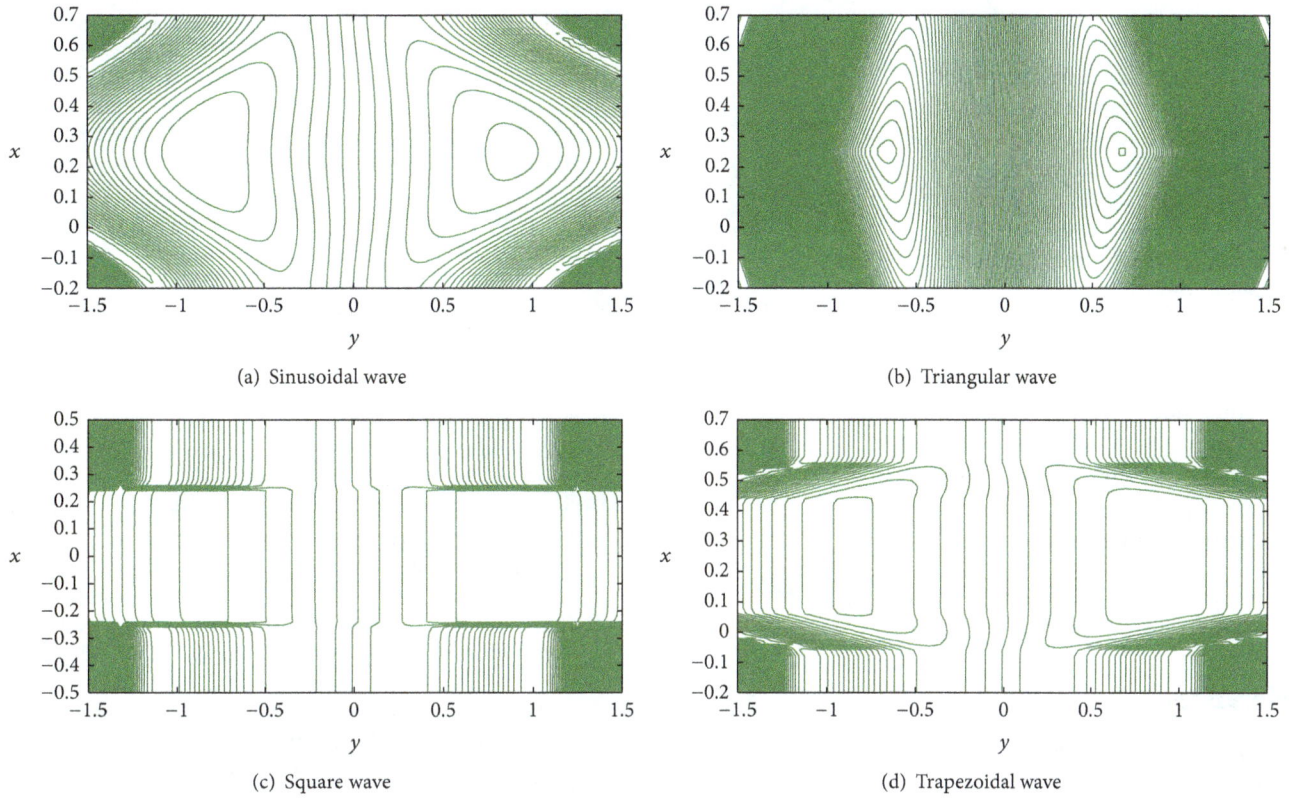

FIGURE 12: Streamlines for various wave forms for fixed values of $a = 0.7$, $b = 0.5$, $d = 1$, $D_a = 0.5$, $M = 1$, $\gamma = 5$, $\phi = \pi/3$, $Gr = 0.5$, $Gc = 0.5$, $Sr = 0.4$, $Sc = 0.4$, $\Theta = 2$, and $\beta = 0$.

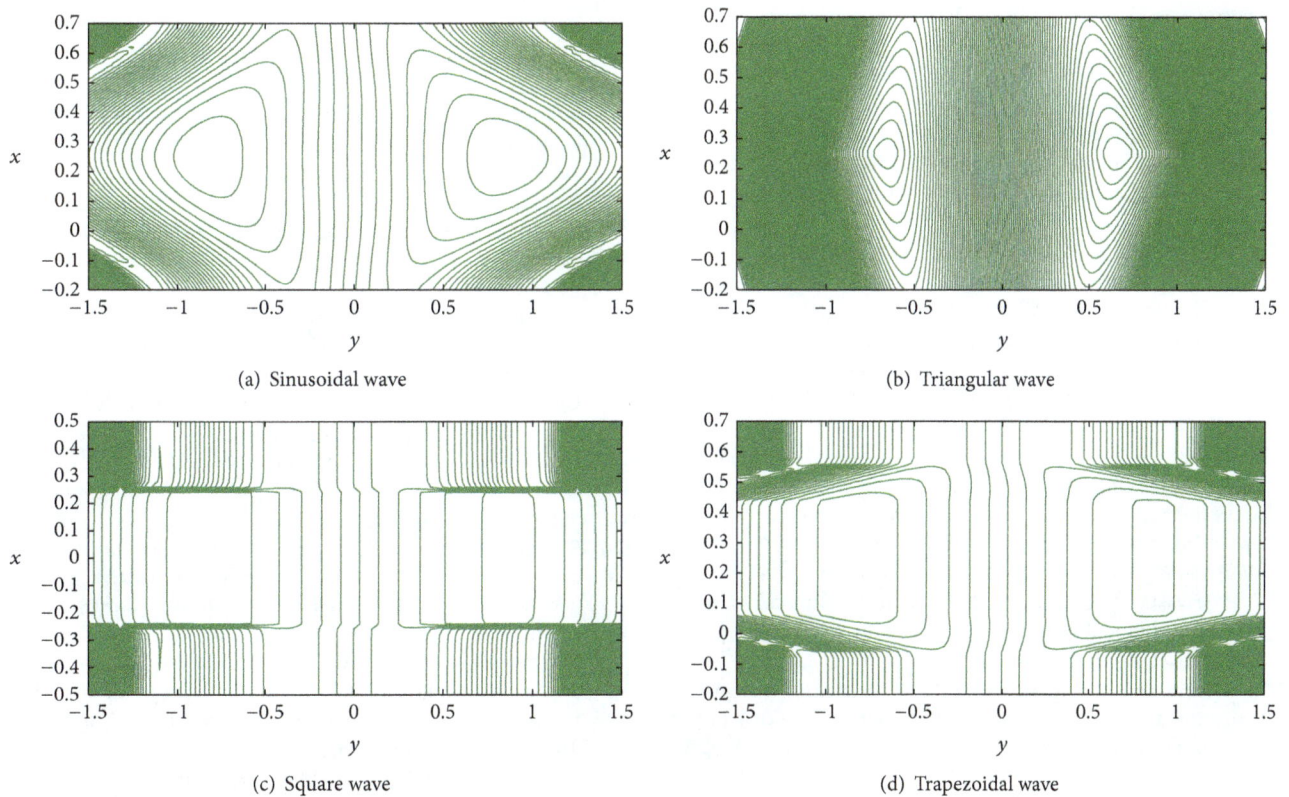

FIGURE 13: Streamlines for various wave forms for fixed values of (a) $a = 0.7$, $b = 0.5$, $d = 1$, $D_a = 0.5$, $M = 1$, $\gamma = 5$, $\phi = \pi/3$, $Gr = 0.5$, $Gc = 0.5$, $Sr = 0.4$, $Sc = 0.4$, $\Theta = 2$, and $\beta = 5$.

M: Hartman number
D_a: Darcy number
γ: Couple stress parameter
Gr: Local temperature Grashof number
Gc: Local concentration Grashof number
Pr: Prandtl number
θ: Dimensionless temperature
Φ: Dimensionless concentration
β: Dimensionless heat generation parameter
Sc: Schmidt number
Sr: Soret number
ψ: Stream function
Θ: Time mean flow rate in the fixed frame
F: Time mean flow rate in the wave frame
Q: Volume flow rate in the fixed frame
q: Volume flow rate in the wave frame.

Conflict of Interests

The authors declare that there is no conflict of interests regarding the publication of this paper.

References

[1] V. K. Stokes, "Couple stresses in fluids," *Physics of Fluids*, vol. 9, no. 9, pp. 1709–1715, 1966.

[2] V. K. Stokes, *Theories of Fluids with Microstructure*, Springer, New York, NY, USA, 1984.

[3] M. Devakar and T. K. V. Iyengar, "Run up flow of a couple stress fluid between parallel plates," *Nonlinear Analysis: Modelling and Control*, vol. 15, no. 1, pp. 29–37, 2010.

[4] M. Devakar and T. K. V. Iyengar, "Stokes' problems for an incompressible couple stress fluid," *Nonlinear Analysis: Modelling and Control*, vol. 1, no. 2, pp. 181–190, 2008.

[5] T. Hayat, M. Mustafa, Z. Iqbal, and A. Alsaedi, "Stagnation-point flow of couple stress fluid with melting heat transfer," *Applied Mathematics and Mechanics*, vol. 34, no. 2, pp. 167–176, 2013.

[6] N. S. Akbar and S. Nadeem, "Intestinal flow of a couple stress nanofluid in arteries," *IEEE Transactions on NanoBioscience*, vol. 12, no. 4, pp. 332–339, 2013.

[7] D. Srinivasacharya, N. Srinivasacharyulu, and O. Odelu, "Flow and heat transfer of couple stress fluid in a porous channel with expanding and contracting walls," *International Communications in Heat and Mass Transfer*, vol. 36, no. 2, pp. 180–185, 2009.

[8] R. Muthuraj, S. Srinivas, and D. Lourdu Immaculate, "Heat and mass transfer effects on MHD fully developed flow of a couple stress fluid in a vertical channel with viscous dissipation and oscillating wall temperature," *International Journal of Applied Mathematics and Mechanics*, vol. 9, no. 7, pp. 95–117, 2013.

[9] V. P. Srivastava and M. Saxena, "A two-fluid model of non-Newtonian blood flow induced by peristaltic waves," *Rheologica Acta*, vol. 34, no. 4, pp. 406–414, 1995.

[10] M. Mishra and A. Ramachandra Rao, "Peristaltic transport of a Newtonian fluid in an asymmetric channel," *Zeitschrift für angewandte Mathematik und Physik*, vol. 54, no. 3, pp. 532–550, 2003.

[11] S. K. Pandey and M. K. Chaube, "Study of wall properties on peristaltic transport of a couple stress fluid," *Meccanica*, vol. 46, no. 6, pp. 1319–1330, 2011.

[12] N. Ali and T. Hayat, "Peristaltic flow of a micropolar fluid in an asymmetric channel," *Computers & Mathematics with Applications*, vol. 55, no. 4, pp. 589–608, 2008.

[13] P. Naga Rani and G. Sarojamma, "Peristaltic transport of a Casson fluid in an asymmetric channel," *Australasian Physical and Engineering Sciences in Medicine*, vol. 27, no. 2, pp. 49–59, 2004.

[14] T. Hayat, A. Afsar, and N. Ali, "Peristaltic transport of a Johnson-Segalman fluid in an asymmetric channel," *Mathematical and Computer Modelling*, vol. 47, no. 3-4, pp. 380–400, 2008.

[15] T. Hayat and M. Javed, "Exact solution to peristaltic transport of power-law fluid in asymmetric channel with compliant walls," *Applied Mathematics and Mechanics*, vol. 31, no. 10, pp. 1231–1240, 2010.

[16] D. Tripathi, "Study of transient peristaltic heat flow through a finite porous channel," *Mathematical and Computer Modelling*, vol. 57, no. 5-6, pp. 1270–1283, 2013.

[17] D. Tripathi, "Peristaltic hemodynamic flow of couple-stress fluids through a porous medium with slip effect," *Transport in Porous Media*, vol. 92, no. 3, pp. 559–572, 2012.

[18] D. Tripathi and O. A. Bég, "A numerical study of oscillating peristaltic flow of generalized Maxwell viscoelastic fluids through a porous medium," *Transport in Porous Media*, vol. 95, no. 2, pp. 337–348, 2012.

[19] Y. Abd elmaboud and Kh. S. Mekheimer, "Non-linear peristaltic transport of a second-order fluid through a porous medium," *Applied Mathematical Modelling*, vol. 35, no. 6, pp. 2695–2710, 2011.

[20] S. Noreen, "Mixed convection peristaltic flow with slip condition and induced magnetic field," *The European Physical Journal Plus*, vol. 129, no. 2, article 33, 2014.

[21] T. Hayat, F. M. Mahomed, and S. Asghar, "Peristaltic flow of a magnetohydrodynamic Johnson-Segalman fluid," *Nonlinear Dynamics*, vol. 40, no. 4, pp. 375–385, 2005.

[22] Y. Wang, N. Ali, and T. Hayat, "Peristaltic motion of a magneto-hydrodynamic generalized second-order fluid in an asymmetric channel," *Numerical Methods for Partial Differential Equations*, vol. 27, no. 2, pp. 415–435, 2011.

[23] S. Nadeem and S. Akram, "Peristaltic flow of a couple stress fluid under the effect of induced magnetic field in an asymmetric channel," *Archive of Applied Mechanics*, vol. 81, no. 1, pp. 97–109, 2011.

[24] S. Nadeem and S. Akram, "Influence of inclined magnetic field on peristaltic flow of a Williamson fluid model in an inclined symmetric or asymmetric channel," *Mathematical and Computer Modelling*, vol. 52, no. 1-2, pp. 107–119, 2010.

[25] T. Hayat, M. Umar Qureshi, and Q. Hussain, "Effect of heat transfer on the peristaltic flow of an electrically conducting fluid in a porous space," *Applied Mathematical Modelling*, vol. 33, no. 4, pp. 1862–1873, 2009.

[26] S. Srinivas and R. Muthuraj, "Effects of chemical reaction and space porosity on MHD mixed convective flow in a vertical asymmetric channel with peristalsis," *Mathematical and Computer Modelling*, vol. 54, no. 5-6, pp. 1213–1227, 2011.

[27] M. Mustafa, S. Hina, T. Hayat, and A. Alsaedi, "Influence of wall properties on the peristaltic flow of a nanofluid: analytic and numerical solutions," *International Journal of Heat and Mass Transfer*, vol. 55, no. 17-18, pp. 4871–4877, 2012.

[28] Y. Abd elmaboud, Kh. S. Mekheimer, and A. I. Abdellateef, "Thermal properties of couple-stress fluid flow in an asymmetric channel with peristalsis," *Journal of Heat Transfer*, vol. 135, no. 4, Article ID 044502, 2013.

[29] S. Nadeem and N. S. Akbar, "Effects of induced magnetic field on peristaltic flow of Johnson-Segalman fluid in a vertical symmetric channel," *Applied Mathematics and Mechanics—English Edition*, vol. 31, no. 8, pp. 969–978, 2010.

[30] S. Srinivas and M. Kothandapani, "Peristaltic transport in an asymmetric channel with heat transfer—a note," *International Communications in Heat and Mass Transfer*, vol. 35, no. 4, pp. 514–522, 2008.

[31] S. Nadeem and S. Akram, "Magnetohydrodynamic peristaltic flow of a hyperbolic tangent fluid in a vertical asymmetric channel with heat transfer," *Acta Mechanica Sinica*, vol. 27, no. 2, pp. 237–250, 2011.

[32] O. U. Mehmood, N. Mustapha, and S. Shafie, "Heat transfer on peristaltic flow of fourth grade fluid in inclined asymmetric channel with partial slip," *Applied Mathematics and Mechanics. English Edition*, vol. 33, no. 10, pp. 1313–1328, 2012.

[33] N. S. Akbar, T. Hayat, S. Nadeem, and S. Obaidat, "Peristaltic flow of a Williamson fluid in an inclined asymmetric channel with partial slip and heat transfer," *International Journal of Heat and Mass Transfer*, vol. 55, no. 7-8, pp. 1855–1862, 2012.

[34] S. Srinivas, R. Gayathri, and M. Kothandapani, "The influence of slip conditions, wall properties and heat transfer on MHD peristaltic transport," *Computer Physics Communications*, vol. 180, no. 11, pp. 2115–2122, 2009.

[35] R. Ellahi, M. Mubashir Bhatti, and K. Vafai, "Effects of heat and mass transfer on peristaltic flow in a non-uniform rectangular duct," *International Journal of Heat and Mass Transfer*, vol. 71, pp. 706–719, 2014.

[36] S. Noreen, "Mixed convection peristaltic flow of third order nanofluid with an induced magnetic field," *PLoS ONE*, vol. 8, no. 11, Article ID e78770, 2013.

[37] M. Saleem and A. Haider, "Heat and mass transfer on the peristaltic transport of non-Newtonian fluid with creeping flow," *International Journal of Heat and Mass Transfer*, vol. 68, pp. 514–526, 2014.

[38] S. Nadeem and N. S. Akbar, "Influence of heat and mass transfer on the peristaltic flow of a Johnson Segalman fluid in a vertical asymmetric channel with induced MHD," *Journal of the Taiwan Institute of Chemical Engineers*, vol. 42, no. 1, pp. 58–66, 2011.

[39] T. Hayat and S. Hina, "The influence of wall properties on the MHD peristaltic flow of a Maxwell fluid with heat and mass transfer," *Nonlinear Analysis: Real World Applications*, vol. 11, no. 4, pp. 3155–3169, 2010.

[40] T. Hayat, S. Noreen, M. S. Alhothuali, S. Asghar, and A. Alhomaidan, "Peristaltic flow under the effects of an induced magnetic field and heat and mass transfer," *International Journal of Heat and Mass Transfer*, vol. 55, no. 1-3, pp. 443–452, 2012.

[41] S. Hina, T. Hayat, S. Asghar, and A. A. Hendi, "Influence of compliant walls on peristaltic motion with heat/mass transfer and chemical reaction," *International Journal of Heat and Mass Transfer*, vol. 55, no. 13-14, pp. 3386–3394, 2012.

[42] T. Hayat, S. Noreen, and M. Qasim, "Influence of heat and mass transfer on the peristaltic transport of a phan-thien-tanner fluid," *Zeitschrift fur Naturforschung A*, vol. 68, no. 12, pp. 751–758, 2013.

2

Numerical Characterization of the Performance of Fluid Pumps Based on a Wankel Geometry

Stephen Wan, Jason Leong, Te Ba, Arthur Lim, and Chang Wei Kang

Institute of High Performance Computing, 1 Fusionopolis Way, No. 16-16 Connexis, Singapore 138632

Correspondence should be addressed to Stephen Wan; wansym@ihpc.a-star.edu.sg

Academic Editor: Yanzhong Li

The performance of fluid pumps based on Wankel-type geometry, taking the shape of a double-lobed limaçon, is characterized. To the authors' knowledge, this is the first time such an attempt has been made. To this end, numerous simulations for three different pump sizes were carried out and the results were understood in terms of the usual scaling coefficients. The results show that such pumps operate as low efficiency (<30%) valveless positive displacements pumps, with pump flow-rate noticeably falling at the onset of internal leakage. Also, for such pumps, the mechanical efficiency varies linearly with the head coefficient, and, within the onset of internal leakage, the capacity coefficient holds steady even across pump efficiency. Simulation of the flow field reveals a structure rich in three-dimensional vortices even in the laminar regime, including Taylor-like counterrotating vortex pairs, pointing towards the utility of these pumps in microfluidic applications. Given the planar geometry of such pumps, their applications as microreactors and micromixers are recommended.

1. Introduction

The present study is part of a larger effort aimed at exploring applications of fluid pumps based on Wankel-type geometry and is focused on the performance characterization of the simplest of such pumps.

Such Wankel-type pumps are essentially rotary positive displacement pumps, which operate by having an inner rotor orbit inside a chamber. The rotor path, determined by the chamber profile, creates a trapped fluid volume which is displaced through the chamber. In contrast to rotodynamic pumps, the trapped fluid is continually compressed to a high pressure without being imparted high kinetic energies. As a positive displacement pump, it has characteristics similar to the reciprocating positive displacement pump and, hence, would generate the same flow at a given speed (RPM) regardless of the discharge pressure, that is, a flat H-Q curve. However, a rotary pump is more susceptible to internal flow leakages especially at high pump heads, leading to a significant reduction in efficiency. The advantages of rotary pumps are that, as well as being able to deliver a flow that is less pulsatile compared with reciprocating piston pumps, they are more compact in design and capable of valveless operation.

There are a number of rotary pump types that have been well established and have found industrial application, such as the Gear Pump, Lobe Pump, Sliding-Vane Pump, Screw Pump, and Progressive-Cavity Pump [1], but so far, to the authors' knowledge, the Wankel-type design for fluid pump applications has not been exactly established. Although rarely reported, it is known that applications of fluid pumps based on Wankel-type geometry are not new. For example, Monties et al. [2] developed a valveless blood pump comprising a double-lobed limaçon-shaped chamber and an elliptically shaped rotor rotating on an eccentric gear. It had a single inlet and outlet and delivered pulsatile flow.

Mathematically, the perimeter of Wankel-type geometry is an epitrochoid, the parametric form of which is given by

$$x = e \cdot \cos(m\theta) + R \cos\theta, \qquad (1a)$$

$$y = e \cdot \sin(m\theta) + R \sin\theta. \qquad (1b)$$

In the case of the blood pump mentioned above and in the pumps examined in the present study, the pump chamber

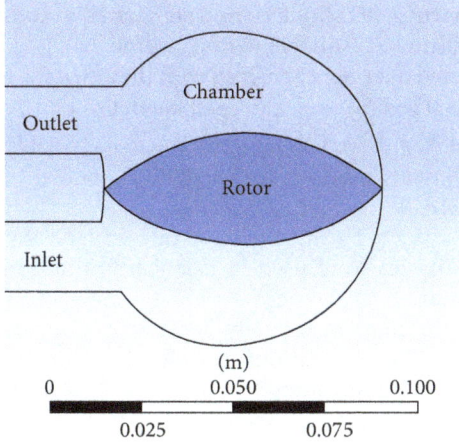

FIGURE 1: Estimated geometry of blood pump used in the calibration simulation.

geometry is obtained by setting $m = 2$ (see Figure 1), which is therefore the simplest of such geometries. (The well-known Wankel rotary engine chamber is generated by setting $m = 3$.)

In the present study, numerous computer simulations for three different pump sizes were carried out using commercially available computational fluid dynamics (CFD) codes (after calibrating a typical setup against experimental data). To the authors' knowledge, it appears that, to date, there has been no such previous attempt for these Wankel-type pumps, although there are numerous studies incorporating computational fluid dynamics (CFD) analyses, reported in the open literature, on particular and general aspects of various other pump types, including both rotodynamic and positive displacement pump types.

We list but several recent references, in the case of positive displacement type pumps, as follows. For gear pumps, see Riemslagh et al. [3] and Houzeaux and Codina [4]; for gerotor pumps, see Ruvalcaba and Hu [5]; for piston pumps, see Casoli et al. [6]; for progressive cavity pumps, see Paladino et al. [7]; for twin screw pumps, see Kovačević et al. [8]; and for vane pumps, see Takemori et al. [9].

In the case of rotodynamic pump types, for centrifugal pumps, see Gao et al. [10], Zhou et al. [11], Stel et al. [12], and Mihalić et al. [13]; for a mixed flow pump, see Liu et al. [14]; for a radial flow pump, see J.-H. Kim and K.-Y. Kim [15]; and for American Petroleum Institute (API) pumps, see Benigni et al. [16].

Numerical results generated in the present study were expressed in terms of the usual scaling coefficients. To conclude, we present details of the flow structure within the chamber for a pump sized for microfluidic applications.

2. Numerical Method

The simulations were carried out in a proprietary finite volume CFD code, ANSYS CFX v 15, using the immersed solid method to model the motion of the rotor.

2.1. Turbulence Model. To solve for the flow field, the shear-stress transport (SST) turbulence model of Menter [17]

was invoked to close the Reynolds averaged continuity and momentum equations (in standard Cartesian tensor notation):

$$\frac{\partial \rho}{\partial t} + \frac{\partial \left(\rho U_j \right)}{\partial x_j} = 0,$$

(2)

$$\frac{\partial \left(\rho U_i \right)}{\partial t} + \frac{\partial \left(\rho U_i U_j \right)}{\partial x_j} = -\frac{\partial P}{\partial x_i} + \frac{\left(\tau_{ij} - \rho \overline{u_i u_j} \right)}{\partial x_j} + S_M.$$

Details of the specific implementation of the SST model can be found in the ANSYS CFX manual [18]. Very briefly, the SST model calculates the Reynolds stresses, $-\rho \overline{u_i u_j} = \mu_t((\partial U_i/\partial x_j) + (\partial U_j/\partial x_i)) - (2/3)\rho k \delta_{ij}$, by solving a transport equation for the turbulent kinetic energy,

$$\frac{\partial \left(\rho k \right)}{\partial t} + \frac{\partial \left(\rho k u_i \right)}{\partial x_i} = \frac{\partial}{\partial x_j} \left(\Gamma_k \frac{\partial k}{\partial x_j} \right) + G_k - Y_k, \quad (3)$$

and another transport equation of the specific dissipation rate, ω, the ratio of energy dissipation rate, ε, to the turbulent kinetic energy, k:

$$\frac{\partial \left(\rho \omega \right)}{\partial t} + \frac{\partial \left(\rho \omega u_i \right)}{\partial x_i} = \frac{\partial}{\partial x_j} \left(\Gamma_\omega \frac{\partial \omega}{\partial x_j} \right) + G_\omega - Y_\omega + D_\omega. \quad (4)$$

The terms G_k and G_ω represent, respectively, the production of the turbulent kinetic energy, k, and the specific dissipation rate, ω, while the terms Y_k and Y_ω represent the dissipation of the turbulent kinetic energy, k, and the specific dissipation rate, ω. The cross-diffusion term, D_ω, arises from the transformation of the $k - \varepsilon$ model into the $k - \omega$ form. The effective diffusivities, $\Gamma_k = \mu + (\mu_t/\sigma_k)$ and $\Gamma_\omega = \mu + (\mu_t/\sigma_\omega)$, are obtained from turbulent Prandtl numbers for k and ω given by $\sigma_k = 1/(F_1/\sigma_{k,1} + (1 - F_1)/\sigma_{k,2})$ and $\sigma_\omega = 1/(F_1/\sigma_{\omega,1} + (1 - F_1)/\sigma_{\omega,2})$ (where F_1 is a blending function; and $\sigma_{k,1}, \sigma_{k,2}, \sigma_{\omega,1}$, and $\sigma_{\omega,2}$ are some constants). The value of the turbulent viscosity, μ_t, finally calculated from $\mu_t = a(\rho k/\omega)$, is limited by an expression, a, containing another blending function, F_2.

2.2. The Immersed Solid Technique. The immersed solid technique, implemented within the finite volume code of CFX [18], treats the presence of a solid object by applying a source term, S_M, to the momentum equation, in the fluid volume geometrically occupied by the solid object in the fluid domain. To track the movement of the immersed solid, the solver updates the mesh positions of the immersed solid at the beginning of each time step and applies the immersed solid sources, S_M, to the fluid nodes that lie within the immersed solid in an attempt to match the fluid velocity with the immersed solid velocity.

2.3. Simulation Calibration. The simulation setup for the pump performance characterization work was calibrated against the experimental data from [2, 19, 20]. To the authors' knowledge, it appears that, to date, these are the only sources of experimental data in the open literature for such pump types.

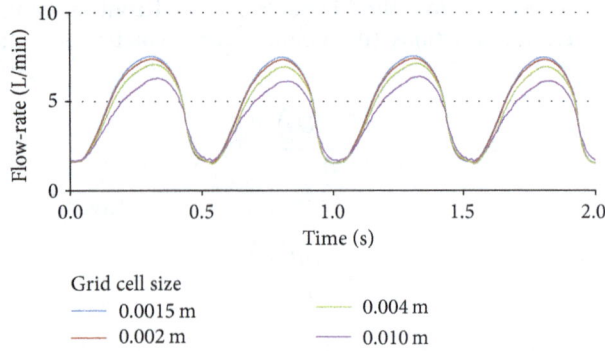

FIGURE 2: Instantaneous flow-rates obtained for various grid cell sizes in grid sensitivity study.

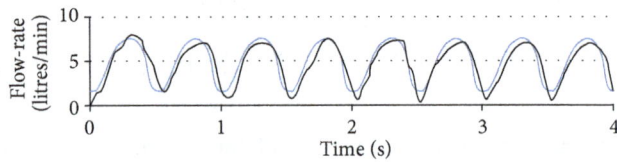

FIGURE 3: Predicted instantaneous flow-rate (blue curve) overlaid on experimental flow-rate (black line) from [1].

Direct comparison was not possible as the exact dimensions of the blood pump and fluid properties were not reported. By trial and error and by examination of the information available, the parameters of $R = 0.039$ m and $e = 0.0114$ m (which generated the size of the pump as shown in Figure 1) and a chamber thickness of 0.015 m were estimated.

A mesh grid sensitivity study was carried out by comparing instantaneous pump flow-rates obtained for different grid cell sizes. As shown in Figure 2, a cell size of 0.0015~0.0020 m would give reasonably accurate results. Using a time step of at least 1/500th of a period of shaft revolution and a mesh cell size of 0.002 m (yielding a total cell count of 55,000), we achieved reasonable agreement with the experimental data of Monties et al. [2], in terms of the instantaneous flow-rate time profile and magnitude, as indicated in Figure 3.

The uncertainty of the numerical results could not be assessed as Monties et al. [2] did not furnish information on the accuracy of and did not include error bounds on their experimental data. The mesh grids for the rotor and pump cavity for the calibration simulation are shown in Figure 4. A blood viscosity of 0.0035 Pa s and a density of $1.06E3$ kg/m^3 [21] were assumed. A first order upwind scheme was used to discretize the convective terms. Temporal discretization was performed with the first order backward Euler scheme performing 10 outward iterations per time step.

3. Pump Performance Scaling Studies

For the scaling studies, pumps of three different sizes of $R = 0.05$ m, $R = 0.1$ m, and $R = 0.2$ m were simulated and rotor tip clearances of $\Delta c = 0$ m (zero clearance), $\Delta c = 0.0025$ m, $\Delta c = 0.0064$ m, and $\Delta c = 0.0105$ m. Owing to the large

number of runs, 2D simulations were carried to complete the studies within a reasonable period of time.

Collected data were analyzed with the aid of the following dimensional groups: capacity coefficient, $C_Q = Q/NR^3$; head coefficient, $C_H = gH/N^2R^2$; efficiency, $\eta_P = \rho QgH/NT_R$; device Reynolds number, Re $= \rho NR^2/\mu$; and tip clearance ratio, $\epsilon = \Delta c/R$.

Time averaged values of flow-rate and reactive torque (over one cycle) were used in calculating the dimensional groups' values.

From the plots of capacity coefficient, C_Q, versus head coefficient, C_H, for various tip clearance ratios, ϵ, at a particular device Reynolds number, Re $= 1E6$ (Figure 5), we observe that the capacity coefficient, C_Q, remains constant up to a certain value of the head coefficient, C_H, beyond which it begins to drop noticeably, particularly for the largest tip clearance ratio, ϵ. These are typical of positive displacements pumps, where internal leakage would cause a deterioration of pump flow-rate at high pump heads. For a particular clearance ratio, we find that the pump efficiency varies fairly linearly with the head coefficient, C_H, across the three pump sizes studied here. See Figure 6.

It appears therefore that the capacity coefficient, C_Q, holds quite steady over variations in pump efficiency, η, as well. To confirm this, we ran simulations at a low and a higher efficiency and indeed the flow-rate, Q, scales linearly with NR^3 on a single line at these two disparate efficiencies ($\eta_P \sim 0.008$ and $\eta_P \sim 0.26$) as shown in Figure 7.

4. Pump Chamber 3D Flow Field

While the results of the above scaling studies are useful for a preliminary assessment of a suitable pump size for a particular application, detailed examination of the flow field in the pump chamber would be necessary to see if such pumps could be exploited for certain biomedical, microfluidic, or microreactor applications.

Hence, in this section, we report on a 3D simulation of pump scales to $R = 0.01$ m (which would be typical of an active microfluidic device) operating at a shaft speed of 200 rad/s on water against zero heads.

By plotting surface streamlines on a horizontal plane bisecting the thickness of the pump chamber as well as on a vertical plane along the axis of symmetry of the pump chamber for various rotor positions, as shown in Figure 8, it can already be observed that the flow pattern within the pump chamber is very complex.

Delving deeper, surface streamlines at various vertical planes across the pressure side of the pump chamber, for a particular rotor position, were then plotted. As shown in Figure 9, pairs of lateral vortices appear to first emerge near the tip of the rotor as indicated on plane P1. As the flow traverses the pump chamber, the vortices develop further while being stretched due to an increase in cross-sectional area in the main flow direction, from plane P2 to P5. By the time the flow reaches the outlet, the vortices appear to have more or less disappeared. A velocity vector plot on one of the planes, plane P2, as shown in Figure 10, clearly

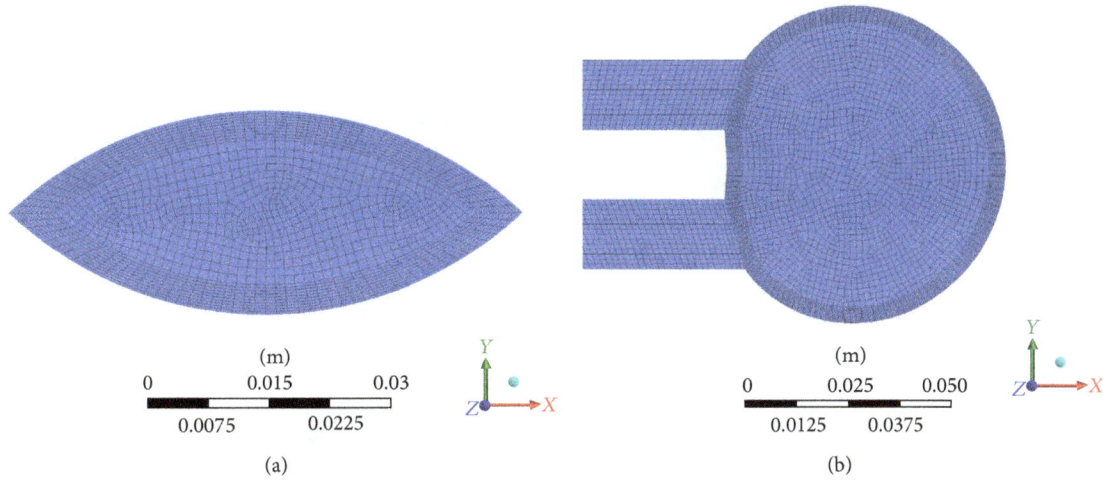

FIGURE 4: Mesh grid for the (a) rotor and (b) pump cavity.

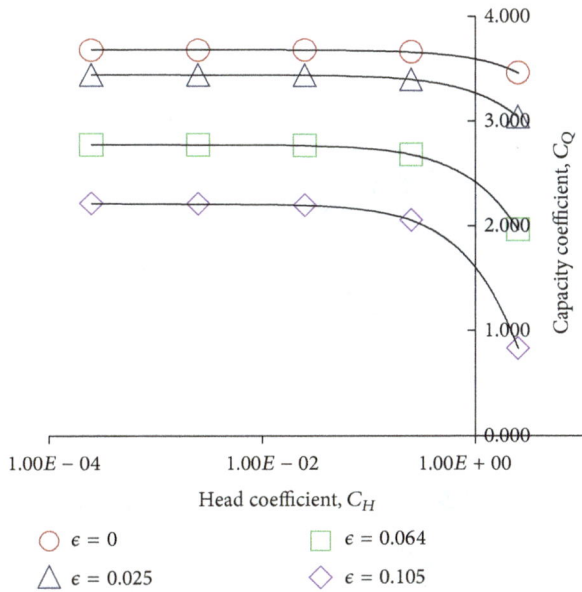

FIGURE 5: Plots of capacity coefficient, C_Q, versus head coefficient, C_H, for various rotor tip clearance ratios, ϵ, for a particular device Reynolds number, Re = 1E6.

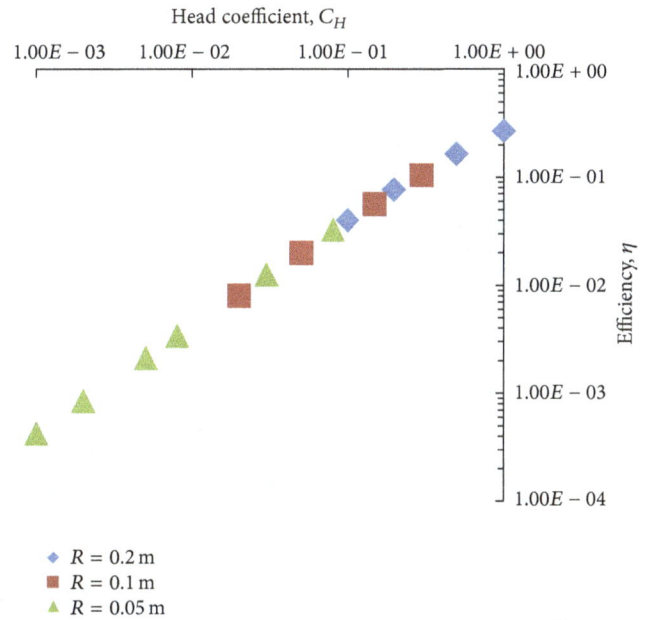

FIGURE 6: Variation of head coefficient, C_H, with pump efficiency, η_P, across three different pump sizes.

shows that these vortices resemble counterrotating Taylor vortices. The suction side of the pump chamber, on the other hand, is dominated by a lateral vortex in the horizontal plane (see again Figure 8). In essence, as the flow is drawn into the pump chamber, it sees a backward facing step, and the well-recognized recirculation zone is produced. A vector plot on a horizontal plane, as shown in Figure 11, confirms this observation.

Thus as the flow transits from the suction to the pressure sides, it is being continuously stretched from a pattern characterized by a dominant, horizontal lateral vortex to one characterized by pairs of vertical, lateral Taylor-like counterrotating vortices that are continuously deformed as the flow is being evacuated.

The resulting complex flow pattern, rich in three-dimensional vortices, suggests that pumps based on Wankel geometries have a good potential for mixing applications.

5. Concluding Remarks

We conclude this technical brief with the following salient points.

(i) A fluid pump based on the simplest Wankel geometry (double-lobed limaçon) operates as a positive displacement pump, without requiring the use of valves.

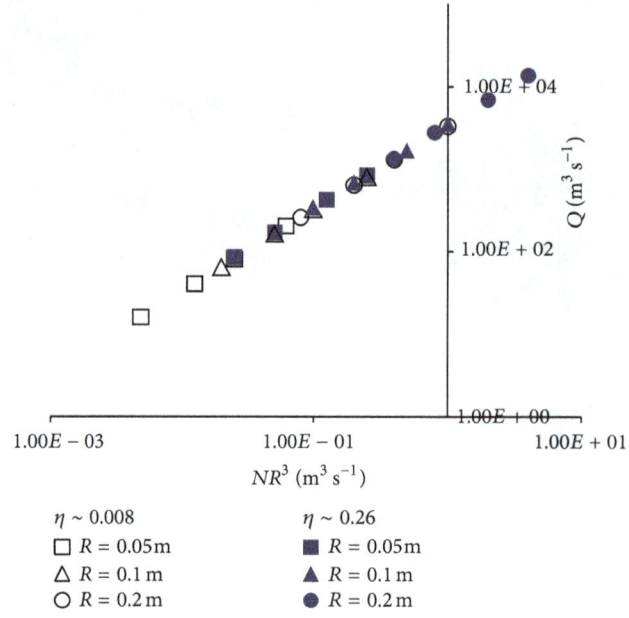

FIGURE 7: Linear scaling of pump flow-rate, Q, with NR^3 for two disparate pump efficiencies.

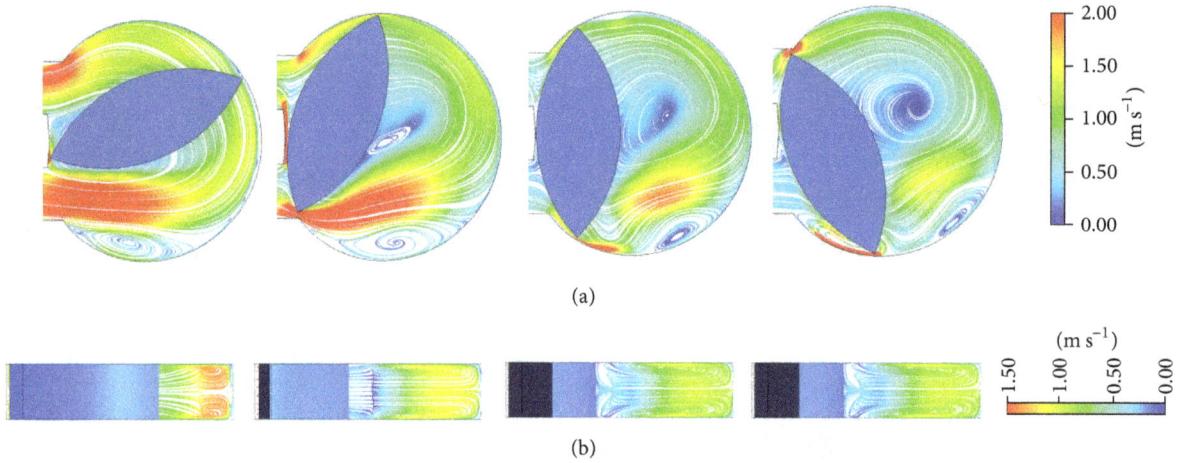

(a)

(b)

FIGURE 8: (a) Surface streamlines taken on a plane bisecting the thickness of the pump chamber for various rotor positions; (b) surface streamlines on a plane along the axis of symmetry of the pump chamber.

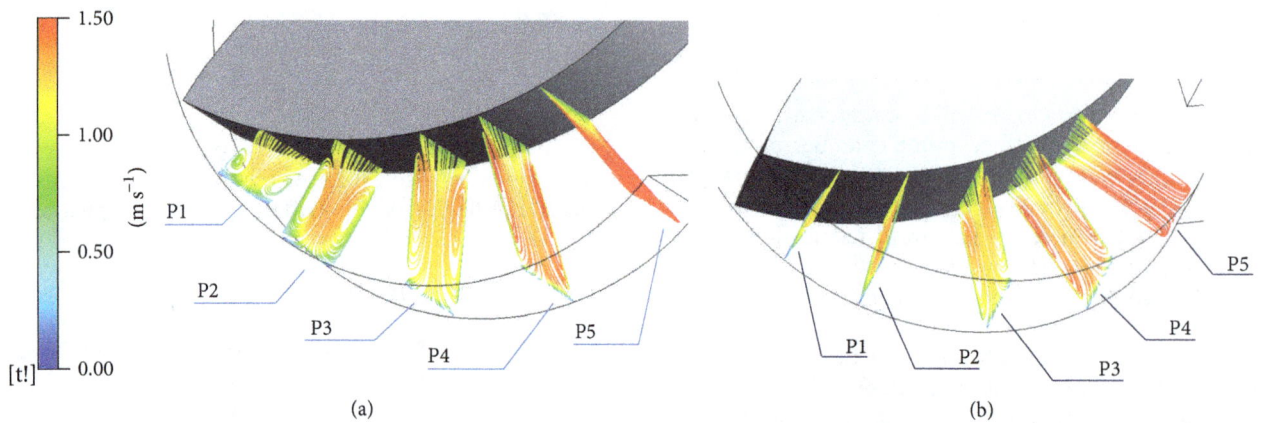

(a)   (b)

FIGURE 9: Two views of 2D streamlines on various cut-planes in the pressure side of the pump chamber, showing Taylor-like vortices being stretched as they traverse to the pump discharge.

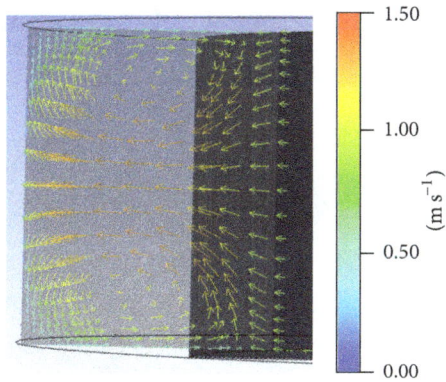

FIGURE 10: Velocity vectors plotted on one of the cut-planes in Figure 9, confirming the existence of Taylor-like counterrotating vortex pairs.

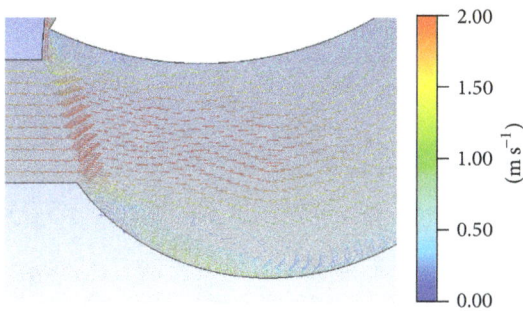

FIGURE 11: Velocity vectors plotted for one of the rotor positions in Figure 8, confirming a backward facing-type recirculation zone in the suction side of the pump.

(ii) As with positive displacement pumps in general, the capacity coefficient, C_Q, falls noticeably at the onset of fluid internal leakage.

(iii) For a particular set of dimensionless groups, pump efficiency, η_P, varies linearly with head coefficient, C_H.

(iv) Within the onset of internal leakage, the capacity coefficient, C_Q, holds steady even across pump efficiency, η_P, thus scaling pump flow-rate, Q, with shaft rotor speed, N, and the cube of rotor size, R (rotor centre-to-tip distance).

(v) Maximum pump efficiency is in the region of 30%.

(vi) Even in the laminar flow regime, flow structure within the pump chamber is complex and very rich in three-dimensional vortices.

(vii) Detailed examination of the flow structure reveals a transition from a structure dominated by a backward facing step type recirculation zone to one characterized by Taylor-like counterrotating vortex pairs that are continuously stretched, as the trapped fluid traverses from the suction side to the pressure side of the pump chamber.

(viii) It appears that a fluid pump based on the simplest Wankel geometry is best suited for applications in which pump efficiency is not an overriding issue; valveless operation is an advantage and complex flow patterns in the pump chamber are exploited to serve a particular function. Given the planar geometry of such pumps, their application as microreactors and micromixers is recommended.

Nomenclature

e: Epitrochoid eccentricity
g: Acceleration due to gravity
k: Turbulent kinetic energy
m: Epitrochoid shape parameter
C_H: Pump head coefficient
C_Q: Pump capacity coefficient
H: Pump head
N: Shaft speed
Q: Pump flow-rate
R: Rotor tip-to-centroid distance
Re: Device Reynolds number
S_M: Momentum source term
T_R: Reactive torque
ϵ: Ratio of rotor tip clearance to rotor tip-to-centroid distance
η_P: Pump mechanical efficiency
μ: Molecular viscosity
μ_T: Eddy viscosity
ρ: Fluid density
ω: Turbulent energy specific dissipation rate
Δc: Rotor tip clearance.

Conflict of Interests

The authors declare that there is no conflict of interests regarding the publication of this paper.

Acknowledgment

The authors would like to acknowledge the support provided by "Down-Hole Multiphase Flow Equipment Design & Analysis"-SERC TSRP Programme of Agency for Science, Technology and Research (A∗STAR) in Singapore (Reference no. 102-164-0077).

References

[1] K. Arnold and M. Stewarts, "Pumps," in *Surface Production Operations: Design of Oil-Handling System and Facilities*, vol. 1, pp. 333–354, 1999.

[2] J. R. Monties, P. Havlik, T. Mesana, J. Trinkl, J. L. Tourres, and J. L. Demunck, "Development of the Marseilles pulsatile rotary blood pump for permanent implantable left ventricular assistance," *Artificial Organs*, vol. 18, no. 7, pp. 506–511, 1994.

[3] K. Riemslagh, J. Vierendeels, and E. Dick, "An arbitrary Lagrangian-Eulerian finite-volume method for the simulation of rotary displacement pump flow," *Applied Numerical Mathematics*, vol. 32, no. 4, pp. 419–433, 2000.

[4] G. Houzeaux and R. Codina, "A finite element method for the solution of rotary pumps," *Computers and Fluids*, vol. 36, no. 4, pp. 667–679, 2007.

[5] M. A. Ruvalcaba and X. Hu, "Gerotor fuel pump performance and leakage study," in *Proceedings of the ASME International Mechanical Engineering Congress & Exposition*, pp. 807–815, Denver, Colo, USA, November 2011.

[6] P. Casoli, A. Vacca, G. Franzoni, and G. L. Berta, "Modelling of fluid properties in hydraulic positive displacement machines," *Simulation Modelling Practice and Theory*, vol. 14, no. 8, pp. 1059–1072, 2006.

[7] E. E. Paladino, J. A. Lima, P. A. S. Pessoa, and R. F. C. Almeida, "A computational model for the flow within rigid stator progressing cavity pumps," *Journal of Petroleum Science and Engineering*, vol. 78, no. 1, pp. 178–192, 2011.

[8] A. Kovačević, N. Stošić, and I. K. Smith, "Numerical simulation of combined screw compressor-expander machines for use in high pressure refrigeration systems," *Simulation Modelling Practice and Theory*, vol. 14, no. 8, pp. 1143–1154, 2006.

[9] C. K. Takemori, E. E. Paladino, and L. Lessa, "Numerical simulation of oil flow in a power steering pump," SAE Technical Paper 2005-01-4061, SAE, 2005.

[10] Z. Gao, W. Zhu, L. Lu, J. Deng, J. Zhang, and F. Wuang, "Numerical and experimental study of unsteady flow in a large centrifugal pump with stay vanes," *Journal of Fluids Engineering, Transactions of the ASME*, vol. 136, no. 7, Article ID 071101, 2014.

[11] L. Zhou, W. Shi, W. Li, and R. Agarwal, "Numerical and experimental study of axial force and hydraulic performance in a deep-well centrifugal pump with different impeller rear shroud radius," *Journal of Fluids Engineering*, vol. 135, no. 10, Article ID 104501, 2013.

[12] H. Stel, G. D. L. Amaral, C. O. R. Negrão, S. Chiva, V. Estevam, and R. E. M. Morales, "Numerical analysis of the fluid flow in the first stage of a two-stage centrifugal pump with a vaned diffuser," *Journal of Fluids Engineering*, vol. 135, no. 7, Article ID 071104, 2013.

[13] T. Mihalić, Z. Guzović, and A. Predin, "Performances and flow analysis in the centrifugal vortex pump," *Journal of Fluids Engineering*, vol. 135, no. 1, Article ID 011002, 2013.

[14] J. Liu, Z. Li, L. Wang, and L. Jiao, "Numerical simulation of the transient flow in a radial flow pump during stopping period," *Journal of Fluids Engineering, Transactions of the ASME*, vol. 133, no. 11, Article ID 111101, 7 pages, 2011.

[15] J.-H. Kim and K.-Y. Kim, "Analysis and optimization of a vaned diffuser in a mixed flow pump to improve hydrodynamic performance," *Journal of Fluids Engineering*, vol. 134, no. 7, Article ID 71104, 2012.

[16] H. Benigni, H. Jaberg, H. Yeung, T. Salisbury, O. Berry, and T. Collins, "Numerical simulation of low specific speed American petroleum institute pumps in part-load operation and comparison with test rig results," *Journal of Fluids Engineering*, vol. 134, no. 2, Article ID 024501, 2012.

[17] F. R. Menter, "Two-equation eddy-viscosity turbulence models for engineering applications," *AIAA Journal*, vol. 32, no. 8, pp. 1598–1605, 1994.

[18] ANSYS, *CFX User's Manual, Version 5*, ANSYS, Cecil Township, Pa, USA, 2014.

[19] J.-R. E. Monties, J. Trinkl, T. Mesana, P. J. Havlik, and J.-L. Y. Demunck, "Cora valveless pulsatile rotary pump: new design and control," *Annals of Thoracic Surgery*, vol. 61, no. 1, pp. 463–468, 1996.

[20] T. Mesana, S. Morita, J. Trinkl et al., "Experimental use of a semipulsatile rotary blood pump for cardiopulmonary bypass," *Artificial Organs*, vol. 19, no. 7, pp. 734–738, 1995.

[21] "The Physics Hypertextbook," http://physics.info/.

3

Separation Criteria for Off-Axis Binary Drop Collisions

Mary D. Saroka[1] and Nasser Ashgriz[2]

[1]*United Technologies Research Center, 411 Silver Lane, MS 129-19, East Hartford, CT 06108, USA*
[2]*Department of Mechanical and Industrial Engineering, University of Toronto, Toronto, ON, Canada M5S 3G8*

Correspondence should be addressed to Nasser Ashgriz; ashgriz@mie.utoronto.ca

Academic Editor: Robert Spall

Off-axis collisions of two equal size droplets are investigated numerically. Various governing processes in such collisions are discussed. Several commonly used theoretical models that predict the onset of separation after collision are evaluated based on the processes observed numerically. A separation criterion based on droplet deformation is found. The numerical results are used to assess the validity of some commonly used phenomenological models for drop separation after collision. Also, a critical Weber number for the droplet separation after grazing collision is reported. The effect of Reynolds number is investigated and regions of permanent coalescence and separation are plotted in a Weber-Reynolds number plane for high impact parameter collisions.

1. Introduction

Many engineering applications and natural process involve drop collisions in which the final outcome has a direct impact on the eventual success or failure of the application or process. In sprays, for example, the behavior is characterized by many small liquid drops with a particular size distribution and with each individual drop moving along a particular trajectory with a prescribed velocity. By singling out and focusing on just two drops within the spray, one can readily see that the interaction between these drops plays a part in the overall evolutional characteristics of the spray. One can further imagine that the likelihood of two drops colliding along the same axis of trajectory is less likely to occur than the situation in which the two drops collide along different trajectories. It is the latter case, typically referred to as off-axis collisions, that is the subject of this investigation.

Over the years, many researchers have studied the collision dynamics of the binary drops. The early studies on drop collision were motivated by understanding of the physical processes that occur during rain fall [1–12]. Rain drops may collide with each other and break into smaller drops or coalesce and generate larger drops. For instance, Park [5] did experiments on water drops in humid environment to determine conditions for the drop separation after collision. Several phenomenological models for the drop separation after collision are provided, for instance, the rotational energy

model [7] and the variation principal model for the minimum potential energy for the stretching separation [12]. The later studies were motivated by drop collision in various spray systems [13–28]. Collisions of many other liquids, such as hydrocarbons [13–18], heavy oils [19, 20], and mercury [21], in addition to water, have also been studied. The bulk of quantitative information on the drop collision is obtained through experiments [21–28], while numerical studies are used to provide detailed understanding of this complex process. The front tracking techniques [29–32] and a combination of level set and VOF methods are used to solve the dynamics of the free surfaces [33–40]. A detailed review of the droplet collision process is provided by Brenn [41]. Review of this body of research indicates that, depending upon the initial energy of the two drops and the angle at which they collide, different types of outcomes will occur. These outcomes can be categorized as bouncing, partial coalescence, permanent coalescence, separation, or shattering. There are also several subcategories for each outcome. Of these five, permanent coalescence and separation are the most frequently observed outcomes, while the other three, namely, bouncing, partial coalescence, and shattering, represent special cases in which the drops have either very small velocities, large size differences, or very high velocities.

The present study is on the off-axis collision of two equal size droplets. This study is a continuation of a previous

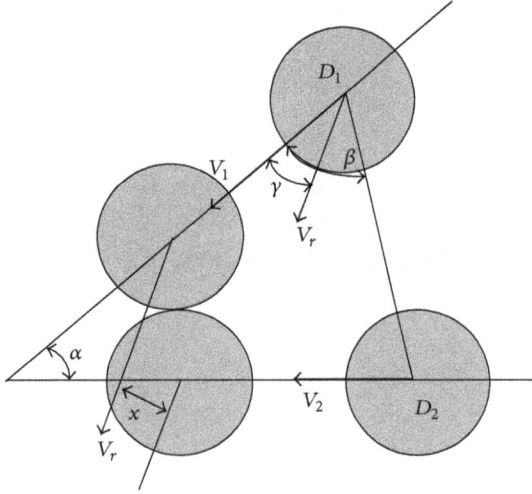

FIGURE 1: General orientation and parameter definition for binary droplet collisions.

study by the authors on the head-on collision of two drops [40]. Droplet collisions in dynamically inert surroundings are considered. Therefore, bouncing collisions, which are due to air entrapment between the two approaching droplets, are not observed. In the next sections, first the problem statement and the numerical methods are described, followed by the results and analysis.

2. Problem Statement

A general schematic depicting the positions and velocity vectors of two drops just prior to contact is shown in Figure 1. Suppose that one of the drops has a velocity defined as V_1, while the other has a velocity of V_2. These individual velocities are used to define a relative velocity, V_r, of the interacting drop. This is done by combining the drop velocities and their impact angle, α, through the following relation: $V_r = \sqrt{V_1^2 + V_2^2 - 2V_1V_2 \cos \alpha}$. Another important parameter necessary to define a particular collision involves the distance between the drops' centers. This distance is referred to as the impact parameter, noted as x. As the impact parameter increases the amount of "interaction" is reduced, while the converse is true. Using this information, it is convenient to define various dimensionless numbers to represent the phenomena under consideration. Three dimensionless parameters, namely, a Weber number, a nondimensional impact parameter, and a drop diameter ratio, are defined based on the drop diameter, liquid density, ρ, and the coefficient of surface tension, σ:

$$\text{We} = \frac{\rho D V_r^2}{\sigma},$$

$$X = \frac{x \sin |\beta - \gamma|}{D}, \tag{1}$$

where

$$\sin \gamma = \frac{V_1}{V_r} \sin \alpha. \tag{2}$$

In addition, Reynolds and Ohnesorge numbers are used to correlate effect of viscosity, μ, and its relationship with surface tension:

$$\text{Re} = \frac{\rho D V_r}{\mu},$$

$$\text{Oh} = \frac{\sqrt{\text{We}}}{\text{Re}} = \frac{\mu}{\sqrt{\rho D \sigma}}. \tag{3}$$

The Ohnesorge number is unique in the fact that it is only a function of fluid properties and the drop size, while both the Weber and Reynolds numbers are functions of the fluid properties, drop geometry, and impact velocity. In the present study, We, Re, and Oh numbers are calculated based on the liquid properties.

In order to discuss different types of collision outcomes, some collision images obtained by Ashgriz and Poo [22], but not published earlier, are provided here. The experimental setup and conditions are all as described in Ashgriz and Poo [22] and will not be repeated here. Figure 2(a) shows a close to head-on collision of two water droplets. After the collision, drops spread radially on each other until the kinetic energy of the drops is transformed into the surface energy. The surface energy is large enough to push back the fluid and form a ligament. The ligament stretches until it breaks into two droplets. This is referred to as a "reflexive separation." Ashgriz and Poo [22] have provided a theoretical model to predict the onset of this type of separation. If the collision impact parameter is increased enough, the impact process becomes complicated. Part of the drop fluid goes through the spreading and reflexive action, and another part of the drop fluid tends to stretch and pull the drops away from each other. If the collision kinetic energy is just enough to have reflexive separation, as shown in Figure 2(a), then a small increase in the impact parameter reduces this reflexive energy, and drops may not separate and stay coalesced. This process is shown in Figure 2(b), where the coalesced drop goes through significant deformation and oscillations, but drops remain coalesced. If the impact parameter is further increased, then the part of each drop which remains uninteracted with the other drop tends to stretch and pull the drops apart. This is observed in Figures 2(c), 2(d), and 2(e), and it is referred to as stretching separation. The present study only addresses stretching separation.

3. Numerical Methodology

We consider two drops composed of the same fluid with constant properties. The flow inside the drops is solved using the standard conservation of mass and momentum equations:

$$\nabla \cdot \mathbf{V} = 0,$$

$$\rho \frac{\partial \mathbf{V}}{\partial t} + \rho \left(\mathbf{V} \cdot \nabla \right) \mathbf{V} = -\nabla p + \mu \nabla^2 \mathbf{V} + \mathbf{F_b}, \tag{4}$$

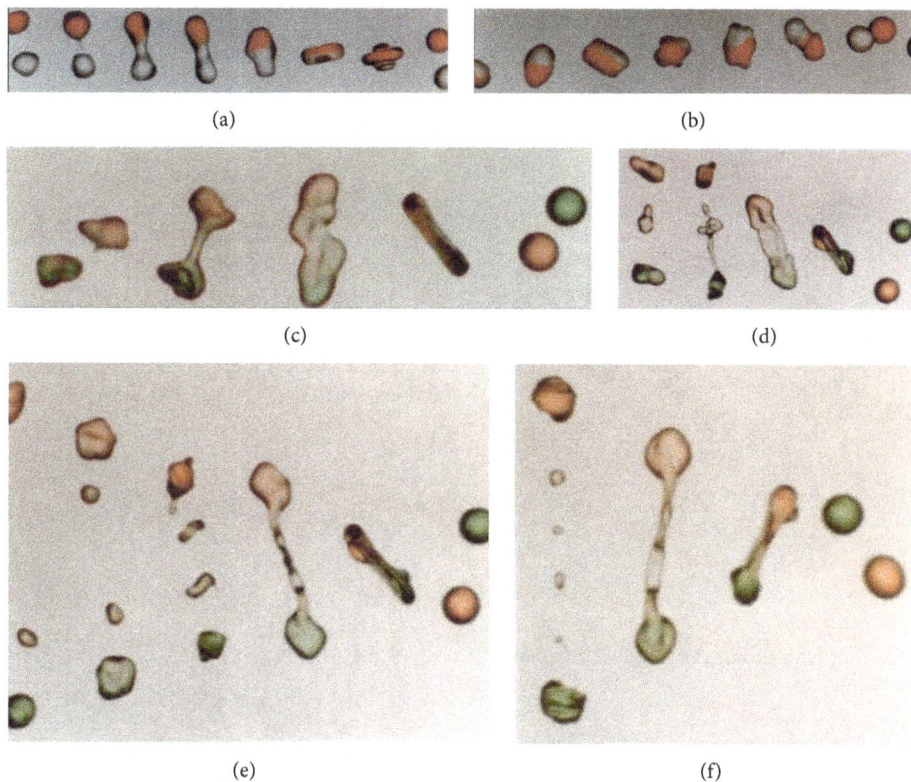

FIGURE 2: Binary collision of water droplets. (a) Reflexive separation due to near head-on collision; (b) coalescence collision; (c) separation collision resulting in two drops; (d) separation collision resulting in three drops; (e) separation collision resulting in four drops; and (f) separation collision resulting in six drops.

where $\mathbf{V}$ is the velocity vector, p is the pressure, and $\mathbf{F_b}$ is any volume force acting on the drops. At the free surface, the boundary condition requires that both mass and momentum are conserved. For drops surrounded by a less dense/viscous fluid or gas, the viscous stresses at the surface are zero. Additionally, assume a constant surface tension coefficient allows the stress boundary condition to reduce to Laplace's equation $p_s = \sigma\kappa$, where p_s is the pressure jump across the surface and κ is the surface curvature. Details of the numerical simulation have been discussed in earlier publications [40, 42, 43] and are not repeated here.

Two drops with a prescribed distance separating their respective centers are considered. In all cases, this distance was defined to occur in one direction only. That is to say that for drops translating in one direction (i.e., z-direction) their centers were placed in the same plane (called the defining plane) composed of this direction and one of the other two remaining directions (i.e., y-z or x-z plane). The third direction (i.e., x or y, resp.) was held constant for both drops. One might imagine that this third direction could also represent an off-set, thus indicating a second impact parameter. However, this case was not considered during this investigation. Figure 3 provides an example of the defining plane for an off-axis collision. This plane represents the y-z plane and is useful in observing the internal flow patterns as the two drops collide and evolve. Although similar planes such as X-Z or X-Y planes could be defined, they would only

represent a specific region of the collision and are therefore not as comprehensive.

The distance between the drops' centers was used to compute the impact parameter. For this investigation, impact parameters ranging from 0 to 1 were considered, with the bulk of the simulations focusing on collisions for $X > 0.5$ (i.e., high impact parameters). In order to cause the drops to collide, each drop was given a nonzero initial velocity. These individual drop velocities were subsequently used to compute the relative velocity, V_r. All simulations were conducted using water drops for Weber numbers ranging from 10 to 60 and Reynolds numbers ranging from 135 to 1639. Since various drop sizes ranging from $10\,\mu$m to 2 mm were used in the course of this investigation, a wide range of relative velocities ranging from 0.5 m/s to 21 m/s were considered. For very small drops of 10 μm, relative velocities as high as 21 m/s were used, while for the larger drops of 2 mm, velocities less than 1 m/s were predominately used. The fluid properties, density, ρ, viscosity, μ, and the coefficient of the surface tension, σ, were kept constant at 1000 kg/m^3, 10^{-6} m^2/s, and 0.073 N/m, respectively.

In order to resolve the flow within the drops, 10 cells per radius were used. A computational grid of $70 \times 70 \times 180$ cells in the x-, y-, and z-directions was used to allow the collision to evolve without hitting any walls. The use of 180 cells in the z-direction was necessary, since this was the primary direction in which the drops approached each other. Higher

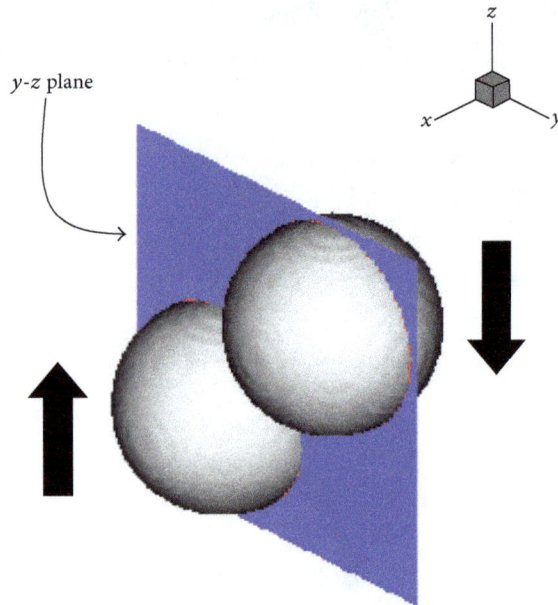

FIGURE 3: Definition of the *y-z* plane. Original two drops translating in the ± *z*-direction.

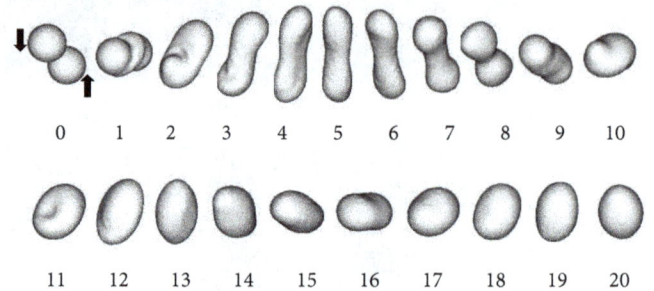

FIGURE 4: Coalescence collision of two 300 μm water droplets with We = 25, Re = 740, and X = 0.7. Numbers in the figure indicate time = $(0.1 \, \text{ms})i$, where i = image number.

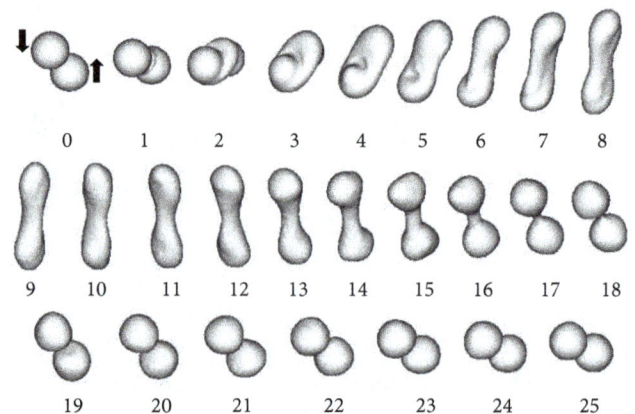

FIGURE 5: Separation collision of two 300 μm water droplets with We = 30, Re = 810, and X = 0.7. The numbers in figure indicate time = $(0.05 \, \text{ms})i$, where i = image number.

Weber numbers than those considered were not simulated, since these types of collisions required a larger computational grid to fully capture the evolutional dynamics of the collision.

4. Results and Discussion

Figures 4–6 show typical surface evolutions of water drops colliding at impact parameters greater than zero. Figure 4 shows a coalescence collision of two 300 μm drops colliding at We = 25, Re = 740, and X = 0.7. Figure 5 shows results for a similar case as in Figure 4 but with a slightly larger relative velocity, where two 300 μm drops collide at We = 30, Re = 811, and X = 0.7. It is shown that the boundary between the coalescence and separation at the impact parameter of X = 0.7 occurs at a Weber number between 25 < We < 30. A small increase in the impact We number causes the drops to separate after collision. Comparing Figures 4 and 5 with each other, it is clear that, in Figure 4, the surface energy pulls the ligament formed at stage 5 back together, and drops coalesce. In Figure 5, the ligament formed at stage 10 is longer than that in Figure 4. In this case, the pinching effect, which is governed by the Rayleigh type effect, is faster than the forces that tend to coalesce the drops. Therefore, the ligament in Figure 5 breaks before it comes back forming a coalesced droplet.

A general description of an off-axis collision process is as follows. As the drops come into contact with one another, a small region of interaction develops, while the remaining portion of the drops continues on their original course. This stretches the collided region. As the combined mass continues to stretch, the interaction region allows a portion of the fluid at the ends to drain back towards the middle of the system. This in turn causes the entire mass to rotate in a clockwise direction when the top drop is to the right of the

bottom drop (opposite orientation results in counterclockwise rotation). Fluid within the mass continues to adjust itself in this manner until a liquid bridge between the two ends is formed. Depending upon the amount of initial energy present in the system and the magnitude of the impact parameter, the rotating mass will either contract resulting in permanent coalescence or separate without or with satellite drops. Figure 6 shows collision of two 700 μm drops colliding at We = 60, Re = 1751, and X = 0.7. In this case, a long ligament is formed, which later results in a satellite droplet in addition to the two main droplets. It can be seen in images 16 to 24 that the rate of contraction of the inner ligament is faster than the end pinching of the drops. Therefore, only one satellite is formed. If the inner ligament was slightly longer, end pinching could result in three satellite droplets.

4.1. Internal Flow Patterns. There has been a lot of debate about the existence of internal circulations in droplets of a spray. Since it is very difficult to see the internal flow of a moving droplet, the main way to determine this effect has been through numerical modeling. The present study puts forward a concept for the formation of the internal circulations in the droplets. Figure 7 shows the internal flow patterns for the drop collision of Figure 6 (We = 60,

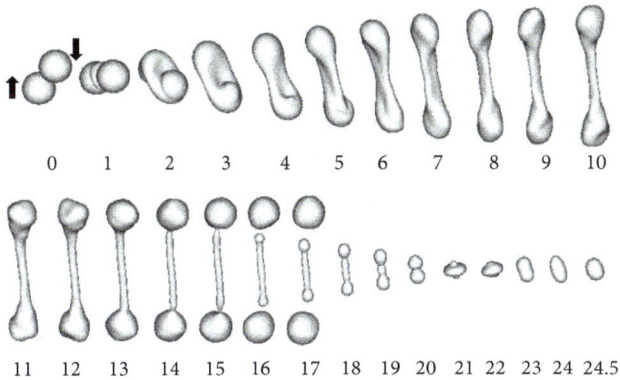

FIGURE 6: Stretching separation with one satellite droplet for the collision of two 700 μm water droplets with We = 60, Re = 1751, and X = 0.7. The numbers in the figure indicate time = (0.2 ms)i, where i = image number.

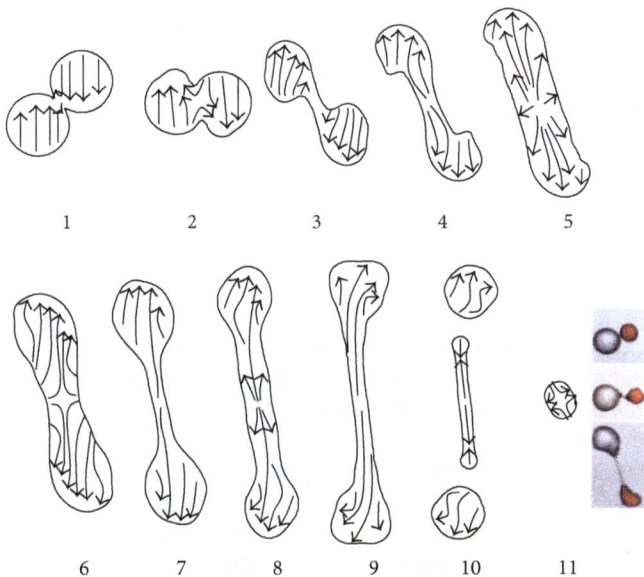

FIGURE 7: Internal flow patterns for the y-z plane. Results are for a binary water drop collision with a diameter of 700 μm and We = 60, Re = 1751.

TABLE 1: Upper critical Weber numbers for two equal size water droplets (X = 1).

Diameter μm	We$_{cr}$	Re
10	32.5	154
75	22	347
150	21	480
300	20.3	665
700	18.3	966
2000	18	1622

the streamlines in images 7–9. Image 10 indicates that separation has occurred with the production of two parent drops and a ligament. It is interesting to note several observations at this point. First, review of early images does not provide any evidence of the pinching mechanism in this defining plane. Therefore, the pinching phenomena are the result of radial type flow. Second, the parent drops exhibit flow patterns representing translating flow. Unlike the scenario for the head-on collisions [40] these parent drops reflect similar flow patterns of the original two drops which causes them to continue to translate after separation occurs. Lastly, the ligament collapses into a satellite drop producing the symmetrical flow pattern observed. Therefore, main droplets after collision have unidirectional internal flows, whereas satellite droplets may have multidirectional and internal flow circulations.

4.2. The Upper Boundary between Coalescence and Separation. The upper boundary between a coalescence and separation collision occurs for grazing collisions, that is, X = 1. As soon as the surfaces of two drops touch, the liquid flows between them. Depending on the collision conditions, the two drops may touch and temporarily coalesce, and then separate, or remain coalesced. The We number corresponding to this condition is referred to as the upper critical We number, We$_{cr}$. Table 1 shows the numerically determined upper critical Weber number for equally sized water drops as a function of drop diameter. It is clear that the upper critical Weber number increases with decreasing diameter. The pressure inside smaller drops is larger than those in larger drops; therefore, the flow is pushed into the liquid bridge between the two drops faster in the smaller drops than for the larger drops. The larger the mass of the bridge, the more difficult it is to break the drops. Therefore, the separation occurs at higher Weber numbers for the smaller drop collisions. The curve fit to the data of Table 1 is We$_{cr}$ = −2.7 lnD + 36. As the drop size increases, the effect of change in drop size becomes less important. The upper critical We number approaches We = 18 for large drop sizes.

4.3. Stretching Separation Boundary. The boundary between permanent coalescence and separation for off-axis collision of two equal size droplets for different collision impact parameters is obtained. Figure 8 presents the results of this investigation for equal size drop collisions with a diameter

Re = 1751, and X = 0.7) and for the plane described in Figure 3. The first image in Figure 7 shows the two drops just after contact. Immediately after the collision a small region of interaction is produced. In this region the fluid is redirected away from the collision point. As this occurs, the outer (or noninteracting) portion continues to translate unimpeded in the original direction (as shown in images 2–4). In the fifth image, in the center of the combined mass some of the flow begins to be redirected "radially" while the bulk of the motion still occurs in original translating direction. However, surface tension compensates and the radial flow is redirected, thus causing the overall flow to be directed upward/downward and away from the center. This action causes the stretching phenomena observed in off-axis collisions. The bulk of the fluid eventually ends up in the bulbous ends as indicated by

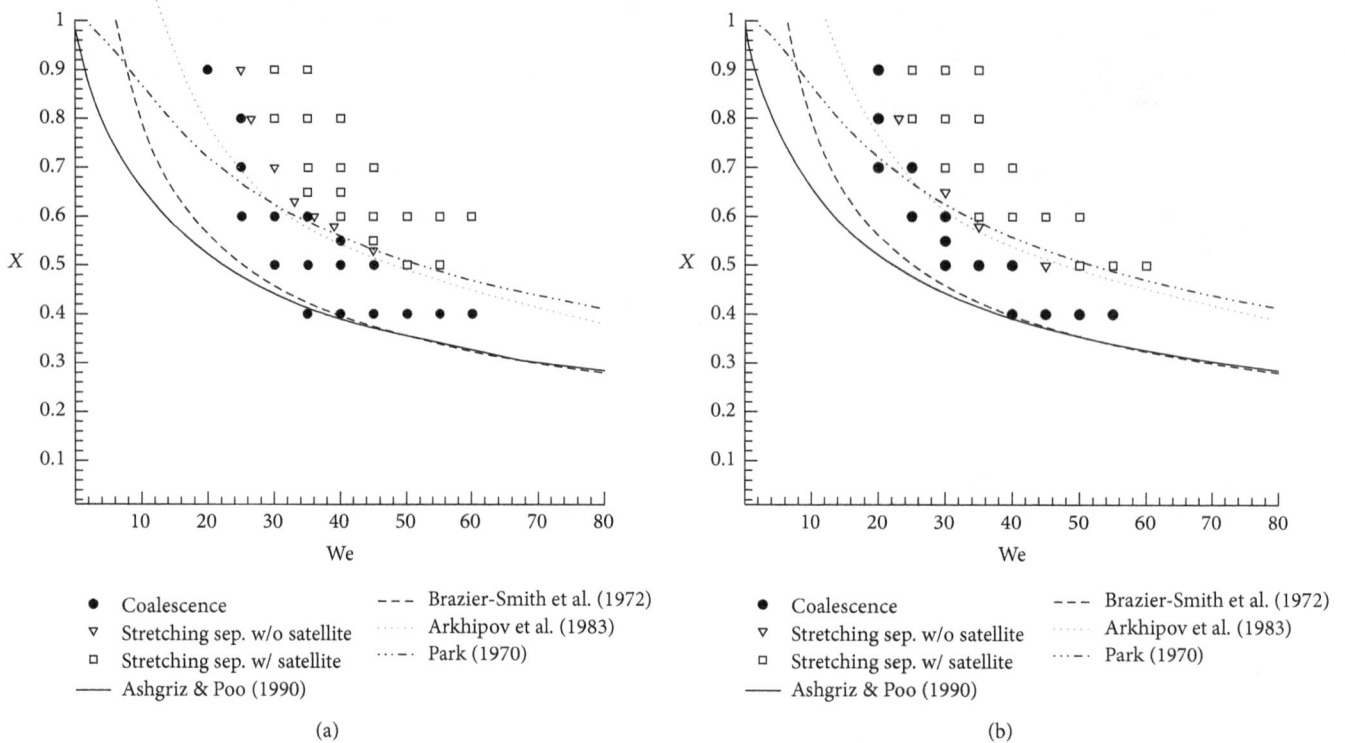

FIGURE 8: Coalescence-fragmentation curves: (a) 300 μm drop collisions and (b) 700 μm drop collisions. Symbols represent numerical data while lines represent boundary models presented by previous researchers.

of 300 μm (Figure 8(a)) and 700 μm (Figure 8(b)). Outcomes of permanent coalescence, separation without satellite production, and separation with one satellite produced are plotted for a range of Weber numbers and impact parameters. Also included in the figures are the boundaries defined by the analytical models proposed by Ashgriz and Poo [22], Arkhipov et al. [12], Brazier-Smith et al. [7], and Park [5]. The model by Ashgriz and Poo [22] is referred to as AP/EXP model for the remainder of the discussion. Similarly, the other models are referred to as A, B-S, and P, respectively. The results show that, for a constant impact parameter, as the Weber number is increased, the outcome changes from coalescence to separation without satellites to separation with a satellite. Similarly, as the Weber number is held constant, the same progression is observed provided there is enough energy initially present in the drops (i.e., at low Weber numbers, regardless of impact parameter, coalescence will always occur).

The numerical division between the coalescence and separation occurs at higher Weber and impact parameter combinations than those observed by the AP/EXP and B-S models. A possible explanation for this behavior is attributed to the existence of trapped air in the gap between the colliding drops. In the experiments, the impact parameter is determined by measuring the lines of trajectory and the corresponding separating angle [22]. Depending upon when this is done (i.e., a few diameters from the drop generators or just prior to impact) the value of the impact parameter can be quite different. As the two drops travel through air,

their trajectories remain relatively constant. However, as they come closer and closer, the surrounding gas becomes trapped in between them, which results in the slippage of the drops on each other. This may alter the original impact parameter based on the drop trajectories prior to collision. The effect of such a slippage is shown in Figure 9 by a shift in data. In this figure, the solid line represents the boundary using the AP/EXP model while the symbols represent the numerical results for the stretching separation without satellite production. The dashed line represents a linear translation of the AP/EXP model and is meant to convey the concept of impact parameter slippage. For example, at We = 31, the impact parameter on the AP/EXP boundary is $X = 0.44$, while that numerically is $X_s = 0.64$. Thus, due to the presence of the trapped air, the impact parameter at the time of collision could actually be higher than that previously thought.

The presence of the trapped air has three different effects on the drops, which ultimately cause an increase (or slippage) in the effective impact parameter. These effects are (i) altered trajectories, (ii) altered internal flow fields, and (iii) shape deformation. The trapped air conveys an external force on the drops, which may alter their trajectories, as shown in Figure 10. The second effect is shown in Figure 11, where drops moving in air may develop internal flows. The internal flow changes the drop dynamics after collision and may change the collision outcome. Finally, the surrounding air may cause drop deformation, as shown in Figure 12. As a result of this deformation the region of contact may increase as opposed to the case in which the drops stay

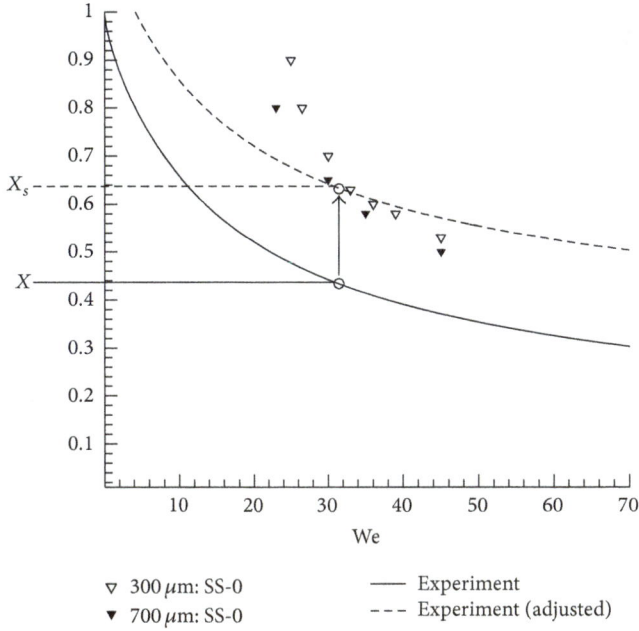

FIGURE 9: Impact parameter deviation for the same Weber numbers. X is a point on the experimental boundary, while X_s represents the impact parameter observed numerically. The solid line represents the boundary model by Ashgriz and Poo [22], while the dashed line represents a linear translation added to their model.

FIGURE 10: Effect of trapped air between colliding drops. Experimental photo from Ashgriz (unpublished material from Ashgriz and Poo [22]). The trapped air alters the collision impact parameter based on droplet trajectories.

perfectly spherical. This increased contact region may result in a larger interaction region and thus more mixing. It is, therefore, concluded that, in determining the collision impact parameter, one has to be worried about the slippage effect that may occur due to the air entrapment.

Figure 8 also compares the numerical results with models by Arkhipov et al. (A) [12] and Park (P) [5]. Arkhipov assumed that after the collision a spherical drop is formed that rotates with a constant angular velocity, while Park assumed that the combined mass rotates as two connected

spheres. The numerical results, as evident in Figure 5, show that, for the drops to separate after collision, they rotate as two drops that are connected, and not as a combined spherical mass. Therefore, Park's model is closer to numerical observation of the collision dynamics. However, the drop shapes and masses are altered and are not exactly the same as the original droplets. This becomes particularly true as the Weber number is increased. However, at low Weber numbers, We < 25, the numerical results do not correlate with any of the models considered. This is most likely explained by an incorrect assessment of the amount of angular velocity produced during the evolution. One could imagine that the amount of angular velocity would in fact not be constant (as assumed by Arkhipov et al.) and be a function of the impact parameter instead.

4.4. Deformation. The deformation variations in each direction for an off-axis collision of two water drops with a Weber number of 30 and a Reynolds number of 810 are provided in Figures 13–16. For clarity, the deformations are grouped according to either low impact parameters, $X < 0.5$, or high impact parameters, $X > 0.5$. It is convenient to nondimensionalize the amount of deformation by dividing by the drop's initial diameter. In doing so, the following definitions are used to represent the dimensionless deformation in each direction:

$$\eta = \frac{L_x}{D},$$

$$\gamma = \frac{L_y}{D}, \tag{5}$$

$$\psi = \frac{L_z}{D},$$

where, L_x is the amount of deformation in the x-direction, L_y is the amount of deformation in the y-direction, and L_z is the amount of deformation in the z-direction. Since off-axis collisions occur on different trajectories, the evolution process occurs in all three dimensions, resulting in separate behavior for each direction. In the next three subsections, the behavior in each direction is analyzed.

As observed with head-on collisions [40], three major periods of the collision process are spreading and production of a disk, period I, contraction of the disk, period II, and stretching along the collision axis (z-direction), period III.

4.5. x-Direction (Minor Axis 1). Figure 13 contains the dimensionless deformation in the x-direction. Review of this figure during period I provides the following information. First, it is observed that, at $t^* = 0$, η increases as the impact parameter, X, increases. This behavior is a direct consequence of the impact parameter definition, since the drop's separating distance is defined in this direction. Second, as the impact parameter is increased the amount of spread decreases. Therefore, as the region of interaction becomes smaller due to the increased impact parameter, less fluid is allowed to flow in this direction. Additionally, at $X \sim 0.6$, no spread is observed and the deformation actually becomes drainage or contraction for higher impact parameters. This

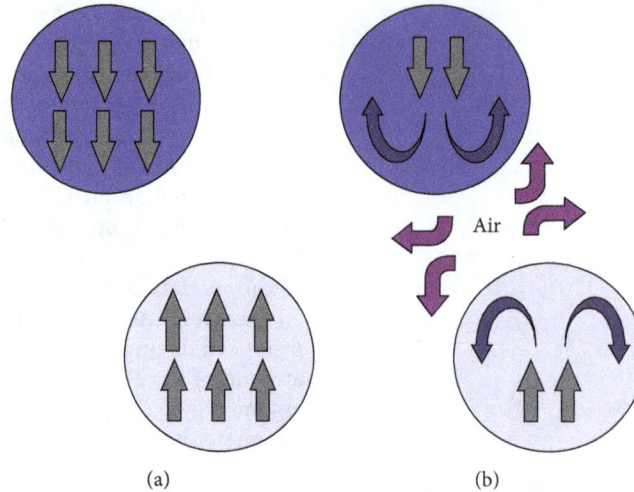

(a)   (b)

FIGURE 11: Internal flow of two drops approaching each other in (a) vacuum and (b) in air.

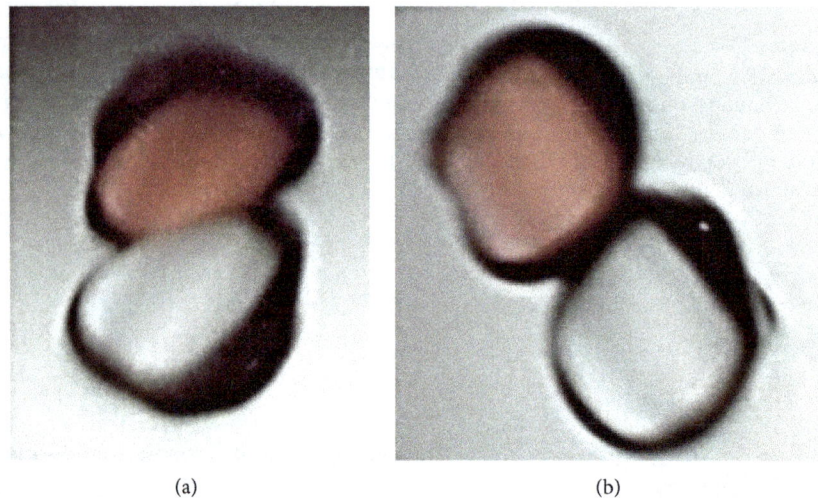

(a)   (b)

FIGURE 12: Two cases show the effect of trapped air in deforming the drop prior to collision and merging [unpublished photos from Ashgriz and Poo experiments [22]].

would seem to imply that, at this (X) and higher impact parameters, fluid is actually pulled out of this direction. The amount of time this period takes to occur is relatively constant regardless of impact parameter.

Period II was defined to represent the action of fluid contraction of the disk formed during period I. When the variation of impact parameter is considered, the following observations can be made. First, the amount of contraction is relatively constant until a value of $\eta \sim 1.3$ is obtained at the end of this period for $X < 0.3$. Essentially, this means that, for low impact parameters, the redirected fluid momentum in this direction dominates the flow which in turn causes the disk to contract inward. Second, as the impact parameter is increased until $X \sim 0.6$, more contraction occurs until a minimum value of $\eta \sim 0.75$ is reached. In this range of $0.3 < X < 0.6$, a competition between the amount of fluid momentum and the surface tension effect becomes evident, with the fluid momentum effect decreasing

as the impact parameter is increased (i.e., due to the reduction in the interaction region). For impact parameters greater than 0.7, this minimum deformation value is maintained. Therefore, at these high impact parameters, the amount of deformation does not change since the surface tension is now the controlling factor.

4.6. y-Direction (Minor Axis 2). The dimensionless deformation in the y-direction is provided in Figure 14. As with the x-direction variation, period I behavior still exists regardless of impact parameter variation. However, the amount of spread decreases as the impact parameter increases. Again, this is seen as a consequence of the reduction of the interaction region. Unlike in the x-direction, spread is observed up until $X \sim 0.8$ (instead of $X \sim 0.6$ as seen in the x-direction). At higher impact parameters, the amount of deformation shifts to contraction instead. This indicates that more mixing occurs in the y-direction than in the x-direction, since

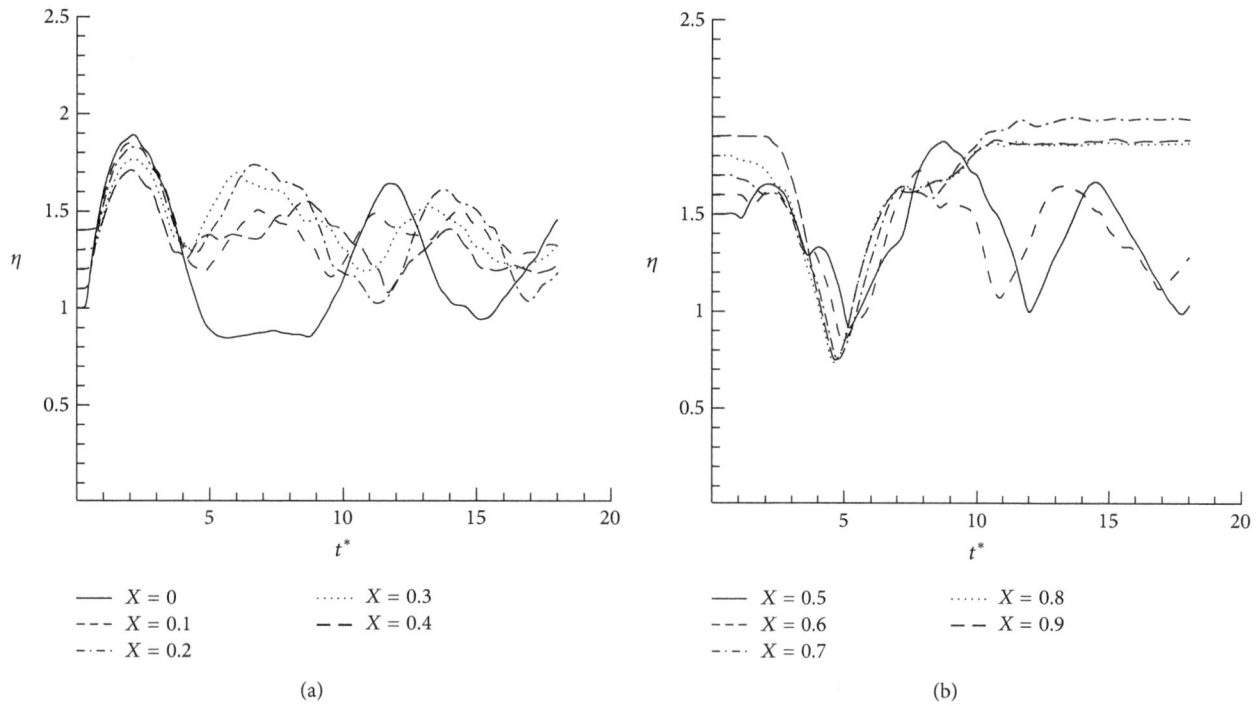

FIGURE 13: x-direction deformation for We $= 30$ and Re $= 810$. (a) Low impact parameters, $X < 0.5$, and (b) high impact parameters, $X \geq 0.5$.

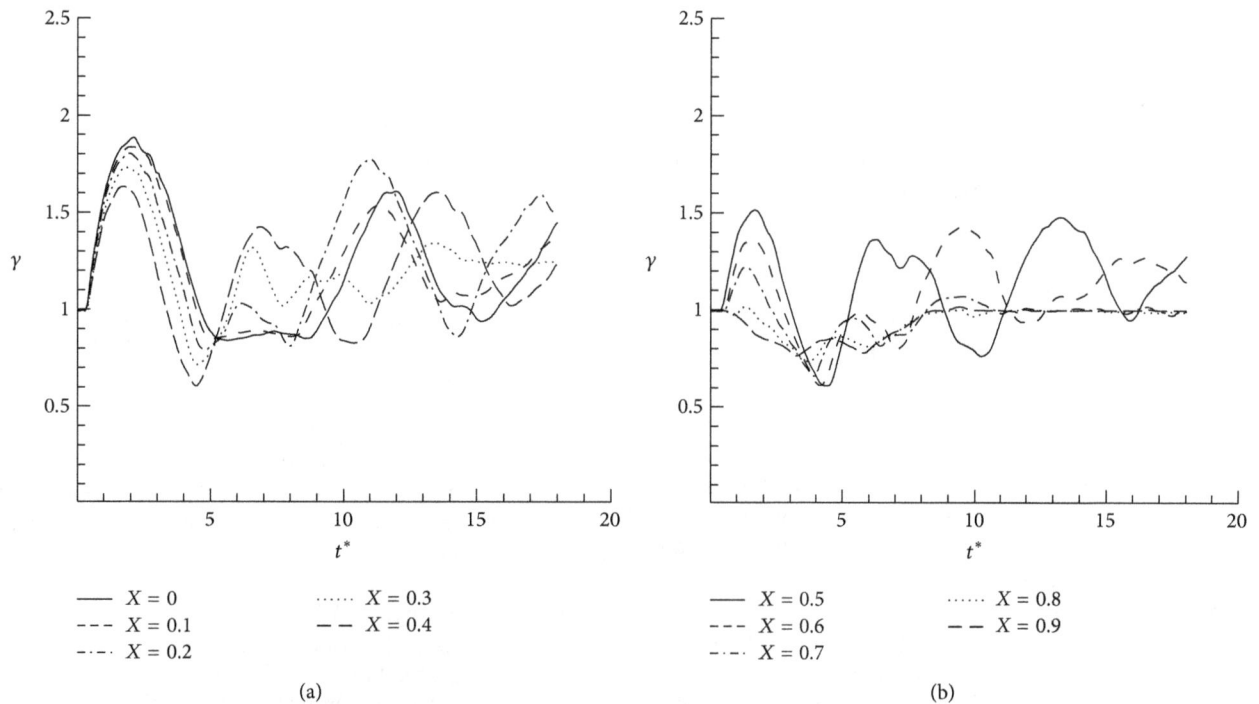

FIGURE 14: y-direction deformation for We $= 30$ and Re $= 810$. (a) Low impact parameters, $X < 0.5$, and (b) high impact parameters, $X \geq 0.5$.

the fluid is still able to spread outward for higher impact parameters.

During period II the overall amount of contraction remains constant (since similar slopes are observed) for low impact parameters of $X < 0.3$ (i.e., γ at the end of this period

decreases from 0.85 to 0.7). Therefore, even though more deformation is observed at the end of this period, the control parameter is still fluid momentum as it is for the x-direction case. The smaller γ values are the consequence of smaller $\gamma_{\max}$ values observed during period I. For impact parameters in

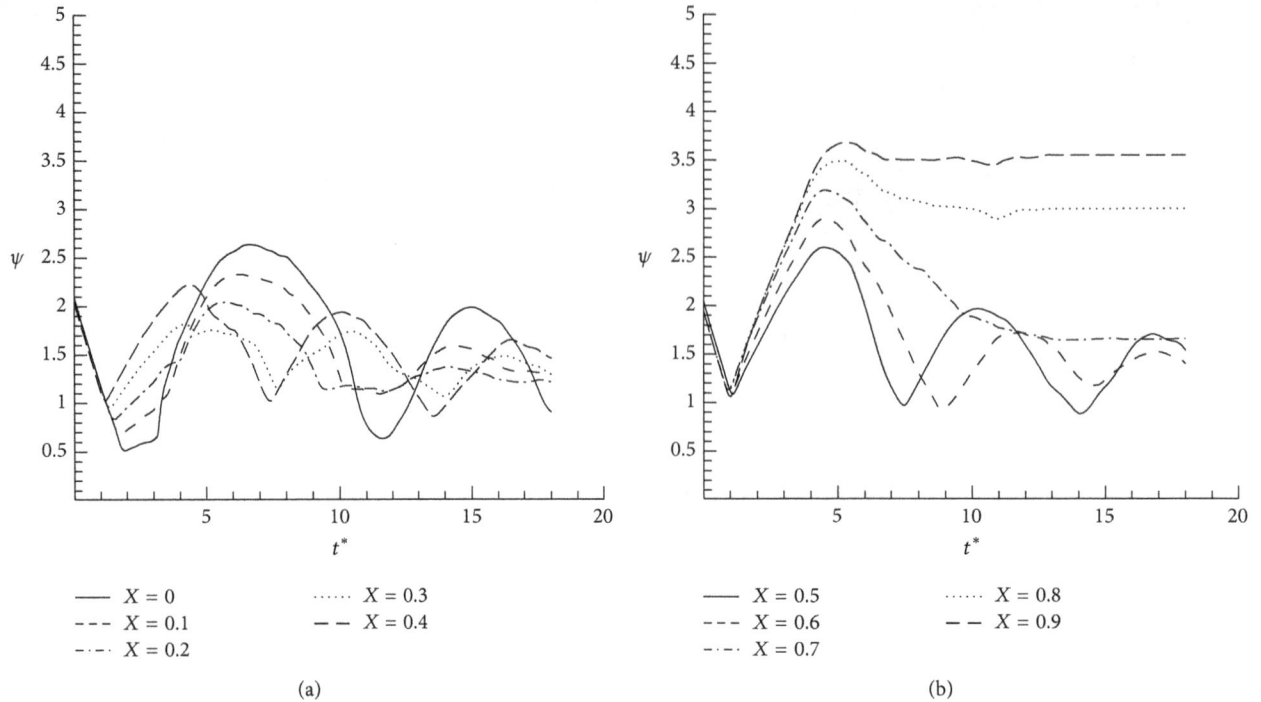

FIGURE 15: z-direction deformation for We = 30 and Re = 810. (a) Low impact parameters, $X < 0.5$, and (b) high impact parameters, $X \geq 0.5$.

the range of 0.4–0.6, the amount of deformation at the end of this period is constant and has a value of $\gamma \sim 0.6$ indicating a shift in dominance from fluid momentum to surface tension. As the impact parameter is further increased, $X > 0.7$, less overall deformation occurs since the amount of initial spread seen during period I has decreased and surface tension effect is in control.

4.7. z-Direction (Major Axis). In general, as the impact parameter increases, the region of interaction between the two drops decreases causing the intensity of deformation to shift from the x- and y-directions to the z-direction. This is clearly observed in the z-deformation curves of Figure 15. For an impact parameter of 0, the maximum deformation obtained is 2.7 times the drop's initial diameter and occurs at $t_c^* = 6.7$. As the impact parameter is increased the maximum height obtained decreases and occurs earlier in time until an impact parameter of 0.3 is reached. At this impact parameter the maximum deformation is 1.8 times the diameter and occurs at $t_c^* = 4.1$. As the impact parameter is increased further the maximum deformation increases linearly until $\psi_{max} = 3$ times the diameter is obtained. As with head-on collisions, this value defines a limiting value in which either permanent coalescence or separation occurs. This is observed at impact parameters of 0.8 and 0.9, in which the combined mass separates by the pinching mechanism explained earlier. In the case of the impact parameter of 0.7, the contraction rate is greater than the pinching mechanism and stretching-coalescence is observed.

The behavior of this maximum deformation parameter, ψ_{max}, as a function of impact parameter and Weber number

TABLE 2: Slopes of lines in Figure 16.

X	$m_{D=300\,\mu m}$	$m_{D=700\,\mu m}$	δm
0.4	0.018	n/a	—
0.5	0.038	0.046	0.008
0.6	0.043	0.051	0.008
0.7	0.058	0.063	0.005
0.8	0.073	0.078	0.005
0.9	0.088	0.087	0.001
0.95	n/a	0.096	—

is plotted in Figure 16. Figure 16(a) represents the results for collisions with drop diameters of 300 μm and Figure 16(b) is for collision with diameters of 700 μm. Open symbols indicate the collision outcome is coalescence while solid symbols indicate separation collision. Review of this figure indicates several aspects. First, the behavior of ψ_{max} is linear over the range of Weber numbers simulated and for illustrative purposes a straight line is fitted through each set of impact data. The slope of each impact parameter data set, m, is computed and presented in Table 2. The subscripts 300 and 700 refer to the drop size. Also included in the table is the change in slope, δm, which indicates the difference in the slope as the drop size is decreased from 700 μm to 300 μm.

Regardless of drop size, the slope of the lines increases with the increase in the impact parameter. This is attributed to the reduction in the interaction region which in turn causes the bulk of the fluid motion to continue traveling in the original z-direction. It is also interesting to note that the drop size has a small impact on the final stretch length. This is

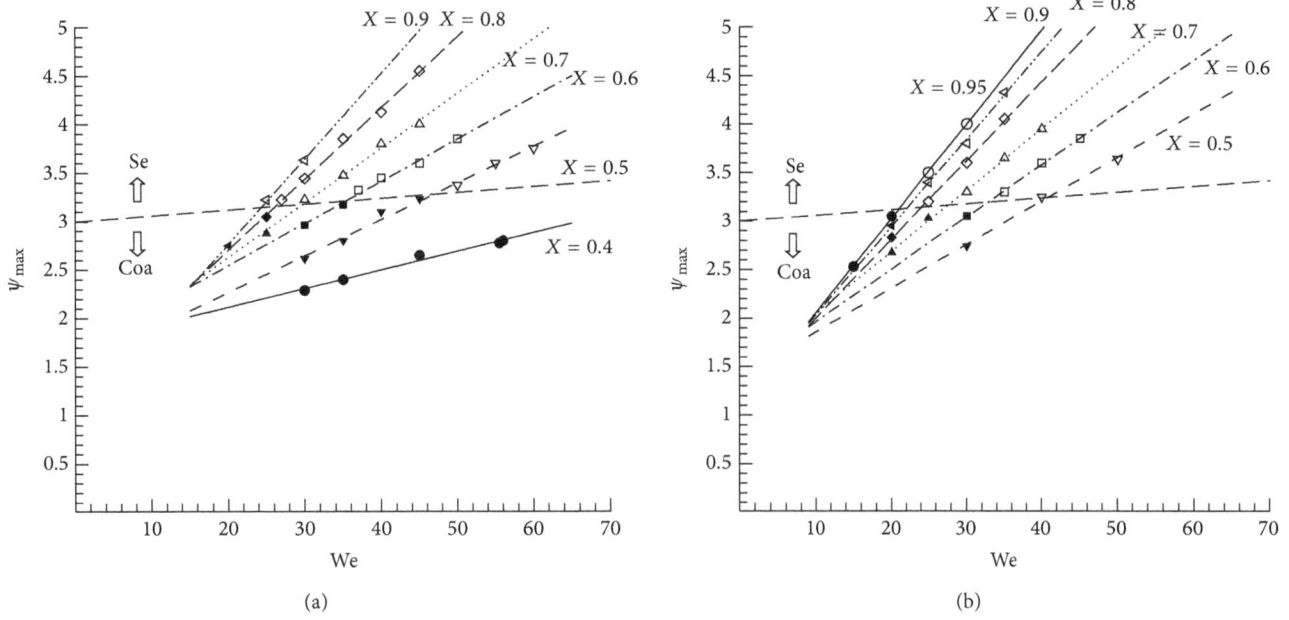

FIGURE 16: Maximum z-deformation for various Weber numbers and impact parameters. (a) 300 μm drops and (b) 700 μm drops. Solid symbols, coalescence, and open symbols, separation. Boundary line, $\psi_{\max} = 0.006\,\mathrm{We} + 3$.

evident by the increased slope values for the 700 μm drop collisions as opposed to 300 μm drop collision occurring at the same impact parameter. Additionally, the slopes for $X = 0.9$ are very similar and as a result this indicates that the size effect is no longer a factor in the overall stretching process.

A last comment concerning Figure 16 concerns the boundary line (represented by a thick dashed line) indicating the regions of coalescence and separation. According to Lord Rayleigh's capillary instability model [44], this boundary has a value of $\psi_{\max} = \pi$. However, for off-axis collisions, this boundary was empirically determined to be

$$\psi_{\max} = 0.006\,\mathrm{We} + 3, \tag{6}$$

where the slope reflects the impact parameter effect.

4.8. Dissipation Based Models. At the beginning of the collision, both drops have a known amount of surface, SE_o, and kinetic energies, KE_o. These energies are computed as follows:

$$\mathrm{SE}_o = 2\pi D_i^2 \sigma,$$
$$\mathrm{KE}_o = \frac{\pi}{24}\rho D_i^3 V_r^2, \tag{7}$$

where the ratio of the kinetic energy to the surface energy is defined as $\mathrm{We}^* = \mathrm{We}/48$ [22]. Each of these energies is computed at each time step using the following definitions:

$$\mathrm{SE}_n = \sigma \sum_{p=1}^{N_{\text{cells}}} \mathrm{SA}_p,$$
$$\mathrm{KE}_n = \frac{1}{2}\sum_{ijk=1}^{N_{\text{cells}}} \left(\frac{f_{ijk}\rho}{v_{\text{cell}}}\right)\left(u_{ijk}^2 + v_{ijk}^2 + w_{ijk}^2\right), \tag{8}$$

where SA_p is the polygon area in the computational domain and n is the time index, N_{cells} is number of computational cells, f_{ijk} is volume fraction of the liquid in the cell, and u_{ijk}, v_{ijk}, and w_{ijk} are the velocity components in each direction. Therefore, using these relationships each energy variation is computed at each time step during the collisional process. Additionally, by using these energies and conservation of energy an estimation of the change in total energy or dissipation energy, ϕ, is obtained as follows:

$$\phi = \mathrm{KE}_o + \mathrm{SE}_o - \mathrm{KE}_n - \mathrm{SE}_n. \tag{9}$$

The kinetic and surface energies are scaled with the initial surface energy as $\alpha = \mathrm{KE}_n/\mathrm{SE}_o$ and $\beta = \mathrm{SE}_n/\mathrm{SE}_o$. The initial surface energy is chosen as the scaling energy since for a same-size drop collision, it is independent of the impact energy. Since the amount of dissipated energy is simply the combined transient kinetic and surface energy subtracted from the initial total energy, a nondimensional dissipation energy is defined as

$$\Phi = 1 - \left(\frac{\mathrm{KE}_n + \mathrm{SE}_n}{\mathrm{KE}_o + \mathrm{SE}_o}\right). \tag{10}$$

4.8.1. Kinetic Energy. The dimensionless kinetic energy variation for the binary collision of two water drops for $\mathrm{We} = 30$ and $\mathrm{Re} = 810$ is provided in Figure 17. As the impact parameter is increased, the maximum kinetic energy, $\alpha_{\max}$, decreases from 0.25 for $X = 0$ to 0.22 for $0 < X < 0.6$. It should be noted that, for this range of impact parameters, the final outcome was coalescence, while higher impact parameters ($X > 0.7$) resulted in separation. This maximum kinetic energy decrease is the result of more energy conversion to surface energy (due to the increased

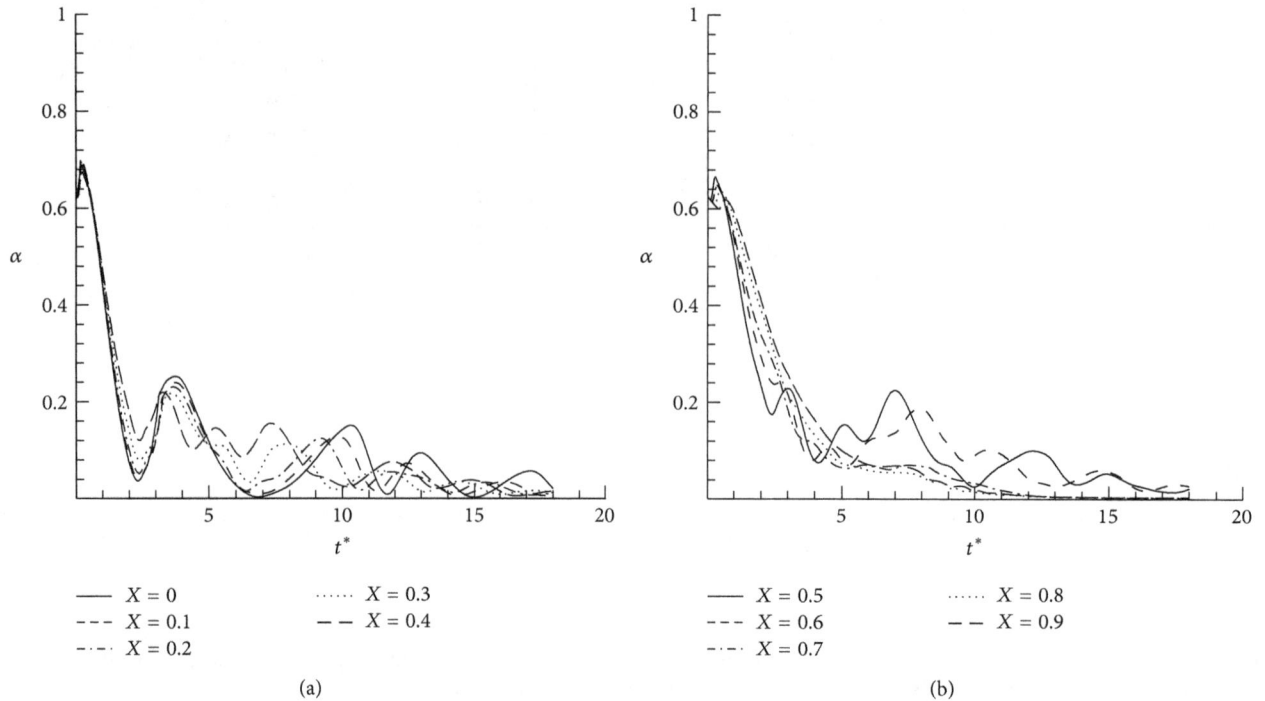

FIGURE 17: Dimensionless kinetic energy, $\alpha = KE_n/SE_o$ for We = 30 and Re = 810. (a) Low impact parameters, $X < 0.5$, and (b) high impact parameters, $X \geq 0.5$.

surface area accompanying an increase in impact parameter). Additionally, whereas head-on collisions exhibit defined peak and valley behavior corresponding to times of maximum spread and contraction, an increase in the impact parameter causes the change in kinetic energy to undergo a more chaotic variation. It is not until the impact parameter is increased to a value greater than 0.6 that no oscillatory behavior (of any kind) is observed. Therefore, if the kinetic energy does not peak early in the evolutional sequence, separation will likely occur.

4.8.2. Surface Energy. Figure 18 provides the dimensionless surface energy variations for the binary collision case under consideration. The maximum surface energy, β_{max}, decreases from a value of 1.325 for $X = 0$ to 1.025 for the range, $0 < X < 0.6$. This behavior parallels the behavior seen for the kinetic energy. At impact parameters greater than 0.6, the absence of this peak behavior at this time in the evolution is conspicuous. In fact at this time, $\beta_{max} \sim 1$ and as noted earlier in the previous discussion, all the remaining collisions at higher impact parameters resulted in separation. Additionally, the variation in surface energy becomes more oscillatory or chaotic for impact parameters up to 0.6. At these low and intermediate impact parameters, various surface deformations occur as the combined mass tries to stabilize itself to its minimum energy state.

4.9. Reynolds Number Effect. The Reynolds number effect is presented based on the collision of two drops with impact parameters of $X = 1$ and $X = 0.9$. For the grazing collision ($X = 1$), 25 different water drop collisions with drop sizes

of 10 μm, 75 μm, 150 μm, 300 μm, 700 μm, and 2 mm are simulated and plotted in Figure 19(a). Note that the Weber and Reynolds numbers are computed using fluid properties, relative velocity, and drop size. By holding the fluid properties and relative velocity constant, the only way to produce different We/Re combinations is to model different drop diameters. Therefore, for this range of drop diameters, the Ohnesorge number varies from 0.0370 to 0.00443. Similarly, the results of 26 different water drop collisions with an impact parameter of 0.9 are provided in Figure 19(b).

A clear boundary between coalescence and separation is observed for both impact parameters. For the grazing collisions ($X = 1$), the boundary was produced by fitting a power function between the data points. This boundary indicates that as the Re number approached zero, the We number approached We $\rightarrow$ 60. This implies that all collisions with this impact parameter and a Weber number greater than 60 would result in separation. By lowering the impact parameter to 0.9 (and thus creating a region of interaction), a minimum Reynolds number between 126 and 100 is observed as the We number approaches zero. Therefore, this interacting region (which allows portions of each drop to mix) increases the viscous forces while reducing the surface tension forces, such that permanent coalescence will occur. Also, the We number approaches 22 for $X = 0.9$ and 18 for $X = 1$ as the Re number is increased. For head-on collision, it was shown that as the Re number increases, the We number approached We $\rightarrow$ 35 [40]. Therefore, the Reynolds number effect is only important for binary drop collisions with a Weber number less than 35 for $X = 0$ and decreases as the impact parameter is increased.

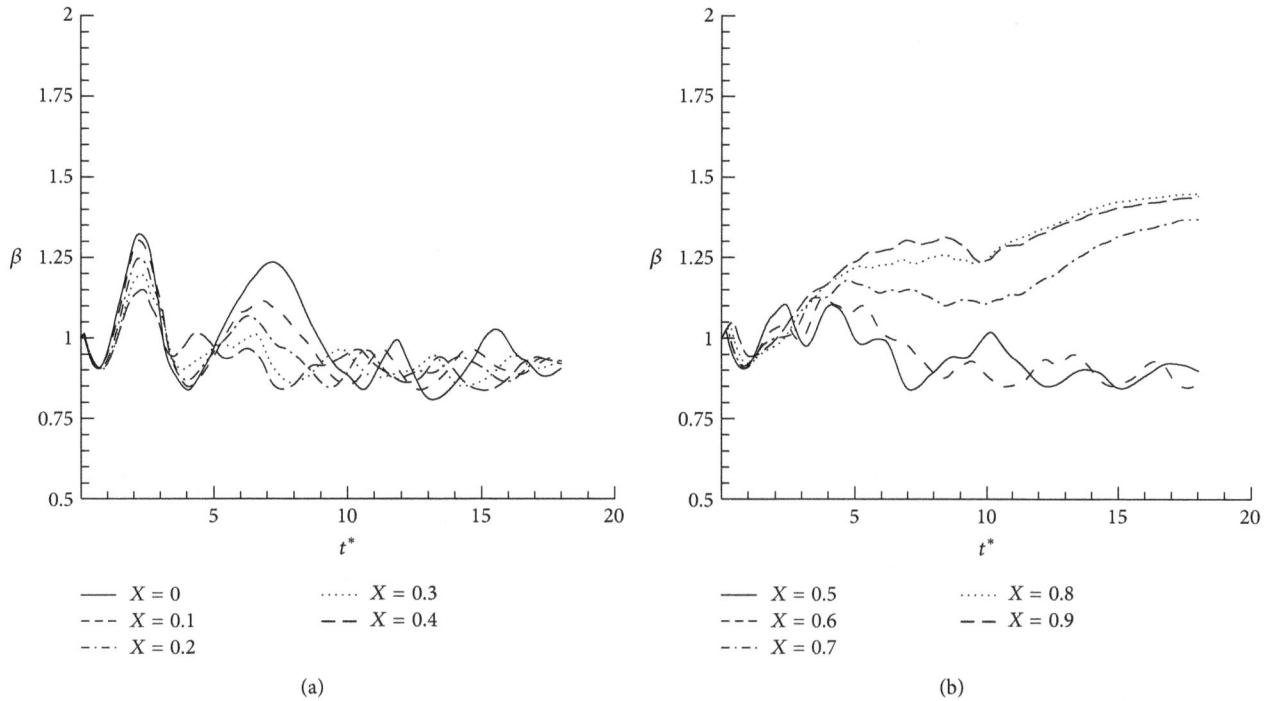

FIGURE 18: Dimensionless surface energy, $\beta = SE_n/SE_o$ for We = 30 and Re = 810. (a) Low impact parameters, $X < 0.5$, and (b) high impact parameters, $X \geq 0.5$.

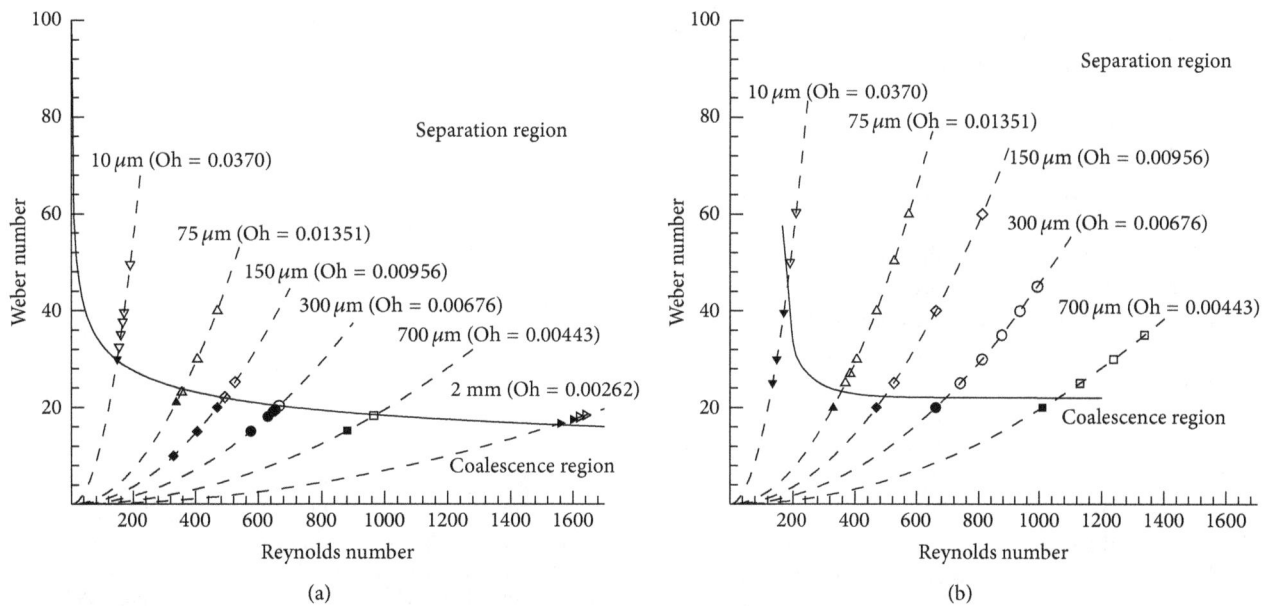

FIGURE 19: Weber versus Reynolds number. Water drop collisions. (a) $X = 1$, boundary between coalescence and separation is depicted using a power fitted line. (b) $X = 0.9$, in this case the boundary between coalescence and separation is shown using a spline fitted line.

5. Conclusion

An investigation of off-axis collisions of water drops across various impact parameters is presented. The results show that as the impact parameter is increased to values greater than zero (off-axis collision), the region of interaction between the drops is reduced. Two different types of break-up

mechanisms are observed. The ligament holding the two drops may go through pinching process governed by the Rayleigh instability, or the ligament may be pulled until it is broken. The pulling (tearing) breakup occurs at high impact parameter and high Weber number combinations. In these situations, the region of interaction is insufficient to allow adequate mixing between the two colliding drops. Therefore, the unmixed portions of the drops continue their travel in their respective original directions, thus tearing the ligament holding the two drops.

A study of the internal flow field of droplets shows that the droplets generated after collision may have different types of internal flows. Collision with high impact parameters may result in main drops and satellite drops. The internal flow field of the main drops seems to be similar to those of the parent drops, generally unidirectional. However, the internal flow field of the satellite drops may have internal circulations and may be multidirectional. Therefore, the internal flow field of droplets in dilute sprays, which do not have significant droplet collisions, may be unidirectional. However, the internal flow fields of dense sprays, which may have significant number of high impact droplet collisions and formation of satellite droplets, may be multidirectional and have internal circulations.

The upper critical Weber number, describing the boundary between the coalescence and separation for grazing collisions ($X = 1$), is determined for water drop collisions with drop diameters ranging from $10\,\mu$m to $2\,$mm. These critical Weber numbers are found to be a function of the drop diameter. The smaller drops have higher We number values, that is, $\mathrm{We_{cr}} = 32.5$ for $D = 10\,\mu$m, whereas $\mathrm{We_{cr}} = 18$ for $D = 2\,$mm. This is due to the higher pressures that exist in smaller drops, which quickly form a liquid bridge between the drops as soon as they touch. Smaller drops form relatively larger bridges, which requires higher impact We numbers to cause separation. However, as the drop size increases, the effect of change in drop size becomes less important. The upper critical We number approaches We = 18 as the drop size is increased. It should be noted that the above conditions are relevant for cases in which the surrounding gas is dynamically neutral, or collisions occur in vacuum. Otherwise, in determining the collision impact parameter, one has to be worried about the slippage effect that may occur due to the air entrapment in between the two drops.

A deformation criterion based on the drop deformation after collision is obtained. This criterion indicates that the drops separate if the length of the ligament formed after collision is approximately 3 times the diameter of the original drop. The maximum length of the ligament formed after the collision is found to vary linearly with the impact parameter.

Plots of the collisional outcomes in the Re-We plane are provided for the grazing collisions ($X = 1$) and collisions with an impact parameter of $X = 0.9$. As the Weber number increases, the Re for the separation approaches zero for an impact parameter of 1 and approached Re $\rightarrow$ 100 for collisions with an impact parameter of 0.9. It was shown [40] that Re $\rightarrow$ 1000 for an impact parameter of $X = 0$. Therefore, it can be concluded that collisions below this critical Reynolds number will always result in coalescence regardless of the Weber number. Also, the Reynolds number does have an effect on the collisional outcome as the impact parameter is reduced. However, this latter influence is small for drops with diameters larger than $75\,\mu$m and for impact parameters of 1 and 0.9.

Conflict of Interests

The authors declare that there is no conflict of interests regarding the publication of this paper.

References

[1] O. W. Jayaratne and B. J. Mason, "The coalescence and bouncing of water drops at an air/water interface," *Proceedings of the Royal Society of London A*, vol. 280, pp. 545–565, 1964.

[2] R. Gunn, "Collision characteristics of freely falling water drops," *Science*, vol. 150, no. 3697, pp. 695–701, 1965.

[3] D. J. Ryley and B. N. Bennett-Cowell, "The collision behaviour of steam-borne water drops," *International Journal of Mechanical Sciences*, vol. 9, no. 12, pp. 817–826, 1967.

[4] J. R. Adam, N. R. Lindblad, and C. D. Hendricks, "The collision, coalescence, and disruption of water droplets," *Journal of Applied Physics*, vol. 39, no. 11, pp. 5173–5180, 1968.

[5] R. W. Park, *Behavior of water drops colliding in humid nitrogen [Ph.D. thesis]*, Department of Chemical Engineering, The University of Wisconsin, 1970.

[6] D. M. Whelpdale and R. List, "The Coalescence process in raindrop growth," *Journal of Geophysical Research*, vol. 76, no. 12, pp. 2836–2856, 1971.

[7] P. R. Brazier-Smith, S. G. Jennings, and J. Latham, "The interaction of falling water drops: coalescence," *Proceedings of the Royal Society of London A*, vol. 326, no. 1566, pp. 393–408, 1972.

[8] S. G. Bradley and C. D. Stow, "Collisions between liquid drops," *Philosophical Transactions of the Royal Society of London A*, vol. 287, no. 1349, pp. 635–678, 1978.

[9] S. G. Bradey and C. D. Stow, "On the production of satellite droplets during collisions between water drops falling in still air," *Journal of the Atmospheric Sciences*, vol. 36, pp. 494–500, 1979.

[10] T. B. Low and R. List, "Collision, coalescence and breakup of raindrops. Part I: experimentally established coalescence efficiencies and fragment size distributions in breakup," *Journal of the Atmospheric Sciences*, vol. 39, no. 7, pp. 1591–1606, 1982.

[11] T. B. Low and R. List, "Collision, coalescence and breakup of raindrops. Part II: parameterization and fragment size distributions," *Journal of the Atmospheric Sciences*, vol. 39, no. 7, pp. 1607–1618, 1982.

[12] V. A. Arkhipov, I. M. Vasenin, and V. F. Trofimov, "Stability of colliding drops of ideal liquid, Tomsk," *Zhurnal Prikladnoi Mekhaniki i Tekhnicheskoi Fiziki*, no. 3, pp. 95–98, 1983.

[13] Y. J. Jiang, A. Umemura, and C. K. Law, "An experimental investigation on the collision behaviour of hydrocarbon droplets," *Journal of Fluid Mechanics*, vol. 234, pp. 171–190, 1992.

[14] J. Qian and C. K. Law, "Regimes of coalescence and separation in droplet collision," *Journal of Fluid Mechanics*, vol. 331, pp. 59–80, 1997.

[15] N. Ashgriz and P. Givi, "Binary collision dynamics of fuel droplets," *International Journal of Heat and Fluid Flow*, vol. 8, no. 3, pp. 205–210, 1987.

[16] N. Ashgriz and P. Givi, "Coalescence efficiencies of fuel droplets in binary collisions," *International Communications in Heat and Mass Transfer*, vol. 16, no. 1, pp. 11–20, 1989.

[17] G. Brenn and A. Frohn, "Collision and merging of two equal droplets of propanol," *Experiments in Fluids*, vol. 7, no. 7, pp. 441–446, 1989.

[18] J.-P. Estrade, H. Carentz, G. Lavergne, and Y. Biscos, "Experimental investigation of dynamic binary collision of ethanol droplets—a model for droplet coalescence and bouncing," *International Journal of Heat and Fluid Flow*, vol. 20, no. 5, pp. 486–491, 1999.

[19] K. D. Willis and M. Orme, "Viscous oil droplet collisions in a vacuum," *Experiments in Fluids*, vol. 29, no. 4, pp. 347–358, 2000.

[20] M. Orme, "Experiments on droplet collisions, bounce, coalescence and disruption," *Progress in Energy and Combustion Science*, vol. 23, no. 1, pp. 65–79, 1997.

[21] A. Menchaca-Rocha, F. Huidobro, A. Martinez-Davalos et al., "Coalescence and fragmentation of colliding mercury drops," *Journal of Fluid Mechanics*, vol. 346, pp. 291–318, 1997.

[22] N. Ashgriz and J. Y. Poo, "Coalescence and separation in binary collisions of liquid drops," *Journal of Fluid Mechanics*, vol. 221, pp. 183–204, 1990.

[23] G. Brenn, S. Kalenderski, and I. Ivanov, "Investigation of the stochastic collisions of drops produced by Rayleigh breakup of two laminar liquid jets," *Physics of Fluids*, vol. 9, no. 2, pp. 349–364, 1997.

[24] G. Brenn, D. Valkovska, and K. D. Danov, "The formation of satellite droplets by unstable binary drop collisions," *Physics of Fluids*, vol. 13, no. 9, pp. 2463–2477, 2001.

[25] I. V. Roisman, "Dynamics of inertia dominated binary drop collisions," *Physics of Fluids*, vol. 16, no. 9, pp. 3438–3449, 2004.

[26] Y. Yoon, M. Borrell, C. C. Park, and L. G. Leal, "Viscosity ratio effects on the coalescence of two equal-sized drops in a two-dimensional linear flow," *Journal of Fluid Mechanics*, vol. 525, pp. 355–379, 2005.

[27] T.-C. Gao, R.-H. Chen, J.-Y. Pu, and T.-H. Lin, "Collision between an ethanol drop and a water drop," *Experiments in Fluids*, vol. 38, no. 6, pp. 731–738, 2005.

[28] S. Guido and M. Simeone, "Binary collision of drops in simple shear flow by computer-assisted video optical microscopy," *Journal of Fluid Mechanics*, vol. 357, pp. 1–20, 1998.

[29] M. R. Nobari, Y.-J. Jan, and G. Tryggvason, "Head-on collision of drops—a numerical investigation," *Physics of Fluids*, vol. 8, no. 1, pp. 29–42, 1996.

[30] S. O. Unverdi, "A front-tracking method for viscous, incompressible, multifluid flows," *Journal of Computational Physics*, vol. 100, no. 1, pp. 25–37, 1992.

[31] M. R. H. Nobari, Y.-J. Jan, and G. Tryggvason, "Head-on collisions of drops—a numerical investigation," *Physics of Fluids*, vol. 8, pp. 29–42, 1996.

[32] M. R. H. Nobari and G. Tryggvason, "Numerical simulations of three-dimensional drop collisions," *AIAA Journal*, vol. 34, no. 4, pp. 750–755, 1996.

[33] M. Rieber and A. Frohn, "Navier-Stokes simulation of droplet collision dynamics," in *Proceedings of the 7th International Symposium on Computational Fluid Dynamic*, pp. 520–525, Beijing, China, 1997.

[34] S. L. Post and J. Abraham, "Modeling the outcome of drop-drop collisions in Diesel sprays," *International Journal of Multiphase Flow*, vol. 28, no. 6, pp. 997–1019, 2002.

[35] F. Mashayek, N. Ashgriz, W. J. Minkowycz, and B. Shotorban, "Coalescence collision of liquid drops," *International Journal of Heat and Mass Transfer*, vol. 46, no. 1, pp. 77–89, 2003.

[36] Y. Morozumi, H. Ishizuka, and J. Fukai, "Criterion between permanent coalescence and separation for head-on binary droplet collision," *Atomization and Sprays*, vol. 15, no. 1, pp. 61–80, 2005.

[37] N. Nikolopoulos, K.-S. Nikas, and G. Bergeles, "A numerical investigation of central binary collision of droplets," *Computers and Fluids*, vol. 38, no. 6, pp. 1191–1202, 2009.

[38] G. H. Ko and H. S. Ryou, "Modeling of droplet collision-induced breakup process," *International Journal of Multiphase Flow*, vol. 31, no. 6, pp. 723–738, 2005.

[39] B. Lafaurie, C. Nardone, R. Scardovelli, S. Zaleski, and G. Zanetti, "Modelling merging and fragmentation in multiphase flows with SURFER," *Journal of Computational Physics*, vol. 113, no. 1, pp. 134–147, 1994.

[40] M. D. Saroka, N. Ashgriz, and M. Movassat, "Numerical investigation of head-on binary drop collisions in a dynamically inert environment," *Journal of Applied Fluid Mechanics*, vol. 5, no. 1, pp. 23–37, 2012.

[41] G. Brenn, "Droplet collision," in *Handbook of Atomization and Sprays*, N. Ashgriz, Ed., chapter 7, pp. 157–181, Springer, New York, NY, USA, 2011.

[42] C. F. Hsu and N. Ashgriz, "Impaction of a droplet on an orifice plate," *Physics of Fluids*, vol. 16, no. 2, pp. 400–411, 2004.

[43] N. Hsu and N. Ashgriz, "Nonlinear penetration of liquid drops into radial capillaries," *Journal of Colloid and Interface Science*, vol. 270, no. 1, pp. 146–162, 2004.

[44] L. Rayleigh, "On the instability of jets," *Proceedings of the London Mathematical Society*, vol. 10, no. 1, pp. 4–13, 1878.

4

Viscous Flows Driven by Uniform Shear over a Porous Stretching Sheet in the Presence of Suction/Blowing

Samir Kumar Nandy[1] and Swati Mukhopadhyay[2]

[1] *Department of Mathematics, A.K.P.C Mahavidyalaya, Bengai, Hooghly 712 611, India*
[2] *Department of Mathematics, The University of Burdwan, West Bengal 713104, India*

Correspondence should be addressed to Samir Kumar Nandy; nandysamir@yahoo.com

Academic Editor: Boming Yu

An analysis is carried out to study the steady two-dimensional flow of an incompressible viscous fluid past a porous deformable sheet, which is stretched in its own plane with a velocity proportional to the distance from the fixed point subject to uniform suction or blowing. A uniform shear flow of strain rate β is considered over the stretching sheet. The analysis of the result obtained shows that the magnitude of the wall shear stress increases with the increase of suction velocity and decreases with the increase of blowing velocity and this effect is more pronounced for suction than blowing. It is seen that the horizontal velocity component (at a fixed streamwise position along the plate) increases with the increase in the ratio of shear rate β and stretching rate (c) (i.e., β/c) and there is an indication of flow reversal. It is also observed that this flow reversal region increases with the increase in β/c.

1. Introduction

Suction or injection (blowing) of a fluid through the bounding surface can significantly change the flow field and consequently affects the heat transfer rate from the surface. Injection or withdrawal of fluid through a porous bounding heated/cooled wall is of great general interest in practical problems such as cooling of films, cooling of wires, and coating of polymer fiber. The process of suction or blowing has its importance in many engineering and industrial activities such as in the design of thrust bearing and radial diffusers and thermal oil recovery. Suction is also applied to chemical processes to remove reactants whereas blowing is used to add reactants, prevent corrosion or scaling, and reduce the drag.

During the last decades, the study of flow over a stretching surface has attracted much more interest of the researchers due to its various industrial applications such as extrusion of polymer sheets, continuous stretching, manufacturing plastic films, and artificial fibers. In a melt-spinning process, the extrudate from the die is generally drawn and simultaneously stretched into a sheet which is then solidified through quenching or gradual cooling by direct contact with the water. The mechanical properties of the final product depend crucially on the rate of cooling/heating along the surface while being stretched. Sakiadas [1] was the first to study boundary layer flow due to a rigid flat continuous surface moving in its own plane. Erickson et al. [2] analyzed the boundary layer flow due to the motion of a porous flat plate when the transverse velocity at the surface is nonzero. A detailed analysis of the boundary layer flow due to a stretching sheet has been carried out by Danberg and Fansler [3].

Later, Crane [4] gave an exact similarity solution in closed analytical form for steady boundary layer flow of an incompressible viscous fluid caused solely by the stretching of an elastic flat sheet which moves in its own plane with a velocity varying linearly with distance from a fixed point. Wang [5] investigated a uniform shear flows over a quiescent liquid and he showed that an interfacial boundary layer develops both in the air and the liquid. P. S. Gupta and A. S. Gupta [6] investigated the heat and mass transfer corresponding to the similarity solution for the boundary layer flow over a stretching sheet subject to uniform suction or blowing. Rajagopal et al. [7] studied the boundary layer flow over a stretching surface for second-order fluid and obtained similarity solutions of the boundary layer equations. Dandapat and Gupta [8] extended the same problem with heat transfer and found exact analytical solutions.

Shear driven flows, namely, wall driven Couette flow, wind driven Ekman flow, and so forth are the classical topics of fluid mechanics. Due to their wide range of applications shear driven flows attract the attention of the researchers. The study of boundary layer flow driven by uniform shear is seen to have fewer contributors in fluid mechanics. Rajagopal et al. [9] studied the nonsimilar boundary layer flow of a second-order fluid over a stretching sheet in the presence of a uniform shear flow. Weidman et al. [10] reported a similarity solution of the boundary layer flow over a flat impermeable plate with free shear flows driven by rotational velocities. The extension of the same problem to a permeable flat plate was analyzed by Magyari et al. [11]. Cossali [12] gave the similarity solutions of the energy and momentum boundary layer equations for a power-law shear driven flow over a semi-infinite flat plate. The thermal boundary layer beneath an external uniform shear flow was considered by Magyari and Weidman [13]. Weidman et al. [14] also investigated the effects of plate extension and transpiration on uniform shear flow over a semi-infinite flat plate by considering the boundary layer approximations. Due to the presence of shear in the free stream, the free stream is no longer rotation-free. As a result the flow behaviours are quite different from the rotation-free flow. In most of the works the flat surface was kept stationary. But the interaction of shear flow and wall stretching affects the flow significantly. Fang [15] analyzed the heat transfer characteristics for boundary layer flow past a stretching sheet in presence of uniform shear-free stream. Xu [16] obtained the analytic solution in case of boundary layer flow driven by power-law shear. Very recently, the heat transfer characteristics of a viscous incompressible fluid over a stretching/shrinking sheet in a uniform shear flow with a convective surface boundary condition were analyzed by Aman et al. [17].

Motivated by the above investigations, in this paper, the steady, two-dimensional incompressible viscous fluid flow past a porous stretching sheet in presence of uniform suction/blowing has been investigated. A uniform shear flow of strain rate β is considered over the stretching surface. The behaviours of the horizontal component of the flow velocity are explained and it is seen that there is a flow reversal in case of suction or blowing at the sheet.

2. Flow Analysis

Consider the steady two-dimensional flow of a viscous fluid towards a stretching surface coinciding with the plane $y = 0$, the flow being confined to the region $y > 0$. We choose the coordinate system such that the x-axis is along the sheet, y-axis is normal to the sheet, and the origin of the coordinate system is located at a finite position on the sheet. Two equal and opposing forces are applied on the stretching surface along the x-axis so that the surface is stretched keeping the origin fixed. Here we consider a uniform shear flow of strain rate β over the stretching surface. The free stream velocity is

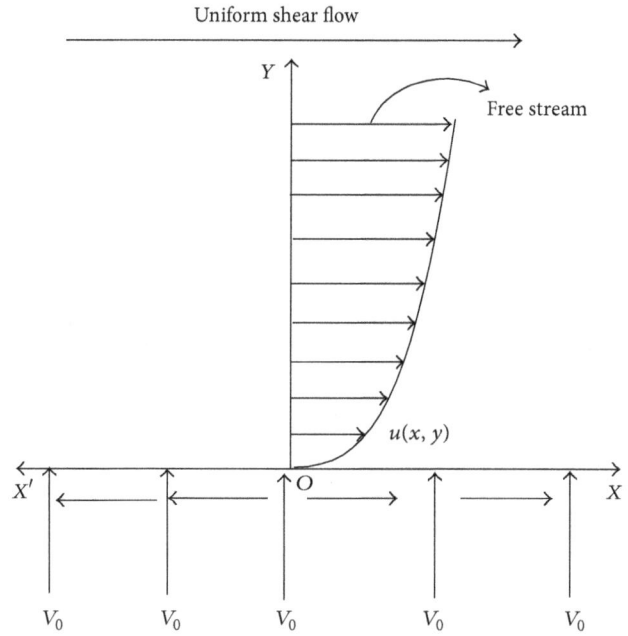

FIGURE 1: A sketch of the physical problem and the coordinate system involved.

uniform shear and the surface is subject to uniform suction or blowing. It is to be noted that, due to the presence of blowing or suction, a streamwise pressure gradient is required to maintain the flow. The flow configuration is shown in Figure 1.

The inviscid version of the present steady flow is given in terms of the stream function ψ_0 as

$$\psi_0 = \beta y \left(\frac{y}{2} - \delta_1 \right) + \delta_2 x, \tag{1}$$

where δ_1 and δ_2 are two constants. Here δ_1 is the displacement thickness arising out of the boundary layer on the stretching surface and δ_2 is the parameter which controls the horizontal pressure gradient that produces the shear flow. So the parameters δ_1 and δ_2 have some effects on the flow field (see Drazin and Riley [18]). The velocity components along x and y directions, respectively, for the flow described by (1) are

$$U_1 = \beta (y - \delta_1), \qquad V_1 = -\delta_2. \tag{2}$$

The boundary conditions at the stretching surface are

$$u = cx, \qquad v = v_w \quad \text{at } y = 0, \tag{3}$$

where $c > 0$ is a constant. The boundary conditions at infinity are

$$u \longrightarrow U_1(x, y), \qquad v \longrightarrow V_1(x, y) \quad \text{as } y \longrightarrow \infty, \tag{4}$$

where U_1 and V_1 are given by (2). Near the stretching surface, we take the stream function in the form:

$$\frac{\psi}{\nu} = \xi F(\eta) + W(\eta), \tag{5}$$

where

$$\xi = \left(\frac{c}{v}\right)^{1/2} x, \qquad \eta = \left(\frac{c}{v}\right)^{1/2} y, \qquad (6)$$

and v is the kinematic viscosity. The velocity components u and v, along x and y directions, are given by

$$u = \frac{\partial \psi}{\partial y}, \qquad v = -\frac{\partial \psi}{\partial x}. \qquad (7)$$

Hence the dimensionless velocity components U and V are obtained from (5) and (6) as

$$U = \xi F'(\eta) + W'(\eta)$$
$$V = -F(\eta), \qquad (8)$$

where $U = u/(cv)^{1/2}$ and $V = v/(cv)^{1/2}$. Substituting (8) into the Navier-Stokes equations we get

$$-\frac{1}{\rho}\frac{\partial p}{\partial x} = v^2 \left(\frac{c}{v}\right)^{3/2}$$
$$\times \left[\xi\left(F'^2 - FF'' - F'''\right) + \left(F'W' - FW'' - W'''\right)\right]$$
$$-\frac{1}{\rho}\frac{\partial p}{\partial y} = v^2 \left(\frac{c}{v}\right)^{3/2} \left[FF' + F''\right], \qquad (9)$$

where a prime denotes differentiation with respect to η. Eliminating p between (9) and equating the coefficients of ξ^0 and ξ^1, we get upon integration

$$F'^2 - FF'' - F''' = C_1,$$
$$F'W' - FW'' - W''' = C_2, \qquad (10)$$

where C_1 and C_2 are the constants of integration.
The boundary conditions (3) and (4) become

$$F(0) = -V_0, \qquad F'(0) = 1, \qquad F'(\infty) = 0, \qquad (11)$$

$$W(0) = 0, \qquad W'(0) = 0,$$
$$W'(\eta) = \frac{\beta}{c}(\eta - d_1) \quad \text{as } \eta \longrightarrow \infty, \qquad (12)$$

where $V_0 = v_w/(cv)^{1/2}$ and $d_1 = (c/v)^{1/2}\delta_1$. It is to be noted that the boundary condition $W(0) = 0$ is taken due to the steam function. Let us assume that $F(\eta) \to d_2$ as $\eta \to \infty$, where $d_2 = \delta_2/(cv)^{1/2}$. Here d_2 is the dimensionless shear rate parameter linked to the shear flow and d_1 is the dimensionless displacement thickness parameter. The constants C_1 and C_2 are now obtained from (10) using (11) and (12) as $C_1 = 0$ and $C_2 = -(\beta/c)d_2$. Hence (10) becomes

$$F'^2 - FF'' - F''' = 0, \qquad (13)$$

$$F'W' - FW'' - W''' = -\frac{\beta}{c}d_2. \qquad (14)$$

Here the dimensionless shear rate parameter d_2 is obtained by integrating (13) numerically by using the boundary condition (11). When (13) and (14) are substituted into (9), we get the dimensionless pressure distribution $P(\xi, \eta)$ after integration as

$$-P = \frac{1}{2}F^2 + F' - \frac{\beta}{c}d_2\xi + \text{constant}, \qquad (15)$$

where $P = p(x, y)/(\rho c v)$. The dimensionless wall shear stress T is given as

$$T = \xi F''(0) + W''(0). \qquad (16)$$

The dimensionless stream function for the flow can be obtained as

$$\psi(\xi, \eta) = \xi F(\eta) + \int_0^\eta W'(\eta)\, d\eta, \qquad (17)$$

where ξ is the dimensionless distance along the surface.

3. Method of Numerical Solution

In the absence of an analytic solution of a problem, a numerical solution is indeed an obvious and natural choice. Thus the governing equations (13) and (14) along with the boundary conditions (11)-(12) are solved numerically by fourth-order Runge-Kutta method with shooting technique. To do this, we first transform the nonlinear differential equation (13) to a system of three first-order differential equations as

$$y_1' = y_2, \qquad y_2' = y_3, \qquad y_3' = y_2^2 - y_1 y_3, \qquad (18)$$

where $y_1 = F(\eta)$, $y_2 = F'(\eta)$, $y_3 = F''(\eta)$, and a prime denotes differentiation with respect to the independent variable η. The boundary conditions (11) become

$$y_1 = -V_0, \qquad y_2 = 1 \quad \text{at } \eta = 0,$$
$$y_2 \longrightarrow 0 \quad \text{as } \eta \longrightarrow \infty. \qquad (19)$$

For a given value of V_0, the values of y_1 and y_2 are known at the starting point $\eta = 0$. Now the value of y_2 as $\eta \to \infty$ is replaced by y_2 at a finite value $\eta = \eta_\infty$ to be determined later. The value of y_2 at $\eta = 0$ is guessed in order to initiate the integration scheme. Starting from the given values of y_1 and y_2 at $\eta = 0$ and the guessed value of y_3 at $\eta = 0$, we integrate the system of first-order equations (18) by using a fourth-order Runge-Kutta method up to the end-point $\eta = \eta_\infty$. The computed value of y_2 at $\eta = \eta_\infty$ is then compared with y_2 at $\eta = \eta_\infty$. The absolute difference between these two values should be as small as possible. To this end we use a Newton-Raphson iteration procedure to assure quadratic convergence of the iterations. The value of η_∞ is then increased till y_2 attains the value zero asymptotically.

Using the numerical values of $F(\eta)$ obtained from the solutions of (11) and (13), (14) along with the boundary conditions (12) is solved numerically using the same method as described above to obtain $W(\eta)$.

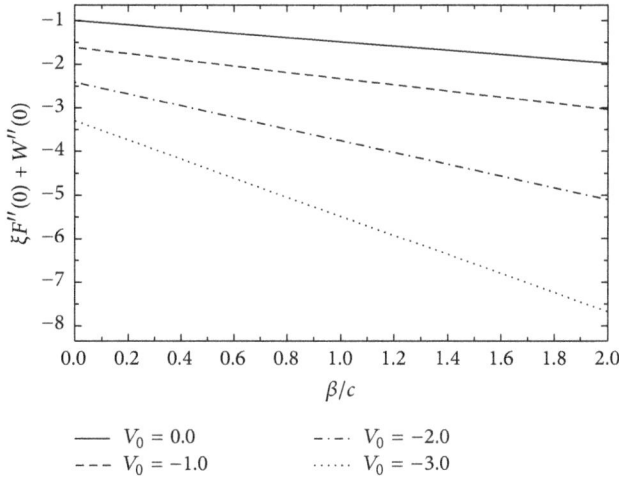

FIGURE 2: Variation of wall shear stress with β/c for several values of $V_0(\leq 0)$ (i.e., suction at the plate) when $\xi = 1.0$ and $d_1 = 1.0$.

Legend:
— $V_0 = 0.0$
--- $V_0 = -1.0$
-·-· $V_0 = -2.0$
······ $V_0 = -3.0$

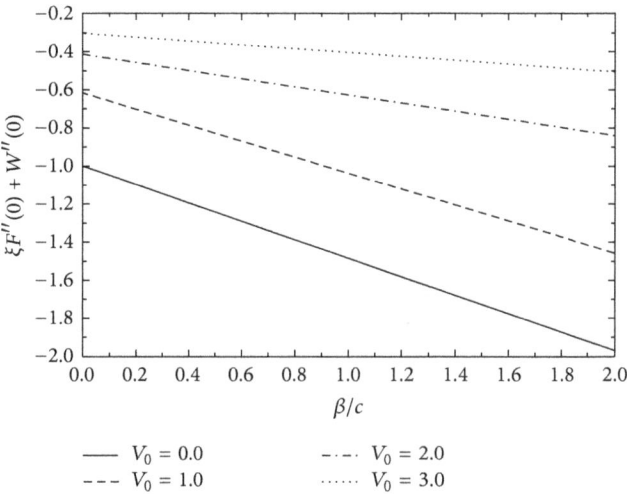

FIGURE 3: Variation of wall shear stress with β/c for several values of $V_0(\geq 0)$ (i.e., blowing at the plate) when $\xi = 1.0$ and $d_1 = 1.0$.

Legend:
— $V_0 = 0.0$
--- $V_0 = 1.0$
-·-· $V_0 = 2.0$
······ $V_0 = 3.0$

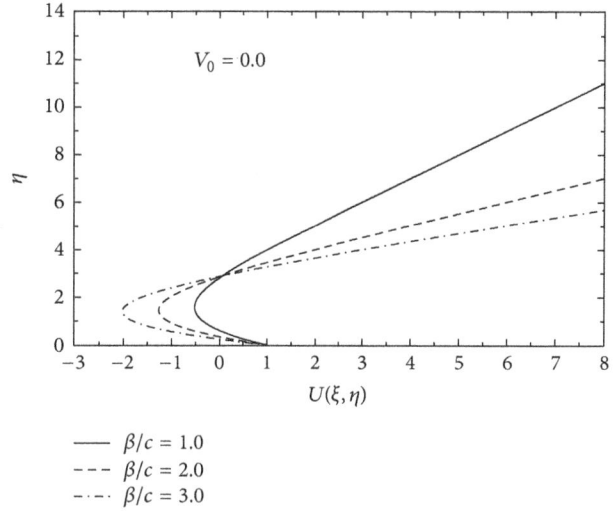

FIGURE 4: Variation of $U(\xi, \eta)$ with η for several values of β/c with $\xi = 1.0, d_1 = 3.0$, and $V_0 = 0$ (i.e., no suction or blowing at the plate).

Legend:
— $\beta/c = 1.0$
--- $\beta/c = 2.0$
-·-· $\beta/c = 3.0$

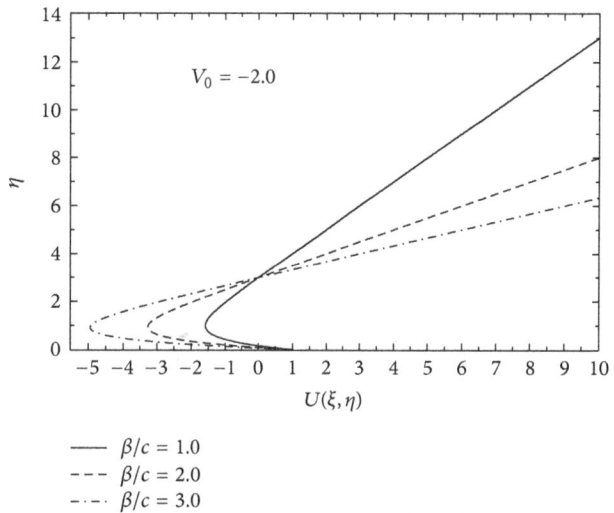

FIGURE 5: Variation of $U(\xi, \eta)$ with η for several values of β/c with $\xi = 1.0, d_1 = 3.0$, and $V_0 = -2.0$ (i.e., for suction at the plate).

Legend:
— $\beta/c = 1.0$
--- $\beta/c = 2.0$
-·-· $\beta/c = 3.0$

4. Results and Discussion

Figures 2 and 3 show the variation of dimensionless wall shear stress with β for several values of V_0 keeping the values of the other parameters fixed. Note that $V_0 = 0$ corresponds to the case when there is no suction or blowing at the sheet, $V_0 < 0$ corresponds to suction, and $V_0 > 0$ corresponds to blowing at the sheet. Figure 2 reveals that as the suction velocity increases, the magnitude of the wall shear stress increases. On the other hand, Figure 3 indicates that with the increase of blowing velocity at the plate, the magnitude of the wall shear stress decreases. From these figures, it is also noticed that the effect of wall shear stress is more pronounced for suction than blowing at the stretching sheet.

Figure 4 shows the variation of $U(\xi, \eta)$, the horizontal component of velocity, with η for several values of β/c when there is no suction or blowing velocity at the sheet. Figure 5 shows the same behaviour for suction ($V_0 = -2.0$) at the

stretching sheet and Figure 6 shows for blowing ($V_0 = 2.0$) at the stretching sheet keeping other parameters fixed. From these figures it is observed that $U(\xi, \eta)$ increases with the increase of β/c except in a small region near the sheet. The figures indicate that there is a flow reversal very near the sheet. This region of flow reversal gradually increases with the increase of β/c.

Figures 7 and 8 show the variation of $U(\xi, \eta)$ with η for several values of the suction and blowing at the plate, respectively. Figure 7 reveals that as $|V_0|$ increases (i.e., suction velocity increases), the horizontal velocity at a point decreases except in a small region near the sheet. Figure 8 shows that as the blowing velocity increases, velocity at a point increases. Notice that there is an indication of flow reversal. From these two figures, it can also be concluded that,

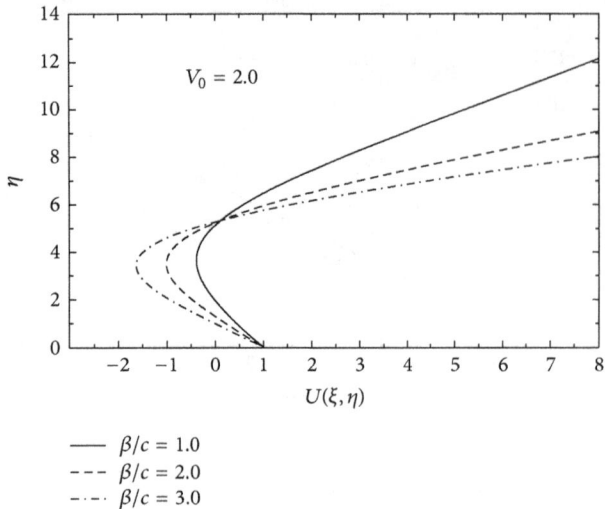

FIGURE 6: Variation of $U(\xi, \eta)$ with η for several values of β/c with $\xi = 1.0$, $d_1 = 3.0$, and $V_0 = 2.0$ (i.e., for blowing at the plate).

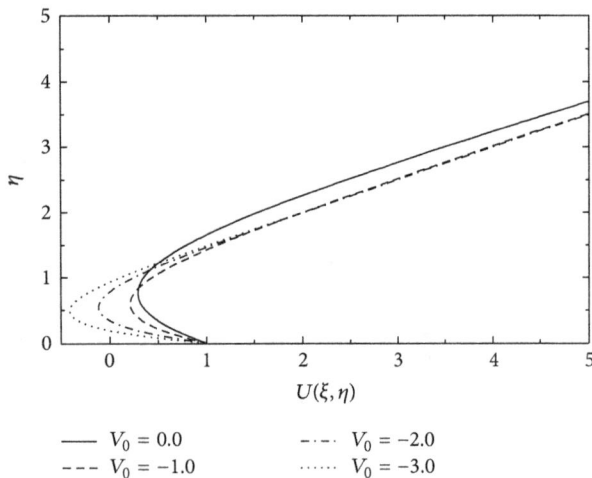

FIGURE 8: Variation of $U(\xi, \eta)$ with η for several values of $V_0 (\geq 0)$ (i.e., blowing at the plate) when $\beta/c = 2.0$, $\xi = 1.0$, and $d_1 = 1.0$.

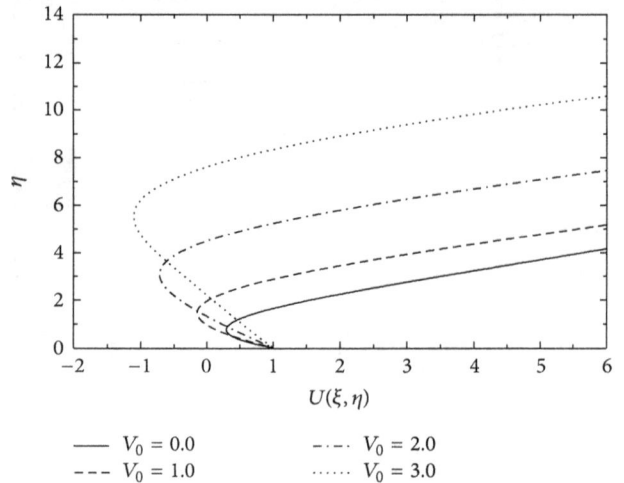

FIGURE 7: Variation of $U(\xi, \eta)$ with η for several values of $V_0 (\leq 0)$ (i.e., suction at the plate) when $\beta/c = 2.0$, $\xi = 1.0$, and $d_1 = 1.0$.

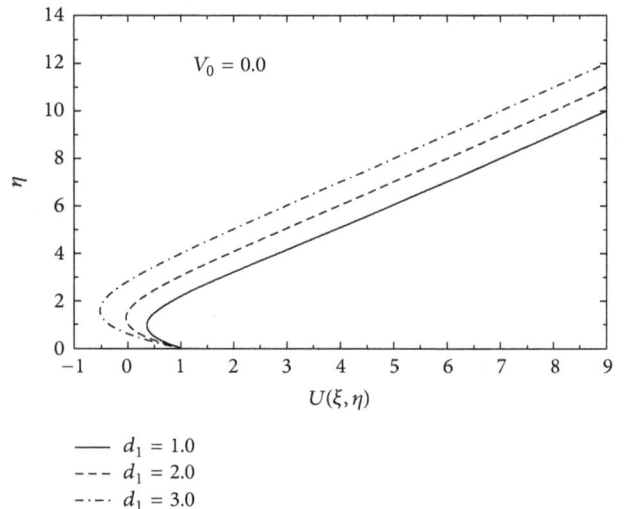

FIGURE 9: Variation of $U(\xi, \eta)$ with η for several values of d_1 with $\xi = 1.0$, $\beta/c = 2.0$, and $V_0 = 0$ (i.e., no suction or blowing at the plate).

with the increase of V_0 (magnitude), the flow reversal region increases.

Figure 9 shows the variation of $U(\xi, \eta)$ with η for several values of the shear rate parameter d_1 in the absence of suction or blowing at the sheet. Figure 10 shows the same for suction and Figure 11 stands for blowing at the sheet for the same set of parameters. From these figures it is seen that, with the increase of shear rate parameter d_1, the region of flow reversal increases. The tendency of flow reversal is more prominent in case of suction than blowing. It is also observed that as the shear rate parameter d_1 increases, velocity at a point also increases. Figure 12 depicts the variation of the horizontal component of velocity, $U(\xi, \eta)$, with η for several values of ξ (negative, zero, and positive) in the presence of blowing at the sheet. The figure reveals that as ξ increases, velocity at a point decreases.

Presentation of full stability analysis is beyond the scope of the present work since a stability analysis requires an unsteady flow, whereas our problem is a steady one. But as we know that, in boundary layer flow, the reverse profiles suggest temporal instability so one can expect that, in the case of full Navier-Stokes solution, the same feature holds good. This can be verified using the stability analysis by adopting the techniques of Merkin [19] and Mahapatra et al. [20].

5. Concluding Remark

We have investigated the steady two-dimensional flow of incompressible viscous fluid past a porous deformable sheet which is stretched in its own plane with velocity cx, x being the distance along the sheet from the fixed point with uniform suction or blowing. A uniform shear flow of

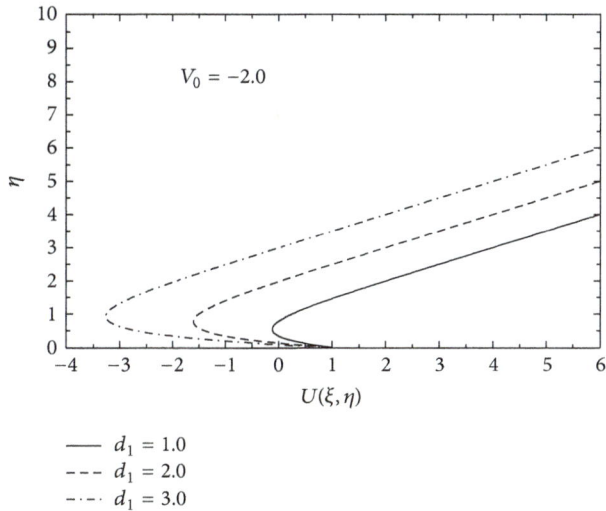

FIGURE 10: Variation of $U(\xi, \eta)$ with η for several values of d_1 with $\xi = 1.0$, $\beta/c = 2.0$, and $V_0 = -2.0$ (i.e., for suction at the plate).

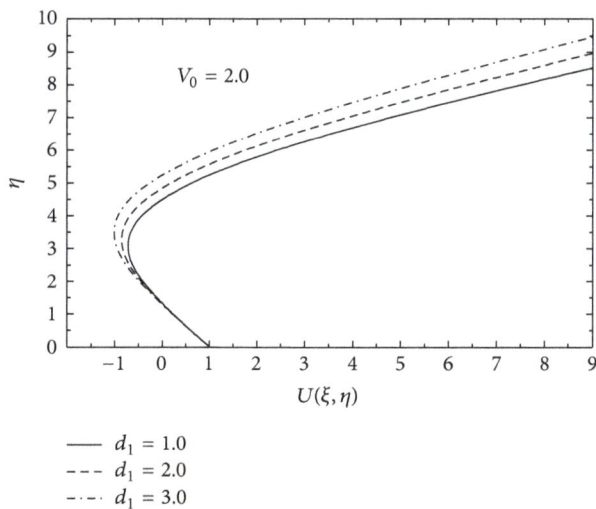

FIGURE 11: Variation of $U(\xi, \eta)$ with η for several values of d_1 with $\xi = 1.0$, $\beta/c = 2.0$, and $V_0 = 2.0$ (i.e., for blowing at the plate).

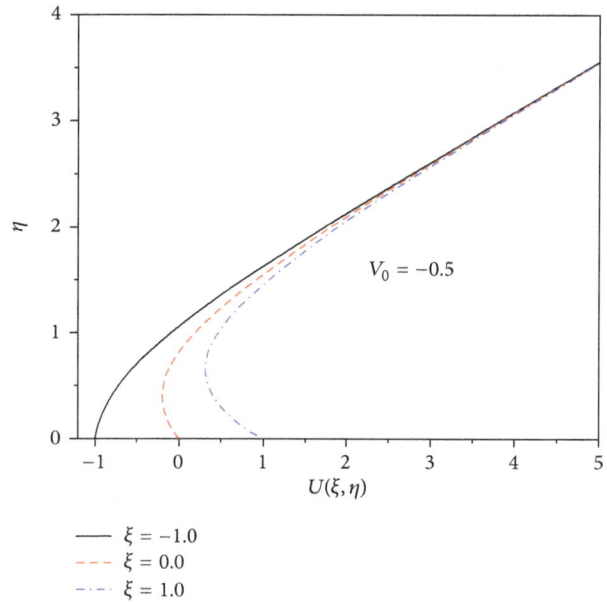

FIGURE 12: Variation of $U(\xi, \eta)$ with η for several values of ξ with $d_1 = 1.0$, $\beta/c = 2.0$, and $V_0 = -0.5$ (i.e., for suction at the plate).

Acknowledgments

The author Samir Kumar Nandy is grateful to the University Grants Commission, New Delhi, India, for providing financial support under a minor research project Grant no. F. PSW-002/13-14 to carry out the work. Authors would like to thank referees for the valuable comments which enhanced the quality of the paper.

References

[1] B. C. Sakiadas, "Boundary layer behaviour on continuous solid surface," *AIChE Journal*, vol. 7, pp. 26–28, 1961.

[2] L. E. Erickson, L. T. Fan, and V. G. Fox, "Heat and mass transfer on a moving continuous flat plate with suction or injection," *Industrial and Engineering Chemistry Fundamentals*, vol. 5, no. 1, pp. 19–25, 1966.

[3] J. E. Danberg and K. S. Fansler, "A non-similar moving-wall boundary-layer problem," *Quarterly of Applied Mathematics*, vol. 34, pp. 305–309, 1976.

[4] L. J. Crane, "Flow past a stretching plate," *Zeitschrift für Angewandte Mathematik und Physik*, vol. 21, no. 4, pp. 645–647, 1970.

[5] C. Y. Wang, "The boundary layers due to shear flow over a still fluid," *Physics of Fluids A*, vol. 4, no. 6, pp. 1304–1306, 1992.

[6] P. S. Gupta and A. S. Gupta, "Heat and mass transfer on a stretching sheet with suction or blowing," *The Canadian Journal of Chemical Engineering*, vol. 55, pp. 744–746, 1977.

[7] K. R. Rajagopal, T. Y. Na, and A. S. Gupta, "Flow of a viscoelastic fluid over a stretching sheet," *Rheologica Acta*, vol. 24, no. 2, pp. 213–215, 1984.

[8] B. S. Dandapat and A. S. Gupta, "Flow and heat transfer in a viscoelastic fluid over a stretching sheet," *International Journal of Non-Linear Mechanics*, vol. 24, no. 3, pp. 215–219, 1989.

strain rate β is considered here over the stretching surface. It is seen that the magnitude of wall skin friction increases with the increase of suction velocity (magnitude) but it decreases with the increase of blowing velocity and this effect is more pronounced in suction than blowing. The horizontal component of velocity at a fixed point increases with the increase of β/c and there is a flow reversal. This region of flow reversal increases with the increase of β/c. The behaviour of the horizontal velocity component is shown for different parameters.

Conflict of Interests

The authors declare that there is no conflict of interests regarding the publication of this paper.

[9] K. R. Rajagopal, T. Y. Na, and A. S. Gupta, "A non-similar boundary layer on a stretching sheet in a non-Newtonian fluid with uniform free stream," *Journal of Mathematical and Physical Sciences*, vol. 21, pp. 189–200, 1987.

[10] P. D. Weidman, D. G. Kubitschek, and S. N. Brown, "Boundary layer similarity flow driven by power-law shear," *Acta Mechanica*, vol. 120, pp. 199–215, 1997.

[11] E. Magyari, B. Keller, and I. Pop, "Boundary-layer similarity flows driven by a power-law shear over a permeable plane surface," *Acta Mechanica*, vol. 163, no. 3-4, pp. 139–146, 2003.

[12] G. E. Cossali, "Similarity solutions of energy and momentum boundary layer equations for a power-law shear driven flow over a semi-infinite flat plate," *European Journal of Mechanics— B/Fluids*, vol. 25, no. 1, pp. 18–32, 2006.

[13] E. Magyari and P. D. Weidman, "Heat transfer on a plate beneath an external uniform shear flow," *International Journal of Thermal Sciences*, vol. 45, no. 2, pp. 110–115, 2006.

[14] P. D. Weidman, A. M. J. Davis, and D. G. Kubitschek, "Crocco variable formulation for uniform shear flow over a stretching surface with transpiration: multiple solutions and stability," *Zeitschrift für Angewandte Mathematik und Physik*, vol. 58, no. 2, pp. 313–332, 2008.

[15] T. Fang, "Flow and heat transfer characteristics of the boundary layers over a stretching surface with a uniform-shear free stream," *International Journal of Heat and Mass Transfer*, vol. 51, no. 9-10, pp. 2199–2213, 2008.

[16] H. Xu, "Homotopy analysis of a self-similar boundary-flow driven by a power-law shear," *Archive of Applied Mechanics*, vol. 78, no. 4, pp. 311–320, 2008.

[17] F. Aman, A. Ishak, and I. Pop, "Heat transfer at a stretching/shrinking surface beneath an external uniform shear flow with a convective boundary condition," *Sains Malaysiana*, vol. 40, no. 12, pp. 1369–1374, 2011.

[18] P. G. Drazin and N. Riley, *The Navier-Stokes Equations: A Classification of Flows and Exact Solutions*, Cambridge University Press, Cambridge, UK, 2006.

[19] J. H. Merkin, "Mixed convection boundary layer flow on a vertical surface in a saturated porous medium," *Journal of Engineering Mathematics*, vol. 14, no. 4, pp. 301–313, 1980.

[20] T. R. Mahapatra, S. K. Nandy, and A. S. Gupta, "Dual solution of MHD stagnation point flow towards a stretching surface," *Engineering*, vol. 2, pp. 299–305, 2010.

5

Phase Separation Behavior and System Properties of Aqueous Two-Phase Systems with Polyethylene Glycol and Different Salts: Experiment and Correlation

Haihua Yuan,[1] **Yang Liu,**[1] **Wanqian Wei,**[1] **and Yongjie Zhao**[2]

[1]*Department of Biology and Guangdong Provincial Key Laboratory of Marine Biotechnology, Shantou University, Shantou, Guangdong 515063, China*
[2]*Department of Mechanical Engineering, College of Engineering, Shantou University, Shantou, Guangdong 515063, China*

Correspondence should be addressed to Yang Liu; liuyanglft@stu.edu.cn

Academic Editor: Robert M. Kerr

The phase separation behaviors of PEG1000/sodium citrate, PEG4000/sodium citrate, PEG1000/ammonium sulfate, and PEG4000/ammonium sulfate aqueous two-phase systems were investigated, respectively. There are two distinct situations for the phase separation rate in the investigated aqueous two-phase systems: one state is top-continuous phase with slow phase separation rate and strong bottom-continuous phase with fast phase separation rate and weak volume ratio dependence. The system properties such as density, viscosity, and interfacial tension between top and bottom phases which have effects on the phase separation rate of aqueous two-phase systems were measured. The property parameter differences between the two phases increased with increasing tie line length and then improved the phase separation rate. Moreover, a modified correlation equation including the phase separation rate, tie line length, and physical properties of the four aqueous two-phase systems has been proposed and successfully tested in the bottom-continuous phase, whose coefficients were estimated through regression analysis. The predicted results of PEG1000/sodium citrate aqueous two-phase systems were verified through the stationary phase retention in the cross-axis countercurrent chromatography.

1. Introduction

In the 1960s, aqueous two-phase systems (ATPS) have been exploited to process different biological sources for the recovery of biological products, such as amino acids, proteins, nucleic acids, antibodies, cells, and organelles [1–5]. ATPS have also become an attractive separation technology due to the mild conditions, the low cost of the phase forming materials, the simplicity of the process, and the easy scale-up to an industrial level. Recently, polymer/salt ATPS, especially polyethylene glycol (PEG)/salt ATPS, have been extensively applied in the separation of many biological products [6–10]. PEG/salt ATPS have advantages of low cost, low viscosity of both the phases, and rapid phase separation rate, so this kind of ATPS has been used for the production of biological molecules on an industrial scale. For example, the scale-up isolation of formate dehydrogenase from 30–50 Kg of wet

cells of *Candida boidinii* was investigated by Kroner et al. [11]. The separation of engineered protein was also successfully scaled up to 1200L by Selber et al. [12]. It displays that ATPS have potential for the large-scale implementation in biotechnology.

Multistage extraction process such as high-speed countercurrent chromatography (HSCCC) has been applied further to improve the target product purity compared with the single-stage extraction. Organic/aqueous solutions system has been widely used in HSCCC due to the short phase separation time (<30 s) [13], while ATPS application in HSCCC is limited due to the long phase separation time (>50 s). The phase separation time of ATPS increases to 3–5 min with the increasing components concentrations [14, 15]. Moreover, the emulsion phenomenon in ATPS is usually serious due to the long phase separation time in the continuous process, which makes it difficult to separate the target products thoroughly.

Many works focused on HSCCC instrument design for ATPS separation to eliminate the emulsion [16–18]. There are few reports about the phase separation rate and physical properties of ATPS [14, 15] for HSCCC.

Much of the research [19, 20] in ATPS has found that polyethylene glycol/organic salts ATPS could be employed as a viable and potentially useful tool for separating proteins instead of the conventional PEG/inorganic systems. The main advantages of PEG/sodium citrate ATPS are the biodegradability and nontoxicity of the organic anion when comparing with the high eutrophication potential of inorganic ions [21] and its harmful impact on the environment. In our present work, the physical properties and phase separation behaviors of the ATPS including PEG/organic salts (PEG/sodium citrate) and PEG/inorganic salts (PEG/ammonium sulfate) systems were investigated for comparisons. Here, ammonium sulfate was selected as the inorganic salt due to its frequent use in protein precipitation. The phase separation rate and the physical properties of ATPS such as density, viscosity, and interfacial tension were measured and correlated with the tie line length (TLL) and ATPS V_r which can contribute to the HSCCC application using ATPS for protein separation. Finally, the stationary phase retention of PEG/sodium citrate ATPS in the cross-axis countercurrent chromatography (X-axis CCC) was investigated based on the above correlated models, in which the X-axis CCC was designed and fabricated in our laboratory.

2. Materials and Methods

2.1. Chemicals. PEG1000 and PEG4000 (GR, $\geq$ 95% mass purity) were purchased from Merck (Shanghai, China). Sodium citrate (GR, $\geq$ 99.5% mass purity) and ammonium sulfate (GR, $\geq$ 99.5% mass purity) were of analytical grade from local sources. Aqueous solutions were prepared with deionized and doubly distilled water.

2.2. Preparation of ATPS. ATPS were prepared from stock solutions of PEG (50% w/w), sodium citrate (30% w/w), and ammonium sulfate (40% w/w). Phase boundaries of all ATPS were obtained by cloud-point measurements in a thermostatic bath (DC-0506, Shanghai Jingtian Electronic Instrument Co. Ltd) of 25°C. The temperature was maintained with an uncertainty of ±0.05°C. The PEG/salt ATPS with five tie lines lengths (30, 35, 40, 45, and 50) were equilibrated for 12 h at 25 ± 0.05°C before use. The compositions of the tie lines were determined by the V_r and the lever rule [22]. The V_r was determined in graduated centrifuge tubes with an uncertainty of ±0.1 mL.

2.3. Density Measurements. The top and bottom phase densities were measured using a 10 mL density bottle with an uncertainty of ±0.1 kg·m^{-3} at 25 ± 0.05°C. The measurements were done in triplicate, and the average value was reported.

2.4. Viscosity Measurements. The viscosity of all ATPS was measured using digital rotary viscosimeter (NDJ-5S, Shanghai Jingtian Electronic Instrument Co. Ltd) with a precision

of ±0.0001 Pa·s. The samples were first maintained at working temperature in thermostatic bath with an uncertainty of ±0.05°C for 10 min, and measurements were done in triplicate.

2.5. Interfacial Tension Measurements. The interfacial tension was measured by a drop volume method [23, 24] at 25°C. Some burettes having capillaries with different outer diameters were used, since the interfacial tension changed with the composition of ATPS. Interfacial tension between the two phases was determined by the numbers and volumes of fallen drops from the top phase solution and the density difference between the top and bottom phases. The interfacial tension is then given by

$$\gamma = \frac{\Delta \rho V g}{2\pi r \psi},$$

$$\psi = 0.9045 - 0.7294 \left(\frac{r}{V^{1/3}} \right) + 0.4293 \left(\frac{r}{V^{1/3}} \right)^2,$$

$$\left(0.3 < \frac{r}{V^{1/3}} < 1.2 \right), \quad (1)$$

$$\psi = 1.007 - 1.479 \left(\frac{r}{V^{1/3}} \right) + 1.829 \left(\frac{r}{V^{1/3}} \right)^2,$$

$$\left(\frac{r}{V^{1/3}} \leq 0.3 \right),$$

where γ is the interfacial tension between the phases of ATPS, $\Delta \rho$ is the density difference of the two phases, V is the average volume of drops, r is the radius of the tip of burette, g is the local value for the acceleration due to gravity, and ψ is the correction factors determined by the values of V and r. The uncertainty of interfacial tension measurement was ±0.0001 mN·m^{-1}. All the measurements were done in triplicate.

2.6. Phase Separation Rate. ATPS mixing was performed in centrifuge tubes with 10 mL volume. The time of mixing was 10 min by manual operation. For all the experiments, the time for phase separation was defined as the time needed for most bulk of the phases to separate and a horizontal interface was formed. Some small drops of one phase which remained in the other phase were ignored. The measurements were done in triplicate with a precision of ±0.1 s.

Kaul et al. found that the kinetic behavior depends greatly on which of the phases is continuous and that the properties of the continuous phase strongly influence the movement of the drops of the dispersed phase [25–27]. Two typical batches' separation is shown in Figure 1. For the top-continuous ATPS, the drops of bottom phase descend in centrifuge tubes and form a settling front; for the bottom-continuous ATPS, the drops of top phase ascend in centrifuge tubes and form a rising front. It can be seen that the process of coalescence is slower than droplet descent and ascent due to the existence of the front.

The problems of phase continuity and phase inversion are important for the phase separation behavior and phase

FIGURE 1: Two typical batch separations for PEG4000/ammonium sulfate ATPS. Left: top-continuous phase; right: bottom-continuous phase.

separation rate. It has been shown that the phase continuity and phase inversion occurred as the V_r changed according to Kaul et al.'s methods [28]. During the ATPS phase separation process, when the descending droplets in the dispersive region were observed, it means a top-continuous phase, while ascending droplets means a bottom-continuous phase. For the same TLL of ATPS, different composition concentrations with various V_r make different continuous phase. Hence, both the top-continuous region and bottom-continuous region can be seen in the same phase diagram. A locus of phase inversion points is found in the phase diagram between top-continuous region and bottom-continuous region, creating a phase inversion band. To determine this band, four ATPS with five different TLL were chosen, and the separation behavior of the ATPS at TLL with different V_r was investigated.

An equation can express the correlation between the phase separation rates and the physical properties of PEG/salt ATPS. The comprehensive efficiency of the physical properties including density, viscosity, and interface tension can be represented by Morton number (M_0) [29–31]. Hence, a generalized correlation equation given by Mistry and Golob [4, 27] was developed for the ATPS with different PEG molar masses and salt relating the Morton number (M_0), the ratio of interfacial tension to surface tension, V_r, and TLL:

$$\frac{dh}{dt}\left(\frac{\mu_c}{\lambda}\right) = aM_0{}^b \left(\frac{\gamma}{\sigma_c}\right)^c \left(\frac{V_{tp}}{V_{bp}}\right)^d (TLL)^e, \qquad (2)$$

$$M_0 = \frac{\Delta\mu^4 g}{\gamma^3 \Delta\rho}, \qquad (3)$$

where a, b, c, d, e are the coefficients of the equation, dh/dt is the phase separation rate, μ_c is the viscosity of the continuous phase, σ_c is the surface tension of continuous phase, V_{tp} is the volume of top phase, V_{bp} is the volume of bottom phase, and $\Delta\mu$ is the viscosity difference between top and bottom phases.

2.7. Stationary Phase Retention in the Cross-Axis Countercurrent Chromatography.

The cross-axis countercurrent chromatography (X-axis CCC) was designed and fabricated in our laboratory. The X-axis CCC apparatus is displayed as Figure 2, which includes the six separation columns at the total column capacity of 71 mL. Each column was tightly coiled by the polytetrafluoroethylene (PTFE) tube with 2.0 mm inner diameter. In view of the best experimental partition efficiencies of ATPS in X-axis CCC [32], the measurement of stationary phase retention of ATPS in our X-axis CCC was designed as follows: the bottom phase was selected as mobile phase, X-axis CCC revolution direction was counterclockwise, the elution mode was from head to tail and from inward to outward, and the flow rate was 0.5 mL/min. The specific operations for the stationary phase retention measurement were the same as Shinomiya's methods [33].

3. Results and Discussion

3.1. The Effect of TLL on ATPS Density, Viscosity, and Interfacial Tension.

The effect of TLL on the density can be seen in Figure 3. It has been shown that the densities of both top and bottom phases are increasing as the TLL becomes longer. Compared with the influence of TLL on the density of the top phase, the variation in density of the bottom phase is very significant. This is because the influence of the composition of salt on the density is larger than the influence of the composition of PEG. Furthermore, the bottom is rich in salt and the top phase is rich in PEG. As a result, the density difference between the bottom phase and the top phase increases with the increasing TLL. Moreover, it can be seen that the PEG molecular weight has no obvious effect on the density of the PEG-rich phase of ATPS compared to Figures 3(a) and 3(b) or Figures 3(c) and 3(d).

The viscosities of the top and bottom phases as a function of the TLL are shown in Figure 4. Figure 4 shows that the viscosity of the top phase increases slightly with the increasing TLL. Furthermore, there is a marked change in viscosity of PEG4000 ATPS as TLL increasing compared with PEG1000 ATPS. It is because the viscosities of ATPS increase with the increasing molar mass of PEG at the same PEG concentration. Perumalsamy and Murugesan have measured the viscosity of PEG solutions at 25°C and observed the same result [34].

Figure 5 shows that the interfacial tension increases with the increasing TLL for all the ATPS. Mishima et al. used the drop volume technique and observed the same trend [35]. The compositions' concentrations play a key role in the solution's interfacial tension. In ATPS, the compositions' concentrations increase with the increase of TLL in both the top phase and the bottom phase. Thus, the interfacial tensions in various ATPS are significantly enhanced with the increasing TLL.

3.2. The Separation Behavior of ATPS.

The separation behaviors of four ATPS are shown in Figure 6. As can be seen in Figure 6, to the left of the solid points, the top phase of ATPS is continuous. To the right of the hollow points, the bottom

(a)   (b)

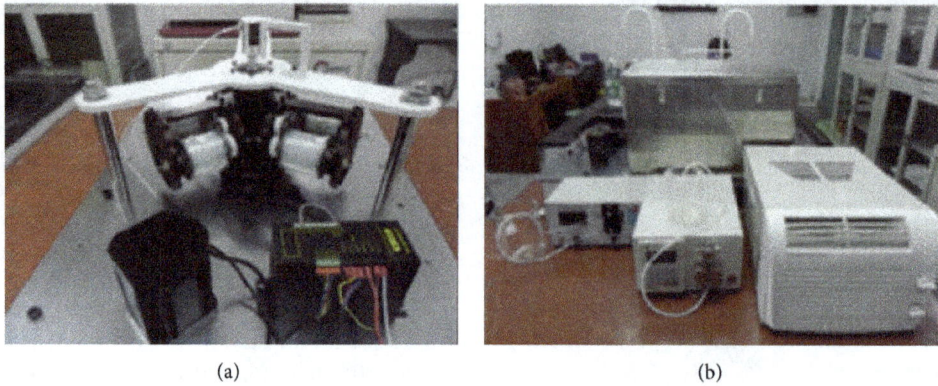

FIGURE 2: Photographs of our cross-axis countercurrent chromatography without the cover case (a) and the overall appearance (b).

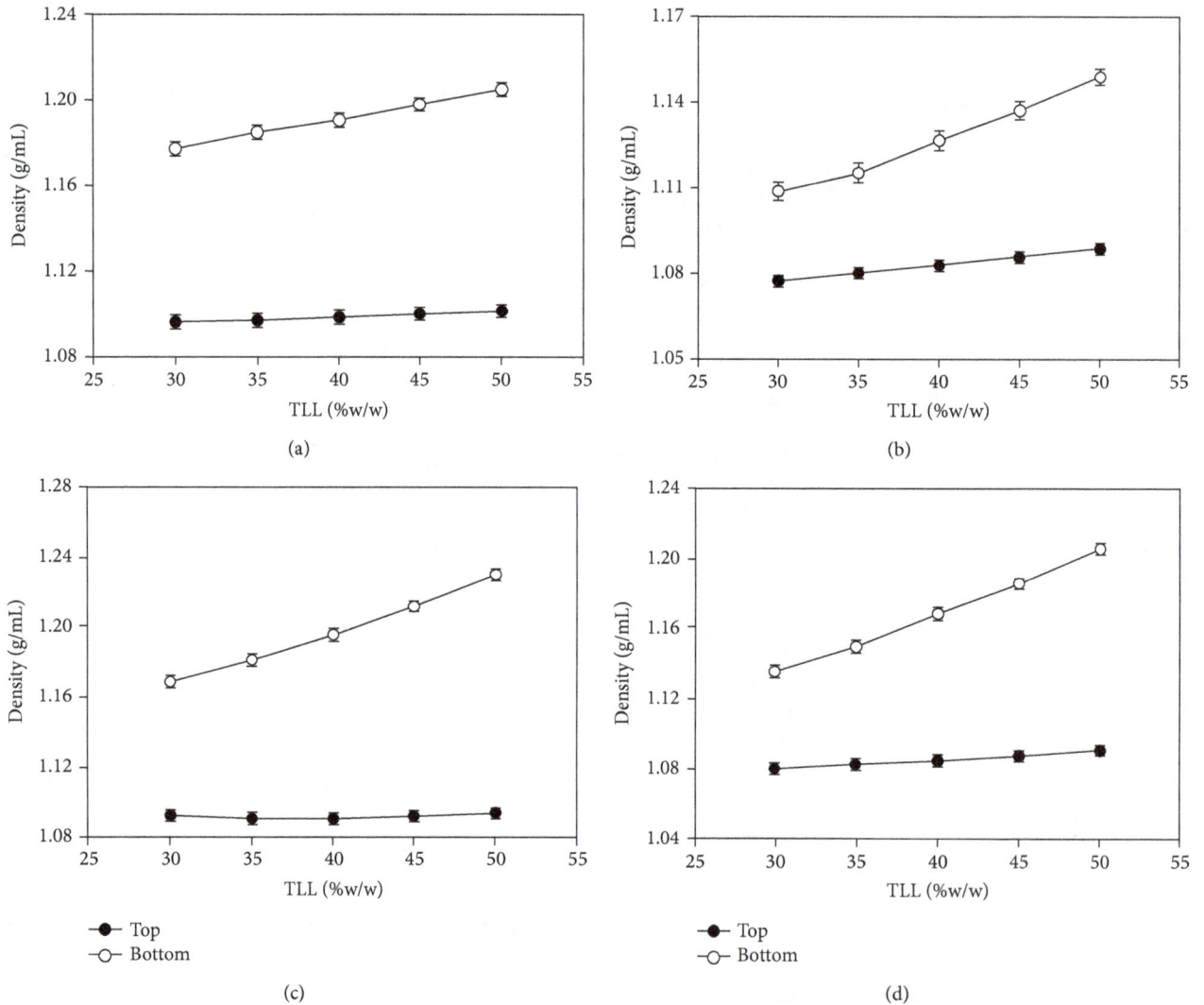

(a)

(b)

(c)

(d)

FIGURE 3: Densities of the top phase and the bottom phase as a function of TLL at 25°C. (a) PEG1000/ammonium sulfate ATPS, (b) PEG4000/ammonium sulfate ATPS, (c) PEG1000/sodium citrate ATPS, and (d) PEG4000/sodium citrate ATPS. The error bars represent the standard deviations in triplicate experiments.

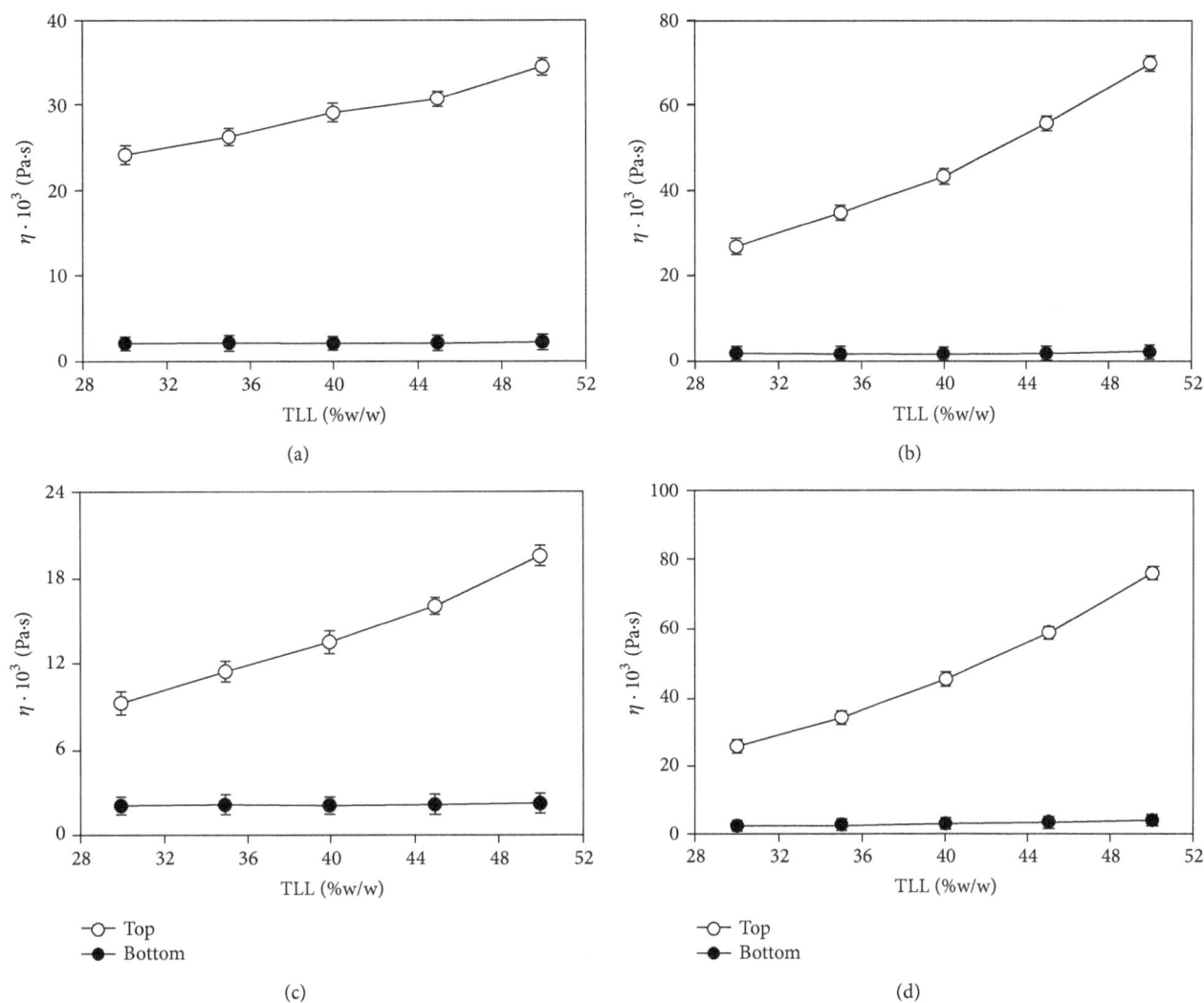

FIGURE 4: Viscosities of the top phase and the bottom phase as a function of TLL at 25°C. (a) PEG1000/ammonium sulfate ATPS, (b) PEG4000/ammonium sulfate ATPS, (c) PEG1000/sodium citrate ATPS, and (d) PEG4000/sodium citrate ATPS. The error bars represent the standard deviations in triplicate experiments.

phase of ATPS is continuous. The region which is called phase inversion band between the solid points and hollow points on the same TLL is a range of ambiguity. Both top-continuous phase and bottom-continuous phase can be seen in this region of ATPS. Within this region, the continuity of the phase is affected not only by composition of the mixture, but also by the fluid dynamics. The region of ambiguity becomes large with the increasing TLL and locates nearly at the constant salt concentration line. In Figure 6, the region of bottom-continuous phase is almost 4 times larger than the region of top-continuous phase. Furthermore, the middle points of TLL locate in the bottom-continuous phase. This indicates that the four investigated ATPS of $V_r = 1$ were operated in the region of bottom-continuous phase.

Figure 7 shows the profile of the phase separation time changes with the increasing top phase ratio ($V_t = 1/(V_r + 1)$). Phase inversion takes place near the point of $V_t = 0.8$ ($V_r = 4$), where the continuous phase can change into the dispersed

phase, and a sudden change of phase separation behavior is observed. The bottom phase is continuous when the V_t is less than 0.8, and short phase separation times are observed with the smooth variation. The top phase is continuous when the V_t is more than 0.8, and the dramatic increase of phase separation time with the increasing V_t is observed. It indicates that ATPS would be better to operate in the bottom-continuous region rather than top-continuous region.

3.3. Correlation Equations of Phase Separation. Since the ATPS of $V_r = 1$ have been widely applied in the separation of biological products, the phase separation rate of ATPS of $V_r = 1$ has been investigated. The surface tension of water (σ_w) can be used in the equation rather than the surface tension of continuous phase (σ_c). Hence, (2) can be modified as

$$\frac{dh}{dt}\left(\frac{\mu_c}{\lambda}\right) = aM_0{}^b\left(\frac{\gamma}{\sigma_w}\right)^c (\text{TLL})^d. \tag{4}$$

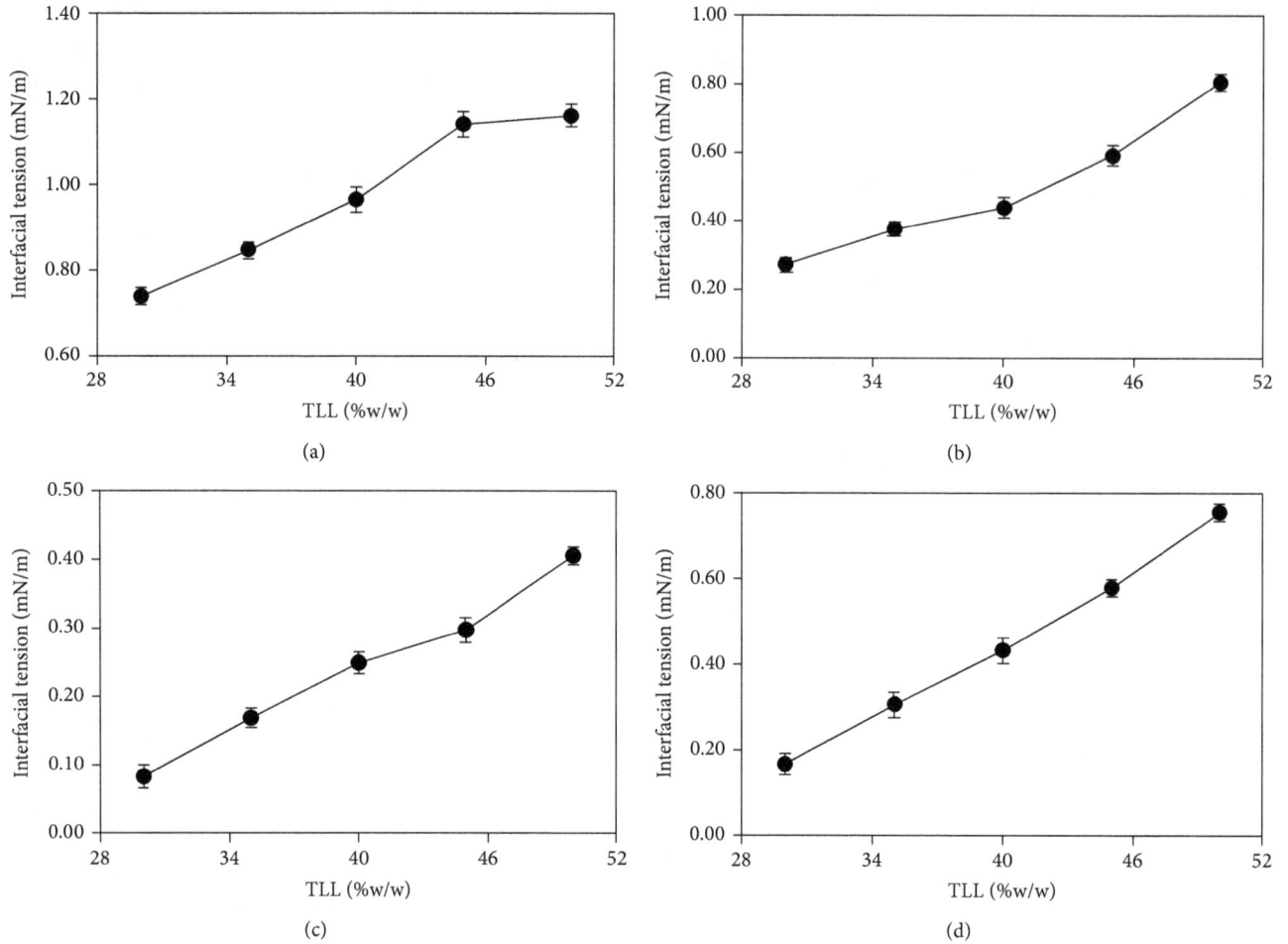

FIGURE 5: Interfacial tension between the top phase and the bottom phase as a function of TLL at 25°C. (a) PEG1000/ammonium sulfate ATPS, (b) PEG4000/ammonium sulfate ATPS, (c) PEG1000/sodium citrate ATPS, and (d) PEG4000/sodium citrate ATPS. The error bars represent the standard deviations in triplicate experiments.

The value of the coefficients (a, b, c, d) can be calculated with the experimental values (Tables 1 and 2). This was performed by nonlinear regression using 1stOpt 1.5 software (China). Using these values for the bottom-continuous region, (4) can be written as (5) and (6).

For the bottom-continuous region of PEG/ammonium sulfate ATPS, the correlation equation can be expressed:

$$\frac{dh}{dt}\left(\frac{\mu_b}{\gamma}\right) = 3.58 \times 10^{-3} M_0^{0.0080}\left(\frac{\gamma}{\sigma_w}\right)^{-1.8688} (\text{TLL})^{0.7650}.$$

(5)

For the bottom-continuous region of PEG/sodium citrate ATPS, the correlation equation can be expressed:

$$\frac{dh}{dt}\left(\frac{\mu_b}{\gamma}\right) = 1.05 \times 10^{-6} M_0^{0.1735}\left(\frac{\gamma}{\sigma_w}\right)^{-1.5403} (\text{TLL})^{2.5750}.$$

(6)

The R^2 (correlation index) of the two equations is 0.960 for PEG/ammonium sulfate ATPS and 0.971 for PEG/sodium citrate ATPS, respectively. Equation (5) was used to predict

the phase separation rate of the bottom-continuous region of PEG/ammonium sulfate ATPS of $V_r = 1$ with an average absolute relative deviation (AARD) value of 7.05%. Equation (6) was used to predict the phase separation rate of the bottom-continuous region of PEG/sodium citrate ATPS of $V_r = 1$ with an AARD value of 6.81%. The values of the coefficient indicate the difference of the phase separation rates between the two ATPS. The two equations show that the TLL has a greater effect on the PEG/sodium citrate ATPS than the PEG/ammonium sulfate ATPS. The M_0 has a similar effect on the phase separation rates of both ATPS.

Both (5) and (6) have predicted the phase separation rates of ATPS with $V_r = 1$. For the batch separation, the results of prediction can provide a reference since the stable ATPS with $V_r = 1$ are widely used. For the continuous separation, such as HSCCC, the phase separation rate is the most important factor for the choice of solvent systems [36]. Hence, it is necessary to investigate the relationship between phase separation rate and physical properties of ATPS. In this paper, as phase diagrams show in Figure 6, the bottom-continuous regions of the four investigated ATPS are fit for continuous separation due to the fast phase separation rates and weak V_r dependence.

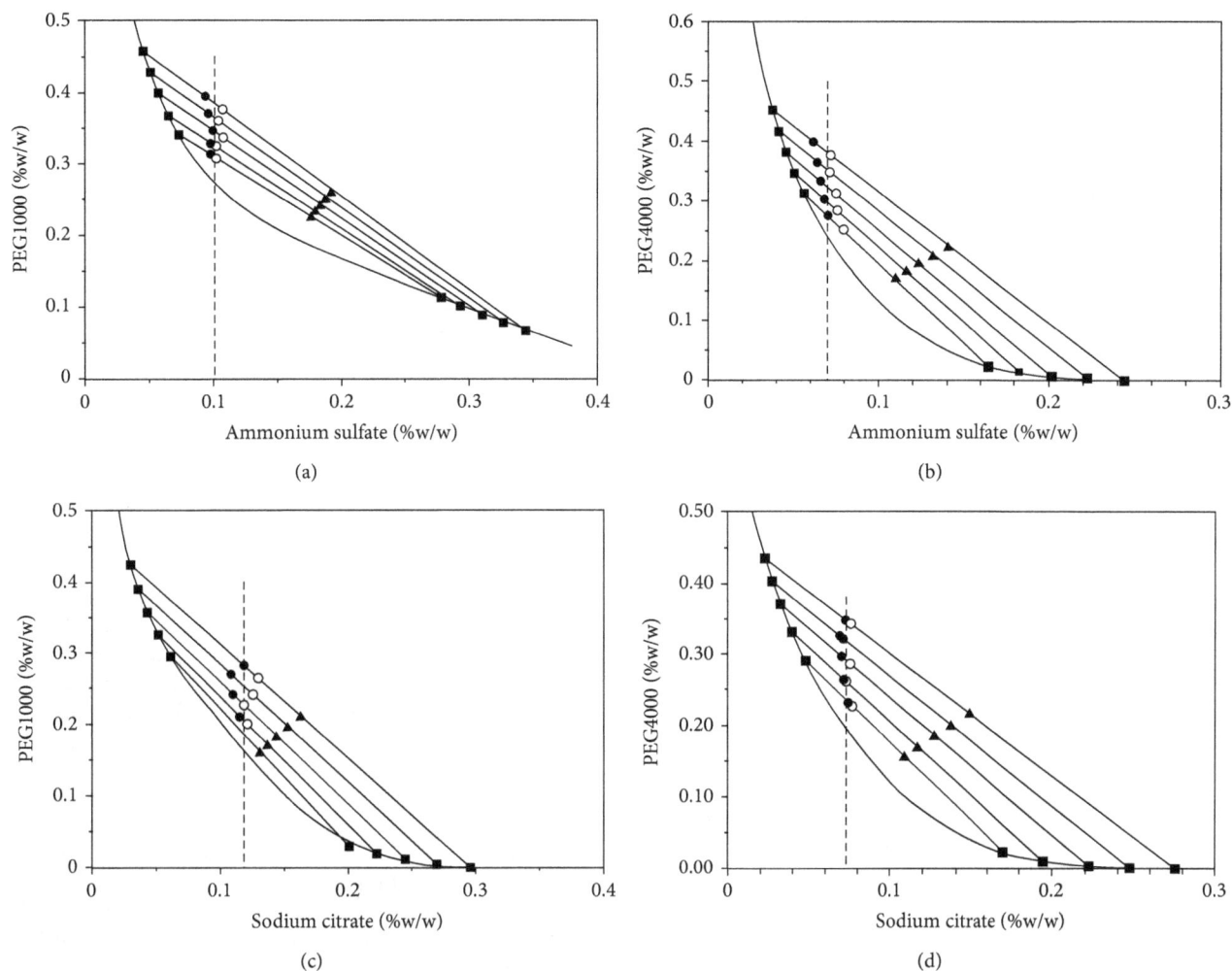

FIGURE 6: Experimental equilibrium data obtained for the ATPS at 25°C, phase inversion band between the points at the same TLL. (a) PEG1000/ammonium sulfate ATPS, (b) PEG4000/ammonium sulfate ATPS, (c) PEG1000/sodium citrate ATPS, and (d) PEG4000/sodium citrate ATPS.

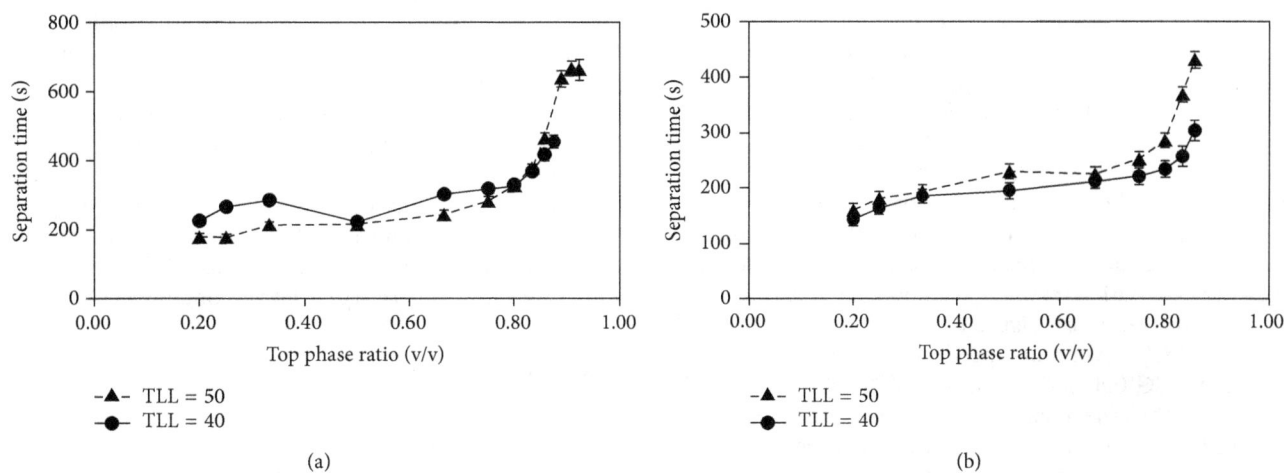

FIGURE 7: Effect of the change in percentage top phase ($V_t = 1/(V_r + 1)$) on phase separation time for the ATPS at 25°C. (a) PEG4000/ammonium sulfate ATPS and (b) PEG4000/sodium citrate ATPS. The error bars represent the standard deviations in triplicate experiments.

TABLE 1: Experimental properties parameters of PEG1000/ammonium sulfate ATPS and PEG4000/ammonium sulfate ATPS at 25°C.

	TLL [%w/w]	Δt [s]	σ_w [mN/m]	γ [mN/m]	$\Delta \mu$ [mPa·s]	μ_b [mPa·s]
	30	90.38	71.97	0.739	22.130	2.120
PEG1000/ ammonium sulfate	35	93.46	71.97	0.846	24.158	2.193
	40	97.52	71.97	0.965	27.053	2.148
	45	96.44	71.97	1.140	28.595	2.205
	50	101.04	71.97	1.161	32.305	2.295
	30	280.62	71.97	0.272	25.010	1.990
PEG4000/ ammonium sulfate	35	247.02	71.97	0.376	33.085	1.865
	40	218.88	71.97	0.439	41.625	1.775
	45	188.00	71.97	0.592	54.005	1.895
	50	164.10	71.97	0.805	67.790	2.210

TABLE 2: Experimental properties parameters of PEG1000/sodium citrate ATPS and PEG4000/sodium citrate ATPS at 25°C.

	TLL [%w/w]	Δt [s]	σ_w [mN/m]	γ [mN/m]	$\Delta \mu$ [mPa·s]	μ_b [mPa·s]
	30	156.18	71.97	0.083	5.650	3.650
PEG1000/ sodium citrate	35	133.48	71.97	0.169	7.780	3.720
	40	122.02	71.97	0.249	9.500	4.050
	45	111.98	71.97	0.297	11.68	4.420
	50	111.32	71.97	0.405	14.365	5.235
	30	279.93	71.97	0.167	23.165	2.435
PEG4000/ sodium citrate	35	247.85	71.97	0.305	31.185	2.715
	40	244.29	71.97	0.433	42.110	2.940
	45	267.05	71.97	0.579	55.145	3.355
	50	266.48	71.97	0.755	71.985	4.015

3.4. Stationary Phase Retention of PEG1000/Sodium Citrate ATPS in the X-Axis CCC. PEG1000/sodium citrate ATPS were selected to investigate the positive effect of fast phase separation rate on the stationary phase retention in the cross-axis countercurrent chromatography. According to the above correlation equation (6), the shorter TLL of PEG1000/sodium citrate ATPS could obtain the faster phase separation rate of the two phases. Thus, PEG1000/sodium citrate ATPS with TLL = 30 (16.54 w/w% PEG1000 and 13.96 w/w% sodium citrate) was determined as the investigated ATPS for the measurement of the stationary phase retention in our X-axis CCC at various rotation rates. Figure 8 displays that the effect of the X-axis CCC revolution speed on the stationary phase retention of PEG1000/sodium citrate ATPS with TLL = 30. It can be seen that the maximum value of stationary phase retention was 54.83% at 800 rpm, which was slightly higher than the reported stationary phase retention of ATPS in other X-axis CCCs [32]. It indicated that the phase separation rate of ATPS played an important role in the stationary phase retention in HSCCC.

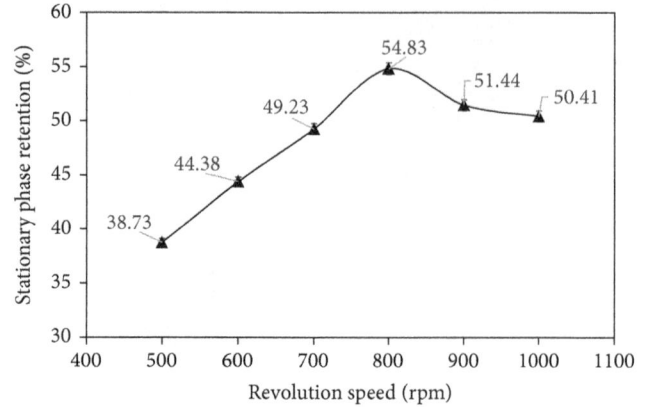

FIGURE 8: Effect of revolution speed on stationary phase retention of PEG1000/sodium citrate in the X-axis CCC. The error bars represent the standard deviations in triplicate experiments.

4. Conclusion

The physical properties of PEG/ammonium sulfate and PEG/sodium citrate ATPS had been studied. The density of bottom phase is larger than that of top phase and increases with increasing TLL. The viscosity of the top phase is 3–35 times larger than the bottom phase and increases with the increasing TLL. The interfacial tension between the phases increases with the increasing TLL.

The bottom phase is continuous at high V_r value, and fast separation rate has been observed as well as the smooth variation. The top phase is continuous at low V_r value, and the dramatic increase of phase separation time with the increasing V_r value was observed. There is a phase inversion band between the top-continuous region and bottom-continuous region. The phase inversion band is located at the constant salt concentration line in the phase diagram. Within this region, the continuity of the phase is affected not only by the composition of the ATPS, but also by the fluid dynamics.

The phase separation rate was correlated using a modified correlation equation and the coefficients were found. It showed that the R^2 of the two equations were 0.960 for PEG/ammonium sulfate ATPS and 0.971 for PEG/sodium citrate ATPS, respectively. The correlation equations gave good results for the bottom-continuous ATPS with $V_r = 1$. The results may be useful for the choice of optimal conditions of ATPS no matter in batch separation or continuous separation process. The correlation equations were applied to estimate the fast phase separation rate of PEG1000/sodium citrate ATPS and further to measure the stationary phase retention of PEG/sodium citrate ATPS in our X-axis CCC, in which the higher stationary phase retention was obtained at 800 rpm.

Nomenclature

R^2: Correlation index,

$$R^2 = \frac{\left(\sum_{i=1}^n x_i y_i - \sum_{i=1}^n x_i \sum_{i=1}^n y_i / n \right)^2}{\left[\sum_{i=1}^n x_i^2 - \left(\sum_{i=1}^n x_i \right)^2 / n \right] \left[\sum_{i=1}^n y_i^2 - \left(\sum_{i=1}^n y_i \right)^2 / n \right]},$$

$(*)$

where x_i, y_i, and n are the experimental, calculated, and number of data points, respectively

AARD: Average absolute relative deviation,

$$\text{AARD} = \left(\sum_{i=1}^{n} \left| \frac{y_{\text{exp},i} - y_{\text{cal},i}}{y_{\text{exp},i}} \times 100\% \right| \right) \times \left(\frac{1}{n} \right), \quad (**)$$

where $y_{\text{exp},i}$, $y_{\text{cal},i}$, and n are the experimental, calculated, and number of data points, respectively

Δh: Height of the interphase

Δt: Time of phase separation

V_{tp}: Volume of top phase

V_{bp}: Volume of bottom phase

M_0: Morton number.

Greek Letters

$\Delta \rho$: Density difference between the top and bottom phases

ρ_c: Density of the continuous phase

γ: Interfacial tension between the top and bottom phases

σ_w: Surface tension of the air-water interface at 25°C

σ_c: Surface tension between air-continuous phase interface

μ_c: Viscosity of the continuous phase

μ_d: Viscosity of the dispersed phase

μ_b: Viscosity of the bottom-continuous phase

$\Delta \mu$: Viscosity difference between the phases.

Conflict of Interests

The authors declare that there is no conflict of interests regarding the publication of this paper.

Acknowledgments

This work is supported by the National Natural Science Foundation of China (no. 21476135), Outstanding Young Teachers Training Program in Guangdong Higher Education Institutions (no. Yq2013076), Science and Technology Planning Project of Guangdong Province, China (no. 2012B060400006), and Education Department Projects of Guangdong Province, China (nos. 2012KJCX0052 and 920-38030337).

References

[1] J. Benavides and M. Rito-Palomares, "Practical experiences from the development of aqueous two-phase processes for the recovery of high value biological products," *Journal of Chemical Technology and Biotechnology*, vol. 83, no. 2, pp. 133–142, 2008.

[2] J. Benavides, O. Aguilar, B. H. Lapizco-Encinas, and M. Rito-Palomares, "Extraction and purification of bioproducts and nanoparticles using aqueous two-phase systems strategies," *Chemical Engineering and Technology*, vol. 31, no. 6, pp. 838–845, 2008.

[3] D. F. C. Silva, A. M. Azevedo, P. Fernandes, V. Chu, J. P. Conde, and M. R. Aires-Barros, "Design of a microfluidic platform for monoclonal antibody extraction using an aqueous two-phase system," *Journal of Chromatography A*, vol. 1249, pp. 1–7, 2012.

[4] S. L. Mistry, A. Kaul, J. C. Merchuk, and J. A. Asenjo, "Mathematical modelling and computer simulation of aqueous two-phase continuous protein extraction," *Journal of Chromatography A*, vol. 741, no. 2, pp. 151–163, 1996.

[5] P. A. Albertsson, *Partition of Cell Particles and Macromolecules*, Wiley-Interscience, New York, NY, USA, 1986.

[6] B. Y. Zaslavsky, *Aqueous Two-Phase Partitioning: Physical Chemistry and Bioanalytical Applications*, Marcel Dekker, New York, NY, USA, 1995.

[7] M. Rito-Palomares, "Practical application of aqueous two-phase partition to process development for the recovery of biological products," *Journal of Chromatography B: Analytical Technologies in the Biomedical and Life Sciences*, vol. 807, no. 1, pp. 3–11, 2004.

[8] E. Huenupi, A. Gomez, B. A. Andrews, and J. A. Asenjo, "Optimization and design considerations of two-phase continuous protein separation," *Journal of Chemical Technology & Biotechnology*, vol. 74, no. 3, pp. 256–263, 1999.

[9] S. Saravanan, J. R. Rao, T. Murugesan, B. U. Nair, and T. Ramasami, "Recovery of value-added globular proteins from tannery wastewaters using PEG—salt aqueous two-phase systems," *Journal of Chemical Technology & Biotechnology*, vol. 81, no. 11, pp. 1814–1819, 2006.

[10] P. A. J. Rosa, I. F. Ferreira, A. M. Azevedo, and M. R. Aires-Barros, "Aqueous two-phase systems: a viable platform in the manufacturing of biopharmaceuticals," *Journal of Chromatography A*, vol. 1217, no. 16, pp. 2296–2305, 2010.

[11] K. H. Kroner, H. Schütte, W. Stach, and M.-R. Kula, "Scale-up of formate dehydrogenase by partition," *Journal of Chemical Technology and Biotechnology*, vol. 32, no. 1, pp. 130–137, 1982.

[12] K. Selber, F. Tjerneld, A. Collén et al., "Large-scale separation and production of engineered proteins, designed for facilitated recovery in detergent-based aqueous two-phase extraction systems," *Process Biochemistry*, vol. 39, no. 7, pp. 889–896, 2004.

[13] Y. Zeng, G. Liu, Y. Ma, X. Y. Chen, and Y. Ito, "Organic high ionic strength aqueous two-phase solvent system series for separation of ultra-polar compounds by spiral high-speed counter-current chromatography," *Journal of Chromatography A*, vol. 1218, no. 48, pp. 8715–8717, 2011.

[14] M. H. Salamanca, J. C. Merchuk, B. A. Andrews, and J. A. Asenjo, "On the kinetics of phase separation in aqueous two-phase systems," *Journal of Chromatography B: Biomedical Applications*, vol. 711, no. 1-2, pp. 319–329, 1998.

[15] A. V. Narayan, M. C. Madhusudhan, and K. S. M. S. Raghavarao, "Demixing kinetics of phase systems employed for liquid-liquid extraction and correlation with system properties," *Food and Bioproducts Processing*, vol. 89, no. 4, pp. 251–256, 2011.

[16] K. Shinomiya, H. Kobayashi, N. Inokuchi et al., "New small-scale cross-axis coil planet centrifuge. Partition efficiency and application to purification of bullfrog ribonuclease," *Journal of Chromatography A*, vol. 1151, no. 1-2, pp. 91–98, 2007.

[17] N. Mekaoui, K. Faure, and A. Berthod, "Advances in countercurrent chromatography for protein separations," *Bioanalysis*, vol. 4, no. 7, pp. 833–844, 2012.

[18] K. Shinomiya, H. Kobayashi, N. Inokuchi, K. Nakagomi, and Y. Ito, "Partition efficiency of high-pitch locular multilayer coil for countercurrent chromatographic separation of proteins using small-scale cross-axis coil planet centrifuge and application to purification of various collagenases with aqueous-aqueous polymer phase systems," *Journal of Liquid Chromatography & Related Technologies*, vol. 34, no. 3, pp. 182–194, 2011.

[19] G. Tubio, G. A. Picó, and B. B. Nerli, "Extraction of trypsin from bovine pancreas by applying polyethyleneglycol/sodium citrate aqueous two-phase systems," *Journal of Chromatography B: Analytical Technologies in the Biomedical and Life Sciences*, vol. 877, no. 3, pp. 115–120, 2009.

[20] R. L. Pérez, D. B. Loureiro, B. B. Nerli, and G. Tubio, "Optimization of pancreatic trypsin extraction in PEG/citrate aqueous two-phase systems," *Protein Expression and Purificatio*, vol. 106, pp. 66–71, 2015.

[21] M. Perumalsamy and T. Murugesan, "Prediction of liquid-liquid equilibria for PEG 2000-sodium citrate based aqueous two-phase systems," *Fluid Phase Equilibria*, vol. 244, no. 1, pp. 52–61, 2006.

[22] Y. Liu, Y. Q. Feng, and Y. J. Zhao, "Liquid–liquid equilibrium of various aqueous two-phase systems: experiment and correlation," *Journal of Chemical & Engineering Data*, vol. 58, no. 10, pp. 2775–2784, 2013.

[23] M. C. Wikinson, "Extended use of, and comments on, the drop-weight (drop-volume) technique for the determination of surface and interfacial tensions," *Journal of Colloid and Interface Science*, vol. 40, pp. 14–26, 1972.

[24] C. Jho and R. Burke, "Drop weight technique for the measurement of dynamic surface tension," *Journal of Colloid and Interface Science*, vol. 95, no. 1, pp. 61–71, 1983.

[25] J. A. Asenjo, S. L. Mistry, B. A. Andrews, and J. C. Merchuk, "Phase separation rates of aqueous two-phase systems: correlation with system properties," *Biotechnology and Bioengineering*, vol. 79, no. 2, pp. 217–223, 2002.

[26] E. Barnea and J. Mizrahi, "Separation mechanism of liquid-liquid dispersions in a deep-layer gravity settler: part IV-continuous settler characteristics," *Transactions of the Institution of Chemical Engineers*, vol. 56, pp. 83–92, 1975.

[27] J. Golob and R. Modic, "Coalescence of liquid/liquid dispersions in gravity settlers," *Transactions of IChemE*, vol. 55, pp. 207–211, 1977.

[28] A. Kaul, R. A. M. Pereira, J. A. Asenjo, and J. C. Merchuk, "Kinetics of phase separation for polyethylene glycol-phosphate two-phase systems," *Biotechnology and Bioengineering*, vol. 48, no. 3, pp. 246–256, 1995.

[29] M. T. Zafarani-Moattar, R. Sadeghi, and A. A. Hamidi, "Liquid-liquid equilibria of an aqueous two-phase system containing polyethylene glycol and sodium citrate: experiment and correlation," *Fluid Phase Equilibria*, vol. 219, no. 2, pp. 149–155, 2004.

[30] V. H. Nagaraja and R. Lyyaswami, "Phase demixing studies in aqueous two-phase system with polyethylene glycol (PEG) and sodium citrate," *Chemical Engineering Communications*, vol. 200, no. 10, pp. 1293–1308, 2013.

[31] T. Murugesan and I. Regupathi, "Prediction of continuous phase axial mixing in rotating disc contactors," *Journal of Chemical Engineering of Japan*, vol. 37, no. 10, pp. 1293–1302, 2004.

[32] K. Shinomiya, K. Yanagidaira, and Y. Ito, "New small-scale cross-axis coil planet centrifuge: the design of the apparatus and its application to counter-current chromatographic separation of proteins with aqueous-aqueous polymer phase systems," *Journal of Chromatography A*, vol. 1104, no. 1-2, pp. 245–255, 2006.

[33] Y. Shibusawa, Y. Eriguchi, and Y. Ito, "Purification of lactic acid dehydrogenase from bovine heart crude extract by counter-current chromatography," *Journal of Chromatography B: Biomedical Applications*, vol. 696, no. 1, pp. 25–31, 1997.

[34] M. Perumalsamy and T. Murugesan, "Phase compositions, molar mass, and temperature effect on densities, viscosities, and liquid-liquid equilibrium of polyethylene glycol and salt-based aqueous two-phase systems," *Journal of Chemical & Engineering Data*, vol. 54, no. 4, pp. 1359–1366, 2009.

[35] K. Mishima, K. Matsuyama, M. Ezawa, Y. Taruta, S. Takarabe, and M. Nagatani, "Interfacial tension of aqueous two-phase systems containing poly(ethylene glycol) and dipotassium hydrogenphosphate," *Journal of Chromatography B: Biomedical Applications*, vol. 711, no. 1-2, pp. 313–318, 1998.

[36] Y. Ito, "Golden rules and pitfalls in selecting optimum conditions for high-speed counter-current chromatography," *Journal of Chromatography A*, vol. 1065, no. 2, pp. 145–168, 2005.

6

Magnetohydrodynamic Mixed Convection Stagnation-Point Flow of a Power-Law Non-Newtonian Nanofluid towards a Stretching Surface with Radiation and Heat Source/Sink

Macha Madhu and Naikoti Kishan

Department of Mathematics, Osmania University, Hyderabad, Telangana 500007, India

Correspondence should be addressed to Macha Madhu; madhumaccha@gmail.com

Academic Editor: Robert Spall

Two-dimensional MHD mixed convection boundary layer flow of heat and mass transfer stagnation-point flow of a non-Newtonian power-law nanofluid towards a stretching surface in the presence of thermal radiation and heat source/sink is investigated numerically. The non-Newtonian nanofluid model incorporates the effects of Brownian motion and thermophoresis. The basic transport equations are made dimensionless first and the complete nonlinear differential equations with associated boundary conditions are solved numerically by finite element method (FEM). The numerical calculations for velocity, temperature, and nanoparticles volume fraction profiles for different values of the physical parameters to display the interesting aspects of the solutions are presented graphically and discussed. The skin friction coefficient, the local Nusslet number and the Sherwood number are exhibited and examined. Our results are compatible with the existing results for a special case.

1. Introduction

Many of the non-Newtonian fluids encountered in chemical engineering processes are known to follow the empirical Ostwald-de Waele power-law model. The concept of boundary layer was applied to power-law fluids by Schowalter [1]. Acrivos [2] investigated the boundary layer flows for such fluids in 1960; since then, a large number of related studies have been conducted because of their importance and presence of such fluids in chemicals, polymers, molten plastics, and others. The theory of non-Newtonian fluids offers mathematicians, engineers, and numerical specialists varied challenges in developing analytical and numerical solutions for the highly nonlinear governing equations. However, due to the practical significance of these non-Newtonian fluids, many authors have presented various non-Newtonian fluid models like Elbashbeshy et al. [3], Nadeem et al. [4], Nadeem et al. [5], Nadeem et al. [6], Nadeem and Akbar [7], Nadeem and Ali [8], Buongiorno [9], and Łukaszewicz [10]. Many interesting applications of non-Newtonian power-law fluids were presented by Shenoy [11]. Details of the behavior of

non-Newtonian fluids for both steady and unsteady flow situations, along with mathematical models, are studied by Astarita and Marrucci [12], Bohme [13], and Kishan and Shashidar Reddy [14].

Nanotechnology has immense applications in industry since materials with sizes of nanometers exhibit unique physical and chemical properties. Fluids with nano-scaled particles interaction are called nanofluid. It represents the most relevant technological cutting edge currently being explored. Nanofluid heat transfer is an innovative technology which can be used to enhance heat transfer. Nanofluid is a suspension of solid nanoparticles (1–100 nm diameters) in conventional liquids like water, oil, and ethylene glycol. Depending on shape, size, and thermal properties of the solid nanoparticles, the thermal conductivity can be increased by about 40% with low concentration (1%–5% by volume) of solid nanoparticles in the mixture. The nanoparticles used in nanofluid are normally composed of metals, oxides, carbides, or carbon nanotubes. Water, ethylene glycol, and oil are common examples of base fluids. Nanofluids have their major applications in heat transfer, including microelectronics, fuel cells, pharmaceutical processes, and hybrid-powered engines, domestic refrigerator, chiller, nuclear reactor coolant, grinding, space technology, and boiler flue gas temperature reduction. They demonstrate enhanced thermal conductivity and convective heat transfer coefficient counterbalanced to the base fluid. Nanofluids have been the core of attention of many researchers for new production of heat transfer fluids in heat exchangers, plants, and automotive cooling significations, due to their enormous thermal characteristics, Nadeem et al. [15]. The nanofluid is stable; it introduces very little pressure drop, and it can pass through nanochannels (e.g., see Zhou [16]). The word nanofluid was coined by Choi [17]. Xuan and Li [18] pointed out that, at higher nanoparticle volume fractions, the viscosity increases sharply, which suppresses heat transfer enhancement in the nanofluid. Therefore, it is important to carefully select the proper nanoparticle volume fraction to achieve heat transfer enhancement. Buongiorno noted that the nanoparticles' absolute velocity can be viewed as the sum of the base fluid velocity and a relative velocity (which he called the slip velocity). He considered in turn seven slip mechanisms: inertia, Brownian diffusion, thermophoresis, diffusiophoresis, Magnus effect, fluid drainage, and gravity settling.

Forced convective heat transfer can be enhanced effectively by using nanofluids, a type of fluid adding different suspending nanoparticles into the conventional base liquid (Pak and Cho [19], Wen and Ding [20], and Ding et al. [21]). However, the characteristics of nanofluids and the mechanism of the enhancement of the forced convective heat transfer of nanofluids are still not clear. Recently, nanofluids have attracted much attention since anomalously large enhancements in effective thermal conductivities were reported over a decade ago (Masuda et al. [22] and Keblinski et al. [23]). Subsequent studies by various groups have reported that nanofluids also have other desirable properties and behaviours such as enhanced wetting and spreading (Wasan and Nikolov [24] and Chengara et al. [25]), as well as increased critical heat fluxes under boiling condition (You et al. [26]).

Boundary layer flow and heat transfer over a continuously stretched surface have received considerable attention in recent years. This is because of the various possible engineering and metallurgical applications such as hot rolling, metal and plastic extrusion, wire drawing, glass fibre production, continuous casting, crystal growing, and paper production. Crane [27] was the first to investigate the boundary layer flow caused by a stretching sheet moving with linearly varying velocity from a fixed point whilst the heat transfer aspect of the problem was investigated by Carragher and Crane [28] under the conditions that the temperature difference between the surface and the ambient fluid was proportional to the power of the distance from a fixed point. The behaviour of non-Newtonian nanofluids could be useful in evaluating the possibility of heat transfer enhancement in various processes of these industries. Several investigators have studied non-Newtonian nanofluid transport in various geometries under various boundary conditions in porous or nonporous media. Ellahi et al. [29] have elaborated that non-Newtonian nanofluids have potential roles in physiological transport as

biological solutions and also in polymer melts, paints, and so forth.

The effect of heat source/sink is very important in cooling process industries. Effects of heat source/sink on the boundary layer flow over a stretching sheet were studied by Cortell [30], Abel et al. [31], Dessie and Kishan [32], Tufail et al. [33], Elbashbeshy and Bazid [34], and Chen [35]. In fact, heat source/sink concepts in fluids have relevance in problems dealing with chemical reactions geonuclear repositions and these concerned with dissociating fluids. Transport phenomena associated with magnetohydrodynamics arise in physics, geophysics, astrophysics, and many branches of chemical engineering which include crystal magnetic damping control, hydromagnetic chromatography, conducting flow in trickle-bed reactors, and enhanced magnetic filtration control (Prasad et al. [36]).

In the present paper, our aim is to investigate the effects of MHD, heat source/sink, and thermal radiation on heat and mass transfer by mixed convection boundary layer stagnation point flow of non-Newtonian power-law fluid towards a stretching surface with a nanofluid. The effects of Brownian motion and thermophoresis are included for the nanofluid. Numerical solutions of the boundary layer equations are obtained and a discussion is provided for several values of the nanofluid parameters governing the problem. The dimensionless profiles of velocity, temperature, and nanoparticle volume fraction as well as the skin friction coefficient, local Nusselt number, and Sherwood number for the different flow parameters have been discussed.

2. Mathematical Formulation

Consider steady, laminar, heat, and mass transfer by mixed convection, boundary layer stagnation-point flow of an electrically conducting, optically dense, and non-Newtonian power-law fluid obeying the Ostwald-de Waele model (see [37]) past a heated or cooled stretching vertical surface in the presence of thermal radiation. It is assumed that the stretching velocity is given by $u_w(x) = cx$, and the velocity distribution in frictionless potential flow in the neighborhood of the stagnation point at $x = y = 0$ is given by $U(x) = ax$. We assumed that the uniform wall temperature T_w and nanoparticles volume fraction C_w are higher than those of their full stream values T_∞, C_∞. A uniform magnetic field is applied in the y-direction normal to the flow direction. The magnetic Reynolds number is assumed to be small so that the induced magnetic field is neglected. In addition, the Hall effect and the electric field are assumed to be negligible. The small magnetic Reynolds number assumption uncouples the Navier-Stokes equations from Maxwell's equations. All physical properties are assumed to be constant except for the density in the buoyancy force term. By invoking all of the boundary layer, Boussinesq and Rosseland diffusion approximations, the governing equations for this investigation can be written as

$$\frac{\partial u}{\partial x} + \frac{\partial v}{\partial y} = 0, \tag{1}$$

$$u\frac{\partial u}{\partial x} + v\frac{\partial u}{\partial y} = U\frac{dU}{dx} + \frac{1}{\rho}\frac{\partial \tau_{xy}}{\partial y} + g\beta(T - T_\infty)$$

$$+ g\beta^*(C - C_\infty) - \frac{\sigma B_0^2}{\rho}(u - U), \tag{2}$$

$$u\frac{\partial T}{\partial x} + v\frac{\partial T}{\partial y} = \alpha_m\frac{\partial^2 T}{\partial y^2} + \tau\left[D_B\frac{\partial C}{\partial y}\frac{\partial T}{\partial y} + \frac{D_T}{T_\infty}\left(\frac{\partial T}{\partial y}\right)^2\right]$$

$$+ \frac{Q_0}{(\rho C)_f}(T - T_\infty) - \frac{1}{\rho C_p}\frac{\partial q_r}{\partial y}, \tag{3}$$

$$u\frac{\partial C}{\partial x} + v\frac{\partial C}{\partial y} = D_B\frac{\partial^2 C}{\partial y^2} + \frac{D_T}{T_\infty}\frac{\partial^2 T}{\partial y^2}. \tag{4}$$

The associated boundary conditions are

$$u = u_w(x) = cx, \quad v = 0, \quad T = T_w, \quad C = C_w$$
$$\text{at } y = 0, \tag{5a}$$

$$u = U(x) = ax, \quad v = -ay, \quad T = T_\infty, \quad C = C_\infty$$
$$\text{at } y \longrightarrow \infty. \tag{5b}$$

u, v, T, and C are the x and y components of velocity, temperature, and nanoparticle volume fraction, respectively. g, ρ, α_m, D_B, D_T, B_0, β, and β^* are the gravitational acceleration, fluid density, thermal diffusivity, Brownian diffusion coefficient, thermophoretic diffusion coefficient, magnetic field, coefficient of thermal expansion, and coefficient of concentration of expansion, respectively. The term $Q_0(T - T_\infty)$ is assumed to be the amount of heat generated or absorbed per unit volume Q_0 as a coefficient constant, which may take on either positive or negative value. When the wall temperature T exceeds the free stream temperature T_∞, the source term $Q_0 > 0$ and heat sink when $Q_0 < 0$. We have $\partial u/\partial y > 0$ when $a/c > 1$ (the ratio of free stream velocity and stretching velocity) which gives the shear stress as

$$\tau_{xy} = K\frac{\partial}{\partial y}\left(\frac{\partial u}{\partial y}\right)^n, \tag{6}$$

where K is the consistency coefficient and n is the power-law fluid. It needs to be mentioned that, for the non-Newtonian power-law model, the case of $n < 1$ is associated with shear-thinning fluids (pseudoplastic fluids); $n = 1$ corresponds to Newtonian fluids and $n > 1$ applies to the case of shear thickening (dilatant).

Using the Rosseland approximation for radiation, the radiative heat flux is simplified as

$$q_r = -\frac{4\sigma_1}{3\chi}\frac{\partial T^4}{\partial y}, \tag{7}$$

where σ_1 and χ are the Stefan-Boltzmann constant and the mean absorption coefficient, respectively. We assume that the temperature differences within the flow, such as the term T^4, may be expressed as a linear function of temperature.

Hence, expanding T^4 in a Taylor series about a free stream temperature T_∞ and neglecting higher-order terms, we get

$$T^4 = 4T_\infty^4 T - 3T_\infty^4. \tag{8}$$

Using (7) and (8) in the last term of (3), we obtain

$$\frac{\partial q_r}{\partial y} = -\frac{16\sigma_1 T_\infty^3}{3\chi}\frac{\partial^2 T}{\partial y^2}. \tag{9}$$

In order to reduce the governing equations into a system of ordinary differential equations, the following dimensionless parameters are introduced

$$\psi = \left(\frac{K/\rho}{c^{1-2n}}\right)^{1/(n+1)}x^{2n/(n+1)}f(\eta), \quad \theta = \frac{T - T_\infty}{T_w - T_\infty},$$

$$\phi = \frac{C - C_\infty}{C_w - C_\infty}, \quad \eta = y\left(\frac{c^{2-n}}{K/\rho}\right)^{1/(n+1)}x^{(1-n)/(1+n)}. \tag{10}$$

It is worth mentioning that the continuity equation (4) is identically satisfied from our choice of the stream function with $u = \partial\psi/\partial y$ and $v = -\partial\psi/\partial x$. Substituting the dimensionless parameters into (2)-(4) gives

$$n(f'')^{(n-1)}f''' + \left(\frac{2n}{n+1}\right)ff'' - f'^2 - Mf'$$

$$+ M\frac{a}{c} + \frac{a^2}{c^2} + \Lambda(\theta + N\phi) = 0,$$

$$\frac{1}{Pr}\left(1 + \frac{4R_d}{3}\right)\theta'' + \left(\frac{2n}{n+1}\right)f\theta' + Nb\theta'\phi' \tag{11}$$

$$+ Nt\theta'^2 + Q\theta = 0,$$

$$\phi'' + \left(\frac{2n}{n+1}\right)Le f\phi' + \frac{Nt}{Nb}\theta'' = 0.$$

The transformed boundary conditions are

$$f(0) = 0, \quad f'(0) = 1, \quad \theta(0) = 1, \quad \phi(0) = 1, \tag{12a}$$

$$f'(\infty) \longrightarrow \frac{a}{c}, \quad \theta(\infty) \longrightarrow 0, \quad \phi(\infty) \longrightarrow 0. \tag{12b}$$

The nine parameters appearing in (11) are defined as follows:

$$M = \frac{\sigma B_0^2}{\rho c}, \quad \Lambda = \frac{g\beta(T_w - T_\infty)x^3/v^2}{u_w^2 x^2/v^2},$$

$$N = \frac{g\beta^*(C_w - C_\infty)}{g\beta(T_w - T_\infty)},$$

$$Pr = \frac{v}{\alpha_m}(c^2 Re_x)^{(n-1)/(n+1)}, \quad R_d = \frac{4\sigma_1 T_\infty^3}{k\chi},$$

$$Nb = \frac{\tau D_B(C_w - C_\infty)}{v}(c^2 Re_x)^{(1-n)/(1+n)}, \tag{13}$$

$$Nt = \frac{\tau D_T(T_w - T_\infty)}{T_\infty v}(c^2 Re_x)^{(1-n)/(1+n)},$$

$$Le = \frac{v}{D_B}(c^2 Re_x)^{(n-1)/(n+1)}, \quad Q = \frac{Q_0}{(\rho C)_f},$$

where $\mathrm{Re}_x = u_w x / v$ is the local Reynolds number based on the stretching velocity $u_w(x)$ and k is the thermal conductivity. It should be noted that $\Lambda > 0$ corresponds to an assisting flow (heated plate), $\Lambda < 0$ corresponds to an opposing flow (cooled plate), and $\Lambda = 0$ yields forced convection flow.

The skin friction coefficient C_{f_x} at the wall is given by

$$C_{f_x} = 2 \left[f''(0) \right]^n \left(\frac{(cx)^{2-n} x^n}{K/\rho} \right)^{-1/(1+n)}, \qquad (14)$$

the local Nusselt number Nu_x is given by

$$\mathrm{Nu}_x = -K \left(\frac{u_w^{2-n}}{K/\rho} \right)^{1/n+1} \left(1 + \frac{4R_d}{3} \right) \theta'(0), \qquad (15)$$

and the local Sherwood number Sh_x is given by

$$\mathrm{Sh}_x = -D \left(\frac{u_w^{2-n}}{K/\rho} \right)^{1/n+1} \phi'(0). \qquad (16)$$

3. Method of Solution

3.1. Finite Element Method. The finite element method is a powerful technique for solving ordinary or partial differential equations. The basic concept of FEM is that the whole domain is divided into smaller elements of finite dimensions called finite elements. This method is the most versatile numerical technique in engineering analysis and has been employed to study diverse problems in heat transfer, fluid mechanics, rigid body dynamics, solid mechanics, chemical processing, electrical systems, acoustics, and many other fields. The steps involved in the finite element analysis are as follows:

(i) discretization of the domain into elements,

(ii) derivation of element equations,

(iii) assembly of element equations,

(iv) imposition of boundary conditions,

(v) solution of assembled equations.

To solve the system of simultaneous nonlinear differential equations (11), with the boundary conditions (12a) and (12b), we assume that

$$f' = h \qquad (17)$$

the system of (11) then reduced to

$$n \left(h' \right)^{(n-1)} h'' + \left(\frac{2n}{n+1} \right) f h' - h^2 - Mh$$

$$+ M \frac{a}{c} + \frac{a^2}{c^2} + \Lambda \left(\theta + N\phi \right) = 0,$$

$$\frac{1}{\mathrm{Pr}} \left(1 + \frac{4R_d}{3} \right) \theta'' + \left(\frac{2n}{n+1} \right) f \theta' + Nb \theta' \phi' \qquad (18)$$

$$+ Nt \theta'^2 + Q\theta = 0,$$

$$\phi'' + \left(\frac{2n}{n+1} \right) \mathrm{Le} f \phi' + \frac{Nt}{Nb} \theta'' = 0$$

and the corresponding boundary conditions now become

$$f(0) = 0, \quad h(0) = 1, \quad \theta(0) = 1, \quad \phi(0) = 1, \qquad (19a)$$

$$h(\infty) \longrightarrow \frac{a}{c}, \quad \theta(\infty) \longrightarrow 0, \quad \phi(\infty) \longrightarrow 0. \qquad (19b)$$

For computational purposes, the ∞ has been shifted to $\eta = 8$, without any loss of generality. The domain is divided into a set of 100 line elements, each element having two nodes.

3.2. Variational Formulation. The variational form associated with (17)-(18) over a typical linear element (η_e, η_{e+1}) is given by

$$\int_{\eta_e}^{\eta_{e+1}} w_1 \left\{ f' - h \right\} d\eta = 0,$$

$$\int_{\eta_e}^{\eta_{e+1}} w_2 \left\{ n \left(h' \right)^{(n-1)} h'' + \left(\frac{2n}{n+1} \right) f h' - h^2 - Mh \right.$$

$$\left. + M \frac{a}{c} + \frac{a^2}{c^2} + \Lambda \left(\theta + N\phi \right) \right\} d\eta = 0, \qquad (20)$$

$$\int_{\eta_e}^{\eta_{e+1}} w_3 \left\{ \frac{1}{\mathrm{Pr}} \left(1 + \frac{4R_d}{3} \right) \theta'' + \left(\frac{2n}{n+1} \right) f \theta' + Nb \theta' \phi' \right.$$

$$\left. + Nt \theta'^2 + Q\theta \right\} d\eta = 0,$$

$$\int_{\eta_e}^{\eta_{e+1}} w_4 \left\{ \phi'' + \left(\frac{2n}{n+1} \right) \mathrm{Le} f \phi' + \frac{Nt}{Nb} \theta'' \right\} d\eta = 0,$$

where w_1, w_2, w_3, and w_4 are weight functions and may be viewed as the variation in f, h, θ, and ϕ, respectively.

3.3. Finite Element Formulation. The finite element model from (20) by substituting finite element approximations of the form

$$f = \sum_{j=1}^{2} f_j \psi_j, \qquad h = \sum_{j=1}^{2} h_j \psi_j, \qquad \theta = \sum_{j=1}^{2} \theta_j \psi_j,$$

$$\phi = \sum_{j=1}^{2} \phi_j \psi_j \qquad (21)$$

with $w_1 = w_2 = w_3 = w_4 = \psi_i$, $(i = 1, 2)$, where ψ_i are the shape functions for a two-noded linear element (η_e, η_{e+1}) and are taken as

$$\psi_1^{(e)} = \frac{\eta_{e+1} - \eta}{\eta_{e+1} - \eta_e}, \quad \psi_2^{(e)} = \frac{\eta - \eta_e}{\eta_{e+1} - \eta_e}, \quad \eta_e \leq \eta \leq \eta_{e+1}. \quad (22)$$

The finite element model of the equations thus formed is given by

$$\begin{bmatrix} \left[K^{11} \right] & \left[K^{12} \right] & \left[K^{13} \right] & \left[K^{14} \right] \\ \left[K^{21} \right] & \left[K^{22} \right] & \left[K^{23} \right] & \left[K^{24} \right] \\ \left[K^{31} \right] & \left[K^{32} \right] & \left[K^{33} \right] & \left[K^{34} \right] \\ \left[K^{41} \right] & \left[K^{42} \right] & \left[K^{43} \right] & \left[K^{44} \right] \end{bmatrix} \begin{bmatrix} \{f\} \\ \{h\} \\ \{\theta\} \\ \{\phi\} \end{bmatrix} = \begin{bmatrix} \{b^1\} \\ \{b^2\} \\ \{b^3\} \\ \{b^4\} \end{bmatrix}, \qquad (23)$$

where $[K^{mn}]$ and $[b^m]$ $(m = 1, 2)$ are defined as

$$K_{ij}^{11} = \int_{\eta_e}^{\eta_{e+1}} \psi_i \frac{d\psi_j}{d\eta} d\eta, \qquad K_{ij}^{12} = -\int_{\eta_e}^{\eta_{e+1}} \psi_i \psi_j d\eta,$$

$$K_{ij}^{13} = K_{ij}^{14} = K_{ij}^{21} = 0,$$

$$K_{ij}^{22} = \int_{\eta_e}^{\eta_{e+1}} \left[-n \left(\overline{h'}\right)^{n-1} \frac{d\psi_i}{d\eta} \frac{d\psi_j}{d\eta} \right.$$
$$\left. + \left(\frac{2n}{n+1}\right) \overline{f} \psi_i \frac{d\psi_j}{d\eta} - \overline{h} \psi_i \psi_j - M \psi_i \psi_j \right] d\eta,$$

$$K_{ij}^{23} = \Lambda \int_{\eta_e}^{\eta_{e+1}} \psi_i \psi_j d\eta, \qquad K_{ij}^{24} = N\Lambda \int_{\eta_e}^{\eta_{e+1}} \psi_i \psi_j d\eta,$$

$$K_{ij}^{31} = K_{ij}^{32} = 0,$$

$$K_{ij}^{33} = \int_{\eta_e}^{\eta_{e+1}} \left[-\frac{1}{Pr} \left(1 + \frac{4R_d}{3}\right) \frac{d\psi_i}{d\eta} \frac{d\psi_j}{d\eta} \right.$$
$$+ \left(\frac{2n}{n+1}\right) \overline{f} \psi_i \frac{d\psi_j}{d\eta} + Nb \overline{\phi'} \psi_i \frac{d\psi_j}{d\eta}$$
$$\left. + Nt \overline{\theta'} \psi_i \frac{d\psi_j}{d\eta} + Q \psi_i \psi_j \right] d\eta,$$

$$K_{ij}^{34} = K_{ij}^{41} = K_{ij}^{42} = 0,$$

$$K_{ij}^{43} = -\frac{Nt}{Nb} \int_{\eta_e}^{\eta_{e+1}} \frac{d\psi_i}{d\eta} \frac{d\psi_j}{d\eta} d\eta,$$

$$K_{ij}^{44} = \int_{\eta_e}^{\eta_{e+1}} \left[-\frac{d\psi_i}{d\eta} \frac{d\psi_j}{d\eta} + Le \left(\frac{2n}{n+1}\right) \overline{f} \psi_i \frac{d\psi_j}{d\eta} \right] d\eta,$$

$$b_i^1 = 0,$$

$$b_i^2 = -\left[\left(\frac{a}{c}\right)^2 + M\left(\frac{a}{c}\right) \right]$$
$$\cdot \int_{\eta_e}^{\eta_{e+1}} \psi_i d\eta - n \left(\overline{h'}\right)^{n-1} \left(\psi_i \frac{dh}{d\eta}\right)_{\eta_e}^{\eta_{e+1}},$$

$$b_i^3 = -\frac{1}{Pr} \left(1 + \frac{4R_d}{3}\right) \left(\psi_i \frac{d\theta}{d\eta}\right)_{\eta_e}^{\eta_{e+1}},$$

$$b_i^4 = -\left(\psi_i \frac{d\phi}{d\eta}\right)_{\eta_e}^{\eta_{e+1}} - \frac{Nt}{Nb} \left(\psi_i \frac{d\theta}{d\eta}\right)_{\eta_e}^{\eta_{e+1}},$$

$$(24)$$

where

$$\overline{f} = \sum_{i=1}^{2} \overline{f_i} \psi_i, \qquad \overline{h} = \sum_{i=1}^{2} \overline{h_i} \psi_i, \qquad \overline{h'} = \sum_{i=1}^{2} \overline{h_i'} \psi_i,$$

$$(25)$$

$$\overline{\theta'} = \sum_{i=1}^{2} \overline{\theta_i'} \psi_i, \qquad \overline{\phi} = \sum_{i=1}^{2} \overline{\phi_i'} \psi_i.$$

Each element matrix is of the order 8×8. The whole domain is divided into 100 linear elements of equal size; after

TABLE 1: Comparison of $f''(0)$ for various values of M at $n = 1$, between analytical solutions obtained by homotopy analysis method and finite element method in the present work in the absence of heat and mass transfer.

M	Mahapatra et al. [38] Analytical results	Present results
0.0	2.0175	2.01753
0.5	2.1363	2.136374
1.0	2.2491	2.249134
1.5	2.3567	2.356684
2.0	2.4597	2.459658
3.0	2.6540	2.65378
5.0	3.0058	3.00392

assembly of all the elements equations, we obtain a matrix of the order 404×404. This system of equations as obtained after assembly of the element equations is nonlinear. Therefore, an iterative scheme must be utilized in the solution. After imposing the boundary conditions only a system of 397 equations remains for the solution, which is solved by the Gauss elimination method maintaining an accuracy of 0.0005.

4. Result and Discussion

The numerical solutions of governing equations (11) with boundary conditions (12a) and (12b) have been solved by using the variational finite element method. To validate our results, the numerical computations of these skin friction coefficients, Nusselt number, and Sherwood number which are, respectively, proportional to $f''(0)$, $-(1 + 4R_d/3)\theta'(0)$, and $-\phi'(0)$ are presented in tabular form and one result is compared with Mahapatra et al. [38]. The validation of present results has been verified with the classical case of Newtonian fluid $(n = 1)$ and there is a good agreement between present and Mahapatra et al. (Table 1) [38].

To analyse the results, numerical computations have been carried out for the dimensionless velocity, temperature, and nanoparticle volume fraction distributions for the flow under considerations and are obtained and their behaviour has been discussed for various governing parameters such as magnetic parameter M, dimensionless mixed convection parameter Λ, concentration to thermal buoyancy ratio N, Prandtl number Pr, radiation parameter R_d, heat source/sink parameter Q, Brownian motion Nb, thermophoresis parameter Nt, and Lewis number Le. Tables 2–5 show the effect of magnetic parameter M, power-law index n, dimensionless mixed convection parameter Λ, Prandtl number Pr, radiation parameter R_d, Brownian motion Nb, thermophoresis parameter Nt, and Lewis number Le on the coefficient of skin friction coefficient $f''(0)$, Nusselt number $-(1 + 4R_d/3)\theta'(0)$, and Sherwood number $-\phi'(0)$, respectively. It can be seen that the effect of magnetic field parameter M is to decrease the values of $f''(0)$, $-(1 + 4R_d/3)\theta'(0)$, and $-\phi'(0)$. The skin friction coefficient $f''(0)$ decreases with the increase of power-law index n.

TABLE 2: Computations of the Nusselt number, skin friction coefficient, and the Sherwood number for various values of n and M.

n	M	$f''(0)$	$-(1 + 4R_d/3)\theta'(0)$	$-\phi'(0)$
	0	3.67583	2.1486	2.5703
0.6	2	3.43416	2.1039	2.5378
	5	3.30083	2.0674	2.5167
	0	2.48416	2.417	2.8058
1.0	2	2.3783	2.3856	2.7928
	5	2.3458	2.3511	2.7822
	0	1.5779	2.5715	2.9596
1.6	2	1.5685	2.5454	2.9584
	5	1.5042	2.5239	2.9573

TABLE 3: Computations of the Nusselt number, skin friction coefficient, and the Sherwood number for various values of n and Nt.

n	Nt	$f''(0)$	$-(1 + 4R_d/3)\theta'(0)$	$-\phi'(0)$
	0.1	3.4350	2.1045	2.5381
0.6	0.4	3.4652	2.0815	2.4863
	0.7	3.4933	2.0591	2.4399
	0.1	2.3783	2.3856	2.7928
1.0	0.4	2.3957	2.3594	2.7357
	0.7	2.4142	2.3332	2.6843
	0.1	1.4992	2.5811	2.9845
1.6	0.4	1.5387	2.5511	2.8973
	0.7	1.5408	2.5217	2.8453

TABLE 4: Computations of the Nusselt number, skin friction coefficient, and the Sherwood number for various values of n and Nb.

n	Nb	$f''(0)$	$-(1 + 4R_d/3)\theta'(0)$	$-\phi'(0)$
0.6	0.2	3.4308	2.0872	2.5483
	0.6	3.4300	2.0219	2.5551
1.0	0.2	2.3758	2.3684	2.8038
	0.6	2.3758	2.3619	2.8114
1.6	0.2	1.4333	2.5683	2.9810
	0.6	1.4333	2.4981	2.9846

TABLE 5: Computations of the Nusselt number, skin friction coefficient, and the Sherwood number for various values of Λ, a/c, Le, Pr, and R_d.

Λ	a/c	Le	Pr	R_d	$f''(0)$	$-(1 + 4R_d/3)\theta'(0)$	$-\phi'(0)$
-1	1.5	10	0.71	5	0.5400	2.1505	2.5521
0	1.5	10	0.71	5	1.0258	2.2182	2.6209
1	1.5	10	0.71	5	1.4925	2.2776	2.6832
3	1.5	10	0.71	5	2.3873	2.3856	2.7928
3	1.8	10	0.71	5	3.0008	2.5332	2.8682
3	2.0	10	0.71	5	3.4425	2.6278	2.9193
3	1.5	20	0.71	5	1.542	0.9756	1.6267
3	1.5	30	0.71	5	1.0333	0.9756	1.9603
3	1.5	10	1	5	2.3475	2.7862	2.7804
3	1.5	10	10	5	2.0817	7.2348	2.5987
3	1.5	10	100	5	1.8883	10.2348	2.5387
3	1.5	10	0.71	10	2.4258	3.3600	2.8093
3	1.5	10	0.71	15	2.4500	4.1597	2.8163

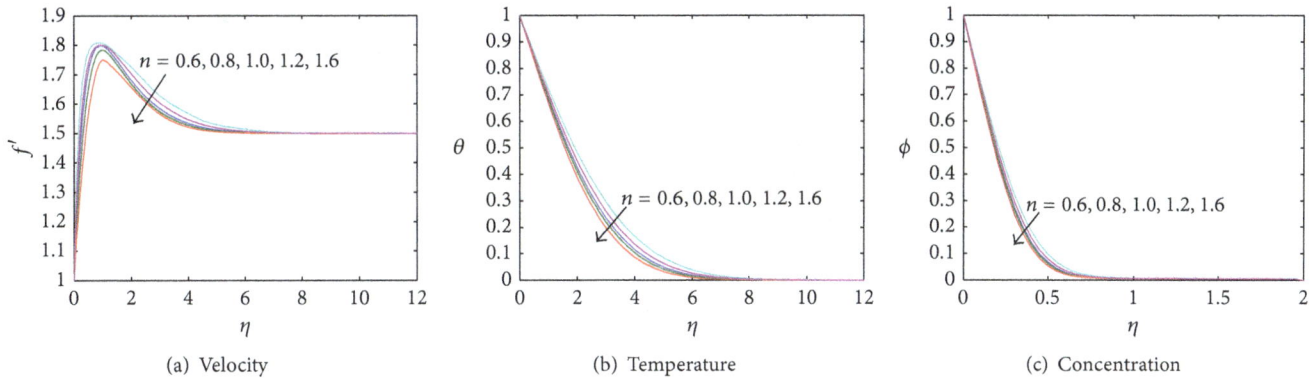

FIGURE 1: Effect of power-law index n on velocity, temperature, and concentration profiles.

It is also observed that Nusselt number $-(1 + 4R_d/3)\theta'(0)$ and Sherwood number $-\phi'(0)$ increase with the increase of power-law index n. The skin friction coefficient $f''(0)$ value increases with the increase of thermophoresis parameter Nt. Nusselt number $-(1 + 4R_d/3)\theta'(0)$ and Sherwood number $-\phi'(0)$ decrease with the increase of thermophoresis parameter Nt. The effect of Brownian motion parameter Nb is to decrease with $f''(0)$ values in case of $n < 1$ and it does not have any effect in case of $n = 1$ and $n > 1$. The effect of Brownian motion parameter (Nb) is to decrease the Nusselt number $-(1 + 4R_d/3)\theta'(0)$ and increase Sherwood number $-\phi'(0)$ for Newtonian and non-Newtonian fluids.

From Table 5 it can be seen that skin friction coefficient $f''(0)$ is to increase with the increase of Λ, a/c, and R_d whereas skin friction coefficient $f''(0)$ values decrease with the increase of Le, Pr. The Nusselt number $-(1 + 4R_d/3)\theta'(0)$ and Sherwood number $-\phi'(0)$ values increase with the increase of Λ, a/c, and R_d. As Le increases, Sherwood number $-\phi'(0)$ increases. The effect of Pr is to increase Nusselt number $-(1+4R_d/3)\theta'(0)$ and decrease the Sherwood number $-\phi'(0)$.

Figures 1(a)–1(c) illustrate the variation of velocity, temperature, and nanoparticles volume fraction profiles, respectively, for different values of power-law index n. The velocity, temperature and nanoparticles volume fraction profiles decrease with the increase of power-law index n from 0.6 to 1.6. The effect of the increased values of n is to reduce the boundary layer thickness. It can be observed from Figure 1(b) that the effect of power-law index n increases from 0.6 to 1.6; the temperature profiles decrease with an increasing viscosity of nanofluid, and thermal diffusion is depressed in the resume which cools the boundary layer and decreases the boundary layer thickness. It can also be seen from Figure 1(c) that the increase of power-law index n from 0.6 to 1.6 decreases the nanoparticle volume fraction which decreases diffusion of nanoparticle volume fraction (concentration) boundary layer thickness.

Figures 2(a)–2(c) are drawn for the velocity, temperature, and concentration profiles for different values of Prandtl number Pr for the cases shear-thinning ($n < 1$), Newtonian ($n = 1$), and shear-thickening ($n > 1$) fluids. The effect of

Prandtl number Pr is to reduce the velocity and temperature profiles for both Newtonian and non-Newtonian fluids. Physically, fluids with smaller Prandtl number Pr have larger thermal diffusivity. Figure 2(c) indicated that increasing Prandtl number Pr leads to increase of the concentration profile for both Newtonian and non-Newtonian fluids.

The effect of thermophoresis parameter Nt is to increase velocity, temperature, and concentration profiles for both Newtonian and non-Newtonian fluids are noticed in Figure 3.

Figure 4 exhibits dimensionless temperature and concentration profiles for various values of Brownian motion parameter Nb. It can be seen that the temperature profile slightly increases with an increase in the value of Brownian motion parameter Nb. The concentration profiles decrease with the value of Brownian motion parameter Nb which is noticed from these Figures 4(a) and 4(b).

Figure 5 shows the effect of thermal buoyancy ratio N on velocity profiles for shear-thinning ($n < 1$), Newtonian ($n = 1$), and shear-thickening ($n > 1$) fluids, respectively. It is noticed that the increase of N values has a tendency to increase the buoyancy effects changing more induced flow along the stretching sheet in the vertical direction reflected by the increase in the fluid velocity. This enhancement in the fluid velocity has more in shear-thinning fluid ($n < 1$) than shear-thickening fluid ($n > 1$).

Figure 6(a) shows the effect of radiation parameter R_d on the velocity profiles for both Newtonian and non-Newtonian fluids. It is noticed from the figure that the velocity of the fluid increases with the increase of radiation parameter R_d values. It can be shown from Figure 6(b) that temperature of the fluid increases with the increase of radiation parameter R_d. As expected, an increase of the radiation parameter R_d has the tendency to increase the effect of conduction as well as increasing the temperature at each point away from the surface. Hence, higher values of radiation parameter R_d imply a higher surface heat flux.

The variations in heat source/sink parameter Q on velocity and temperature profiles are given in Figure 7; from the figure it can be seen that the velocity as well as temperature profiles increases with the increase of heat source/sink parameter Q. A gradually increasing heat source/sink parameter

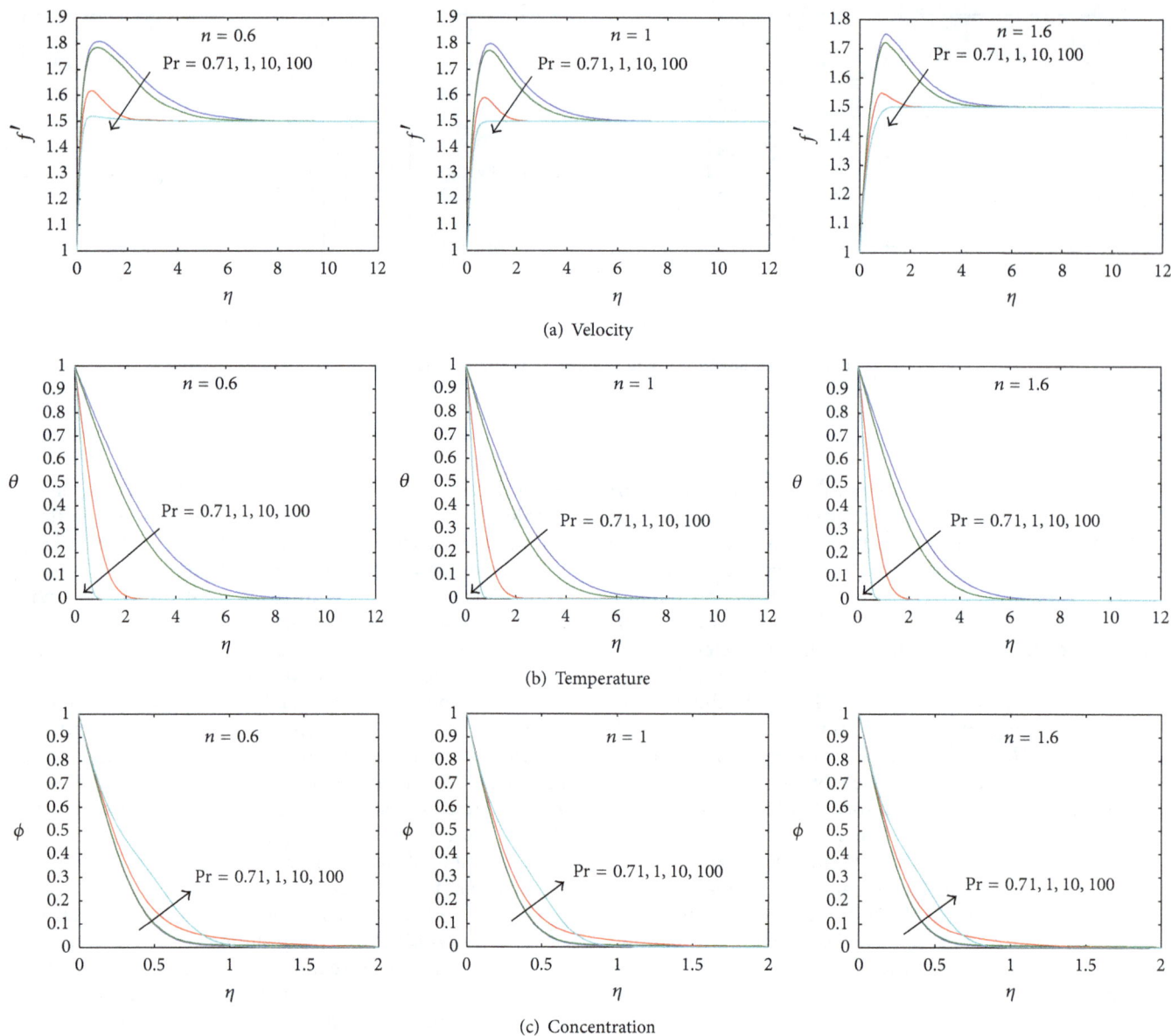

(a) Velocity

(b) Temperature

(c) Concentration

FIGURE 2: Effect of Pr on velocity, temperature, and concentration profiles for pseudoplastic, Newtonian, and dilatant fluids.

increases the thermal boundary layer thickness which physically reveals the fact that an increase in the heat source/sink parameter means an increase in the heat generated inside the boundary layer which leads to higher temperature field. It is noted that the temperature profiles decrease for increasing strength of heat sink and due to increase of heat source strength the temperature increases. So the thickness of the thermal boundary layer reduces for increase with heat source parameter. These results are very much significant for the flow where heat transfer is given prime importance.

Figures 8(a)–8(c) present the changes in the velocity profiles with the effect of magnetic parameter M for shear-thinning ($n < 1$), Newtonian ($n = 1$), and shear-thickening ($n > 1$) fluids, respectively. The velocity profiles f' decrease with the raising of magnetic parameter M. This is due to

magnetic field opposing the transport phenomena, since the variation of magnetic parameter M causes the variation of Lorentz forces. The Lorentz force is a drag like force that produces more resistance to transport phenomena and that causes reduction in the fluid velocity. The effect of magnetic field is more in shear-thinning fluids than shear-thickening fluids. The effect of magnetic fields is very meager on temperature profiles; hence, it is not shown.

Figures 9(a)–9(c) show the effect of dimensionless mixed convection parameter Λ on velocity, temperature, and concentration profiles, respectively. The velocity profiles are increasing with increasing values of Λ whereas temperature and concentration profiles are decreasing with increasing values of Λ. The presence of the thermal buoyancy effects represented by finite values of the mixed parameter has

(a) Velocity

(b) Temperature

(c) Concentration

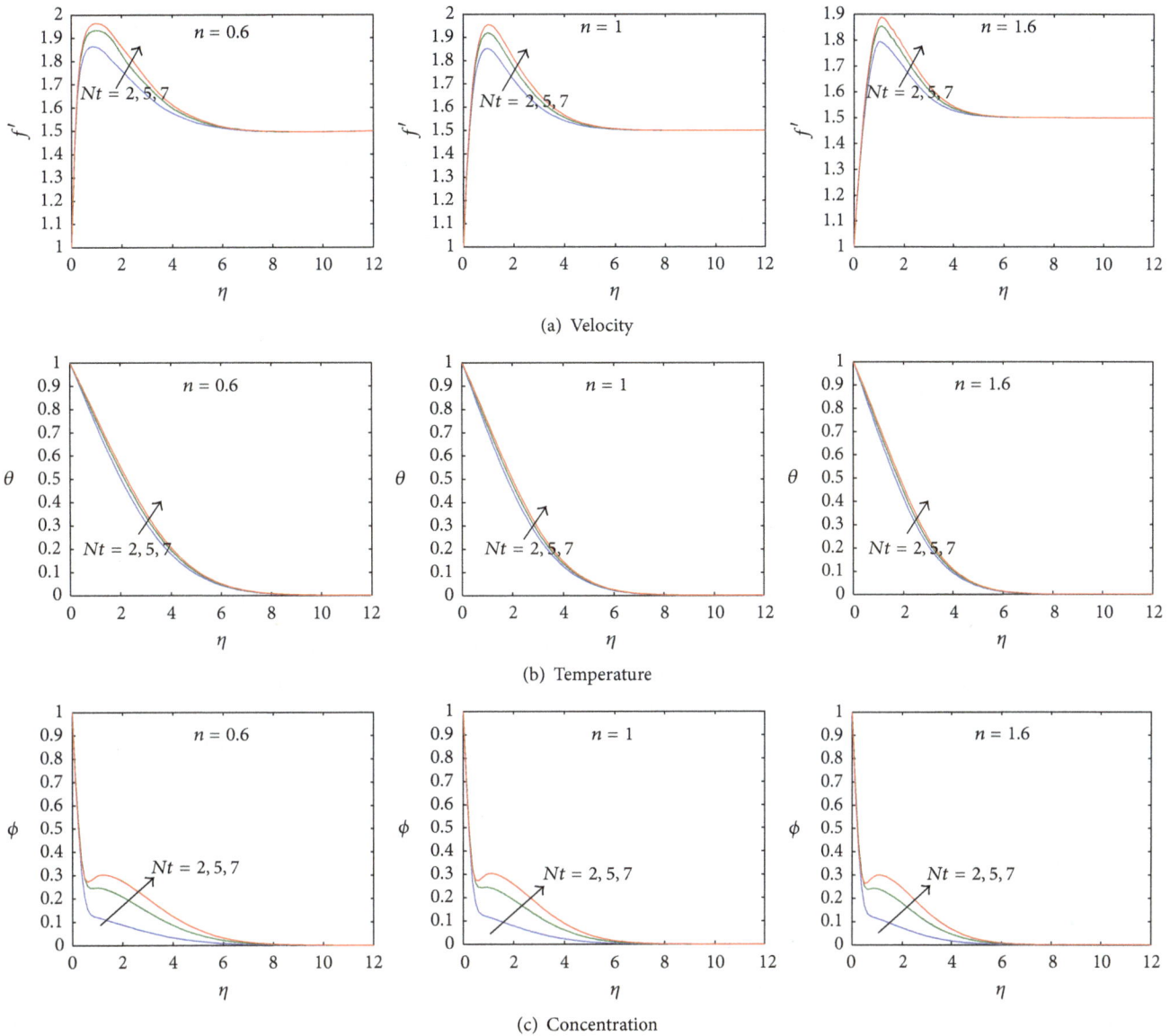

FIGURE 3: Effect of Nt on velocity, temperature, and concentration profiles for pseudoplastic, Newtonian, and dilatant fluids.

the tendency to induce more flow along the surface at the expense of small reductions in the temperature and concentration. Distinctive peaks in the velocity profiles which are characteristics of free convection flows are also observed as Λ increases.

Figure 10 depicts the variation of nanoparticle concentration profiles for Newtonian and non-Newtonian fluids with different values of Lewis number Le. The concentration profile is decreased due to the increase of Lewis number Le. Increase in the Lewis number Le reduces the nanoparticle volume fraction and its boundary layer thickness.

Figures 11(a)–11(c) present the velocity, temperature, and concentration profiles for various values of ratio of velocity parameter a/c. It can be observed that an increase in a/c causes increase in velocity profiles and signifiant decrease on

the temperature and concentration profiles. These behaviours are clearly shown in Figure 11.

5. Conclusions

The influence of Brownian motion and thermophoresis on mixed convection magneto hydrodynamic boundary layer flow of heat and mass transfer stagnation-point flow of power-law non-Newtonian nanofluid towards a stretching surface is investigated. Various thermal radiation and heat source/sink parameters have been considered for the flow, temperature, and nanoparticle volume fraction as well as the skin friction coefficient, Nusselt number, and Sherwood

(a) Temperature

(b) Concentration

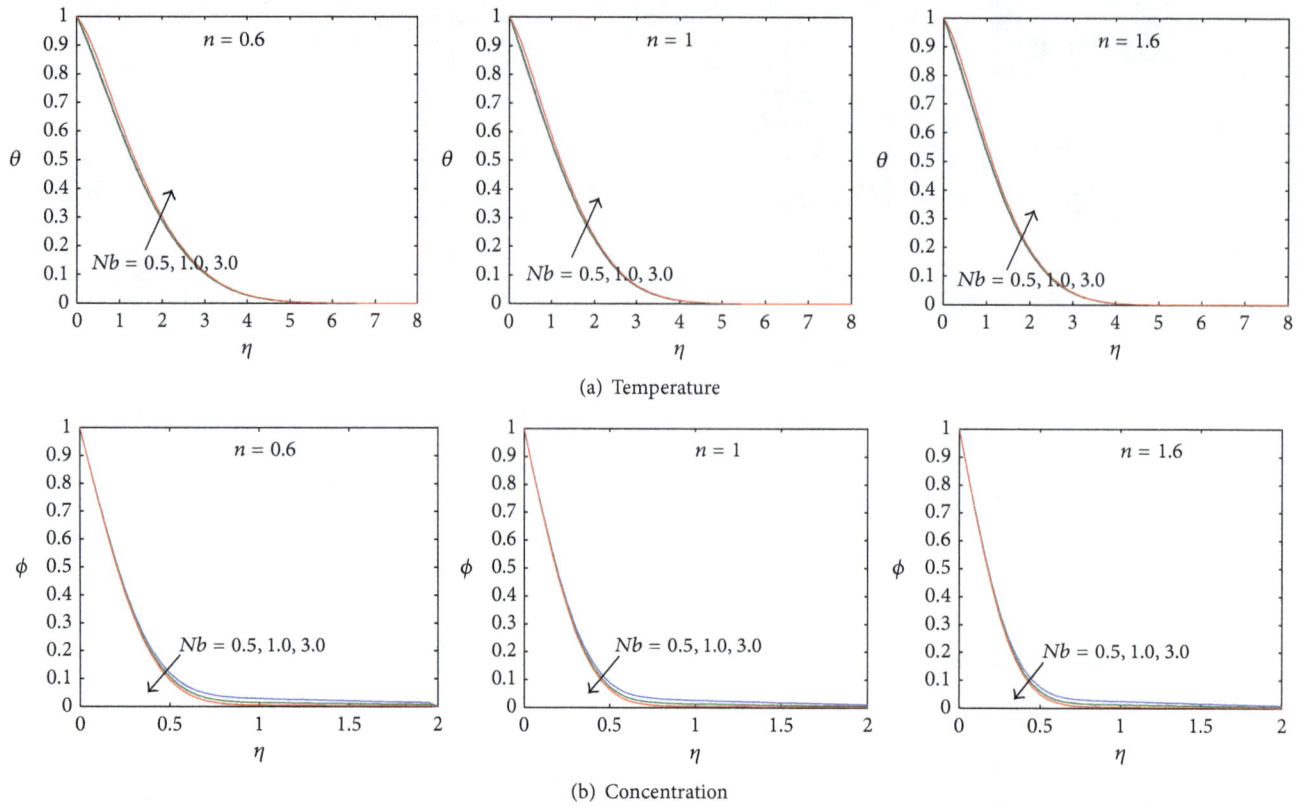

FIGURE 4: Effect of Nb on temperature and concentration profiles for pseudoplastic, Newtonian, and dilatant fluids.

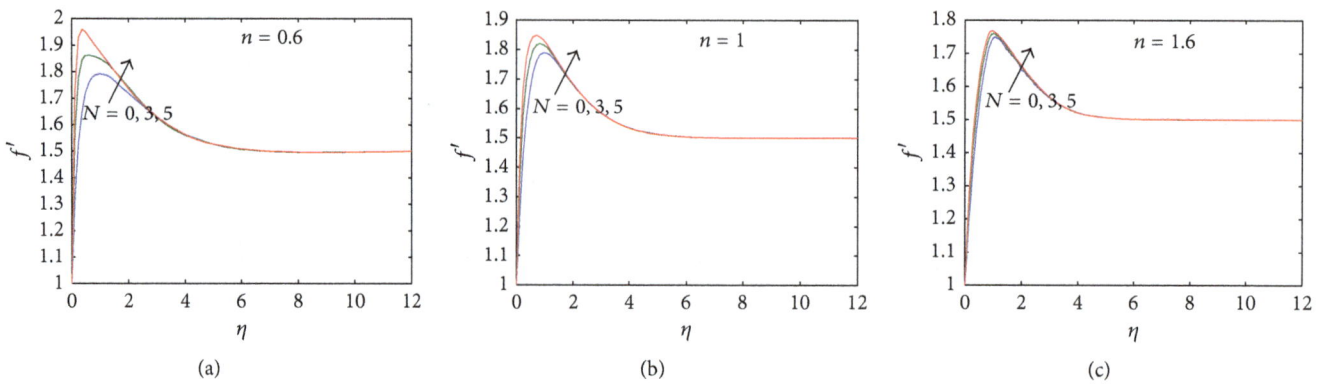

FIGURE 5: Effect of N on velocity profiles for pseudoplastic, Newtonian, and dilatant fluids.

number. The main findings of the present study can be summarised as follows.

(i) The effect of Magnetic field parameter M reduces the velocity profiles.

(ii) The influence of thermophoresis Nt increases the velocity, temperature, and concentration profiles for the cases of pseudoplastic, Newtonian, and dilatant fluids. The effect of Brownian motion Nb is to increase

the temperature profiles and decrease the concentration profiles.

(iii) The velocity and concentration profiles increase with the increase of radiation parameter R_d and heat source/sink parameter Q for both Newtonian and non-Newtonian fluids.

(iv) The effect of mixed convection parameter Λ and velocity ratio parameter a/c is to increase the velocity

(a) Velocity

(b) Temperature

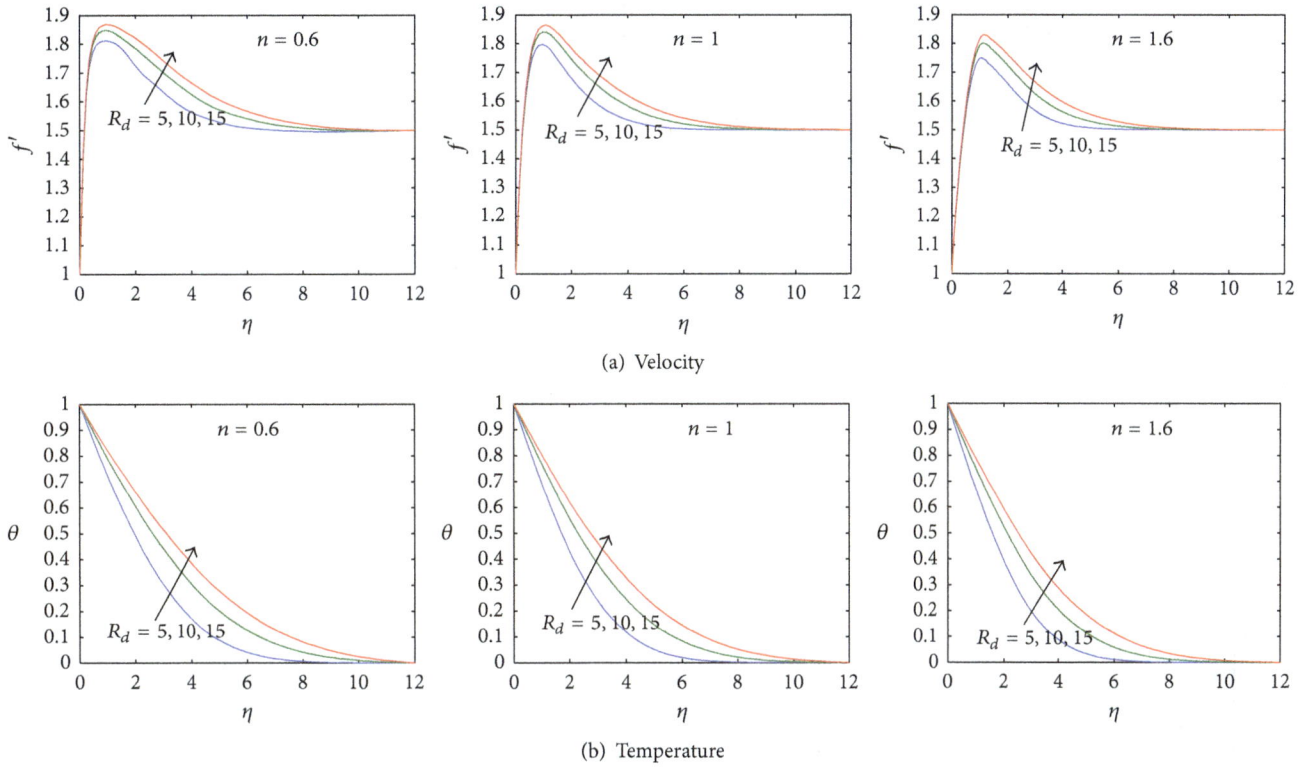

FIGURE 6: Effect of R_d on velocity and temperature profiles for pseudoplastic, Newtonian, and dilatant fluids.

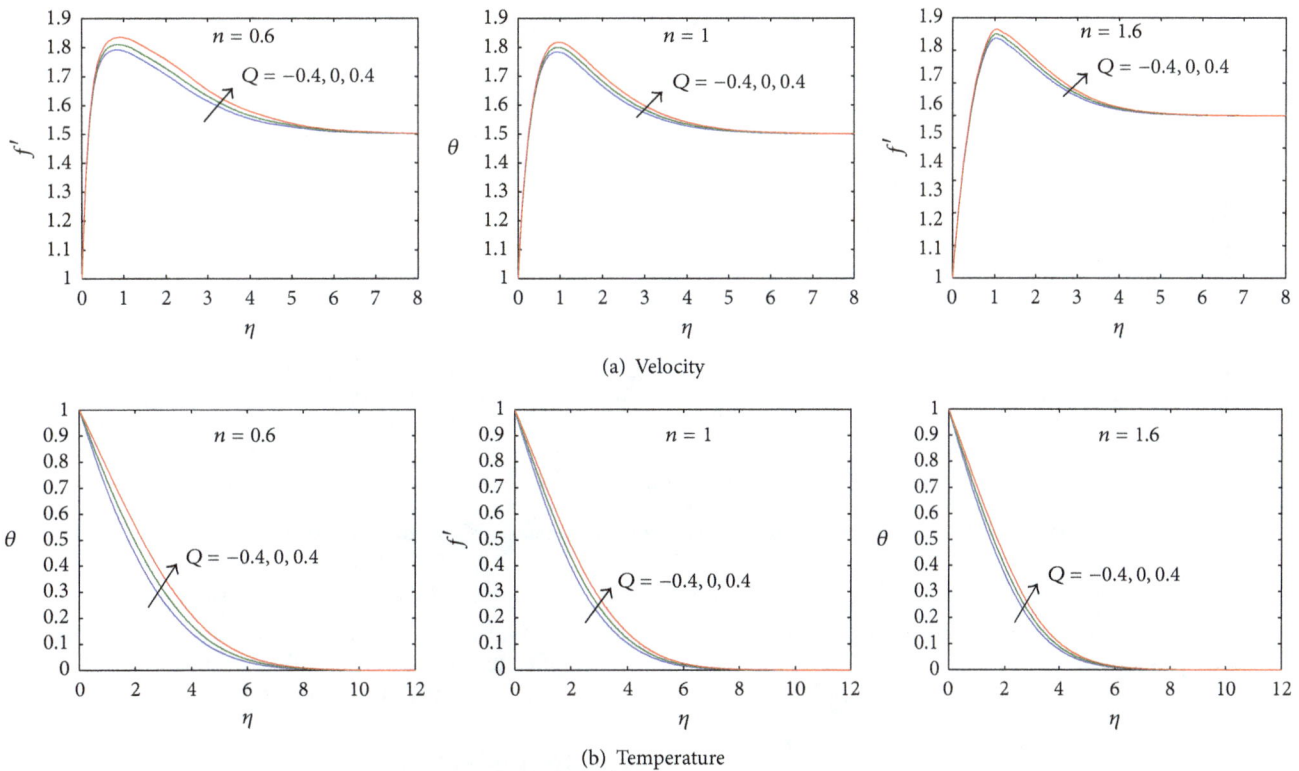

(a) Velocity

(b) Temperature

FIGURE 7: Effect of Q on velocity and temperature profiles for pseudoplastic, Newtonian, and dilatant fluids.

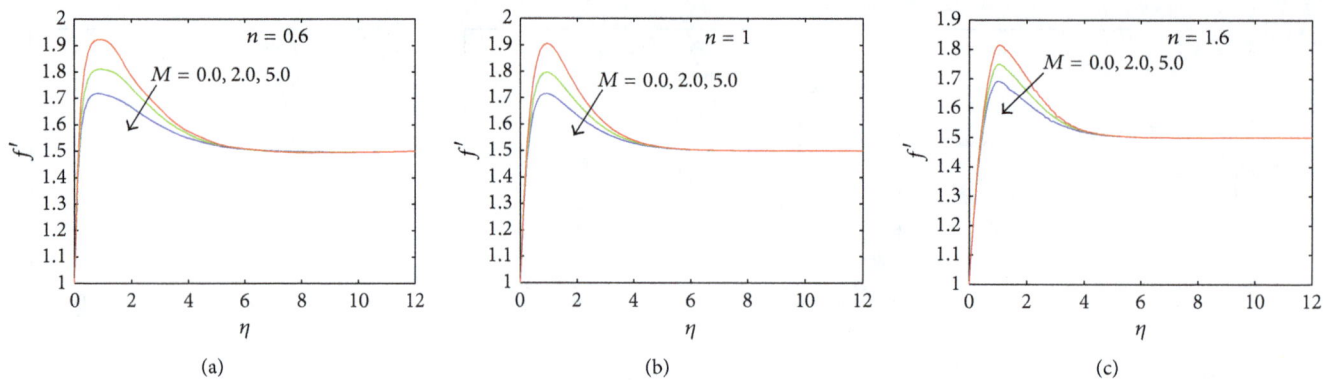

FIGURE 8: Effect of M on velocity profiles for pseudoplastic, Newtonian and dilatant fluids.

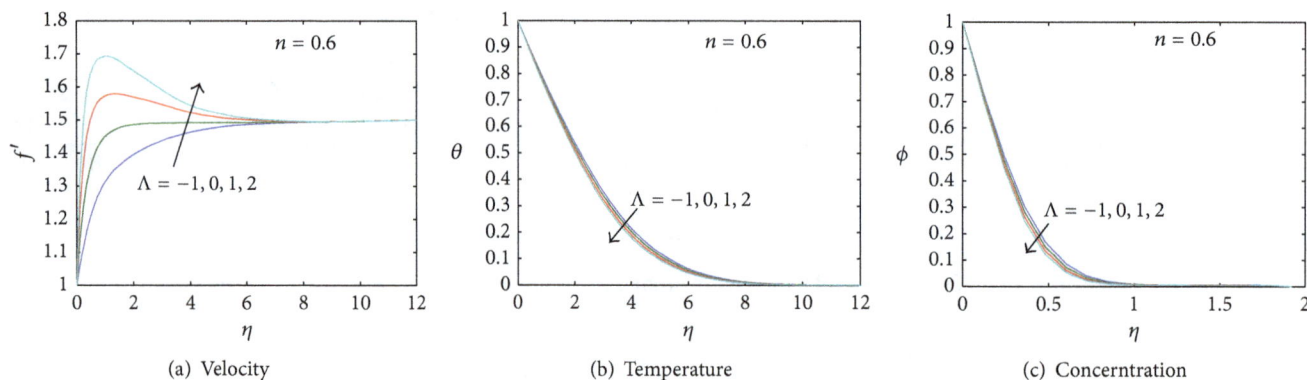

(a) Velocity   (b) Temperature   (c) Concerntration

FIGURE 9: Effect of Λ on the velocity, temperature, and concentration profiles for $n = 0.6$.

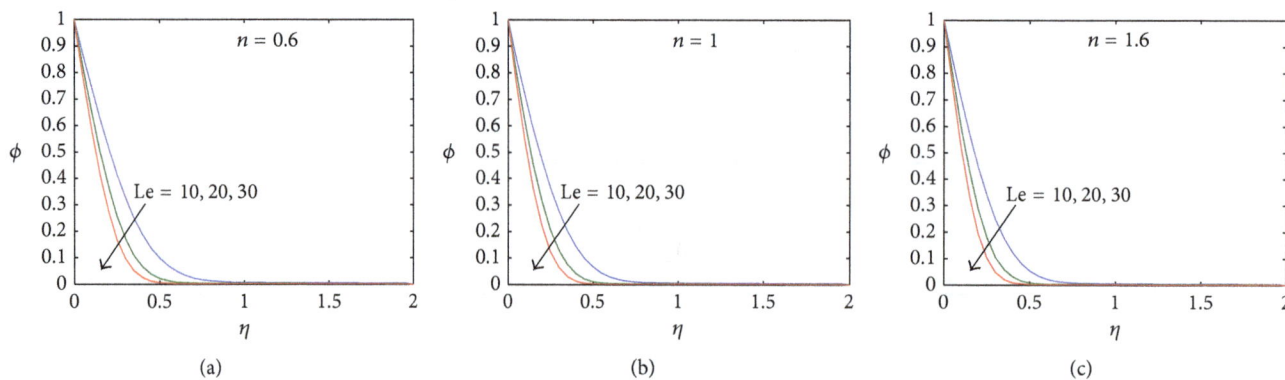

FIGURE 10: Effect of Le on concentration profiles for pseudoplastic, Newtonian, and dilatant fluids.

profiles and reduces the temperature and concentration profiles.

(v) The skin friction coefficient $f''(0)$ increases with the increase of thermophoresis parameter Nt and it decreases with the increase of Brownian motion parameter Nb and the power-law index n. The coefficient of Nusselt number and Sherwood number decreases with the increase of Nt.

Conflict of Interests

The authors declare that there is no conflict of interests regarding the publication of this paper.

Magnetohydrodynamic Mixed Convection Stagnation-Point Flow of a Power-Law Non-Newtonian...

71

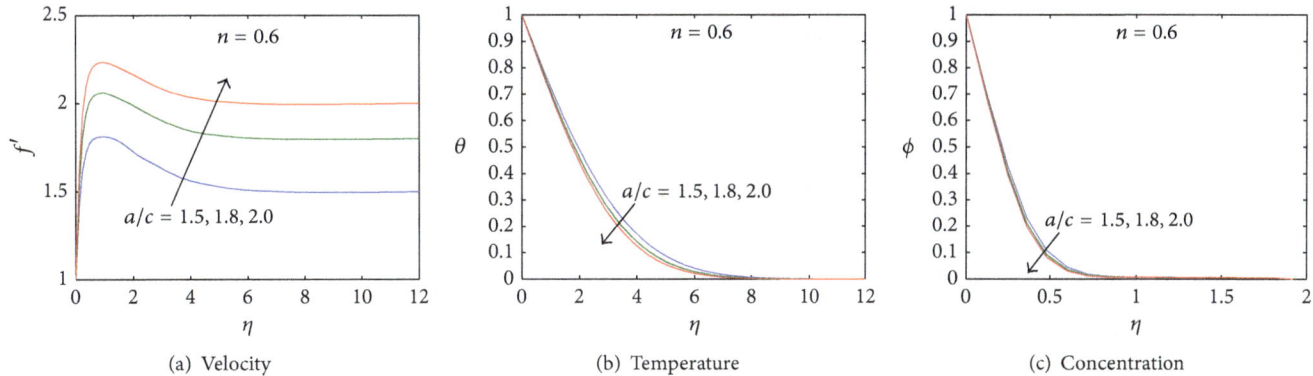

(a) Velocity

(b) Temperature

(c) Concentration

FIGURE 11: Effect of a/c on the velocity, temperature, and concentration profiles for $n = 0.6$.

References

[1] W. R. Schowalter, "The application of boundary-layer theory to power-law pseudoplastic fluids: similar solutions," *AIChE Journal*, vol. 6, no. 1, pp. 24–28, 2004.

[2] A. Acrivos, "On laminar boundary layer flows with a rapid homogeneous chemical reaction," *Chemical Engineering Science*, vol. 13, no. 2, pp. 57–62, 1960.

[3] E. M. A. Elbashbeshy, T. G. Emam, and M. S. Abdel-Wahed, "Three-dimensional flow over a stretching surface with thermal radiation and heat generation in the presence of chemical reaction and suction/injection," *International Journal of Energy Technology*, vol. 16, pp. 1–8, 2011.

[4] S. Nadeem, T. Hayat, M. Y. Malik, and S. A. Rajput, "Thermal radiation effects on the flow by an exponentially stretching surface: a series solution," *Zeitschrift fur Naturforschung—Section*, vol. 65, no. 6-7, pp. 495–503, 2010.

[5] S. Nadeem, A. Hussain, and K. Vajravelu, "Effects of heat transfer on the stagnation flow of a third order fluid over a shrinking sheet," *Zeitschrift für Naturforschung*, vol. 65, no. 11, pp. 969–994, 2010.

[6] S. Nadeem, A. Hussain, and M. Khan, "HAM solutions for boundary layer flow in the region of the stagnation point towards a stretching sheet," *Communications in Nonlinear Science and Numerical Simulation*, vol. 15, no. 3, pp. 475–481, 2010.

[7] S. Nadeem and N. S. Akbar, "Influence of heat transfer on a peristaltic flow of Johnson Segalman fluid in a non uniform tube," *International Communications in Heat and Mass Transfer*, vol. 36, no. 10, pp. 1050–1059, 2009.

[8] S. Nadeem and M. Ali, "Analytical solutions for pipe flow of a fourth grade fluid with Reynold and Vogel's models of viscosities," *Communications in Nonlinear Science and Numerical Simulation*, vol. 14, no. 5, pp. 2073–2090, 2009.

[9] J. Buongiorno, "Convective transport in nanofluids," *Journal of Heat Transfer*, vol. 128, no. 3, pp. 240–250, 2006.

[10] G. Łukaszewicz, "Asymptotic behavior of micropolar fluid flows," *International Journal of Engineering Science*, vol. 41, no. 3–5, pp. 259–269, 2003.

[11] A. V. Shenoy, "Non-Newtonian fluid heat transfer in porous media," *Advances in Heat Transfer*, vol. 24, pp. 101–190, 1994.

[12] G. Astarita and G. Marrucci, *Principles of Non-Newtonian Fluid Mechanics*, McGraw-Hill, New York, NY, USA, 1974.

[13] H. Bohme, *Non-Newtonian Fluid Mechanics*, North-Holland Series in Applied Mathematics and Mechanics, North-Holland, 1987.

[14] N. Kishan and B. Shashidar Reddy, "MHD Effects on Non-Newtonian power-law fluid past a continuously moving porous flat plate with heat flux and viscous dissipation," *International Journal of Applied Mechanics and Engineering*, vol. 18, no. 2, pp. 425–445, 2013.

[15] S. Nadeem, A. Riaz, R. Ellahi, and N. S. Akbar, "Effects of heat and mass transfer on peristaltic flow of a nanofluid between eccentric cylinders," *Applied Nanoscience*, vol. 4, pp. 393–404, 2014.

[16] D. W. Zhou, "Heat transfer enhancement of copper nanofluid with acoustic cavitation," *International Journal of Heat and Mass Transfer*, vol. 47, no. 14–16, pp. 3109–3117, 2004.

[17] S. U. S. Choi, "Enhancing thermal conductivity of fluids with nanoparticles," *ASME Fluids Engineering Division*, vol. 231, pp. 99–105, 1995.

[18] Y. Xuan and Q. Li, "Investigation on convective heat transfer and flow features of nanofluids," *Journal of Heat Transfer*, vol. 125, no. 1, pp. 151–155, 2003.

[19] B. C. Pak and Y. I. Cho, "Hydrodynamic and heat transfer study of dispersed fluids with submicron metallic oxide particles," *Experimental Heat Transfer*, vol. 11, no. 2, pp. 151–170, 1998.

[20] D. S. Wen and Y. L. Ding, "Effective thermal conductivity of aqueous suspensions of carbon nanotubes (Nanouids)," *Journal of Thermophysics and Heat Transfer*, vol. 18, no. 4, pp. 481–485, 2004.

[21] Y. L. Ding, H. Alias, D. S. Wen, and R. A. Williams, "Heat transfer of aqueous suspensions of carbon nanotubes (CNT nanofluids)," *International Journal of Heat and Mass Transfer*, vol. 49, no. 1-2, pp. 240–250, 2006.

[22] H. Masuda, A. Ebata, K. Teramae, and N. Hishinuma, "Alteration of thermal conductivity and viscosity of liquid by dispersing ultra-fine particles (dispersion of G-Al$_2$O$_3$, SiO$_2$ and TiO$_2$ ultra-fine particles)," *Netsu Bussei*, vol. 7, no. 4, pp. 227–233, 1993.

[23] P. Keblinski, S. R. Phillpot, S. U. S. Choi, and J. A. Eastman, "Mechanisms of heat flow in suspensions of nano-sized particles (nanofluids)," *International Journal of Heat and Mass Transfer*, vol. 45, no. 4, pp. 855–863, 2001.

[24] D. T. Wasan and A. D. Nikolov, "Spreading of nanofluids on solids," *Nature*, vol. 423, no. 6936, pp. 156–159, 2003.

[25] A. Chengara, A. D. Nikolov, D. T. Wasan, A. Trokhymchuk, and D. Henderson, "Spreading of nanofluids driven by the structural disjoining pressure gradient," *Journal of Colloid and Interface Science*, vol. 280, no. 1, pp. 192–201, 2004.

[26] S. M. You, J. H. Kim, and K. H. Kim, "Effect of nanoparticles on critical heat flux of water in pool boiling heat transfer," *Applied Physics Letters*, vol. 83, no. 16, pp. 3374–3376, 2003.

[27] L. J. Crane, "Flow past a stretching plate," *Zeitschrift für Angewandte Mathematik und Physik ZAMP*, vol. 21, no. 4, pp. 645–647, 1970.

[28] P. Carragher and L. J. Crane, "Heat transfer on a continuous stretching sheet," *Zeitschrift für Angewandte Mathematik und Mechanik*, vol. 62, pp. 564–573, 1982.

[29] R. Ellahi, M. Raza, and K. Vafai, "Series solutions of non-Newtonian nanofluids with Reynolds' model and Vogel's model by means of the Homotopy analysis method," *Mathematical and Computer Modelling*, vol. 55, no. 7-8, pp. 1876–1891, 2012.

[30] R. Cortell, "Flow and heat transfer of a fluid through a porous medium over a stretching surface with internal heat generation/absorption and suction/blowing," *Fluid Dynamics Research*, vol. 37, no. 4, pp. 231–245, 2005.

[31] M. S. Abel, M. M. Nandeppanavar, and M. B. Malkhed, "Hydro-magnetic boundary layer flow and heat transfer in viscoelastic fluid over a continuously moving permeable stretching surface with nonuniform heat source/sink embedded in fluid-saturated porous medium," *Chemical Engineering Communications*, vol. 197, no. 5, pp. 633–655, 2010.

[32] H. Dessie and N. Kishan, "MHD effects on heat transfer over stretching sheet embedded in porous medium with variable viscosity, viscous dissipation and heat source/sink," *Ain Shams Engineering Journal*, vol. 5, no. 3, pp. 967–977, 2014.

[33] M. N. Tufail, A. S. Butt, and A. Ali, "Heat source/sink effects on non-Newtonian MHD fluid flow and heat transfer over a permeable stretching surface: lie group analysis," *Indian Journal of Physics*, vol. 88, no. 1, pp. 75–82, 2014.

[34] E. M. A. Elbashbeshy and M. A. A. Bazid, "Heat transfer in a porous medium over a stretching surface with internal heat generation and suction or injection," *Applied Mathematics and Computation*, vol. 158, no. 3, pp. 799–807, 2004.

[35] C.-H. Chen, "Magneto-hydrodynamic mixed convection of a power-law fluid past a stretching surface in the presence of thermal radiation and internal heat generation/absorption," *International Journal of Non-Linear Mechanics*, vol. 44, no. 6, pp. 596–603, 2009.

[36] K. V. Prasad, K. Vajravelu, and P. S. Datti, "Mixed convection heat transfer over a non-linear stretching surface with variable fluid properties," *International Journal of Non-Linear Mechanics*, vol. 45, no. 3, pp. 320–330, 2010.

[37] A. B. Metzner, "Heat transfer in non-Newtonian fluid," *Advances in Heat Transfer*, vol. 2, pp. 357–397, 1965.

[38] T. R. Mahapatra, S. K. Nandy, and A. S. Gupta, "Analytical solution of magnetohydrodynamic stagnation-point flow of a power-law fluid towards a stretching surface," *Applied Mathematics and Computation*, vol. 215, no. 5, pp. 1696–1710, 2009.

7

Double Diffusive Convection in a Layer of Maxwell Viscoelastic Fluid in Porous Medium in the Presence of Soret and Dufour Effects

Ramesh Chand[1] and G. C. Rana[2]

[1] Department of Mathematics, Government College Dhaliara, Himachal Pradesh 177103, India
[2] Department of Mathematics, Government College Nadaun, Himachal Pradesh 177033, India

Correspondence should be addressed to Ramesh Chand; rameshnahan@yahoo.com

Academic Editor: Mahmoud Mamou

Double diffusive convection in a horizontal layer of Maxwell viscoelastic fluid in a porous medium in the presence of temperature gradient (Soret effects) and concentration gradient (Dufour effects) is investigated. For the porous medium Darcy model is considered. A linear stability analysis based upon normal mode technique is used to study the onset of instabilities of the Maxwell viscolastic fluid layer confined between two free-free boundaries. Rayleigh number on the onset of stationary and oscillatory convection has been derived and graphs have been plotted to study the effects of the Dufour parameter, Soret parameter, Lewis number, and solutal Rayleigh number on stationary convection.

1. Introduction

Bénard convection originated from the experimental works of Bénard [1] and theoretical analysis of Lord Rayleigh [2]. Lord Rayleigh studied the dynamic origins of convective cells and proposed his theory on the buoyancy driven convection. The detailed study of Bénard convection in Newtonian fluid in nonporous medium under varying assumptions of hydrodynamics and hydromagnetics had been given by Chandrasekhar [3]. Lapwood [4] had studied the stability of convective flow in hydromagnetics in a porous medium using Rayleigh's procedure. The Rayleigh instability of a thermal boundary layer in flow through a porous medium had been considered by Wooding [5]. McDonnel [6] suggested the importance of porosity in the astrophysical context.

Double-diffusive convection is referred to buoyancy-driven flows induced by combined temperature and concentration gradients. The onset of double diffusive convection in a fluid saturated in porous medium is regarded as a classical problem due to its wide range of applications in many engineering fields such as evaporative cooling of high temperature systems, agricultural product storage, soil sciences, enhanced oil recovery, packed-bed catalytic reactors, and the pollutant transport in underground. A detailed review of the literature concerning double diffusive convection in binary fluid in a porous medium was given by Nield and Bejan [7], Trevisan and Bejan [8], and Malashetty and Kollur [9]. Thermal convection in binary fluid driven by the Soret and Dufour effects had been investigated by Knobloch [10] and showed that the equations were identical to the thermosolutal problem except relation between the thermal and solutal Rayleigh numbers. The above literature dealt with Newtonian fluids.

The study of natural convection of non-Newtonian fluids in a porous medium had gained much attention because of its engineering and industrial applications. These applications included design of chemical processing equipment, formation and dispersion of fog, distributions of temperature and moisture over agricultural fields and groves of fruit trees, and damage of crops due to freezing and pollution of the environment.

The fluids that show distinct deviation from "Newtonian hypothesis" (stress on fluid is linearly proportional to strain rate of fluid) are called non-Newtonian fluids. Different models had been proposed to explain the behavior of non-Newtonian fluids. Maxwell model is one of them. These fluids help us to understand the wide variety of fluids that exist in the physical world and characterized by power-law model. The work on viscoelastic fluid appears to be that of Herbert on plane coquette flow heated from below. He found a finite elastic stress in the undistributed state to be required for the elasticity to affect the stability. Using a three constants rheological model due to Oldroyd [11], the author demonstrated, for finite rate of strain, that the elasticity has a destabilizing effect, which results solely from the change in apparent viscosity.

The importance of the study of viscoelastic fluids in a porous medium has been increasing for the last few years. This is mainly due to their applications in petroleum drilling, manufacturing of foods and paper, and many others. The problem of convective instability of viscoelastic fluid heated from below was first studied by Green [12]. Vest and Arpaci [13] investigated problems of overstability in a horizontal layer of a viscoelastic fluid heated from below. Bhatia and Steiner [14] studied the problem of thermal instability of a Maxwellian viscoelastic fluid in the presence of rotation and found that rotation has a destabilizing influence in contrast to its stabilizing effect on a viscous Newtonian fluid. Bhatia and Steiner [15] have also studied the thermal instability of a Maxwellian viscoelastic fluid in hydromagnetic and found that magnetic field had stabilizing effect on the Maxwell fluid, just as in the case of Newtonian fluid. Sharma and Kumar [16] studied the Hall effect on thermosolutal instability in a Maxwellian viscoelastic fluid and found that Hall effect destabilizes the fluid layer while Kirti and Chand [17] studied the combined effect Hall Current, suspended particles, and variable gravity in a layer of Maxwell viscoelastic fluid. Chand and Kango [18], Chand [19–21], and Chand and Kumar [22] studied problems of thermal instability of Maxwell viscoelastic fluid in porous medium under various assumptions. Chand and Rana [23] investigated the Dufour and Soret effects in layer of elasticoviscous fluid in a porous medium and found that Dufour parameter destabilizes the fluid layer while Soret parameter has both the stabilizing and destabilizing effects on fluid layer depending upon certain conditions.

In this paper an attempt has been made to study the Dufour and Soret effects on the onset of instability in a horizontal layer of Maxwell viscoelastic fluid in a porous medium.

2. Mathematical Formulations of the Problem

Consider an infinite horizontal layer of Maxwell viscoelastic fluid of thickness "d," confined between the planes $z = 0$ and $z = d$ in a porous medium of porosity ε and medium permeability k_1 and is acted upon by gravity $\mathbf{g}(0, 0, -g)$. This layer of fluid is heated and soluted in such a way that a constant temperature and concentration distribution is prescribed at the boundaries of the fluid layer. The temperature (T) and concentration (C) are taken to be T_0 and C_0 at $z = 0$ and

T_1 and C_1 at $z = d$, ($T_0 > T_1, C_0 > C_1$). Let ΔT and ΔC be the difference in temperature and concentration across the boundaries.

Let $\mathbf{q}(u, v, w)$, $p, \rho, T, C, \alpha, \alpha', \mu, \kappa$, and κ' be the Darcy velocity vector, hydrostatic pressure, density, temperature, solute concentration, coefficient of thermal expansion, an analogous solvent coefficient of expansion, viscosity, thermal diffusivity, and solute diffusivity of fluid, respectively.

2.1. Assumptions.
The mathematical equations describing the physical model are based upon the following assumptions.

(i) Thermophysical properties expect for density in the buoyancy force (Boussinesq hypothesis) are constant.

(ii) Darcy's model with time derivative is employed for the momentum equation.

(iii) The porous medium is assumed to be isotropic and homogeneous.

(iv) No chemical reaction takes place in a layer of fluid.

(v) The fluid and solid matrix are in thermal equilibrium state.

(vi) Radiation heat transfer between the sides of the wall is negligible when compared with other modes of the heat transfer.

2.2. Governing Equations.
According to the works of Bhatia and Steiner [14, 15], Sharma and Kumar [16], and Chand [19–21] the appropriate governing equations for Maxwell viscoelastic fluid in a porous medium are

$$\nabla \cdot \mathbf{q} = 0,$$

$$0 = \left(1 + \lambda \frac{\partial}{\partial t}\right)$$

$$\times \left(-\nabla p + \rho \left(1 - \alpha (T - T_0) - \alpha' (C - C_0)\right) \mathbf{g}\right) - \frac{\mu}{k_1} \mathbf{q},$$

$$\sigma \frac{\partial T}{\partial t} + \mathbf{q} \cdot \nabla T = \kappa \nabla^2 T + D_{TC} \nabla^2 C,$$

$$\varepsilon \frac{\partial C}{\partial t} + \mathbf{q} \cdot \nabla C = \kappa' \nabla^2 C + D_{CT} \nabla^2 T,$$

$$(1)$$

where D_{TC} and D_{CT} are the Dufour and Soret coefficients; $\sigma = (\rho c_p)_m / (\rho c_p)_f$ is thermal capacity ratio, c_p is specific heat, and the subscripts m and f refer to porous medium and fluid, respectively.

We assume that temperature and concentration are constant at the boundaries of the fluid layer. Therefore, boundary conditions are

$$w = 0, \qquad T = T_0, \qquad C = C_0 \quad \text{at } z = 0,$$
$$w = 0, \qquad T = T_1, \qquad C = C_1 \quad \text{at } z = d. \qquad (2)$$

2.3. Steady State and Its Solutions. The steady state is given by

$$u = v = w = 0, \qquad p = p(z), \qquad T = T_s(z), \qquad C = C_s(z). \tag{3}$$

The solution of steady state is given as

$$T_s = T_0 - \frac{\Delta T}{d} z,$$

$$C_s = C_0 - \frac{\Delta C}{d} z, \tag{4}$$

$$p_s = p_0 - \rho_0 g \left(z + \alpha \frac{\Delta T}{2d} z^2 + \alpha' \frac{\Delta C}{2d} z^2 \right),$$

where subscript 0 shows the value of the variable at boundary $z = 0$.

2.4. Perturbation Solution. To study the stability of the system, we superimposed infinitesimal perturbations on the basic state, which are of the forms

$$q = 0 + q', \qquad T = T_s + T', \qquad C = C_s + C',$$

$$p = p_s + p', \tag{5}$$

where the prime denotes the perturbed quantities. Substituting (5) into (1) and neglecting higher order terms of the perturbed quantities, we get

$$\nabla \cdot q' = 0,$$

$$0 = \left(1 + \lambda \frac{\partial}{\partial t} \right) \left(-\nabla p' + \rho_0 \left(\alpha T' + \alpha' C' \right) \mathbf{g} \right) - \frac{\mu}{k_1} q',$$

$$\sigma \frac{\partial T'}{\partial t} - w' \frac{\Delta T}{d} = \kappa \nabla^2 T' + D_{TC} \nabla^2 C', \tag{6}$$

$$\varepsilon \frac{\partial C'}{\partial t} - w' \frac{\Delta C}{d} = \kappa' \nabla^2 C' + D_{CT} \nabla^2 T'.$$

Introduce the dimensionless variables as

$$\left(x'', y'', z'' \right) = \left(\frac{x', y', z'}{d} \right),$$

$$\left(u'', v'', w'', \right) = \left(\frac{u', v', w'}{\kappa} \right) d,$$

$$t'' = \frac{\kappa}{\sigma d^2} t, \qquad p'' = \frac{k_1 d^2}{\mu \kappa} p', \tag{7}$$

$$T'' = \frac{T'}{\Delta T}, \qquad C'' = \frac{C'}{\Delta C}.$$

Thereafter drop the dashes ($''$) for simplicity.

Equation (6) in nondimensional form can be written as

$$\nabla \cdot \mathbf{q} = 0,$$

$$0 = \left(1 + F \frac{\partial}{\partial t} \right) \left(-\nabla p + \mathrm{Ra} T + \mathrm{Rs} C \right) - \mathbf{q},$$

$$\frac{\partial T}{\partial t} - w = \nabla^2 T + D_f \nabla^2 C, \tag{8}$$

$$\frac{\varepsilon}{\sigma} \frac{\partial C}{\partial t} - w = \frac{1}{\mathrm{Le}} \nabla^2 C + S_r \nabla^2 T.$$

Here nondimensional parameters are as follows.

$\mathrm{Ra} = g\rho_0 \alpha k_1 \Delta T d / \mu \kappa$ is the thermal Rayleigh number, $\mathrm{Rs} = g\rho_0 \alpha' k_1 \Delta C d / \mu \kappa'$ is the solutal Rayleigh number, $\mathrm{Le} = \kappa / \kappa'$ is the Lewis number, $F = (\kappa / \sigma d^2) \lambda$ is the stress relaxation parameter, $D_f = D_{TC} \Delta C / \kappa \Delta T$ is the Dufour parameter, and $S_r = D_{CT} \Delta T / \kappa \Delta C$ is the Soret parameter.

The nondimensional boundary conditions are

$$w = T = C = 0 \quad \text{at } z = 0, \ z = 1. \tag{9}$$

3. Normal Modes and Stability Analysis

Analyze the disturbances into the normal modes and assume that the perturbed quantities are of the form

$$[w, T, C] = [W(z), \Theta(z), \Gamma(z)] \exp \left(i k_x x + i k_y y + nt \right), \tag{10}$$

where k_x, k_y are wave numbers along x and y directions, respectively, and n is growth rate of disturbances.

Using (10), (8) becomes

$$\left(D^2 - a^2 \right) W + (1 + Fn) \left(a^2 \mathrm{Ra} \Theta + a^2 \mathrm{Rs} \Gamma \right) = 0,$$

$$W + \left(D^2 - a^2 - n \right) \Theta + D_f \left(D^2 - a^2 \right) \Gamma = 0, \tag{11}$$

$$W + S_r \left(D^2 - a^2 \right) \Theta + \left(\frac{1}{\mathrm{Le}} \left(D^2 - a^2 \right) - \frac{\varepsilon}{\sigma} n \right) \Gamma = 0,$$

where $D = d/dz$ and $a = \sqrt{k_x^2 + k_y^2}$ is the dimensionless resultant wave number.

The boundary conditions are

$$W = 0, \qquad D^2 W = 0, \qquad \Theta = 0, \qquad \Gamma = 0 \quad \text{at } z = 0,$$

$$W = 0, \qquad D^2 W = 0, \qquad \Theta = 0, \qquad \Gamma = 0 \quad \text{at } z = 1. \tag{12}$$

We assume the solution to W, Θ, and Γ is of the form

$$W = W_0 \sin \pi z, \qquad \Theta = \Theta_0 \sin \pi z, \qquad \Gamma = \Gamma_0 \sin \pi z, \tag{13}$$

which satisfy boundary conditions (12).

Substituting solution (13) in (11), integrating each equation from $z = 0$ to $z = 1$ by parts, we obtain the following matrix equation as

$$\begin{bmatrix} J & -a^2\,(1+Fn)\,\mathrm{Ra} & -a^2\,(1+Fn)\,\mathrm{Rs} \\ -1 & (J+n) & D_f J \\ -1 & S_r J & \left(\dfrac{J}{\mathrm{Le}}+\dfrac{\varepsilon n}{\sigma}\right) \end{bmatrix} \begin{bmatrix} W_0 \\ \Theta_0 \\ \Gamma_0 \end{bmatrix} = \begin{bmatrix} 0 \\ 0 \\ 0 \end{bmatrix}, \quad (14)$$

where $J = \pi^2 + a^2$.

The nontrivial solution of the above matrix requires that

$$\mathrm{Ra} = \frac{\left(J\,(J+n)\,((J/\mathrm{Le})+(\varepsilon n/\sigma)) - S_r D_f J^2\right)}{a^2\,(1+Fn)\,J\left((1/\mathrm{Le})-D_f\right)+(\varepsilon n/\sigma)} \qquad (15)$$
$$+ \frac{S_r J - (J+n)}{J\left((1/\mathrm{Le})-D_f\right)+(\varepsilon n/\sigma)}\mathrm{Rs}.$$

For neutral instability $n = i\omega$, (where ω is real and dimensionless frequency of oscillation) and equating real and imaginary parts of (15), we have

$$J\left(\left(\frac{J^2}{\mathrm{Le}}-\frac{\omega^2\varepsilon}{\sigma}\right)-S_r D_f J^2\right)+a^2\mathrm{Ra}\left(J\left(D_f-\frac{1}{\mathrm{Le}}\right)+\frac{\omega^2\varepsilon F}{\sigma}\right)$$
$$- a^2\mathrm{Rs}\left(J-S_r J-\omega^2 F\right)=0,$$

$$J^2\left(\frac{1}{\mathrm{Le}}+\frac{\varepsilon}{\sigma}\right)+a^2\mathrm{Ra}\left(FJ\left(D_f-\frac{1}{\mathrm{Le}}\right)-\frac{\varepsilon}{\sigma}\right)$$
$$- a^2\mathrm{Rs}\left(JF-S_r JF+1\right)=0. \qquad (16)$$

For stationary convection $\omega = 0$ ($n = 0$), we have

$$\mathrm{Ra} = \frac{\left(\pi^2+a^2\right)^2}{a^2}\left(\frac{D_f S_r \mathrm{Le}-1}{D_f \mathrm{Le}-1}\right)+\frac{(1-S_r)\,\mathrm{Le}}{D_f \mathrm{Le}-1}\mathrm{Rs}. \qquad (17)$$

It is clear from (17) that stationary Rayleigh number Ra is a function of dimensionless wave number a, Dufour parameter D_f, Soret parameter S_r, Lewis number Le and solutal Rayleigh number Rs, and independent of stress relaxation parameter F. Thus for stationary convection the Maxwell viscoelastic fluid behaves like an ordinary Newtonian fluid. This result is the same as obtained by Motsa [24] and Chand and Rana [23].

The critical cell size at the onset of instability is obtained from the condition $(\partial \mathrm{Ra}/\partial a)_{a=a_c} = 0$, which gives $a_c = \pi$.

This result is the same as obtained by Lapwood [4] for Newtonian fluid.

The corresponding critical Rayleigh number Ra_c for steady onset is

$$\mathrm{Ra}_c = 4\pi^2\left(\frac{D_f S_r \mathrm{Le}-1}{D_f \mathrm{Le}-1}\right)+\frac{(S_r-1)\,\mathrm{Le}}{1-D_f \mathrm{Le}}\mathrm{Rs}. \qquad (18)$$

This result is the same as obtained by Motsa [24] and Chand and Rana [23].

If $\mathrm{Rs} = D_f = S_r = 0$ then

$$\mathrm{Ra}_c = 4\pi^2. \qquad (19)$$

This is exactly the same result which was obtained by Nield [25].

4. Result and Discussion

The onset of double diffusive convection in a horizontal layer of Maxwell viscoelastic fluid in the presence of Soret and Dufour in a porous medium is investigated analytically and graphically. The expressions for both the stationary and oscillatory Rayleigh numbers, which characterize the stability of the system, are obtained analytically. The stationary critical Rayleigh number is found to be independent of the viscoelastic parameter F; thus Maxwell viscoelastic binary fluid behaves like ordinary Newtonian binary fluid. The stationary critical Rayleigh number and critical wave number are independent of viscoelastic parameter because of the absence of base flow in the present case. The computations are carried out for different values solutal Rayleigh number Rs, Soret parameter S_r, Dufour parameter D_f, and Lewis number Le. The parameters considered are in the range of $10^2 \le \mathrm{Ra} \le 10^5$ (thermal Rayleigh number), $10^2 \le \mathrm{Rs} \le 10^3$ (solutal Rayleigh number), $0 \le S_r \le 1$ (Soret parameter), $0 \le D_f \le 1$ (Dufour parameter), [24, 26], and $10^{-2} \le \mathrm{Le} \le 1$ (Lewis number) [27, 28], [29]. We choose the values of Dufour parameter D_f and Lewis number Le in such a way that $D_f \mathrm{Le} \ne 1$.

In order to investigate effects of the Dufour parameter D_f, Soret parameter S_r, Lewis number Le, and solutal Rayleigh number Rs on stationary convection, we examine the behavior of $\partial \mathrm{Ra}/\partial D_f, \partial \mathrm{Ra}/\partial S_r, \partial \mathrm{Ra}/\partial \mathrm{Le}$, and $\partial \mathrm{Ra}/\partial \mathrm{Rs}$ analytically.

From (17), we have

(i)

$$\frac{\partial \mathrm{Ra}}{\partial D_f} = \frac{(1-S_r)\,\mathrm{Le}}{\left(D_f \mathrm{Le}-1\right)^2}\left[\frac{\left(\pi^2+a^2\right)^2}{a^2}D_f - \mathrm{Le}\,\mathrm{Rs}\right],$$

$$\frac{\partial \mathrm{Ra}}{\partial D_f} > 0 \quad \text{if } \mathrm{Le}\,\mathrm{Rs} > \frac{\left(\pi^2+a^2\right)^2}{a^2}D_f,$$

$$< 0 \quad \text{if } \mathrm{Le}\,\mathrm{Rs} < \frac{\left(\pi^2+a^2\right)^2}{a^2}D_f, \qquad (20)$$

$$= 0$$

at two specific values of wave number, which depend upon Lewis number Le, solutal Rayleigh number Rs, and Dufour parameter D_f and is independent of Soret parameter S_r.

Thus for the stationary convection Dufour parameter D_f has a stabilizing effect if $\mathrm{Le}\,\mathrm{Rs} > ((\pi^2+a^2)^2/a^2)D_f$ and destabilizing effect if $\mathrm{Le}\,\mathrm{Rs} < ((\pi^2+a^2)^2/a^2)D_f$:

(ii)

$$\frac{\partial Ra}{\partial S_r} = \frac{Le}{D_f Le - 1}\left[\frac{\left(\pi^2 + a^2\right)^2}{a^2}D_f - Rs\right],$$

$$\frac{\partial Ra}{\partial S_r} > 0 \quad \text{if} \quad \frac{\left(\pi^2 + a^2\right)^2}{a^2}D_f < Rs, \qquad (21)$$

$$< 0 \quad \text{if} \quad \frac{\left(\pi^2 + a^2\right)^2}{a^2}D_f > Rs,$$

$$= 0$$

at two specific values of wave number, which depend upon solutal Rayleigh number Rs and Dufour parameter D_f and is independent of Lewis number Le and Soret parameter S_r.

Thus for stationary convection Soret parameter S_r, therefore, has a stabilizing effect if $((\pi^2 + a^2)^2/a^2)D_f < $ Rs and destabilizing effect if $((\pi^2 + a^2)^2/a^2)D_f > $ Rs:

(iii)

$$\frac{\partial Ra}{\partial Le} = \frac{(1 - S_r)}{\left(D_f Le - 1\right)}\left[\frac{\left(\pi^2 + a^2\right)^2}{a^2}D_f - Rs\right],$$

$$\frac{\partial Ra}{\partial Le} > 0 \quad \text{if} \quad \frac{\left(\pi^2 + a^2\right)^2}{a^2}D_f < Rs, \qquad (22)$$

$$< 0 \quad \text{if} \quad \frac{\left(\pi^2 + a^2\right)^2}{a^2}D_f > Rs,$$

$$= 0$$

at two specific values of wave number, which depend upon solutal Rayleigh number Rs and Dufour parameter D_f and is independent of Lewis number Le and Soret parameter S_r.

Thus for stationary convection Lewis number Le has a stabilizing effect if $((\pi^2 + a^2)^2/a^2)D_f < $ Rs and destabilizing effect if $((\pi^2 + a^2)^2/a^2)D_f > $ Rs:

(iv)

$$\frac{\partial Ra}{\partial Rs} = \frac{(1 - S_r)Le}{D_f Le - 1} < 0. \qquad (23)$$

Thus for stationary convection solutal Rayleigh number Rs has a destabilizing effect.

Now we discussed the effects of various parameters on the onset of double diffusive convection of Maxwell viscoelastic fluid in a porous medium for stationary convection graphically. The convection curves for solutal Rayleigh number Rs, Soret parameter S_r, Dufour parameter D_f, and Lewis number Le in the (Ra, a) plane are shown in Figures 1–4.

Figure 1 shows the variation of stationary Rayleigh number with wave number for different values of Dufour parameter and it is found that the Rayleigh number first increases

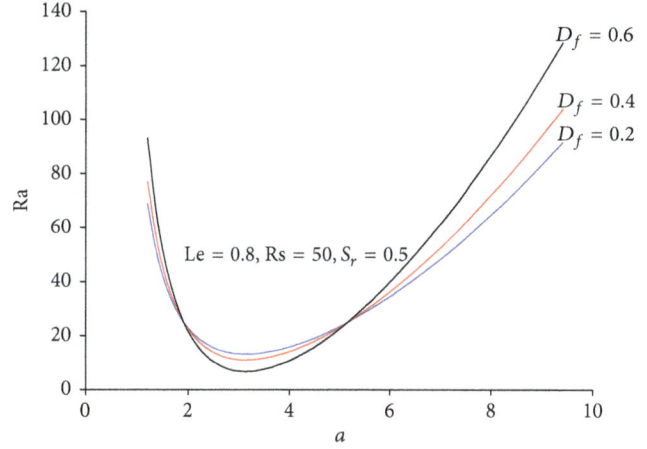

FIGURE 1: Variation of Rayleigh number Ra with wave number a for different values of Dufour parameter D_f.

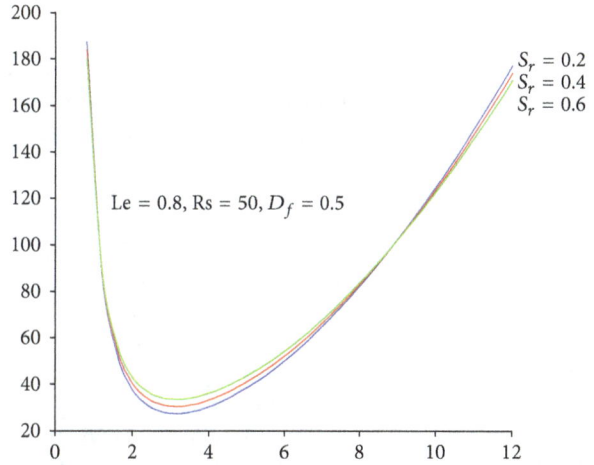

FIGURE 2: Variation of Rayleigh number Ra with wave number a for different values of Soret parameter S_r.

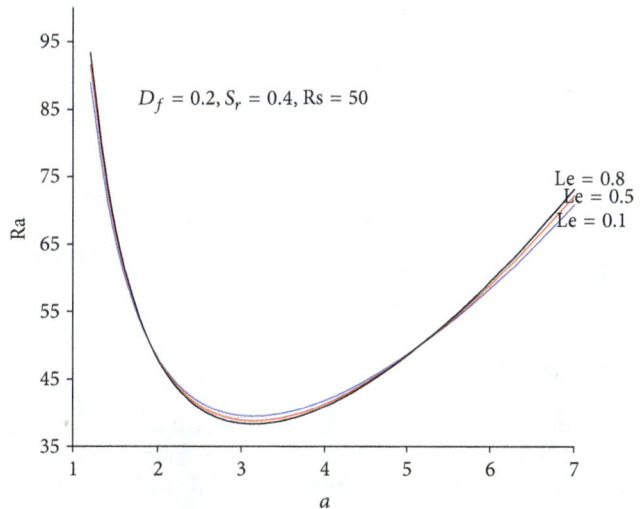

FIGURE 3: Variation of Rayleigh number Ra with wave number a for different values of Lewis number Le.

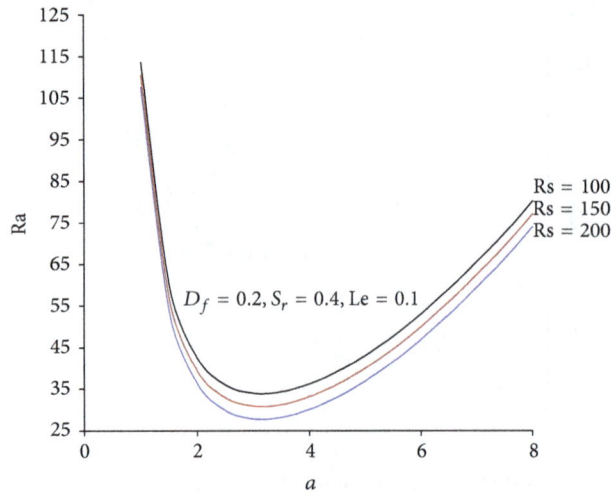

FIGURE 4: Variation of Rayleigh number Ra with wave number a for different values of solutal Rayleigh number Rs.

then decreases and finally increases with increase in the value of Dufour parameter; thus for stationary convection Dufour parameter has both the stabilizing and destabilizing effects depending upon certain conditions.

Figure 2 shows the variation of stationary Rayleigh number with wave number for different values of Soret parameter and it is found that the Rayleigh number first decreases then increases and finally decreases with increase in the value of Soret parameter; thus for stationary convection Soret parameter has both the stabilizing and destabilizing effects depending upon certain conditions.

Figure 3 shows the variation of stationary Rayleigh number with wave number for different values of Lewis number and it is found that the Rayleigh number first increases then decreases and finally increases with increase in the value of Lewis number; thus for stationary convection Lewis number has both the stabilizing and destabilizing effects depending upon certain conditions.

Figure 4 shows the variation of Rayleigh number with wave number for different value of the solutal Rayleigh number Rs and it is found that the Rayleigh number decreases with increase in the value of solutal Rayleigh number Rs; thus solutal Rayleigh number Rs has destabilizing effect on the stationary convection.

Curves in Figures 1–3 intersect two specific values of wave number a which is independent of Lewis number Le and Soret parameter S_r as explained in the analytical part of the results and discussion.

5. Conclusions

A linear stability analysis of double diffusive convection in a horizontal layer of Maxwell viscoelastic fluid in the presence of Soret and Dufour in a porous medium is investigated analytically and graphically. The expressions for both the stationary and oscillatory Rayleigh numbers, which characterize the stability of the system, are obtained.

The main conclusions are as follows.

(i) In stationary convection Maxwell viscoelastic fluid behaves like ordinary Newtonian fluid.

(ii) Dufour parameter, Soret parameter, and Lewis parameter have both stabilizing and destabilizing effects on the stationary convection.

(iii) Solutal Rayleigh number destabilizes the stationary convection.

(iv) In limiting case when Rs $= S_r = D_f = 0$ the critical thermal Rayleigh number obtained is the same as reported by Nield [25].

Nomenclature

a:	Wave number
C:	Solute concentration
c_p:	Heat capacity
D_f:	Dufour parameter
D_{TC}:	Dufour coefficient
D_{CT}:	Soret coefficient
d:	Thickness of fluid layer
F:	Stress relaxation parameter
$\mathbf{g}$:	Acceleration due to gravity
k_1:	Medium permeability
Le:	Lewis number
n:	Growth rate of disturbances
p:	Pressure
$\mathbf{q}$:	Darcy fluid velocity
Ra:	Thermal Rayleigh number
Ra_c:	Critical Rayleigh number
Rs:	Solutal Rayleigh number
S_r:	Soret parameter
t:	Time
T:	Temperature
(u, v, w):	Components of fluid velocity
(x, y, z):	Space coordinates.

Greek Symbols

ρ:	Density of fluid
α:	Coefficient of thermal expansion
α':	Analogous solvent coefficient of expansion
μ:	Viscosity
κ:	Thermal diffusivity
κ':	Solute diffusivity
λ:	Relaxation time
∂:	Curly operator
ε:	Porosity
ω:	Dimensionless frequency of oscillation
σ:	Thermal capacity ratio.

Superscripts

$'$:	Nondimensional variables
$''$:	Perturbed quantity.

Subscripts

0: Value of variables at lower boundary
1: Value of variables at upper boundary
s: Steady state
f: Fluid
m: Porous medium.

Conflict of Interests

The authors declare that there is no conflict of interests regarding the publication of this paper.

Acknowledgment

The authors are grateful to the reviewers for their lucid comments and suggestions which have served to improve the research paper.

References

[1] H. Bénard, "Les tourbillons cellularies dans une nappe liquid," *Revue Generale des Sciences Pures et Appliqués*, vol. 11, pp. 1261–1271, 1900.

[2] L. Rayleigh, "On convective currents in a horizontal layer of fluid when the higher temperature is on the under side," *Philosophical Magazine*, vol. 32, pp. 529–546, 1916.

[3] S. Chandrasekhar, *Hydrodynamic and Hydromagnetic Stability*, Dover Publication, New York, NY, USA, 1961.

[4] E. R. Lapwood, "Convection of a fluid in porous medium," *Mathematical Proceedings of the Cambridge Philosophical Society*, vol. 44, no. 4, pp. 508–519, 1948.

[5] R. A. Wooding, "Rayleigh instability of a thermal boundary layer in flow through a porous medium," *Journal of Fluid Mechanics*, vol. 9, pp. 183–192, 1960.

[6] J. A. M. McDonnel, *Cosmic Dust*, John Wiley & Sons, Toronto, Canada, 1978.

[7] D. A. Nield and A. Bejan, *Convection in Porous Medium*, Springer, New York, NY, USA, 3rd edition, 2006.

[8] O. V. Trevisan and A. Bejan, "Combined heat and mass transfer by natural convection in a porous medium," *Advances in Heat Transfer*, vol. 20, pp. 315–352, 1990.

[9] M. S. Malashetty and P. Kollur, "The onset of double diffusive convection in a couple stress fluid saturated anisotropic porous layer," *Transport in Porous Media*, vol. 86, no. 2, pp. 435–459, 2011.

[10] E. Knobloch, "Convection in binary fluids," *The Physics of Fluids*, vol. 23, no. 9, pp. 1918–1919, 1980.

[11] J. G. Oldroyd, "Non-Newtonian effects in steady motion of some idealized elastico-viscous liquids," *Proceedings of the Royal Society A: Mathematical, Physical and Engineering Sciences*, vol. 245, pp. 278–297, 1958.

[12] T. Green III, "Oscillating convection in an elasticoviscous liquid," *Physics of Fluids*, vol. 11, no. 7, pp. 1410–1412, 1968.

[13] C. M. Vest and V. Arpaci, "Overstability of visco-elastic fluid layer heated from below," *Journal of Fluid Mechanics*, vol. 36, no. 3, pp. 613–623, 1969.

[14] P. K. Bhatia and J. M. Steiner, "Convective instability in a rotating viscoelastic fluid layer," *Zeitschrift für Angewandte Mathematik und Mechanik*, vol. 52, pp. 321–327, 1972.

[15] P. K. Bhatia and J. M. Steiner, "Thermal instability in a viscoelastic fluid layer in hydromagnetics," *Journal of Mathematical Analysis and Applications*, vol. 41, pp. 271–283, 1973.

[16] R. C. Sharma and P. Kumar, "Hall efect on thermosolutal instability in a Maxwellian visco-elastic fluid in porous medium," *Archives of Mechanics*, vol. 48, pp. 199–209, 1996.

[17] K. Prakash and R. Chand, "Thermosolutal instability of Maxwell visco-Elastic fluid with Hall Current, suspended particles and variable gravity in porous medium," *Ganita Sandesh*, vol. 13, no. 1, pp. 1–12, 1999.

[18] R. Chand and S. K. Kango, "Thermosolutal instability of dusty rotating Maxwell visco-elastic fluid in porous medium," *Advances in Applied Science Research*, vol. 2, no. 6, pp. 541–553, 2011.

[19] R. Chand, "Gravitational effect on thermal instability of Maxwell visco-elastic fluid in porous medium," *Ganita Sandesh*, vol. 24, no. 2, pp. 166–170, 2010.

[20] R. Chand, "Effect of suspended particles on thermal instability of Maxwell visco-elastic fluid with variable gravity in porous medium," *Antarctica Journal of Mathematics*, vol. 8, no. 6, pp. 487–497, 2011.

[21] R. Chand, "Thermal instability of rotating Maxwell visco-elastic fluid with variable gravity in porous medium," *The Journal of the Indian Mathematical Society*, vol. 80, no. 1-2, pp. 23–31, 2013.

[22] R. Chand and A. Kumar, "Thermal instability of rotating Maxwell visco-elastic fluid with variable gravity in porous medium," *International Journal of Advances in Applied Mathematics and Mechanics*, vol. 1, no. 2, pp. 30–38, 2013.

[23] R. Chand and G. C. Rana, "Dufour and soret effects on the thermosolutal instability of rivlin-ericksen elastico-viscous fluid in porous medium," *Zeitschrift fur Naturforschung A*, vol. 67, no. 12, pp. 685–691, 2012.

[24] S. S. Motsa, "On the onset of convection in a porous layer in the presence of Dufour and Soret effects," *SAMSA Journal of Pure and Applied Mathematics*, vol. 3, pp. 58–65, 2008.

[25] D. A. Nield, "Convection in a porous medium with inclined temperature gradient," *International Journal of Heat and Mass Transfer*, vol. 34, no. 1, pp. 87–92, 1991.

[26] N. Nithyadevi and R. J. Yang, "Double diffusive natural convection in a partially heated enclosure with Soret and Dufour effects," *International Journal of Heat and Fluid Flow*, vol. 30, no. 5, pp. 902–910, 2009.

[27] J. Martínez-Mardones, R. Tiemann, and D. Walgraef, "Convection in binary viscoelastic fluid," *Revista Mexicana de Fisica*, vol. 48, no. 3, pp. 103–105, 2002.

[28] Z. Yang, S. Wang, M. Zhao, S. Li, and Q. Zhang, "The onset of double diffusive convection in a viscoelastic fluid-saturated porous layer with non-equilibrium model," *PLoS ONE*, vol. 8, no. 11, Article ID e79956, 2013.

[29] D. Srinivasacharya and G. Swamy Reddy, "Double diffusive natural convection in power-law fluid saturated porous medium with Soret and Dufour effects," *Journal of the Brazilian Society of Mechanical Sciences and Engineering*, vol. 34, no. 4, pp. 525–530, 2012.

8

Free Convective MHD Flow Past a Vertical Cone with Variable Heat and Mass Flux

J. Prakash,[1] S. Gouse Mohiddin,[2] and S. Vijaya Kumar Varma[3]

[1] Department of Mathematics, University of Botswana, Private Bag 0022, Gaborone, Botswana
[2] Department of Mathematics, Madanapalle Institute of Technology & Science, Madanapalle 517325, Andhra Pradesh, India
[3] Department of Mathematics, Sri Venkateswara University, Tirupati 517502, Andhra Pradesh, India

Correspondence should be addressed to J. Prakash; prakashj@mopipi.ub.bw

Academic Editor: Mohy S. Mansour

A numerical study of buoyancy-driven unsteady natural convection boundary layer flow past a vertical cone embedded in a non-Darcian isotropic porous regime with transverse magnetic field applied normal to the surface is considered. The heat and mass flux at the surface of the cone is modeled as a power law according to $q_w(x) = x^m$ and $q_w^*(x) = x^m$, respectively, where x denotes the coordinate along the slant face of the cone. Both Darcian drag and Forchheimer quadratic porous impedance are incorporated into the two-dimensional viscous flow model. The transient boundary layer equations are then nondimensionalized and solved by the Crank-Nicolson implicit difference method. The velocity, temperature, and concentration fields have been studied for the effect of Grashof number, Darcy number, Forchheimer number, Prandtl number, surface heat flux power-law exponent (m), surface mass flux power-law exponent (n), Schmidt number, buoyancy ratio parameter, and semivertical angle of the cone. Present results for selected variables for the purely fluid regime are compared with the published results and are found to be in excellent agreement. The local skin friction, Nusselt number, and Sherwood number are also analyzed graphically. The study finds important applications in geophysical heat transfer, industrial manufacturing processes, and hybrid solar energy systems.

1. Introduction

Combined heat and mass transfer in fluid-saturated porous media finds applications in a variety of engineering processes such as heat exchanger devices, petroleum reservoirs, chemical catalytic reactors and processes, and others. A thorough discussion of these and other applications is available in the monographs [1, 2]. Comprehensive reviews of the much of the work communicated in porous media transport phenomena have been presented in [3, 4]. Most studies dealing with porous media have employed the Darcy law. However, for high velocity flow situations, the Darcy law is inapplicable, since it does not account for inertial effects in the porous medium. Such flows can arise, for example, in the near-wellbore region of high capacity gas and condensate petroleum reservoirs and also in highly porous filtration systems under high blowing rates. The most popular approach for simulating high-velocity transport in porous

media is the Darcy-Forchheimer drag force model. This adds a second-order (quadratic) drag force to the momentum transport equation. This term is related to the geometrical features of the porous medium and is independent of viscosity. A seminal study discussing the influence of Forchheimer inertial effects in porous media convection is presented by Vafai and Tien [5]. The mixed convective boundary layer flow from a vertical surface in a fluid-saturated non-Darcian porous medium including Forchheimer inertial effects is studied by C.-H. Chen and C.-K. Chen [6] and Chen et al. [7]. Thermal convection boundary layer flow with buoyancy and suction/blowing effects from a cone with nonuniform surface temperature is studied by Hossain and Paul [8]. The study is extended by Hossain and Paul [9] by considering nonuniform surface heat flux, both studies employing numerical methods. Chamkha et al. [10] studied the double-diffusive convection heat and mass transfer over a cone (and wedge) in Darcy-Forchheimer porous media. Magnetohydrodynamic (MHD) flow and heat transfer is of considerable interest because it can occur in many geothermal, geophysical, technological, and engineering applications such as nuclear reactors and others. The geothermal gases are electrically conducting and are affected by the presence of a magnetic field. Vajravelu and Nayfeh [11] studied hydromagnetic convection from a cone and a wedge with variable surface temperature and internal heat generation or absorption. Thus far the transient thermal convection flow over a cone in Darcy-Forchheimer porous media has not been studied in the literature despite important applications in geothermics, geophysics, and materials processing.

2. Mathematical Model

An axisymmetric unsteady natural convection boundary layer flow past a vertical cone with transverse magnetic field applied normal to the surface with variable heat and mass flux in a Darcy-Forchheimer fluid saturated porous medium in a cartesian (x, y) coordinate system is formulated mathematically in this section. Initially, it is assumed that the cone surface and the surrounding fluid which are at rest possess the same temperature T'_∞ and concentration level C'_∞ everywhere in the fluid. At time $t' > 0$, heat supplied from the cone surface to the fluid and concentration level near the cone surface are raised at a rate of $q_w(x) = x^m$ and $q^*_w(x) = x^n$, respectively, and they are maintained at the same level. It is assumed that the concentration C' of the diffusing species in the binary mixture is very less in comparison to the other chemical species, which are present and hence the Soret and Dufour effects are negligible. We consider viscous flow where pressure work, viscous dissipation, and thermal dispersion effects are neglected. The coordinate system chosen (as shown in Figure 1) is such that the x-direction is measured along the cone surface from the leading edge O, and the y-direction is normal to the cone generator. The cone apex is located at the origin $(x = y = 0)$.

Here ϕ designates the semivertical angle of the cone, and r is the local radius of the cone.

Then under the previous assumptions, the governing boundary layer equations with Boussinesq's approximation are

$$\frac{\partial (ur)}{\partial x} + \frac{\partial (vr)}{\partial y} = 0,$$

$$\frac{\partial u}{\partial t'} + u \frac{\partial u}{\partial x} + v \frac{\partial u}{\partial y}$$
$$= \nu \frac{\partial^2 u}{\partial y^2} - \frac{\sigma B_0^2}{\rho} u + g\beta \cos\phi \left(T' - T'_\infty\right)$$
$$+ g\beta^* \cos\phi \left(C' - C'_\infty\right) - \frac{\nu}{K} u - \frac{b}{K} u^2,$$

$$\frac{\partial T'}{\partial t'} + u \frac{\partial T'}{\partial x} + v \frac{\partial T'}{\partial y} = \alpha \frac{\partial^2 T'}{\partial y^2},$$

$$\frac{\partial C'}{\partial t'} + u \frac{\partial C'}{\partial x} + v \frac{\partial C'}{\partial y} = D \frac{\partial^2 C'}{\partial y^2},$$

(1)

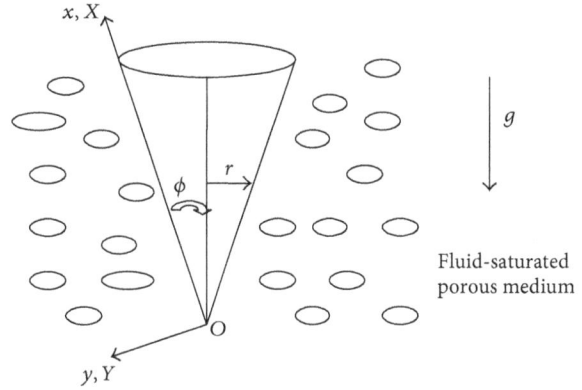

FIGURE 1: Physical Model.

where all terms are defined in the nomenclature. Under the Boussinesq approximation, buoyancy effects are simulated only in the momentum equation, which is coupled to the energy equation, constituting a free convection regime. The corresponding spatial and temporal initial and boundary conditions at the surface and far from the cone take the following form:

$$t' \leq 0: u = 0, \quad v = 0, \quad T' = T'_\infty, \quad C' = C'_\infty \quad \forall\, x, y,$$

$$t' > 0: u = 0, \quad v = 0, \quad \frac{\partial T'}{\partial y} = -\frac{q_w(x)}{k},$$

$$\frac{\partial C'}{\partial y} = -\frac{q^*_w(x)}{D} \quad \text{at } y = 0,$$

$$u = 0, \quad T' = T'_\infty, \quad C' = C'_\infty \quad \text{at } x = 0,$$

$$u \longrightarrow 0, \quad T' \longrightarrow T'_\infty, \quad C' \longrightarrow C'_\infty \quad \text{as } y \longrightarrow \infty,$$

(2)

where all the parameters are defined in the nomenclature.

Equation (1) is highly coupled, parabolic, and nonlinear. An analytical solution is clearly intractable, and in order to facilitate a numerical solution we *nondimensionalize* the model. Proceeding with the analysis, we now introduce the following transformations:

$$X = \frac{x}{L}, \qquad Y = \frac{y}{L}(\mathrm{Gr}_L)^{1/4},$$

$$R = \frac{r}{L}, \quad \text{where } r = x\sin\phi,$$

$$V = \frac{vL}{\nu}(\mathrm{Gr}_L)^{-1/4}, \qquad U = \frac{uL}{\nu}(\mathrm{Gr}_L)^{-1/2},$$

$$t = \frac{\nu t'}{L^2}(\mathrm{Gr}_L)^{1/2}, \qquad M = \frac{\sigma B_0^2 L^2}{\mu}\mathrm{Gr}_L^{-1/2},$$

$$T = \frac{T' - T'_\infty}{[q_w(L)L/k]}, \qquad \mathrm{Gr}_L = \frac{g\beta \cos\phi\, [q_w L/k]\, L^4}{\nu^2},$$

$$\mathrm{Pr} = \frac{\nu}{\alpha}, \quad \mathrm{Da} = \frac{K}{L^2}, \quad C = \frac{C' - C'_\infty}{[q_w^*(L)\, L/D]},$$

$$N = \frac{\beta^*\left(C'_w - C'_\infty\right)}{\beta\left(T'_w - T'_\infty\right)}, \quad \mathrm{Sc} = \frac{\nu}{D}, \quad \mathrm{Fs} = \frac{b}{L}.$$

$$(3)$$

The transport (1) is thereby reduced to the following dimensionless form:

$$\frac{\partial(UR)}{\partial X} + \frac{\partial(VR)}{\partial Y} = 0,$$

$$\frac{\partial U}{\partial t} + U\frac{\partial U}{\partial X} + V\frac{\partial U}{\partial Y}$$

$$= \frac{\partial^2 U}{\partial Y^2} - MU + T\cos\phi + NC\cos\phi$$

$$- \frac{U}{\mathrm{DaGr}_L} - \frac{\mathrm{Fs}}{\mathrm{Da}}U^2$$

$$\frac{\partial T}{\partial t} + U\frac{\partial T}{\partial X} + V\frac{\partial T}{\partial Y} = \frac{1}{\mathrm{Pr}}\frac{\partial^2 T}{\partial Y^2},$$

$$\frac{\partial C}{\partial t} + U\frac{\partial C}{\partial X} + V\frac{\partial C}{\partial Y} = \frac{1}{\mathrm{Sc}}\frac{\partial^2 C}{\partial Y^2}.$$

$$(4)$$

The corresponding nondimensional initial and boundary conditions are given by

$$t \le 0 : U = 0, \qquad V = 0,$$

$$T = 0, \quad C = 0 \quad \forall\, X, Y,$$

$$t > 0 : U = 0, \qquad V = 0, \qquad \frac{\partial T}{\partial Y} = -X^m,$$

$$\frac{\partial C}{\partial Y} = -X^n \quad \text{at } Y = 0,$$

$$U = 0, \quad T = 0, \quad C = 0 \quad \text{at } X = 0,$$

$$U \longrightarrow 0, \quad T \longrightarrow 0, \quad C \longrightarrow 0 \quad \text{as } Y \longrightarrow \infty,$$

$$(5)$$

where again all the parameters are given in the nomenclature. The dimensionless local values of the skin friction, Nusselt number, and the Sherwood number are given by the following expressions:

$$\tau_x = -\left(\frac{\partial U}{\partial Y}\right)_{Y=0}$$

$$\mathrm{Nu}_x = -X\left(\frac{\partial T}{\partial Y}\right)_{Y=0}$$

$$\mathrm{Sh}_x = -X\left(\frac{\partial C}{\partial Y}\right)_{Y=0}.$$

$$(6)$$

3. Numerical Solution

In order to solve the unsteady, nonlinear, coupled (4) under condition (5), an implicit finite difference scheme of Crank-Nicolson type has been employed which is discussed by many researchers [12–15]. The finite difference scheme of dimensionless governing equations is reduced to tridiagonal system of equations and is solved by Thomas algorithm as discussed elsewhere [16]. The region of integration is considered as a rectangle with $X_{\max} = 1$ and $Y_{\max} = 22$ where $Y_{\max}$ and corresponds to $Y = \infty$ which lies very well outside both the momentum and thermal boundary layers. The maximum of Y was chosen as 22, after some preliminary investigation so that the last two boundary conditions of (5) are satisfied within the tolerance limit 10^{-5}. The mesh sizes have been fixed as $\Delta X = 0.05$, $\Delta Y = 0.05$ with time step $\Delta t = 0.01$. The computations are carried out first by reducing the spatial mesh sizes by 50% in one direction, and later in both directions by 50%. The results are compared. It is observed in all cases that the results differ only in the fifth decimal place. Hence, the choice of the mesh sizes seems to be appropriate. The scheme is unconditionally stable. The local truncation error is $O(\Delta t^2 + \Delta Y^2 + \Delta X)$, and it tends to zero as Δt, ΔX and ΔY tend to zero. Hence, the scheme is compatible. Stability and compatibility ensure the convergence. The derivatives involved in (6) are evaluated using five point approximation formula.

4. Results and Discussion

Only selective figures have been reproduced here for brevity. In the numerical computations, the following values for the dimensionless thermophysical parameters are prescribed: Grashof number (Gr_L) = 1.0, Darcy number (Da) = 0.1 (high permeability), Forchheimer number (Fs) = 0.1 (weak quadratic drag), Prandtl number (Pr) = 7.0 (water), Schmidt number (Sc) = 0.6 (oxygen diffusing in air), surface heat flux power law exponent (m) = 0.5, surface mass flux power law exponent (n) = 0.5, buoyancy ratio parameter (N) = 1.0, and semivertical angle of the cone (ϕ) = 20°. All graphs therefore correspond to these values unless otherwise indicated. To test the accuracy of the computations, the local shear stress and local Nusselt number computations for the nonporous case are compared with those of Hossain and Paul [9] for a heat flux gradient of $m = 0.5$ and $X = 1.0$ in the steady state, in Tables 1 and 2, respectively, and are found to be in good agreement.

In Figures 2(a) and 2(b), the influence of Grashof number (Gr_L) on steady state velocity (U) and temperature (T) distributions with Y-coordinate are shown. Free convection, that is, thermal buoyancy effects are analyzed via the Grashof number. For an increasing Gr_L from 0.1 through 1.0, 10.0, 50.0 to 100.0 cooling of the cone by free convection occurs; that is, heat is conducted away from the cone to the surrounding regime.

Figures 3(a) and 3(b) show the effect of Darcy number (Da) on dimensionless velocity (U) and temperature (T) with transformed radial coordinate (Y) close to the leading edge (i.e., cone apex) at $X = 1.0$. To study the influence of

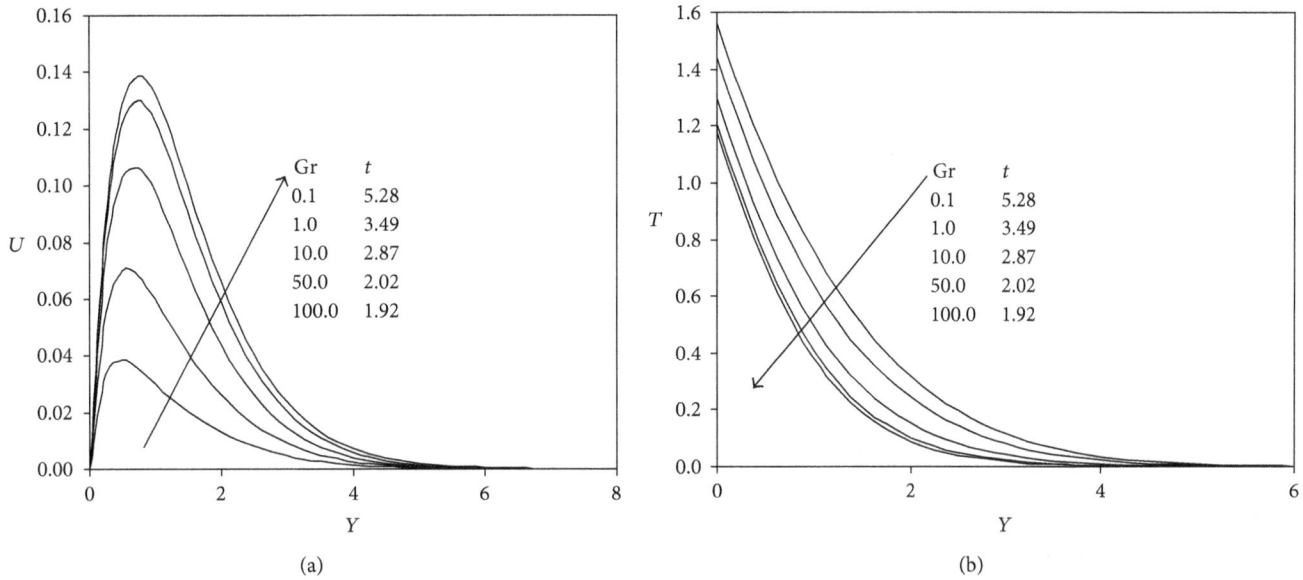

FIGURE 2: (a) Steady state velocity profiles at $X = 1.0$ for different Gr. (b) Steady state temperature profiles at $X = 1.0$ for different Gr.

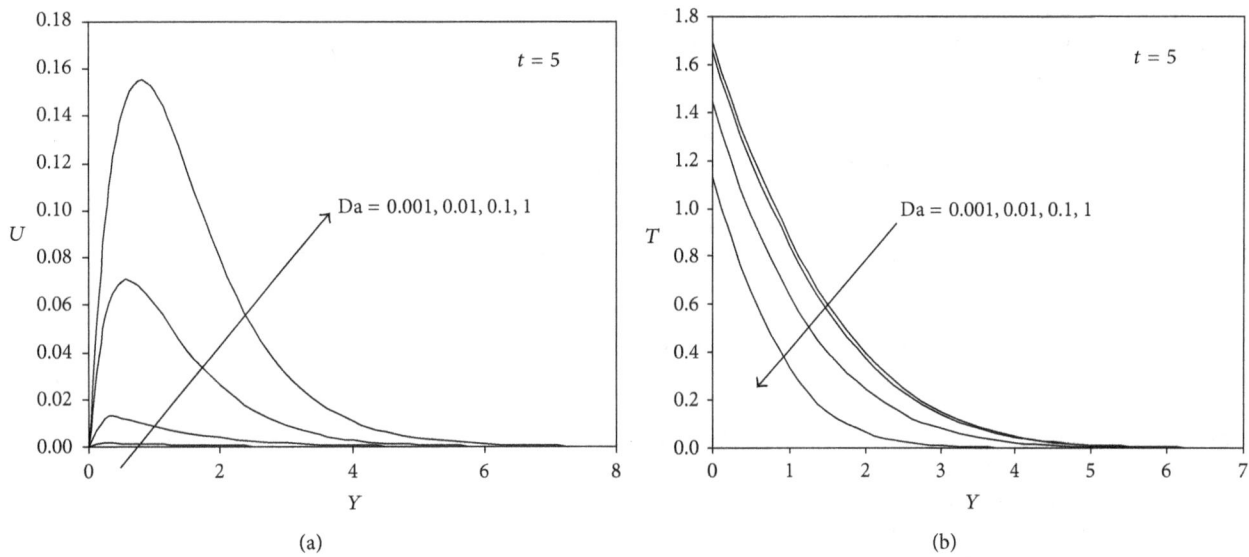

FIGURE 3: (a) Transient velocity profiles at $X = 1.0$ for different Da. (b) Transient temperature profiles at $X = 1.0$ for different Da.

TABLE 1: Comparison of local skin friction values at $X = 1.0$ and $m = 0.5$ with those of Hossain and Paul [9] for steady state purely fluid (Da $\to \infty$ in present model) case.

Pr	Hossain and Paul [9] $\tau_X/Gr_L^{3/5}$	Present results
0.01	5.13457	5.13424
0.05	2.93993	2.93180
0.1	2.29051	2.29044

TABLE 2: Comparison of local Nusselt number values at $X = 1.0$ and $m = 0.5$ with those of Hossain and Paul [9] for steady state purely fluid (Da $\to \infty$ in present model) case.

Pr	Hossain and Paul [9] $Nu_X/Gr_L^{3/5}$	Present results
0.01	0.14633	0.14648
0.05	0.26212	0.26227
0.1	0.33174	0.33648

regime permeability from sparsely packed media to densely packed materials, the following values Da = 1.0, 0.1, 0.01, 0.001 are considered. Da = K/L^2 for a fixed value of the reference length (L) is directly proportional to permeability (K) of the porous regime. Increasing Da increases the porous medium permeability and simultaneously decreases the Darcian impedance since progressively less solid fibers are present in the regime. The flow is therefore accelerated

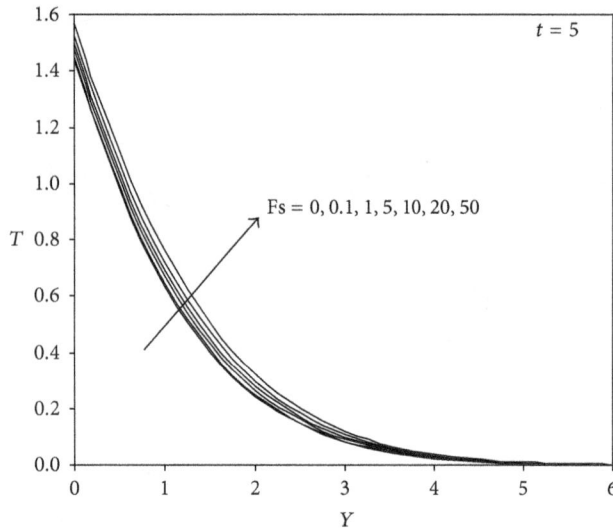

FIGURE 4: Transient temperature profiles at $X = 1.0$ for different Fs.

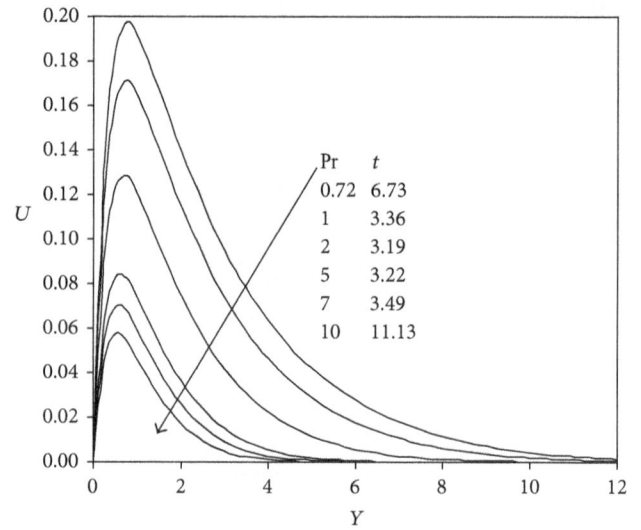

FIGURE 5: Steady state velocity profiles at $X = 1.0$ for different Pr.

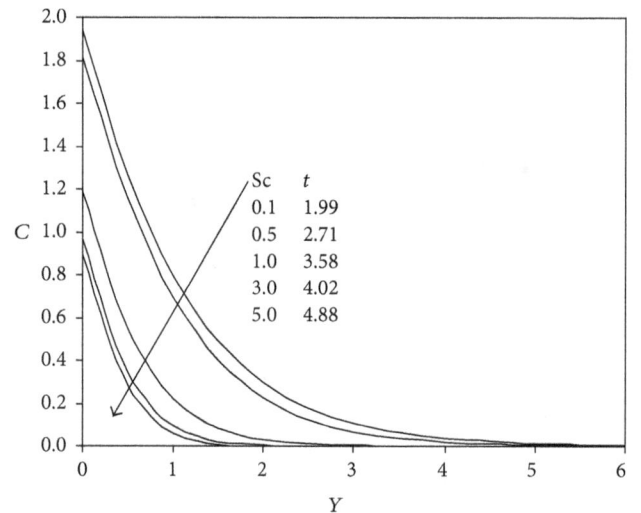

FIGURE 6: Steady state concentration profiles at $X = 1.0$ for different Sc.

for higher Da values causing an increase in the velocity U as shown in Figure 3(a). Maximum effect of rising Darcy number is observed at intermediate distance from the cone surface around $Y \sim 1$. Conversely, temperature T depicted in Figure 3(b) is opposed by increasing Darcy number. The presence of fewer solid fibers in the regime with increasing Da inhibits the thermal conduction in the medium which reduces distribution of thermal energy. The regime is therefore cooled when more fluid is present, and T values in the thermal boundary layer are decreased. Profiles for both velocity and temperature are smoothly asymptotic decays to the free stream indicating that excellent convergence (and stability) is obtained with the numerical method. Velocity boundary layer thickness will be increased with a rise in Da and thermal boundary layer thickness reduced. The effect of the Forchheimer inertial drag parameter (Fs) on dimensionless temperature (T) profiles is shown in Figure 4. The Forchheimer drag force is a second-order retarding force simulated in the momentum conservation equation. Increasing Fs values from 0.0 through 0.1, 1.0, 5.0, 10.0, 20.0, and 50.0 causes a strong increase in Forchheimer drag which decelerates the flow, that is, reduces velocities. For higher values of Fs, it is expected that the porous medium flow becomes increasingly chaotic. Temperature (T) however is slightly increased with a rise in Forchheimer parameter. The effects of the Prandtl number (Pr) on velocity profiles are depicted in Figure 5. Pr encapsulates the ratio of momentum diffusivity to thermal diffusivity. Larger Pr values imply a thinner thermal boundary layer thickness and more uniform temperature distributions across the boundary layer. Hence, thermal boundary layer will be much less thick than the hydrodynamic (translational velocity) boundary layer. Smaller Pr fluids have higher thermal conductivities, so that heat can diffuse away from the cone surface faster than for higher Pr fluids (thicker boundary layers). Physically the lower values of Pr correspond to liquid metals (Pr $\sim$ 0.02, 0.05), Pr = 0.7 is accurate for air or hydrogen and Pr = 7.0

for water. The computations show that translational velocity U is therefore reduced as Pr rises from 0.72 through 1.0, 2.0, 5.0, 7.0 and 10.0 since the fluid is increasingly viscous as Pr rises.

Figure 6 shows the effect of the Schmidt number (Sc) on the dimensionless concentration (C). We note that the Schmidt number (Sc) embodies the ratio of the momentum to the mass diffusivity. Sc therefore quantifies the relative effectiveness of momentum and mass transport by diffusion in the hydrodynamic (velocity) and concentration (species) boundary layers. Smaller Sc values can represent for example hydrogen gas as the species diffusing in air, Sc = 2.0 implies hydrocarbon diffusing in air, and higher values to petroleum derivatives diffusing in fluids (e.g. ethyl benzene) as indicated elsewhere [17]. As Sc increases, Figure 6 shows that C values are strongly decreased as larger values of Sc correspond to

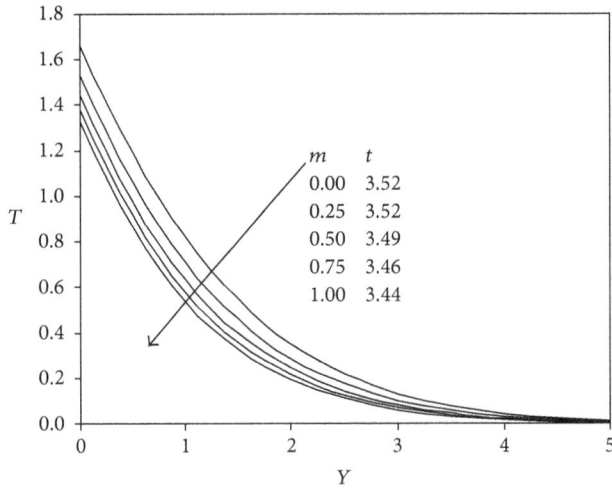

FIGURE 7: Steady state temperature profiles at $X = 1.0$ for different m.

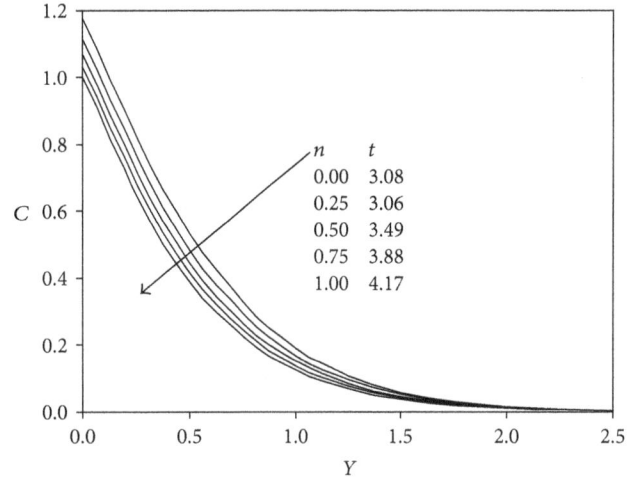

FIGURE 8: Steady state concentration profiles at $X = 1.0$ for different n.

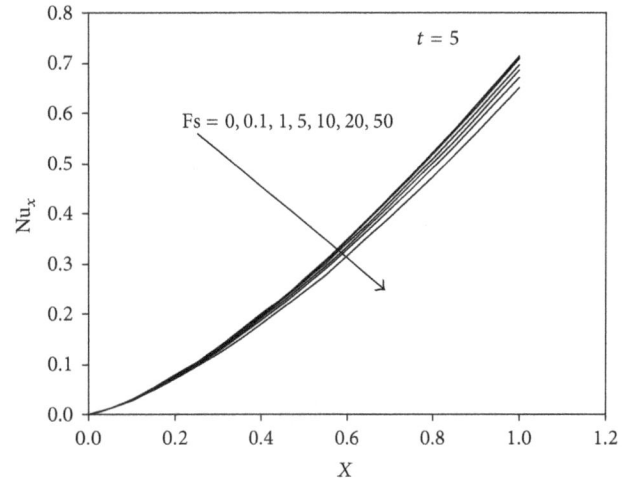

FIGURE 9: Effect of Fs on local Nusselt number at $X = 1.0$.

a decrease in the chemical molecular diffusing that is, less diffusion therefore takes place by mass transport. The dimensionless concentration profiles all decay from a maximum concentration to zero in the freestream. Greater Sc values correspond to lower chemical molecular diffusivity of the parent fluid so that less diffusion of the species occurs in the regime. Concentration boundary layer thickness will therefore be reduced. For low Sc fluid greater species diffusion occurs and concentration boundary layer thickness increased. For Sc = 1, the Concentration and velocity boundary layers will have approximately the same thickness that is, species and momentum will be diffused at the same rates. With lower Sc values the decay of concentration from the cone surface is more controlled, for increasing values of Sc the profiles descend more and more steeply and concentration falls faster from the surface to a short distance into the boundary layer regime.

The effect of surface heat flux power exponent (m) on the steady state temperature (T) is shown in Figure 7. An increase in the value of m reduces the temperature. It is also seen that the time required to reach the steady state temperature is more at lower values of m. Figure 8 depict the distribution of concentration (C) with radial coordinate (Y) for various values of the surface mass flux power law exponent (n). The concentration reduces with the increasing n values from 0.0 through 0.25, 0.50, 0.75 and 1.0. Increasing Fs clearly reduces the local Nusselt number as shown in Figure 9.

A slight increase in local Nusselt number accompanies the increment in Pr as shown in Figure 10. The influence of the concentration to thermal buoyancy ratio parameter (N), on dimensionless temperature (T) with radial coordinate (Y) is shown in Figure 11. $N = 0$ indicates that thermal and species buoyancy forces are both absent. For $N > 0$, thermal and species buoyancy forces aid each other. $N = 1$ implies that both buoyancy forces are of the same order of magnitude. A rise in N from 0.0 through 1.0, 2.0, 3.0 and 5.0 induces a retarding effect on the flow in the porous regime that is,

velocities are decreased. Increasing N (thermal and concentration buoyancy forces assisting each other) decreases temperatures in the regime that is, cools the boundary layer regime. The effect of semivertical angle of the cone (ϕ) on dimensionless temperature (T) with Y-coordinate is shown in Figure 12. It is observed that a rise in ϕ substantially increases the temperature T in the boundary layer regime. And more time is required to reach the steady state. Figure 13 the influence of magnetic parameter (M) versus spanwise spatial distributions of velocity U are depicted. Application of magnetic field normal to the flow of an electrically conducting fluid gives rise to a resistive force that acts in the direction opposite to that of the flow. This force is called the Lorentz force. This resistive force tends to slow down the motion of the fluid along the cone and causes an increase in its temperature and a decrease in velocity as M increases. An increase in M from 1 though 2, 3, 4 clearly reduces streamwise velocity U both in the near-wall regime and far-field regime of the boundary layer.

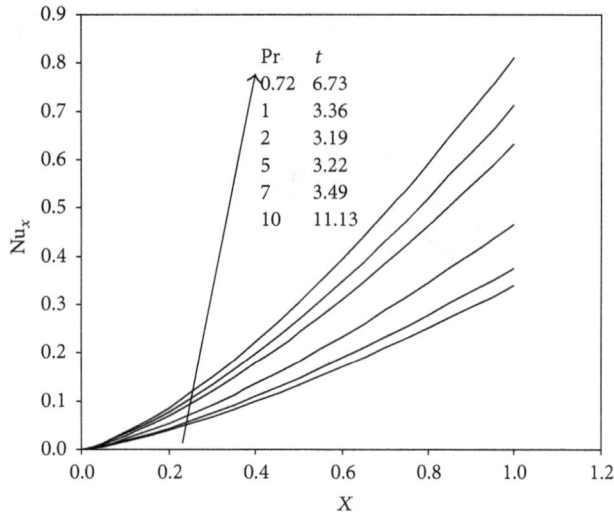

FIGURE 10: Effect of Pr on local Nusselt number at $X = 1.0$.

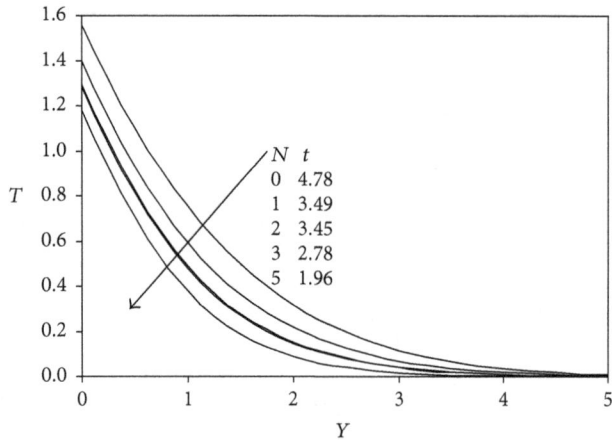

FIGURE 12: Steady state temperature profiles at $X = 1.0$ for different ϕ.

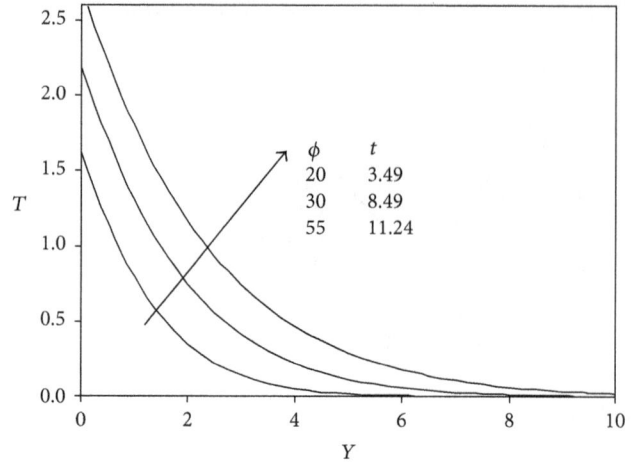

FIGURE 11: Steady state temperature profiles at $X = 1.0$ for different N.

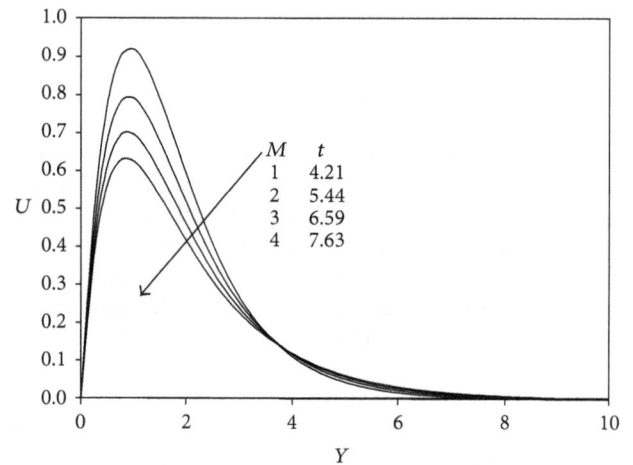

FIGURE 13: Steady state velocity profiles at $X = 1.0$ for different values of M.

5. Conclusions

Numerical solutions have been presented for the buoyancy-driven unsteady natural convection boundary layer flow past a vertical cone embedded in a non-Darcian isotropic porous regime. Present results are compared with those of [8] and found to be in excellent agreement. The following conclusions are drawn.

(i) Increasing Grashof number boosts the translational velocity in the cone surface regime and decreases temperature throughout the flow regime.

(ii) Increasing Darcy number accelerates the flow, that is, increases translational velocities. However, the temperature is reduced with a rise in Darcy number.

(iii) An increase in the Forchheimer inertial drag parameter is observed to slightly increase the temperature but reduces both velocity and local Nusselt number.

(iv) An increase in Prandtl number is observed to decrease both temperature and velocity, but

the concentration is slightly increased. A slight increase in local Nusselt number accompanies the increment in Pr.

(v) The concentration is observed to significantly decrease with an increase in Schmidt number.

(vi) The temperature is observed to decrease with an increase in buoyancy ratio parameter but decrease with an increase in semivertical angle of the cone. The time taken to reach the steady state increases with increasing ϕ.

Nomenclature

x, y: Coordinates along the cone generator and normal to the generator

u, v: Velocity components along the x- and y-directions

g: Gravitational acceleration

r: Local radius of cone

t' :   Time

t: Dimensionless time

T':   Temperature

T: Dimensionless temperature

C':   Concentration

C: Dimensionless concentration

D: Mass diffusion coefficient

K: Permeability of porous medium

q_w: Heat flux (i.e., heat transfer rate per unit area)

q_w^*:   Mass flux (i.e., mass transfer rate per unit area)

k: Thermal conductivity of fluid

L: Reference length

X, Y: Dimensionless coordinates along the cone generator and normal to the generator

U, V: Dimensionless velocity components along the x- and y-directions

b: Forchheimer geometrical constant

Da: Darcy number

Fs: Forchheimer number

Gr_L: Grashof number

M: Magnetic parameter

B_0:   Magnetic field strength

Pr: Prandtl number

N: Buoyancy ratio parameter

Sc: Schmidt number

m: Power-law index for surface heat flux relation

n: Power-law index for surface mass flux relation

Nu_x: Local Nusselt number

Nu_X: Dimensionless local Nusselt number

Sh_x: Local Sherwood number

Sh_X: Nondimensional local Sherwood number

R: Dimensionless local radius of cone.

Greek Symbols

μ: Dynamic viscosity of fluid

v: Kinematic viscosity of fluid

ϕ: Semivertical cone angle

α: Thermal diffusivity

β: Volumetric thermal expansion coefficient

θ: Dimensionless temperature function

τ: Dimensionless time

τ_X: Dimensionless local shear stress function (skin friction).

Subscripts

w: Condition on the wall

∞:   Free stream condition.

References

[1] D. B. Ingham and I. I. Pop, *Transport Phenomena in Porous Media II*, Pergamon Press, Oxford, UK.

[2] D. A. Nield and A. Bejan, *Convection in Porous Media*, Springer, New York, NY, USA, 3rd edition, 2006.

[3] K. Vafai, *Handbook of Porous Media*, CRC Press, Boca Raton, Fla, USA, 2nd edition, 2005.

[4] O. V. Trevisan and A. Bejan, "Combined heat and mass transfer by natural convection in a porous medium," *Advances in Heat Transfer*, vol. 20, pp. 315–352, 1990.

[5] K. Vafai and C. L. Tien, "Boundary and inertia effects on flow and heat transfer in porous media," *International Journal of Heat and Mass Transfer*, vol. 24, no. 2, pp. 195–203, 1981.

[6] C.-H. Chen and C.-K. Chen, "Non-darcian mixed convection along a vertical plate embedded in a porous medium," *Applied Mathematical Modelling*, vol. 14, no. 9, pp. 482–488, 1990.

[7] C. K. Chen, C. Chien-Hsin, W. J. Minkowycz, and U. S. Gill, "Non-Darcian effects on mixed convection about a vertical cylinder embedded in a saturated porous medium," *International Journal of Heat and Mass Transfer*, vol. 35, no. 11, pp. 3041–3046, 1992.

[8] M. A. Hossain and S. C. Paul, "Free convection from a vertical permeable circular cone with non-uniform surface temperature," *Acta Mechanica*, vol. 151, no. 1-2, pp. 103–114, 2001.

[9] M. A. Hossain and S. C. Paul, "Free convection from a vertical permeable circular cone with non-uniform surface heat flux," *Heat and Mass Transfer*, vol. 37, no. 2-3, pp. 167–173, 2001.

[10] A. J. Chamkha, A.-R. A. Khaled, and O. Al-Hawaj, "Simultaneous heat and mass transfer by natural convection from a cone and a wedge in porous media," *Journal of Porous Media*, vol. 3, no. 2, pp. 155–164, 2000.

[11] K. Vajravelu and J. Nayfeh, "Hydromagnetic convection at a cone and a wedge," *International Communications in Heat and Mass Transfer*, vol. 19, no. 5, pp. 701–710, 1992.

[12] R. Muthukumaraswamy and P. Ganesan, "Unsteady flow past an impulsively started vertical plate with heat and mass transfer," *Heat and Mass Transfer*, vol. 34, no. 2-3, pp. 187–193, 1998.

[13] S. Gouse, *Mohiddin Computational Fluid Dynamics*, LAP Lambert Academic Publishing, Saarbrücken, Germany, 2011.

[14] S. G. Mohiddin, V. R. Prasad, S. V. K. Varma, and O. Anwar Bég, "Numericalstudy of unsteady free convective heat and mass transfer in a walters-b viscoelastic flow along a vertical cone," *International Journal of Applied Mathematics and Mechanics*, vol. 6, pp. 88–114, 2010.

[15] P. Bapuji, K. Ekambavanan, and A. J. Chamkha, "Unsteady laminar free convection from a vertical cone with uniform surface heat flux," *Nonlinear Analysis*, vol. 13, pp. 47–60, 2008.

[16] B. Carnahan, H. A. Luther, and J. O. Wilkes, *Applied Numerical Methods*, John Wiley & Sons, New York, NY, USA, 1969.

[17] B. Gebhart, Y. Jaluria, R. L. Mahajan, and B. Sammakia, *Buoyancy—Induced Flows and Transport*, Hemisphere, New York, NY, USA, 1998.

9

Double-Diffusive Convection in Presence of Compressible Rivlin-Ericksen Fluid with Fine Dust

Mahinder Singh[1] and Rajesh Kumar Gupta[2]

[1] *Department of Mathematics, Govt. Post Graduate College, Seema (Rohru), Shimla, India*
[2] *Department of Mathematics, Lovely School of Engineering and Technology, Lovely Professional University, Phagwara, India*

Correspondence should be addressed to Mahinder Singh; mahinder_singh91@rediffmail.in

Academic Editor: Toshiyuki Gotoh

An investigation is made on the effect of suspended particles (fine dust) on double-diffusive convection of a compressible Rivlin-Ericksen elastico-viscous fluid. The perturbation equations are analyzed in terms of normal modes after linearizing the relevant set of equations. A dispersion relation governing the effects of viscoelasticity, compressibility, stable solute gradient, and suspended particles is derived. For stationary convection, Rivlin-Ericksen fluid behaves like an ordinary Newtonian fluid due to the vanishing of the viscoelastic parameter. The stable solute gradient compressibility has a stabilizing effect on the system whereas suspended particles hasten the onset of thermosolutal instability. The Rayleigh numbers and the wave numbers of the associated disturbances for the onset of instability as stationary convection are obtained and the behaviour of various parameters on Rayleigh numbers has been depicted graphically. It has been observed that oscillatory modes are introduced due to the presence of viscoelasticity, suspended particles, and stable solute gradient which were not existing in the absence of these parameters.

1. Introduction

A layer of Newtonian fluid heated from below, under varying assumptions of hydrodynamics, has been treated in detail by Chandrasekhar [1]. Chandra [2] performed careful experiments in an air layer and found contradiction between the theory and the experiment. He found that the instability depended on the depth of the layer. A Bénard-type cellular convection with fluid descending at the cell centre was observed when predicted gradients were imposed, if the layer depth was more than 10 mm. But if the layer of depth was less than 7 mm, convection occurred at much lower gradients than predicted and appeared as irregular strips of elongated cells with fluid rising at the centre. Chandra called this motion columnar instability. The effect of particle mass and heat capacity on the onset of Bénard convection has been considered by Scanlon and Segel [3]. They found that the critical Rayleigh number was reduced solely because the heat capacity of the clean gas was supplemented by that of the particles. The effect of suspended particles was found to destabilize the layer. Palniswamy and purushotham [4] have considered the stability of shear flow of stratified fluids with fine dust and have found the effect of fine dust to increase the region of instability. A study of double-diffusive convection with fine dust has been made by Sharma and Rani [5]. Kumar et al. [6] have studied effect of magnetic field on thermal instability of rotating Rivlin-Ericksen viscoelastic fluid, in which effect of magnetic field has stabilizing as well as destabilizing effect on the system. Also, Rayleigh-Taylor instability of Rivlin-Ericksen elastico-viscous fluid through porous medium has been considered by Sharma et al. [7]. They have studied the stability aspects of the system. The effects of a uniform horizontal magnetic field and a uniform rotation on the problem have also been considered separately. Kumar [8] has also studied the stability of superposed viscous-viscoelastic Rivlin-Ericksen fluids in presence of suspended particles through a porous medium. In one other study, Kumar and Singh [9] have studied the stability of superposed viscous-viscoelastic fluids through porous medium, in which effects of uniform horizontal magnetic field and a uniform rotation are considered. Kumar et al. [10] have also studied hydero-dynamic and hyderomagnetic stability of Rivlin-Ericksen fluid and found that the growth rates decrease as well as increase with the increase in kinematic viscosity and kinematic viscoelasticity in absence and presence of magnetic field. Singh and Gupta [11] have studied thermal instability of Rivlin-Ericksen elastico viscous fluid permeated with suspended particles in hydrodynamics in a porous medium and found that magnetic field has only stabilizing effect whereas medium permeability has a destabilizing effect on the system. EI-Sayed et al. [12] have studied nonlinear Kelvin-Helmholtz instability of Rivlin-Ericksen viscoelastic electrified fluid particle mixtures saturating porous medium and, in one another study, Kumar and Mohan [13] have also studied double-diffusive convection in compressible viscoelastic fluid through Brinkman porous media.

The present paper attempts to study the stability of double-diffusive convection Rivlin-Ericksen elastico-viscous fluids permeated with suspended particles. Viscosity is a function of space and time in a large variety of fluid flows, and its variation can have a dramatic effect on flow stability. In this paper, instability due to double-diffusive effects in viscosity permeated with suspended particles flow has been discussed. Double-diffusive systems are known to display a rich variety of instability behavior in density permeated with suspended particles fluid flow system. In viscosity permeated systems, it was found that stable flow in the context of single component systems becomes unstable due to double-diffusive effect. Many interesting flow patterns arise due to this instability; these aspects form the motivation for the present study.

2. Formulation of the Problem and Perturbation Equations

We have considered an infinite, horizontal, and compressible electrically conducting Rivlin-Ericksen elastico-viscous fluid permeated with suspended particles, bounded by the planes $z = 0$ and $z = d$. This layer is heated from below so that temperature at bottom (at $z = 0$) and at the upper layer (at $z = d$) is T_0 and T_d, respectively, and that a steady adverse temperature gradient $\beta(= |dT/dz|)$ and solute gradient $\beta'(= |dC/dz|)$ are maintained. Here, $\vec{g}(0, 0, -g)$ denotes the acceleration due to gravity. The effect of fluid compressibility, even small in magnitude, is also considered.

Let ρ, μ, μ', and $\vec{u}(u, v, w)$ denote the density, viscosity, viscoelasticity, and velocity of pure fluid, and let $\vec{v}(\vec{x}, t)$ and $N(\vec{x}, t)$ denote the velocity and number density of the suspended particles, $\vec{x}(x, y, z)$ and $\lambda(0, 0, 1)$; $K = 6\pi\mu\eta$, η being particle radius, is the Stokes drag coefficient. Then, the equations of motion and continuity governing the flow are

$$\rho \left[\frac{\partial \vec{u}}{\partial t} + (\vec{u} \cdot \nabla) \vec{u} \right] = -\nabla p + \rho \vec{g} + KN (\vec{v} - \vec{u})$$
$$+ \left(\mu + \mu' \frac{\partial}{\partial t} \right) \nabla^2 \vec{u}, \tag{1}$$

$$\nabla \cdot \vec{u} = 0. \tag{2}$$

Assuming uniform particle size, spherical shape, and small relative velocities between the fluid and particles, the presence of particles adds an extra force term in equation of

motion (1), proportional to the velocity difference between particles and fluid. Since the force exerted by the fluid on the particles is equal and opposite to that exerted by the particles on the fluid, there must be an extra force term, equal in magnitude but opposite in sign, in the equations of motion of the particles. The distance between the particles is assumed quite large compared with their diameters, so that interparticle reactions are ignored. The buoyancy forces on the particles are neglected. If mN is the mass of the particles per unit volume, then the equations of motion and continuity for the particles, under the above assumptions, are

$$mN \left[\frac{\partial \vec{v}}{\partial t} + (\vec{v} \cdot \nabla) \vec{v} \right] = KN (\vec{u} - \vec{v}),$$
$$\frac{\partial N}{\partial t} + \vec{\nabla} \cdot (N\vec{v}) = 0. \tag{3}$$

Let C_v, C_{pt}, C_p, T, and q denote the heat capacity of fluid at constant volume, heat capacity of particles, and heat capacity of fluid at constant pressure, temperature, and effective thermal conductivity of the pure fluid, respectively. Hence, the volume fractions of the particles are assumed to be small; the effective properties of the suspension are taken to be those of the clean fluid. If we assume that the particles and fluid are in the thermal equilibrium, the equation of heat conduction gives

$$\rho C_v \left[\frac{\partial}{\partial t} + \vec{u} \cdot \nabla \right] T + mNC_{pt} \left(\frac{\partial}{\partial t} + \vec{v} \cdot \nabla \right) T = q\nabla^2 T. \tag{4}$$

If C denotes the solute concentration, the equation of solute conduction gives

$$\rho C'_v \left[\frac{\partial}{\partial t} + \vec{u} \cdot \nabla \right] C + mNC'_{pt} \left[\frac{\partial}{\partial t} + \vec{v} \cdot \nabla \right] C = q'^{\nabla^2 C}, \tag{5}$$

where C'_v, C'_{pt}, and q' denote the analogous solute quantities.

Spiegel and Veronis [14] defined f as any one of the state variables (pressure p, density ρ, or temperature T) and expressed these in the form

$$f(x, y, z, t) = f_m + f_0(z) + f'(x, y, z, t), \tag{6}$$

where f_m is the constant space average of f, f_0 is the variation in the absence of motion, and f' is the fluctuation resulting from motion.

The initial state of the system is taken to be quiescent layer (no settling) with a uniform particle distribution N_0, therefore a state in which the density, pressure, temperature, solute concentration, and velocity at any point in the fluid are given by $\rho = \rho(z)$, $p = p(z)$, $T = T(z)$, $C = C(z)$, $\vec{v} = 0$, $\vec{u} = 0$, $N_0 = $ constant, respectively, where

$$T(z) = T_0 - \beta z, \qquad C(z) = C_0 - \beta' z,$$

$$P(z) = p_m - g \int_0^z (\rho_m - \rho_0)\, dz,$$

$$\rho(z) = \rho_m \left[1 - \alpha_m (T - T_m) + \alpha'_m (C - C_m) \right.$$

$$\left. + K_m (p - p_m) \right],$$

$$\alpha_m = -\left(\frac{1}{\rho} \frac{\partial \rho}{\partial T} \right)_m \ (= \alpha \text{ say}),$$

$$\alpha'_m = -\left(\frac{1}{\rho} \frac{\partial \rho}{\partial C} \right)_m \ (= \alpha' \text{ say}),$$

$$K_m = -\left(\frac{1}{\rho} \frac{\partial \rho}{\partial p} \right)_m.$$

(7)

Consider a small perturbation on the steady state solution and let $\delta p, \delta \rho, \theta, \gamma, \vec{u}(u, v, w), \vec{v}(1 \cdot r, s)$, and N denote, respectively, the perturbations in pressure p, density ρ, temperature T, solute concentration C, fluid velocity $\vec{u}(0, 0, 0)$, particle velocity $\vec{v}(0, 0, 0)$, and number density N_0. The change in density $\delta \rho$, caused mainly by the perturbations θ and γ in temperature and solute concentration, is given by

$$\delta \rho = -\rho_m \left(\alpha \theta - \alpha' \gamma \right).$$

(8)

Then the linearized perturbation equations relevant to the problem, Spiegel and Veronis [14], Scanlon and Segel [3], and Rivlin and Ericksen [15], become

$$\frac{\partial \vec{u}}{\partial t} = -\frac{1}{\rho_m} \nabla \delta p + g \left(\alpha \theta - \alpha' \gamma \right) \lambda + \frac{KN}{\rho_m} (\vec{v} - \vec{u})$$

$$+ \left(\nu + \nu' \frac{\partial}{\partial t} \right) \nabla^2 \vec{u},$$

(9)

$$\nabla \cdot \vec{u} = 0,$$

(10)

$$\left(\frac{m}{K} \frac{\partial}{\partial t} + 1 \right) \vec{v} = \vec{u},$$

(11)

$$\frac{\partial N}{\partial t} + \nabla \cdot (N_0 \vec{v}) = 0,$$

(12)

$$(1 + h) \frac{\partial \theta}{\partial t} = \beta \left(\frac{G-1}{G} \right) (w + hs) + \kappa \nabla^2 \theta,$$

(13)

$$\left(1 + h' \right) \frac{\partial \gamma}{\partial t} = \beta' (w + h's) + \kappa' \nabla^2 \gamma,$$

(14)

where $\mu, \mu', \nu = \mu/\rho_m, \nu' = \mu'/\rho_m, \kappa (= q/\rho_m C_v)$, and $\kappa' (= q'/\rho_m C'_v)$ stand for viscosity, viscoelasticity, kinematic viscosity, kinematic viscoelasticity, thermal diffusivity, and analogous solute diffusivity, respectively.

Also, $h = f(C_{pt}/C_v)$, $h' = f(C'_{pt}/C'_v)$, $f = mN_0/\rho_m$, and $G = C_p \beta/g$.

Initially, $\vec{u} = (0, 0, 0)$, $\vec{v} = (0, 0, 0)$, $T = T(z)$, and $N = N_0$, so, (4) yields $0 = 0$, identically.

After perturbation, (4) becomes

$$(\rho_m + \delta \rho) C_v \left(\frac{\partial}{\partial t} + \vec{u} \cdot \nabla \right) (T + \theta) + (mN_0 + mN) C_{pt}$$

$$\times \left(\frac{\partial}{\partial t} + \vec{v} \cdot \nabla \right) (T + \theta) = q \nabla^2 (T + \theta).$$

(15)

Follow Speigal and Veronis [14] where the flow equations are found to be the same as those for incompressible fluids except that the static temperature gradient is replaced by its excess over the adiabatic and C_v is replaced by C_p, that is, β is replaced by $(\beta - (g/C_p))$, and linearizing (4) gives

$$\frac{\partial \theta}{\partial t} + \frac{mN_0}{\rho_m} \frac{C_{pt}}{C_v} \frac{\partial \theta}{\partial t} = \left(\beta - \frac{g}{C_p} \right) (w + hs) + \frac{q}{\rho_m C_v} \nabla^2 \theta,$$

(16)

that is, (13). However, β' remains unaltered and, as above, (5) yields (14).

3. The Dispersion Relation

Analyzing the disturbances into normal modes, we assume that the perturbation quantities are of the form

$$[w, \theta, \gamma] = [W(z), \Theta(z), \Gamma(z)] \exp \left[ik_x x + ik_y y + nt \right],$$

(17)

where k_x and k_y are wave numbers along x- and y-directions, respectively, $k(= \sqrt{k_x^2 + k_y^2})$ is the resultant wave number, and n is the growth rate, which is, in general a complex constant. Using (15), (9)–(14) in nondimensional form become

$$\left[\sigma \left(1 + \frac{M}{1 + \tau_1 \sigma} \right) - (1 + F\sigma) \left(D^2 - a^2 \right) \right] \left(D^2 - a^2 \right) W$$

$$+ \frac{ga^2 d^2}{\nu} \left(\alpha \Theta - \alpha' \Gamma \right) = 0,$$

(18)

$$\left(D^2 - a^2 - Hp_1 \sigma \right) \Theta = -\beta \left(\frac{G-1}{G} \right) \frac{d^2}{\kappa} \frac{(H + \tau_1 \sigma)}{(1 + \tau_1 \sigma)} W,$$

(19)

$$\left(D^2 - a^2 - H'q\sigma \right) \Gamma = -\beta' \frac{d^2}{\kappa'} \frac{(H' + \tau_1 \sigma)}{(1 + \tau_1 \sigma)} W,$$

(20)

where we have put $a = kd$, $\sigma = nd^2/\nu$, $\tau = m/\kappa$, $\tau_1 = \tau \nu/d^2$, $M = mN/\rho_m$, $p_1 = \nu/\kappa$, $q = \nu/\kappa'$, $H = 1 + h$, $H' = 1 + h'$, $F = \nu'/d^2$, and $D = d/dz$.

Eliminating Θ and Γ between (18) and (20), we obtain

$$\left[\sigma\left(1 + \frac{M}{1+\tau_1\sigma}\right) - (1+F\sigma)\left(D^2 - a^2\right)\right]\left(D^2 - a^2 - Hp_1\sigma\right)$$

$$\times \left(D^2 - a^2 - H'q\sigma\right)\left(D^2 - a^2\right)W$$

$$- R\left(\frac{G-1}{G}\right)a^2\left(\frac{H+\tau_1\sigma}{1+\tau_1\sigma}\right)\left(D^2 - a^2 - H'q\sigma\right)W$$

$$+ Sa^2\left(\frac{H'+\tau_1\sigma}{1+\tau_1\sigma}\right)\left(D^2 - a^2 - Hp_1\sigma\right)W = 0,$$

$$\tag{21}$$

where $R = g\alpha\beta d^4/\nu\kappa$ is the thermal Rayleigh number, $S = g\alpha'\beta'd^4/\nu\kappa'$ is the analogous solute Rayleigh number, $p_1 = \nu/\kappa$ is the thermal Prandtl number, and $q = \nu/\kappa'$ is the analogous Schmidt number.

We consider the case where both boundaries are free and perfect conductors of heat and solute, while the adjoining medium is assumed to be electrically nonconducting. The appropriate boundary conditions for the case are

$$W = D^2W = \Theta = \Gamma = DZ = 0 \quad \text{at } z = 0, 1. \tag{22}$$

The case of two free boundaries though little artificial is the most appropriate for stellar atmospheres. Using (22), we can show that all the even order derivatives of W must vanish for $z = 0$ and 1 and hence the proper solution of W characterizing the lowest mode is

$$W = W_0 \sin \pi z, \tag{23}$$

where W_0 is a constant.

Substituting (23) in (21), we obtain the dispersion relation

$$R_1 x = \left(\frac{G}{G-1}\right)$$

$$\times \left[\left\{i\sigma_1\left(1 + \frac{M}{1+i\tau_1\sigma\pi^2}\right) + \left(1 + iF\sigma\pi^2\right)(1+x)\right\}\right.$$

$$\times \left\{\frac{\left(1+i\tau_1\sigma\pi^2\right)(1+x)\left(1+x+iHp_1\sigma\right)}{H+i\tau_1\sigma\pi^2}\right\}$$

$$\left. + \frac{S_1 x\left(H' + i\tau_1\sigma\pi^2\right)(1+x+iHp_1\sigma)}{\left(H+i\tau_1\sigma\pi^2\right)\left(1+x+iH'q\sigma\right)}\right],$$

$$\tag{24}$$

where

$$R_1 = \frac{R}{\pi^4}, \qquad x = \frac{a^2}{\pi^2},$$

$$i\sigma_1 = \frac{\sigma}{\pi^2}, \qquad S_1 = \frac{S}{\pi^4}. \tag{25}$$

Equation (24) is the required dispersion relation studying the effects of suspended particles and compressibility on the thermosolutal convection in Rivlin-Ericksen elastico-viscous fluid.

4. The Stability of the System and Oscillatory Modes

Here, we examine whether the instability can occur as oscillatory modes, if any, on the Rivlin-Ericksen elastico-viscous fluid in the presence of compressibility and suspended particles effects.

Multiplying (18) by W^*, the complex conjugate of W, integrating over the range of z, and making use of (19) and (20) with the help of boundary conditions (22), we obtain

$$\sigma\left(1 + \frac{M}{1+\tau_1\sigma}\right)I_1 + (1+F\sigma)I_2$$

$$- \frac{g\alpha a^2\kappa}{\nu\beta}\left(\frac{G}{G-1}\right)\left(\frac{1+\tau_1\sigma^*}{H+\tau_1\sigma^*}\right)\left(I_3 + Hp_1\sigma^* I_4\right)$$

$$+ \frac{g\alpha' a^2\kappa'}{\nu\beta'}\left(\frac{1+\tau_1\sigma^*}{H'+\tau_1\sigma^*}\right)$$

$$\times \left(I_5 + H'q\sigma^* I_6\right) = 0,$$

$$\tag{26}$$

where

$$I_1 = \int_0^1 \left(|DW|^2 + a^2|w|^2\right),$$

$$I_2 = \int_0^1 \left(\left|D^2W\right|^2 + 2a^2|Dw|^2 + a^4|W|^2\right),$$

$$I_3 = \int_0^1 \left(|D\Theta|^2 + a^2|\Theta|^2\right)dz,$$

$$I_4 = \int_0^1 |\Theta|^2 dz, \tag{27}$$

$$I_5 = \int_0^1 \left(|D\Gamma|^2 + a^2|\Gamma|^2\right)dz,$$

$$I_6 = \int_0^1 |\Gamma|^2 dz.$$

The integrals $I_1 - I_6$ are all positive definite. Putting $\sigma = i\sigma_i$, where σ_i is real and equating the imaginary parts, we obtain

$$\sigma_i\left[\left(1 + \frac{M}{1+\tau_1\sigma_i}\right)I_1 + FI_2 + \frac{g\alpha a^2\kappa}{\nu\beta}\left(\frac{G}{G-1}\right)\right.$$

$$\times \left(\frac{\tau_1(H-1)}{H^2 + \tau_1^2\sigma_i^2}I_3 + \frac{H+\tau_1^2\sigma_i^2}{H^2 + \tau_1^2\sigma_i^2}Hp_1\sigma^* I_4\right)$$

$$- \frac{g\alpha' a^2\kappa'}{\nu\beta'}\left(\frac{\tau_1(H'-1)}{H'^2 + \tau_1^2\sigma_i^2}I_5\right.$$

$$\left.\left. + \frac{H'+\tau_1^2\sigma_i^2}{H'^2 + \tau_1^2\sigma_i^2}H'q\sigma^* I_6\right)\right] = 0.$$

$$\tag{28}$$

Equation (28) implies that $\sigma_i = 0$ or $\sigma_i \neq 0$, which mean that modes may be nonoscillatory or oscillatory.

The oscillatory modes are introduced due to presence of stable solute gradient, which were nonexistent in its absence.

5. The Stationary Convection

When instability sets in as stationary convection, the marginal state will be characterized by $\sigma = 0$. Putting $\sigma = 0$ in (24), the dispersion relation reduces to

$$R_1 = \left(\frac{G}{G-1}\right)\left[\frac{(1+x)^3}{xH} + S_1\frac{H'}{H}\right], \qquad (29)$$

and Rivlin-Ericksen elastico-viscous fluid behaves like an ordinary Newtonian fluid.

To study the effect of stable solute gradient and suspended particles, we examine the behaviour of dR_1/dS_1 and dR_1/dH analytically.

Equation (29) yields

$$\frac{dR_1}{dS_1} = \left(\frac{G}{G-1}\right)\frac{H'}{H}, \qquad (30)$$

which is positive, thereby Rayleigh number increases the with increase in solute parameter. The stable solute gradient, therefore, has a stabilizing effect on the system

$$\frac{dR_1}{dH} = -\left(\frac{G}{G-1}\right)\left[\frac{(1+x)^3}{x} + S_1 H'\right]\frac{1}{H^2}, \qquad (31)$$

which is negative, implying thereby that the Rayleigh number decreases with the increase in the suspended particles number density. Therefore, the effect of suspended particles is to destabilize the system. We studied here these effects graphically as shown in Figure 1.

In Figure 1, as value of stable solute gradient parameter increased, the value of Rayleigh number is increased by taking values of wave number $x(= 1, 2, 3, 4, 5)$, for fixed values $G = 9.8$, $H = 2$, $H' = 10$, and $S_1(= 10, 20, 30)$, respectively. Therefore value of Rayleigh number increased with the increase in wave number showing the stabilizing effect on the system.

In Figure 2, Rayleigh number decreased with the increase in the suspended particles by taking values of wave number $x(= 1, 2, 3, 4, 5)$, for fixed values $G = 9.8$, $S_1 = 10$, $H' = 5$, and $H(= 2, 4, 6)$, respectively. Therefore values of Rayleigh number have increased with the decrease of suspended particles parameter, showing the destabilizing effect on the system.

For fixed S_1, H, and H', let G (accounting for the compressibility effects) also be fixed. Then, we find that

$$\overline{R}_c = \left(\frac{G}{G-1}\right)R_c, \qquad (32)$$

where $\overline{R}_c$ and R_c denote, respectively, the critical Rayleigh number in the presence and absence of compressibility. $G > 1$ is relevant here. The cases $G < 1$ and $G = 1$ correspond to negative and infinite values of the critical Rayleigh number in the presence of compressibility, which are not relevant in the present study. The effect of compressibility is thus to postpone the onset of thermosolutal convection.

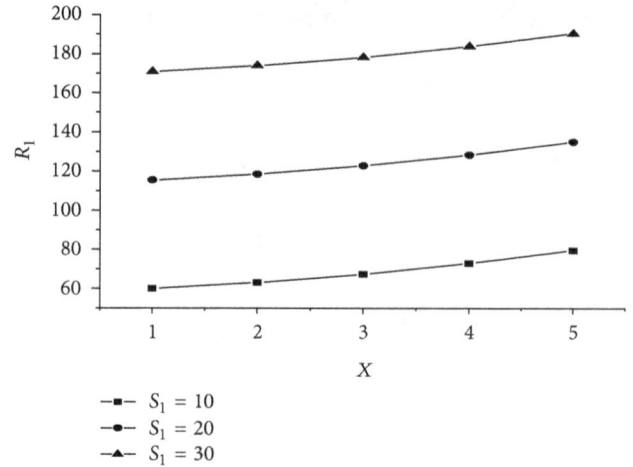

FIGURE 1: The variation of Rayleigh number (R_1) with wave number $x(= 1, 2, 3, 4, 5)$, for $G = 9.8$, $H = 2$, $H' = 10$, and $S_1(= 10, 20, 30)$.

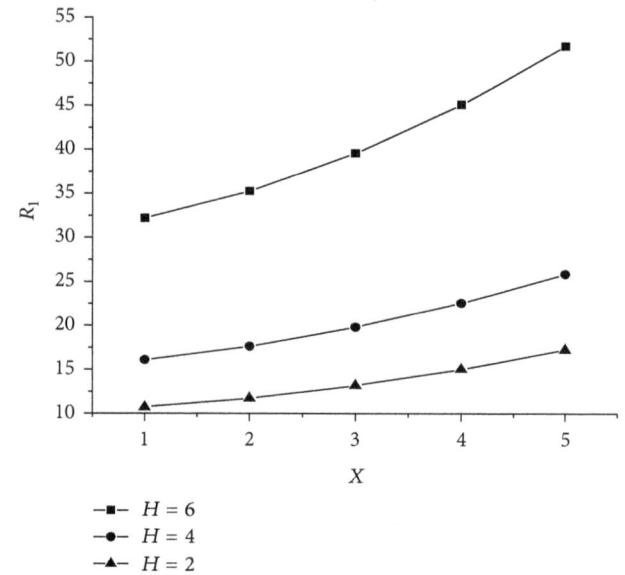

FIGURE 2: The variation of Rayleigh number (R_1) with wave number $x(= 1, 2, 3, 4, 5)$, for $G = 9.8$, $S_1 = 10$, $H' = 5$, and $H(= 2, 4, 6)$.

6. Conclusion

Combined effect of various parameters, that is, compressibility, suspended particles, and stable solute gradient, has been investigated on thermosolutal convection of a Rivlin-Ericksen fluid. The motivation for the present study is due to the fact thata fluid-particle mixture is not commensurate with their scientific and industrial importance. The analysis would be relevant to the stability of some polymer solutions and the problem finds its usefulness in several Geophysical situations and in chemical technology. Hence, a study has been made on thermosolutal convection in presence of compressible fluid with fine dust. For stationary convection, Rivlin-Ericksen fluid behaves like an ordinary Newtonian fluid due to the vanishing of the viscoelastic parameter.

From (32), it is clear that the effect of compressibility is to postpone the onset of instability. To investigate the effects of suspended particles and stable solute gradient, we examined the expressions dR_1/dH and dR_1/dS_1 analytically. Stable solute gradient postpones the onset of instability whereas suspended particles are found to hasten the onset of instability. These results are graphically verified by Figures 1 and 2, respectively. The oscillatory modes are introduced due to the presence of viscoelasticity, suspended particles, and stable solute gradient. In the absence of these, the principle of exchange of stabilities is found to hold good.

Conflict of Interests

The authors declare that there is no conflict of interests regarding the publication of this paper.

Acknowledgment

The authors are thankful to the chief editor and learned referee for his useful technical comments and valuable suggestions, which led to a significant improvement of the paper.

References

[1] S. Chandrasekhar, *Hydrodynamic and Hydromagnetic Stability*, Dover, New York, NY, USA, 1981.

[2] K. Chandra, "Instability of fluid heated from below," *Proceedings of the Royal Society A*, vol. 164, pp. 231–242, 1938.

[3] J. W. Scanlon and L. A. Segel, "Some effects of suspended particles on the onset of Bénard convection," *Physics of Fluids*, vol. 16, no. 10, pp. 1573–1578, 1973.

[4] V. I. Palaniswamy and C. M. Purushotham, "Stability of shear flow of stratified fluids with fine dust," *Physics of Fluids*, vol. 24, no. 7, pp. 1224–1228, 1981.

[5] R. C. Sharma and N. Rani, "Double-diffusive convection with fine dust," *Czechoslovak Journal of Physics B*, vol. 39, no. 7, pp. 710–716, 1989.

[6] P. Kumar, H. Mohan, and R. Lal, "Effect of magnetic field on thermal instability of a rotating Rivlin-Ericksen viscoelastic fluid," *International Journal of Mathematics and Mathematical Sciences*, vol. 2006, Article ID 28042, 10 pages, 2006.

[7] R. C. Sharma, P. Kumar, and S. Sharma, "Rayleigh-Taylor instability of Rivlin-Ericksen elastico-viscous fluid through porous medium," *Indian Journal of Physics B*, vol. 75, no. 4, pp. 337–340, 2001.

[8] P. Kumar, "Stability of superposed viscous-viscoelastic (Rivlin-Ericksen) fluids in the presence of suspended particles through a porous medium," *Zeitschrift fur Angewandte Mathematik und Physik*, vol. 51, no. 6, pp. 912–921, 2000.

[9] P. Kumar and G. J. Singh, "The stability of superposed viscous-viscoelastic fluids through porous medium," *Applications and Applied Mathematics*, vol. 5, no. 1, pp. 110–119, 2010.

[10] P. Kumar, R. Lal, and M. Singh, "Hydrodynamic and hydromagnetic stability of two stratified Rivlin-Ericksen elastico-viscous superposed fluid," *International Journal of Applied Mechanics and Engineering*, vol. 12, no. 3, pp. 645–653, 2007.

[11] M. Singh and R. Gupta, "Thermal instability of Revilin-Ericksen Elastico-Viscous fluid permeated with suspended particles in hydrodynamics in a porous medium," *International Journal of Applied Mechanics and Engineering*, vol. 16, no. 4, pp. 1169–1179, 2011.

[12] M. F. El-Sayed, N. T. Eldabe, M. H. Haroun, and D. M. Mostafa, "Nonlinear Kelvin-Helmholtz instability of Rivlin-Ericksen viscoelastic electrified fluid-particle mixtures saturating porous media," *The European Physical Journal Plus*, vol. 127, article 29, 2012.

[13] P. Kumar and H. Mohan, "Double-diffusive convection in compressible viscoelastic fluid through Brinkman porous media," *The American Journal of Fluid Mechanics*, vol. 2, pp. 1–6, 2012.

[14] E. A. Spiegel and G. Veronis, "On the Boussinesq approximation for a compressible fluid," *The Astrophysical Journal*, vol. 131, pp. 442–447, 1960.

[15] R. S. Rivlin and J. L. Ericksen, "Stress deformation relaxations for isotropic materials," *Journal of Rational Mechanics and Analysis*, vol. 4, pp. 323–425, 1955.

10

Linear Stability Analysis of Thermal Convection in an Infinitely Long Vertical Rectangular Enclosure in the Presence of a Uniform Horizontal Magnetic Field

Takashi Kitaura and Toshio Tagawa

Tokyo Metropolitan University, 6-6, Asahigaoka, Hino-shi 191-0065, Japan

Correspondence should be addressed to Toshio Tagawa; tagawa-toshio@tmu.ac.jp

Academic Editor: Robert Spall

Stability of thermal convection in an infinitely long vertical channel in the presence of a uniform horizontal magnetic field applied in the direction parallel to the hot and cold walls was numerically studied. First, in order to confirm accuracy of the present numerical code, the one-dimensional computations without the effect of magnetic field were computed and they agreed with a previous study quantitatively for various values of the Prandtl number. Then, linear stability analysis for the thermal convection flow in a square horizontal cross section under the magnetic field was carried out for the case of Pr = 0.025. The thermal convection flow was once destabilized at certain low Hartmann numbers, and it was stabilized at high Hartmann numbers.

1. Introduction

Nuclear fusion energy has received considerable attention as one of the environment-friendly in modern society. Heading towards implementation of the use of nuclear fusion energy, experimental facility called ITER (International Thermonuclear Experimental Reactor) is currently under construction in France. ITER has a toroidal shape like a donut, in which the high-temperature plasma enough to induce nuclear fusion reaction is controlled by both operation of magnetic field generated by the superconducting coils disposed around the plasma and the imposed electric current in the plasma [1]. Blankets located close to the plasma side play an important role for cooling, shielding neutrons, and fuel production.

Fusion reactor blanket can be classified into the solid blanket using a solid compound of lithium as fuel production material, or liquid blanket using liquid lithium. Above all, liquid blanket has the advantage of being of relatively simple structure [2], but, on the other hand, it has a serious problem called the MHD pressure loss [3]. The MHD pressure loss obstructs convection of liquid metal as a coolant and it depends on the direction of the magnetic field and the electric conductivity of the wall. To elucidate that problem, researches on thermal convection under the electromagnetic force have been actively conducted not only for the application of fusion reactor blankets but also for crystal growth such as the horizontal Bridgman method.

The effect of the direction of uniform magnetic field on the natural convection in a cubic enclosure heated from a vertical wall and cooled from an opposing vertical wall was numerically studied by Ozoe and Okada [5]. They showed that the horizontal magnetic field parallel to the hot and cold walls (Y-directional magnetic field) has much less influence on the damping of natural convection than the two other directions of magnetic field (X- or Z-directional magnetic field). This finding was confirmed with an experiment using 30 mm × 30 mm × 30 mm cubical box filled with molten gallium by Okada and Ozoe [6]. They noticed the heat transfer rate of natural convection under the Y-directional magnetic field indicated not only weak damping of heat transfer rate at high magnetic fields but also slight enhancement of heat transfer rate at low magnetic fields. However, they were not convinced by this enhancement of heat transfer.

After several years, Tagawa and Ozoe (1997, 1998) [7, 8] carried out both three-dimensional computations with the higher Rayleigh numbers than those by Ozoe and Okada and the molten gallium experiment using 64 mm × 64 mm × 64 mm cubical box for the Y-directional magnetic field in order to confirm the existence of enhancement of heat transfer rate at low magnetic fields. Due to the exploration of the larger Rayleigh numbers, both the numerical and experimental results showed the enhancement of heat transfer rate (the Nusselt number) when the uniform magnetic field was applied in the Y-direction.

The enhancement of Nusselt number in the presence of low magnetic fields has been reported by several other groups. Authié et al. [9] carried out both three-dimensional computations and the corresponding experiment using mercury for the buoyant convection in a long vertical enclosure. Their experiments indicated that the Nusselt number takes its maximum around the Hartmann number 200–250 for the values of the Grashof number from 3×10^7 to 1.5×10^8. Burr et al. [10] carried out the natural convection for the similar configuration under a horizontal magnetic field perpendicular to both the gravity and the applied heat flux. They showed that the Nusselt numbers increased from the value of natural convection without the magnetic field in the range of the Hartmann number $100 < Ha < 225$ with a tendency of higher Hartmann numbers at larger heat fluxes. Zhang et al. [11] carried out an experimental study on the flow structure of a bubble-driven liquid-metal jet in a horizontal magnetic field. They concluded that the application of a moderate magnetic field destabilizes the global flow and gives rise to transient, oscillating flow patterns with predominant frequencies. The electric conducting fluid motion is usually damped by the use of static magnetic field. The above-mentioned literatures concern the enhancement of natural convection heat transfer or destabilizing effect of the electric conducting fluid flow at low Hartmann numbers in the presence of static magnetic field. This interesting finding could be related to the flow instability and/or transition.

Fujimura and Nagata (1998) [12] studied instability of the natural convection in an infinitely long vertical rectangular channel in the presence of a uniform horizontal magnetic field by two-dimensional analysis. Kakutani (1964) [13] studied the hydromagnetic stability on the plane laminar flow between the parallel plates in the presence of a transverse magnetic field. The plane Couette flow is destabilized by the magnetic field in the range of Hartmann number $3.91 < Ha < 5.4$.

The purpose of this research is to clarify the stability of parallel flow of thermal convection in an infinitely long vertical rectangular channel in the presence of a horizontal magnetic field.

2. Numerical Model

The schematic model considered in this study is shown in Figure 1. The fluid treated in this research is assumed to be an incompressible Newtonian fluid, and the Boussinesq approximation is employed. The boundary conditions are

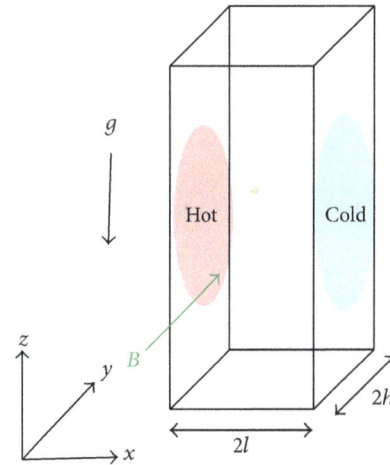

FIGURE 1: Schematic model considered.

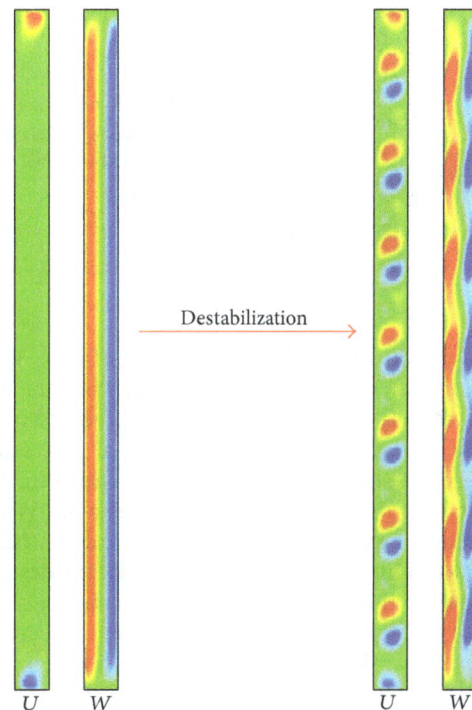

FIGURE 2: An example of destabilization of the thermal convection in a long vertical rectangular channel at Pr = 0.025 and Gr = 900.0.

the no-slip condition for all walls and the heated and cooled walls are isothermal and the other two walls are the thermally insulated. Governing equations are the equations of continuity, the momentum, the energy, Ohm's law, and the conservation of electric charge. We used a finite difference method with a staggered mesh system. By introducing the time derivative term into the governing equations, we obtain approximated solutions under discretization using a second-order or fourth-order accurate central difference method, together with the use of HS-MAC method for the solution of the Poisson equations of the pressure and the electric potential.

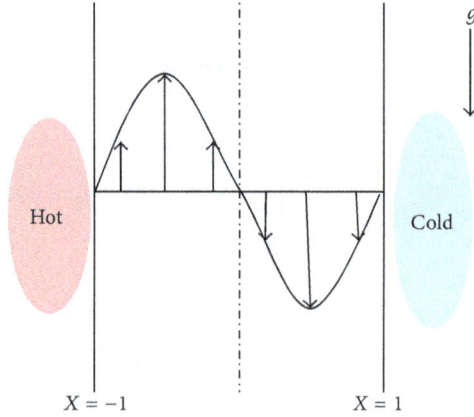

FIGURE 3: Simplified calculation model ($A = h/l \rightarrow \infty$).

TABLE 1: Comparisons of the critical Grashof number and the critical wavenumber between a reference and the present research at Pr = 0.67 and 7.5.

Method	Reference [4] Chebyshev polynomials expansion		Present Fourth-order accurate central difference method with 101 grids	
Pr	0.67	7.5	0.67	7.5
k_c	1.40417	1.38334	1.40419	1.38349
Gr_c	503.293	491.900	503.305	491.829

The instability of the convection of the infinitely long vertical rectangular channel is investigated by using the Grashof number, which is a dimensionless number representing the magnitude of the influence of the inertial and buoyancy forces. Figure 2 shows how the flow of a two-dimensional vertical rectangular channel becomes unstable. When the Grashof number exceeds a certain value, the flow structure is divided into a number of cells. In this study, the destabilized flow in the three-dimensional vertical rectangular channel is assumed to be periodic flow in the Z-axis direction, and then we perform the computation with applying a uniform horizontal magnetic field perpendicular to the temperature gradient.

3. Verification of the Numerical Code

As the verification of the numerical code, preliminary computations were performed for the model of the aspect ratio ($h/l \rightarrow \infty$) without magnetic field, and those results were compared with previous results [4]. The present model and the numerical results are described in Figures 3 and 4, respectively. Figure 4 describes the result performed with the number of grids 101 using the discretization of a fourth-order accurate central difference method.

It is noted that this graph shows result of the case of the standing wave disturbance. It is known that for the lower Prandtl number (Pr < 12.45) [14], the neutral Grashof number in the case of the standing wave disturbance is lower than that in the case of the traveling wave disturbance.

Table 1 shows the numerical results for comparison of the critical Grashof number and the critical wavenumber when air (Pr = 0.67) and water (Pr = 7.5) are assumed as test fluids. The agreement between the reference's result and the present result is negligibly small.

4. Dimensionless Equations

The flow instability in a long vertical enclosure in the presence of a magnetic field applied in the Y-direction is investigated here. In this study, the destabilized flow is assumed to be periodic and stationary in the Z-axis direction, and then the numerical analyses were performed within a horizontal cross section. Due to the computational resources, we focus on the case that the aspect ratio of the horizontal cross section is unity and the Prandtl number is 0.025, which is a typical value of liquid metals.

The dimensionless governing equations are shown below:

$$\frac{\partial \mathbf{V}}{\partial \tau} + (\mathbf{V} \cdot \nabla)\mathbf{V} = -\nabla P + \frac{1}{Gr}\nabla^2 \mathbf{V} + \frac{1}{Gr}\Theta \mathbf{e}_z + \frac{Ha^2}{Gr}\mathbf{J} \times \mathbf{e}_{B_0},$$

$$\frac{\partial \Theta}{\partial \tau} + \mathbf{V} \cdot \nabla \Theta = \frac{1}{GrPr}\nabla^2 \Theta,$$

$$\nabla \cdot \mathbf{V} = 0, \qquad \nabla \cdot \mathbf{J} = 0,$$

$$\mathbf{J} = -\nabla \Phi + \mathbf{V} \times \mathbf{e}_{B_0},$$

$$(1)$$

The dimensionless variables are defined as follows:

$$\tau = \frac{t}{l^2/(Gr\nu)}, \qquad \mathbf{X} = \frac{\mathbf{x}}{l}, \qquad \mathbf{V} = \frac{\mathbf{v}}{Gr\nu/l},$$

$$\Theta = \frac{T - T_0}{\Delta T}, \qquad P = \frac{p}{\rho_0 Gr^2 \nu^2/l^2},$$

$$\mathbf{J} = \frac{\mathbf{j}}{Gr\nu B_0}, \qquad \Phi = \frac{\phi}{Gr\sigma_e \nu B_0/l}, \qquad Gr = \frac{g\beta\Delta T l^3}{\nu^2},$$

$$Pr = \frac{\nu}{\alpha}, \qquad Ha = B_0 l\sqrt{\frac{\sigma_e}{\rho_0 \nu}}, \qquad A = \frac{h}{l}.$$

$$(2)$$

Since we limit ourselves to the case that the uniform magnetic field is applied in the Y-direction, it holds that $\mathbf{e}_{B_0} = \mathbf{e}_y$.

5. Basic Flow

First, the flow of the basic state is computed with the assumption that the rectangular enclosure is long enough to neglect recirculation area near the top and the bottom walls. Therefore, the basic state of temperature field is in heat conduction. Since the Prandtl and Grashof numbers are included in the scaling, the dimensionless basic flow is independent of the Grashof and Prandtl numbers. Dimensionless governing equations used in the computation are shown as follows.

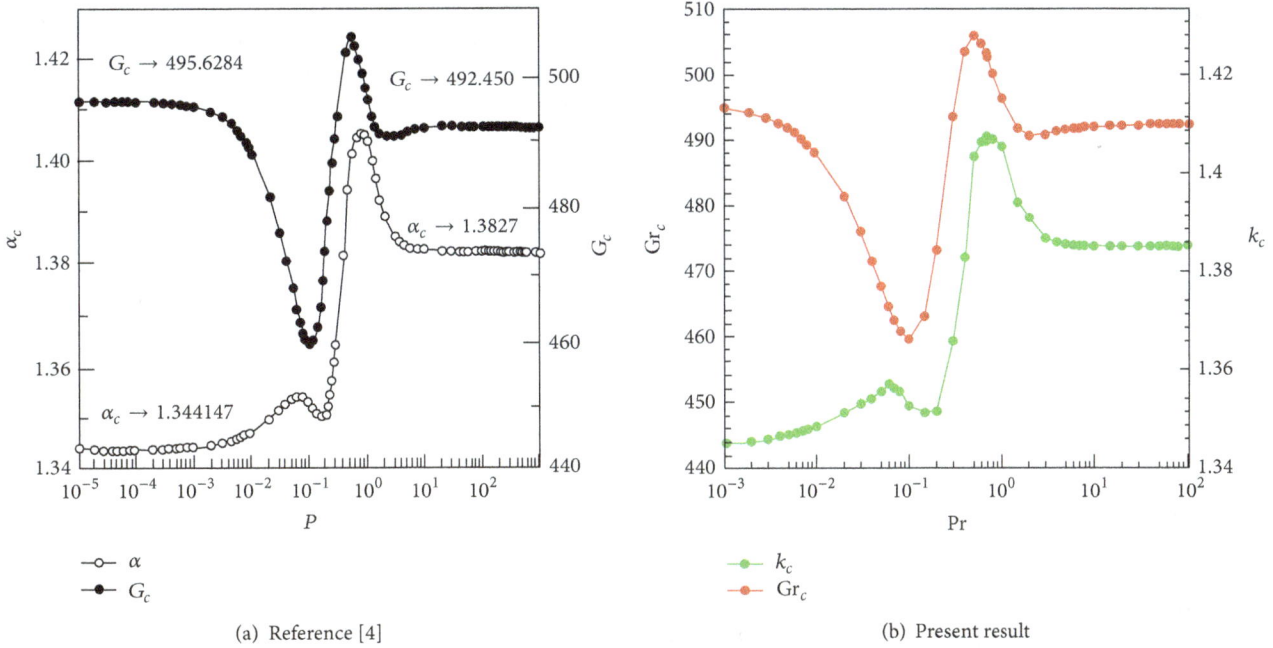

(a) Reference [4]

(b) Present result

FIGURE 4: Variations of the critical Grashof number and the critical wavenumber for the wide range of the Prandtl number. (a) is referred from a previous study [4] and (b) indicates the present result.

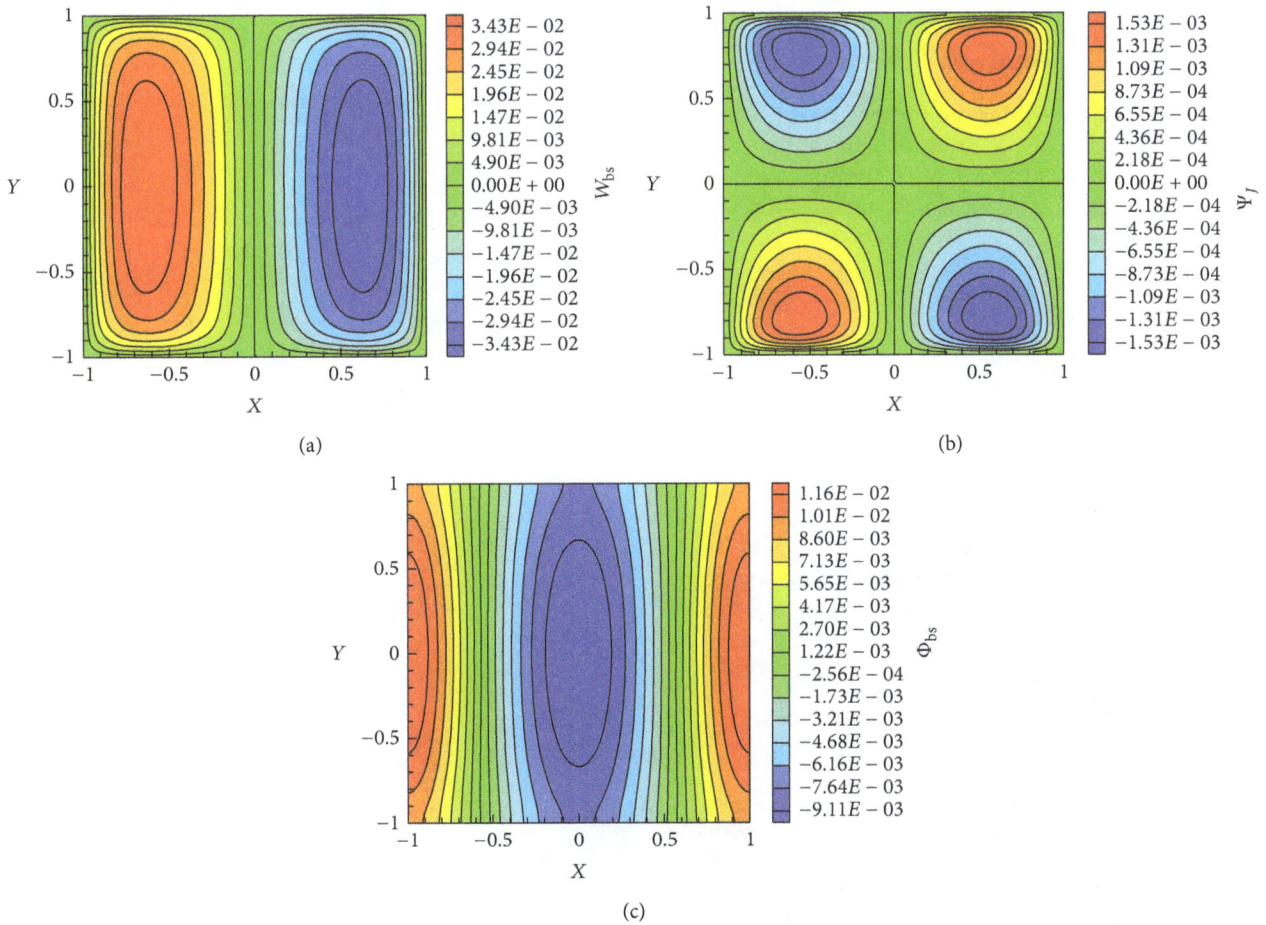

(a)

(b)

(c)

FIGURE 5: Contour maps of the basic flow at Ha = 10.0: (a) vertical velocity W_{bs}, (b) stream line of electric current Ψ_J, and (c) electric potential Φ_{bs}.

In this study, the aspect ratio of the horizontal cross section is limited to $A = h/l = 1$:

$$\frac{\partial^2 W_{bs}}{\partial X^2} + \frac{\partial^2 W_{bs}}{\partial Y^2} + \Theta_{bs} + \mathrm{Ha}^2 Jx_{bs} = 0,$$

$$\Theta_{bs} = -X,$$

$$\frac{\partial Jx_{bs}}{\partial X} + \frac{\partial Jy_{bs}}{\partial Y} = 0, \qquad (3)$$

$$Jx_{bs} = -\frac{\partial \Phi_{bs}}{\partial X} - W_{bs},$$

$$Jy_{bs} = -\frac{\partial \Phi_{bs}}{\partial Y}.$$

Here, the subscript character bs represents the basic state. The vertical component of velocity W, the streamlines of electric current density Ψ_j, and the electric potential Φ of the basic state at the Hartmann number Ha = 10.0 are depicted in Figure 5.

6. Linear Stability Analysis

Disturbance equations are derived for investigating the flow instability. First of all, velocity, pressure, temperature, electric potential, and electric current density are represented with the sum of the solution of the basic state and the infinitesimal disturbances [15]:

$$\mathbf{v} = \mathbf{v}_{bs} + \mathbf{v}', \qquad p = p_{bs} + p',$$

$$T = T_{bs} + T', \qquad \phi = \phi_{bs} + \phi', \qquad (4)$$

$$\mathbf{j} = \mathbf{j}_{bs} + \mathbf{j}'.$$

We assume that the infinitesimal disturbances are represented by the normal mode as follows:

$$\left(\mathbf{v}', p', T', \phi', \mathbf{j}'\right) = \left(\widetilde{\mathbf{v}}, \widetilde{p}, \widetilde{T}, \widetilde{\phi}, \widetilde{\mathbf{j}}\right) \exp\left(iaz + \lambda t\right). \qquad (5)$$

Note that i is the imaginary unit, λ is the complex growth rate which is a complex number, and a is the wavenumber which is a real number here. The dimensionless equations obtained by substituting (4) and (5) to (1) are shown below. It is known that the natural convection flow in a vertical slot becomes unstable for standing wave disturbance rather than for travelling wave disturbance in the range of

TABLE 2: Variation of the neutral Grashof number for several cases of number of grids at Pr = 0.025, k = 0.83, and Ha = 5.0.

Number of meshes	Gr	
	2nd-order	4th-order
30×30	2280.63	2191.22
50×50	2177.64	2152.50
70×70	2153.61	2142.76

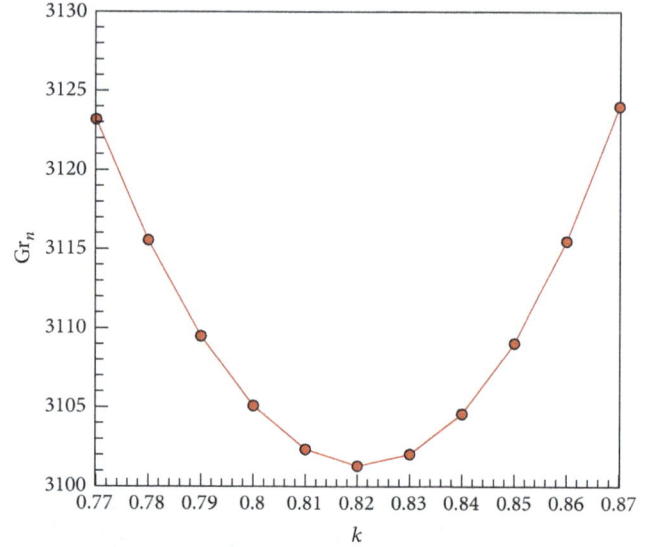

FIGURE 6: Neutral stability curve of the Grashof number at Pr = 0.025 and Ha = 10.0.

Pr < 12.45 [14]. Therefore, we assumed that the neutral stable state of the flow of Pr = 0.025 is taken as $\lambda = 0$:

$$\frac{\partial \widetilde{U}}{\partial X} + \frac{\partial \widetilde{V}}{\partial Y} + ik\widetilde{W} = 0,$$

$$iW_{bs}k\widetilde{U} = -\frac{\partial \widetilde{P}}{\partial X} + \frac{1}{\mathrm{Gr}}\left(\frac{\partial^2}{\partial X^2} + \frac{\partial^2}{\partial Y^2} - k^2\right)\widetilde{U} + \frac{\mathrm{Ha}^2}{\mathrm{Gr}}\left(-\widetilde{J}z\right),$$

$$iW_{bs}k\widetilde{V} = -\frac{\partial \widetilde{P}}{\partial Y} + \frac{1}{\mathrm{Gr}}\left(\frac{\partial^2}{\partial X^2} + \frac{\partial^2}{\partial Y^2} - k^2\right)\widetilde{V},$$

$$\widetilde{U}\frac{\partial W_{bs}}{\partial X} + \widetilde{V}\frac{\partial W_{bs}}{\partial Y} + iW_{bs}k\widetilde{W}$$

$$= -ik\widetilde{P} + \frac{1}{\mathrm{Gr}}\left(\frac{\partial^2}{\partial X^2} + \frac{\partial^2}{\partial Y^2} - k^2\right)\widetilde{W} + \frac{1}{\mathrm{Gr}}\widetilde{\Theta} + \frac{\mathrm{Ha}^2}{\mathrm{Gr}}\widetilde{J}x,$$

$$\frac{\partial \widetilde{J}x}{\partial X} + \frac{\partial \widetilde{J}y}{\partial Y} + ik\widetilde{J}z = 0, \qquad \widetilde{J}x = -\frac{\partial \widetilde{\Phi}}{\partial X} - \widetilde{W},$$

$$\widetilde{J}y = -\frac{\partial \widetilde{\Phi}}{\partial Y}, \qquad \widetilde{J}z = -ik\widetilde{\Phi} + \widetilde{U}.$$

$$(6)$$

Table 2 shows variation of the critical Grashof number on the number of grids at the dimensionless wavenumber k = 0.83 and Hartmann number Ha = 5.0 when discretized with

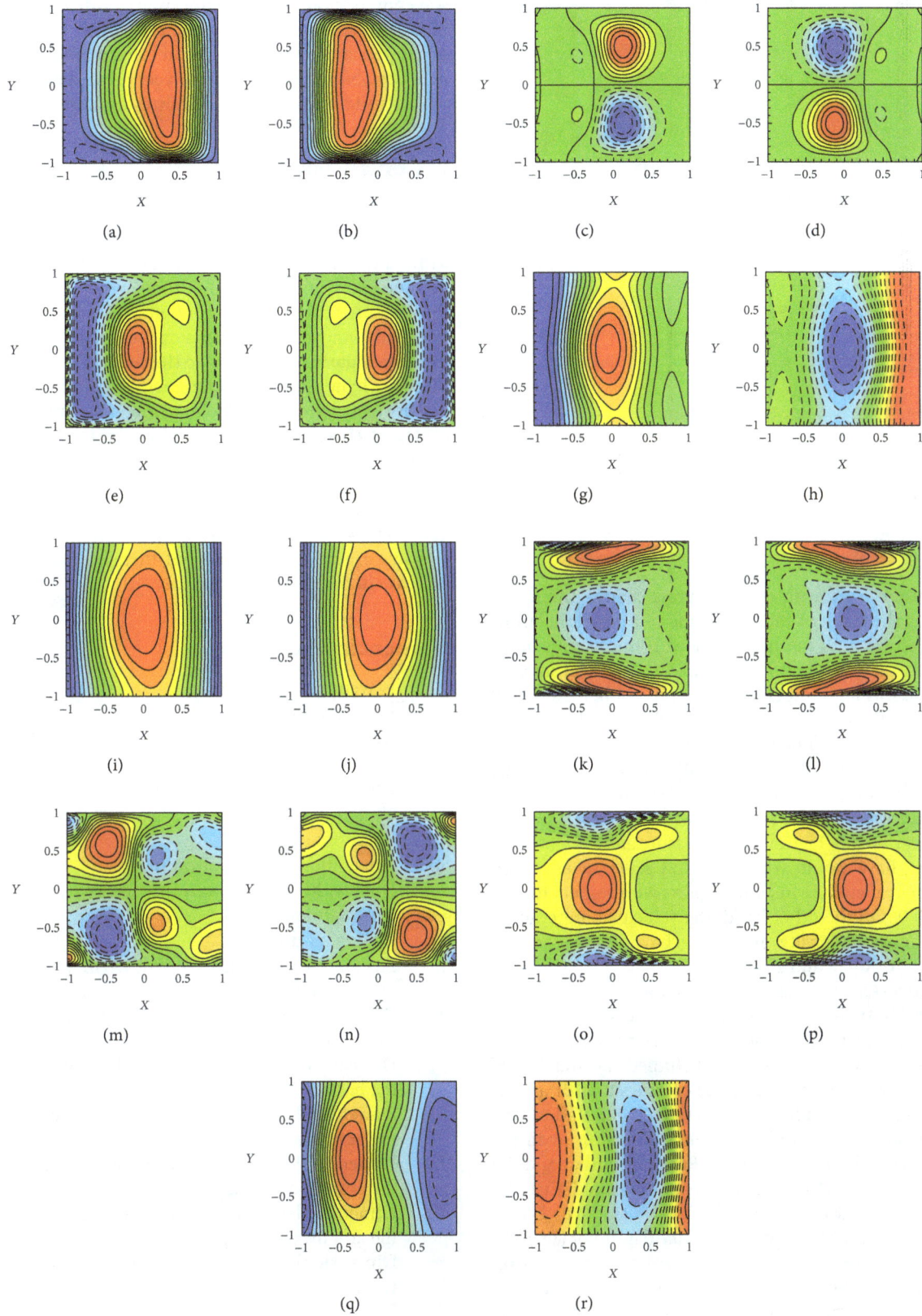

FIGURE 7: Contour maps of the characteristic function at a marginal state at Pr = 0.025, k = 0.83, Gr = 2152.5, and Ha = 5.0: (a) real part of $\widetilde{U}$, (b) imaginary part of $\widetilde{U}$, (c) real part of $\widetilde{V}$, (d) imaginary part of $\widetilde{V}$, (e) real part of $\widetilde{W}$, (f) imaginary part of $\widetilde{W}$, (g) real part of $\widetilde{P}$, (h) imaginary part of $\widetilde{P}$, (i) real part of $\widetilde{\Theta}$, (j) imaginary part of $\widetilde{\Theta}$, (k) real part of $J\widetilde{x}$, (l) imaginary part of $J\widetilde{x}$, (m) real part of $J\widetilde{y}$, (n) imaginary part of $J\widetilde{y}$, (o) real part of $J\widetilde{z}$, (p) imaginary part of $J\widetilde{z}$, (q) real part of $\widetilde{\Phi}$, and (r) imaginary part of $\widetilde{\Phi}$.

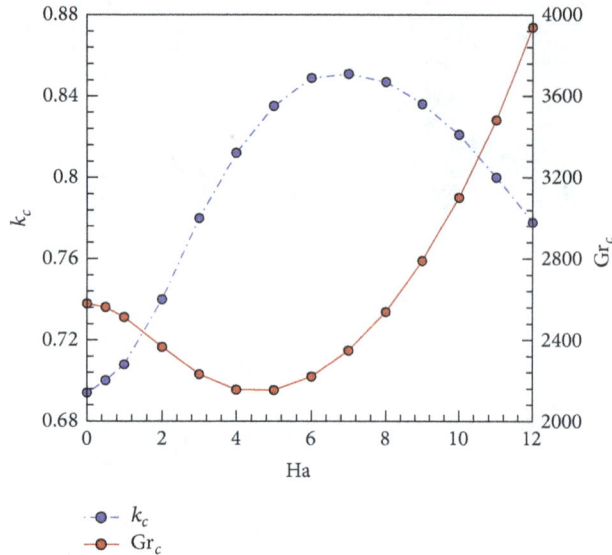

FIGURE 8: Variation of the critical Grashof number and the critical wavenumber against the Hartmann number at Pr = 0.025.

a second-order or a fourth-order accurate central difference. Considering the result of Table 2 and the computational time, we employed 70×70 mesh number with the second-order accurate central difference. For example, the neutral stability curve of the Grashof number at Hartmann number Ha = 10.0 is depicted in Figure 6. The critical Grashof numbers at the wavenumber 0.821 with mesh size 70×70, 90×90 and 120×120 take the value of 3101, 3109, and 3114, respectively. So the dependency of the mesh size on the result is not so significant.

Characteristic functions for various variables at a neutral state are illustrated in Figure 7. The pressure and the potential distribution have point symmetry and have reverse sign between the real and the imaginary parts with respect to $X = 0$ and $Y = 0$. On the other hand, the velocity, temperature, and electric current density distribution have point symmetry with respect to $X = 0$ and $Y = 0$.

The critical Grashof number and the wavenumber plotted against the Hartmann number up to 12 are shown in Figure 8. Concerning the application of a horizontal Bridgeman crystal growth, Lyubimov et al. [16] studied the stability of thermal convection in a rectangular cross section of an oblong channel. They studied two cases of computation, the magnetic field applied in the horizontal direction or in the vertical direction with changing the Prandtl number and the aspect ratio of cross section. It should be noted that the present study could be compared with the study of Lyubimov et al. at the case of zero value of Prandtl number (Pr = 0). In the configuration of horizontal Bridgeman crystal growth, the plane-parallel flow does not exist in nonzero values of Prandtl number, but in the present study the plane-parallel flow always exists irrespective of Prandtl number since the basic thermal state is independent of Prandtl number. According to the results of horizontal magnetic field by Lyubimov et al., the critical Grashof number takes its minimum at a certain low Hartmann number and the critical

wavenumber takes its maximum at another low Hartmann number. This tendency is quite coincident with the finding of the present study. This indicates that a moderate magnetic field destabilizes the natural convection of plane-parallel flow but a strong magnetic field stabilizes the convection.

7. Conclusion

In this study, we studied the linear stability of thermal convection in an infinitely long vertical rectangular channel in the presence of a uniform horizontal magnetic field. We obtained some findings. First, the characteristic function of the pressure and the electric potential distribution have point symmetry and have reverse sign between the real and the imaginary parts with respect to the center of cross section. On the other hand, the velocity, temperature, and electric current density distribution have point symmetry with respect to the center of cross section. Second, except in the range of the Hartmann number of 0 < Ha < 8, increase in the Hartmann number leads to well stabilization. And the critical Grashof number takes a minimum value at the Hartmann number of about Ha = 4.5. Third, the critical wavenumber increases at a low Hartmann number but then decreases for further increase in the Hartmann number.

Nomenclature

a:	Wavenumber [rad/m]
A:	Aspect ratio (= h/l)
B_0:	Applied magnetic field [T]
$\mathbf{e}_{B_0}$:	Unit vector (direction of B_0)
$\mathbf{e}_y$:	Unit vector (direction of y-axis)
$\mathbf{e}_z$:	Unit vector (direction of z-axis)
Gr:	Grashof number
g:	Gravitational acceleration [m/s^2]
h:	Duct width in the y-axis [m]
Ha:	Hartmann number
i:	Imaginary unit
$\mathbf{j}$:	Electric current density vector [A/m^2]
$\mathbf{J}$:	Dimensionless electric current density
Jx_{bs}:	Dimensionless electric current density of the basic flow (along x-axis)
Jy_{bs}:	Dimensionless electric current density of the basic flow (along y-axis)
$J\tilde{x}$:	Dimensionless electric current density of the amplitude (along x-axis)
$J\tilde{y}$:	Dimensionless electric current density of the amplitude (along y-axis)
$J\tilde{z}$:	Dimensionless electric current density of the amplitude (along z-axis)
k:	Dimensionless wavenumber (along z-axis)
l:	Duct width in the x-axis [m]
p:	Pressure [Pa]
P:	Dimensionless pressure
$\bar{P}$:	Dimensionless pressure of the amplitude
Pr:	Prandtl number
T_0:	Reference temperature [K]
T:	Temperature [K]

$\widetilde{U}$:   Dimensionless velocity of the amplitude (along x-axis)

v: Velocity vector [m/s]

V: Dimensionless velocity vector

$\widetilde{V}$:   Dimensionless velocity of the amplitude (along y-axis)

W_{bs}:   Dimensionless velocity of the basic flow (along z-axis)

$\widetilde{W}$:   Dimensionless velocity of the amplitude (along z-axis)

α: Thermal diffusivity [m^2/s]

β: Volumetric thermal expansion coefficient [K^{-1}]

Θ_{bs}:   Dimensionless temperature of the basic flow

$\widetilde{\Theta}$:   Dimensionless temperature of the amplitude

λ: Complex growth rate [rad/s]

ν: Kinematic viscosity [m^2/s]

ρ_0:   Fluid density at reference temperature [kg/m^3]

σ_e: Electric conductivity [$\Omega^{-1}\,\mathrm{m}^{-1}$]

ϕ: Electric potential [V]

Φ_{bs}:   Dimensionless electric potential of the basic flow

$\widetilde{\Phi}$:   Dimensionless electric potential of the amplitude.

Conflict of Interests

The authors declare that there is no conflict of interests regarding the publication of this paper.

References

[1] Y. Shimomura, "The new ITER," *Journal of Plasma and Fusion Research*, vol. 79, no. 9, pp. 949–961, 2000.

[2] C. P. C. Wong, S. Malang, M. Sawan et al., "An overview of dual coolant Pb-17Li breeder first wall and blanket concept development for the US ITER-TBM design," *Fusion Engineering and Design*, vol. 81, no. 1–4, pp. 461–467, 2006.

[3] T. Tanaka, T. Muroga, T. Shikama et al., "Electrical insulating properties of ceramic coating materials for liquid Li blanket system under irradiation," *Journal of Plasma and Fusion Research*, vol. 83, no. 4, pp. 391–396, 2007.

[4] The society of Fluid Mechanics of Japan, *Handbook of Fluid Mechanics (Japanese)*, The society of Fluid Mechanics of Japan, Tokyo, Japan, 2002.

[5] H. Ozoe and K. Okada, "The effect of the direction of the external magnetic field on the three-dimensional natural convection in a cubical enclosure," *International Journal of Heat and Mass Transfer*, vol. 32, no. 10, pp. 1939–1954, 1989.

[6] K. Okada and H. Ozoe, "Experimental heat transfer rates of natural convection of molten gallium suppressed under an external magnetic field in either the X, Y, or Z direction," *Journal of Heat Transfer*, vol. 114, no. 1, pp. 107–114, 1992.

[7] T. Tagawa and H. Ozoe, "Enhancement of heat transfer rate by application of a static magnetic field during natural convection of liquid metal in a cube," *Journal of Heat Transfer*, vol. 119, no. 2, pp. 265–271, 1997.

[8] T. Tagawa and H. Ozoe, "Enhanced heat transfer rate measured for natural convection in liquid gallium in a cubical enclosure under a static magnetic field," *Journal of Heat Transfer*, vol. 120, no. 4, pp. 1027–1032, 1998.

[9] G. Authié, T. Tagawa, and R. Moreau, "Buoyant flow in long vertical enclosures in the presence of a strong horizontal magnetic field. Part 2. Finite enclosures," *European Journal of Mechanics, B/Fluids*, vol. 22, no. 3, pp. 203–220, 2003.

[10] U. Burr, L. Barleon, P. Jochmann, and A. Tsinober, "Magnetohydrodynamic convection in a vertical slot with horizontal magnetic field," *Journal of Fluid Mechanics*, no. 475, pp. 21–40, 2003.

[11] C. Zhang, S. Eckert, and G. Gerbeth, "The flow structure of a bubble-driven liquid-metal jet in a horizontal magnetic field," *Journal of Fluid Mechanics*, vol. 575, pp. 57–82, 2007.

[12] K. Fujimura and M. Nagata, "Degenerate 1 : 2 steady state mode interaction-MHD flow in a vertical slot," *Physica D: Nonlinear Phenomena*, vol. 115, no. 3-4, pp. 377–400, 1998.

[13] T. Kakutani, "The hydromagnetic stability of the modified plane Couette flow in the presence of a transverse magnetic field," *Journal of the Physical Society of Japan*, vol. 19, no. 6, pp. 1041–1057, 1964.

[14] K. Fujimura and J. Mizushima, "Nonlinear equilibrium solutions for travelling waves in a free convection between vertical parallel plates," *European Journal of Mechanics, B/Fluids*, vol. 10, no. 2, pp. 25–31, 1991.

[15] T. P. Lyubimova, D. V. Lyubimov, V. A. Morozov, R. V. Scuridin, H. B. Hadid, and D. Henry, "Stability of convection in a horizontal channel subjected to a longitudinal temperature gradient. Part 1. Effect of aspect ratio and Prandtl number," *Journal of Fluid Mechanics*, vol. 635, pp. 275–295, 2009.

[16] D. V. Lyubimov, T. P. Lyubimova, A. B. Perminov, D. Henry, and H. B. Hadid, "Stability of convection in a horizontal channel subjected to a longitudinal temperature gradient. Part 2. Effect of a magnetic field," *Journal of Fluid Mechanics*, vol. 635, pp. 297–319, 2009.

11

Unsteady/Steady Hydromagnetic Convective Flow between Two Vertical Walls in the Presence of Variable Thermal Conductivity

M. M. Hamza,[1] **I. G. Usman,**[2] **and A. Sule**[2]

[1]*Department of Mathematics, Usmanu Danfodiyo University, PMB 2346, Sokoto, Nigeria*
[2]*Department of Mathematics, Zamfara State College of Education, PMB 1002, Maru, Nigeria*

Correspondence should be addressed to M. M. Hamza; hmbtamb@yahoo.com

Academic Editor: Miguel Onorato

Unsteady as well as steady natural convection flow in a vertical channel in the presence of uniform magnetic field applied normal to the flow region and temperature dependent variable thermal conductivity is studied. The nonlinear partial differential equations governing the flow have been solved numerically using unconditionally stable and convergent semi-implicit finite difference scheme. For steady case, approximate solutions have been derived for velocity, temperature, skin friction, and the rate of heat transfer using perturbation series method. Results of the computations for velocity, temperature, skin friction, and the rate of heat transfer are presented graphically and discussed quantitatively for various parameters embedded in the problem. An excellent agreement was found during the numerical computations between the steady-state approximate solutions and unsteady numerical solutions at steady-state time. In addition, comparison with previously published work is performed and the results agree well.

1. Introduction

In recent years, the interest in the study of hydromagnetic flow in a channel region has been growing rapidly because of its extensive engineering applications. The experimental investigation of modern MHD flow in a laboratory was first carried out by [1]. This study provided the basic knowledge for the development of many MHD devices, such as MHD pumps, MHD generators, brakes, flow meters, plasma studies, and geothermal energy extraction. Unsteady free convection heat transfer with MHD effects in a channel region can be found in [2]. Unsteady hydromagnetic flows in rotating systems have been studied by [3–9]. An exact solution for unsteady hydromagnetic free convection flow with constant heat flux is to be found in [10]. All the above mentioned studies assumed the thermal conductivity of the fluid to be constant. However, it is known that the fluid physical properties may change significantly with temperature changes. To accurately predict the flow behavior and heat transfer rate, it is necessary to take into account the variation of thermal conductivity with temperature (see [11]). Thermal properties, particularly thermal conductivity and diffusivity, are essential materials parameters of bedrock controlling the heat transfer and temperature increases in the vicinity of repository. There has been considerable published work dealing with steady flow with variable thermal conductivity (see [12–18]). Recently (see [19]) studied steady MHD flow with variable thermal conductivity over an inclined radiative isothermal permeable surface.

To the best of our knowledge, the problem of unsteady/steady hydromagnetic convective flow between two vertical walls heated symmetrically/asymmetrically in the presence of variable thermal conductivity has not been studied. The present paper is committed to study unsteady as well as steady natural convection flow of a viscous, incompressible fluid between two parallel vertical walls in the presence of transverse magnetic field and temperature dependent variable thermal conductivity when convection between the vertical parallel walls is set up by a change in the temperature of the walls compared to the fluid temperature.

2. Governing Equations

Consider the unsteady natural convection flow of viscous, incompressible, and electrically conducting fluid between

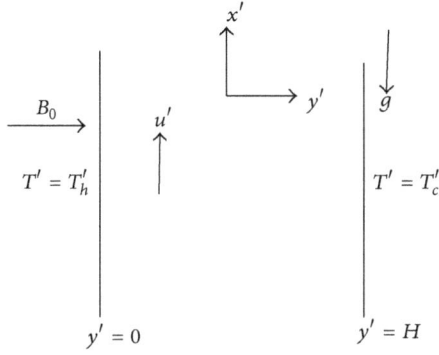

FIGURE 1: Geometry of the problem.

two vertical walls in the presence of a transversely imposed magnetic field of strength B_0. Initially, it is assumed that both the fluid and the walls are at rest and at the same temperature T'_m. At time $t' > 0$, the temperature of the walls $y' = 0$ and $y' = H$ is instantaneously raised or lowered to T'_h and T'_c, respectively, such that $T'_h > T'_c$ which is thereafter maintained constant. We chose a Cartesian coordinate system with x' axis along the upward direction and the y' axis normal to it as shown in Figure 1. Thermal conductivity (k_f) of the fluid is assumed to vary as a linear function of temperature in the form (see [20]), $k_f = k_m[1 + \delta(T' - T'_m)]$, where k_m is the fluid free stream thermal conductivity and δ is a constant depending on the nature of the fluid, where $\delta > 0$ for fluids such as water and air, while $\delta < 0$ for fluids such as lubrication oils (see [20]). The governing equations under Boussinesq's approximation can be written as

$$\frac{\partial u'}{\partial t'} = \nu \frac{\partial^2 u'}{\partial y'^2} + g\beta \left(T' - T'_m\right) - \frac{\sigma B_0^2 u'}{\rho}$$

$$\frac{\partial T'}{\partial t'} = \frac{1}{\rho c_p} \frac{\partial}{\partial y'} \left[k_f \frac{\partial T'}{\partial y'} \right]. \tag{1}$$

The initial and boundary conditions for the present problem are

$$t' \leq 0: u' = 0, \ T' = T'_m, \ \text{for } 0 \leq y' \leq H,$$

$$t' > 0: u' = 0, \ T' = T'_h \ \text{at } y' = 0, \tag{2}$$

$$u' = 0, \ T' = T'_c \ \text{at } y' = H,$$

where β is the coefficient of the thermal expansion, ν is the kinematic viscosity, g is the gravitational force, σ is the conductivity of the fluid, B_0 is the electromagnetic induction, ρ is the density of the fluid, and c_p is the specific heat at constant pressure.

In order to solve (1) to (2), we employ the following dimensionless parameters:

$$y = \frac{y'}{H},$$

$$t = \frac{t'\nu}{H^2},$$

$$U = \frac{u'\nu}{g\beta \left(T'_h - T'_m\right) H^2},$$

$$M^2 = \frac{\sigma B_0^2 H^2}{\nu\rho}, \tag{3}$$

$$\text{Pr} = \frac{\nu\rho c_p}{k_m},$$

$$\theta = \frac{T' - T'_m}{T'_h - T'_m},$$

$$R = \frac{T'_c - T'_m}{T'_h - T'_m},$$

$$\lambda = \delta \left(T'_h - T'_m\right).$$

Using (3), (1) to (2) can take the following form:

$$\frac{\partial U}{\partial t} = \frac{\partial^2 U}{\partial y^2} + \theta - M^2 U,$$

$$\text{Pr}\frac{\partial \theta}{\partial t} = (1 + \lambda\theta)\frac{\partial^2 \theta}{\partial y^2} + \lambda \left(\frac{\partial \theta}{\partial y}\right)^2. \tag{4}$$

The initial and boundary conditions in dimensionless form are

$$t \leq 0: U = 0, \ \theta = 0, \ 0 \leq y \leq 1,$$

$$t > 0: U = 0, \ \theta = 1, \ \text{at } y = 0, \tag{5}$$

$$U = 0, \ \theta = R, \ \text{at } y = 1.$$

3. Approximate Solutions

The approximated solutions played an important role in validating and exploring computer routes of complicated problems. Therefore, we reduce the governing equations of this problem due to its nonlinearity into a form that can be solved analytically. By setting $\partial u/\partial t = 0$, and $\partial\theta/\partial t = 0$ into (4), we get

$$\frac{d^2 U}{dy^2} - M^2 U = -\theta, \tag{6}$$

$$(1 + \lambda\theta)\frac{d^2\theta}{dy^2} + \lambda \left(\frac{d\theta}{dy}\right)^2 = 0. \tag{7}$$

The boundary conditions are

$$U = 0, \ \theta = 1, \ \text{at } y = 0$$

$$U = 0, \ \theta = R, \ \text{at } y = 1. \tag{8}$$

In order to construct an approximate solution to (6) and (7) subject to (8), we employed a regular perturbation method by taking a power series expansion in the variable thermal conductivity parameter λ:

$$U = U_0 + \lambda U_1,$$
$$\theta = \theta_0 + \lambda \theta_1. \qquad (9)$$

Substituting (9) into (6) to (8), the solution of the governing equations are obtained as

$$U = \frac{1}{M^2} \left[1 + (R-1)\, y - \cosh(My) \right.$$
$$+ (k_2 - k_3) \sinh(My) \right] + \lambda \left[A \cosh(My) \right.$$
$$+ B \sinh(My) + k_4 y^3 + k_5 y^2 + k_6 y + k_7 \right], \qquad (10)$$

$$\theta = 1 + (R-1)\, y + \lambda \left[k_1 y - (R-1)\, y^2 \right.$$
$$\left. - (R-1)^2 \frac{y^3}{3} \right].$$

Using (10), we write the steady-state skin friction and rate of heat transfer on the boundaries as follows.

Steady-state skin friction on the boundary plates is

$$\left. \frac{\partial U}{\partial y} \right|_{y=0} = \frac{1}{M^2} \left[(R-1) - M(k_2 - k_3) \right] + \lambda \left[k_6 \right.$$
$$- BM \right]$$

$$\left. \frac{\partial U}{\partial y} \right|_{y=1} = \frac{1}{M^2} \left[(R-1) - M \sinh(M) \right. \qquad (11)$$
$$- M(k_2 - k_3) \cosh(M) \right] + \lambda \left[AM \sinh(M) \right.$$
$$- BM \cosh(M) + 3k_4 + 2k_5 + k_6 \right].$$

The steady-state rate of heat transfer on the boundary plates is

$$\left. \frac{\partial \theta}{\partial y} \right|_{y=0} = (R-1) + \lambda k_1$$

$$\left. \frac{\partial \theta}{\partial y} \right|_{y=1} = (R-1) + \lambda \left[k_1 - 2(R-1) - (R-1)^2 \right], \qquad (12)$$

where

$$k_1 = \frac{(R-1)^2 + 3(R-1)}{3},$$

$$k_2 = \frac{\cosh(M)}{\sinh(M)},$$

$$k_3 = \frac{R}{\sinh(M)},$$

$$k_4 = -\frac{(R-1)^2}{3M^2},$$

$$k_5 = -\frac{(R-1)}{M^2},$$

$$k_6 = \frac{k_1 + 6k_4}{M^2},$$

$$k_7 = \frac{2k_5}{M^2}, \quad A = -k_7,$$

$$B = \frac{k_7 \cosh(M)}{\sinh(M)} - \frac{(k_4 + k_5 + k_6 + k_7)}{\sinh(M)}. \qquad (13)$$

4. Numerical Solutions

The complete forms of (4) are solved numerically using semi-implicit finite difference scheme. We used forward difference formulas for all time derivatives and approximate the spatial derivatives with central difference formula. The semi-implicit finite difference equations corresponding to (4) are as follows:

$$-r_1 U_{j-1}^{(N+1)} + (1 + 2r_1) U_j^{(N+1)} - r_1 U_{j+1}^{(N+1)}$$
$$= r_2 U_{j-1}^{(N)} + (1 - 2r_2 - M^2 \Delta t) U_j^{(N)} + r_2 U_{j+1}^{(N)}$$
$$+ \Delta t \theta_j^N$$

$$-r_1 \theta_{j-1}^{(N+1)} + (\mathrm{Pr} + 2r_1) \theta_j^{(N+1)} - r_1 \theta_{j+1}^{(N+1)} \qquad (14)$$
$$= (r_2 + r_3) \theta_{j-1}^{(N)} + (\mathrm{Pr} - 2r_2 - 2r_3) \theta_j^{(N)}$$
$$+ (r_2 + r_3) \theta_{j+1}^{(N)} + r_4 \left(\theta_{j+1}^{(N)} - \theta_{j-1}^{(N)} \right)^2,$$

where $r_1 = \xi \Delta t / \Delta y^2$, $r_2 = (1-\xi)\Delta t / \Delta y^2$, $r_3 = \lambda \Delta t \theta_j^{(N)} / \Delta y^2$, $r_4 = \lambda \Delta t / 4 \Delta y^2$, and $0 \le \xi \le 1$. We chose $\xi = 1$ so that we are free to choose larger time steps. The approximated solutions displayed in the previous section are used as a check on the accuracy and effectiveness of the numerical scheme. Also, in order to reconfirm accuracy of the scheme, the numerical results for velocity and temperature are compared with the approximated solutions. It has been found that numerical values of the velocity and temperature fields calculated from expressions (10) have matched very well with the numerical solutions obtained from expressions (14) at the steady-state time. See Figure 2 for the graph of the numerical solutions at steady-state and steady-state approximate solutions for velocity and temperature fields. Again, Figures 3 and 4 represent comparisons with the work of [21] when Pr = 0.71, $M = 0$, and $\lambda = 0$. It is clear that excellent agreement between the present numerical solutions and the approximated solutions of [21] exists.

5. Results and Discussion

The numerical results are obtained by solving (14) using the method described in the previous section for various values

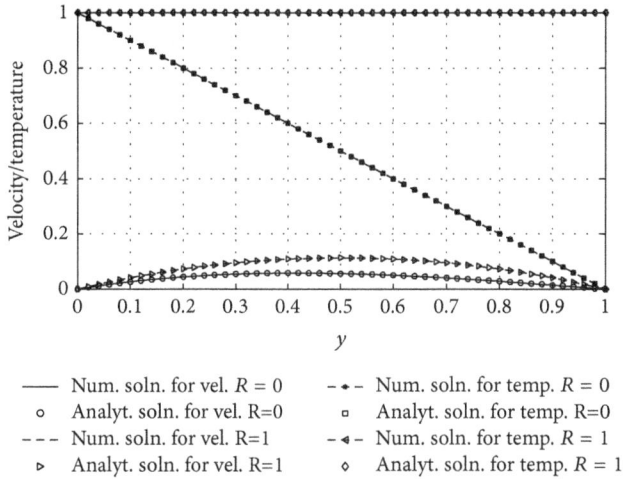

FIGURE 2: Unsteady and steady-state solutions for velocity and temperature profiles.

Legend for Figure 2:
- —— Num. soln. for vel. $R = 0$
- ○ Analyt. soln. for vel. R=0
- - - - Num. soln. for vel. R=1
- ▷ Analyt. soln. for vel. R=1
- - • - Num. soln. for temp. $R = 0$
- ▫ Analyt. soln. for temp. R=0
- - ◂ - Num. soln. for temp. $R = 1$
- ◇ Analyt. soln. for temp. $R = 1$

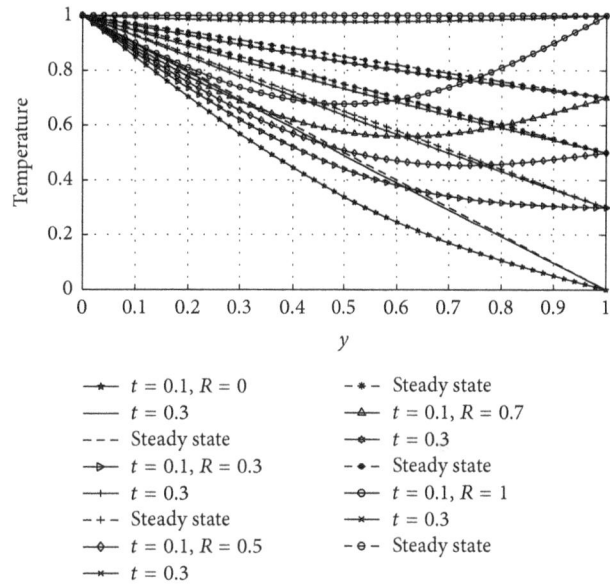

FIGURE 4: Unsteady and steady-state solutions for temperature profiles $M = 0$, Pr $= 0.71$, and $\lambda = 0$.

Legend for Figure 4:
- —★— $t = 0.1, R = 0$
- —— $t = 0.3$
- - - - Steady state
- —▷— $t = 0.1, R = 0.3$
- —+— $t = 0.3$
- - + - Steady state
- —◆— $t = 0.1, R = 0.5$
- —*— $t = 0.3$
- - * - Steady state
- —△— $t = 0.1, R = 0.7$
- —◆— $t = 0.3$
- - • - Steady state
- —○— $t = 0.1, R = 1$
- —×— $t = 0.3$
- - ⊖ - Steady state

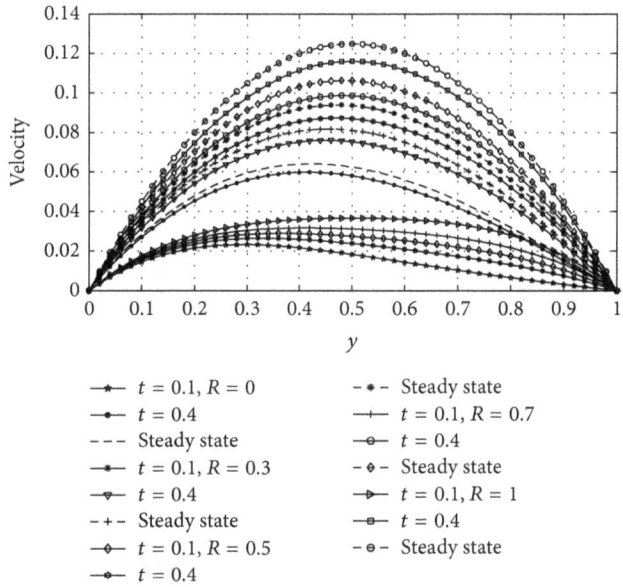

FIGURE 3: Unsteady and steady-state solutions for velocity profiles $M = 0$, Pr $= 0.71$, and $\lambda = 0$.

Legend for Figure 3:
- —★— $t = 0.1, R = 0$
- —●— $t = 0.4$
- - - - Steady state
- —★— $t = 0.1, R = 0.3$
- —▼— $t = 0.4$
- - + - Steady state
- —◆— $t = 0.1, R = 0.5$
- —●— $t = 0.4$
- - * - Steady state
- —+— $t = 0.1, R = 0.7$
- —○— $t = 0.4$
- - ◇ - Steady state
- —▷— $t = 0.1, R = 1$
- —▫— $t = 0.4$
- - ⊖ - Steady state

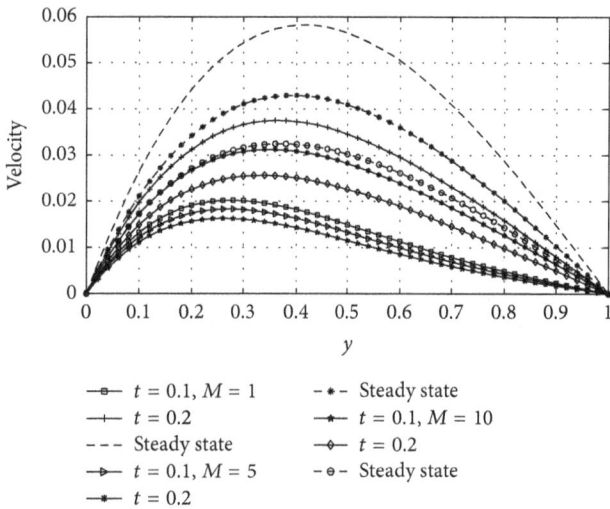

FIGURE 5: Unsteady and steady-state solutions for velocity profiles $R = 0$, Pr $= 1$, and $\lambda = 0.1$.

Legend for Figure 5:
- —▫— $t = 0.1, M = 1$
- —+— $t = 0.2$
- - - - Steady state
- —▷— $t = 0.1, M = 5$
- —*— $t = 0.2$
- - * - Steady state
- —★— $t = 0.1, M = 10$
- —◆— $t = 0.2$
- - ⊖ - Steady state

of physical parameters to describe the physics of the problem. The nondimensional parameters that govern the flow are the Prandtl number (Pr), magnetic parameter (M), variable thermal conductivity parameter (λ), and buoyancy force distribution parameter (R). The value of Prandtl number (Pr) is taken as 1.0, which corresponds to electrolyte solution (see [22, 23]). Results obtained are presented graphically for velocity, temperature, skin friction, and Nusselt number for various flow parameters.

Figures 5 and 6 illustrate the velocity profiles of the fluid for different values of magnetic parameter (M) and nondimensional time (t) for asymmetric and symmetric case, respectively (i.e., $R = 0$ and $R = 1$). From Figure 5 it is observed that velocity of the fluid is maximum near

the heated wall ($y = 0$) and then gradually decreases as it moves towards the cooled wall ($y = 1$). It is clear from Figure 6 that symmetric flow about occurs between the walls for all considered values of M and the nature of the figures is parabolic. In both Figures 5 and 6 it is noted that increasing M decreases the velocity of the flow throughout the channel walls. The physical explanation of this behavior is that the presence of magnetic field produces a resistivity force (Lorentz force) similar to the drag force which retard the velocity.

Figures 7 and 8, respectively, show the response of the fluid temperature to variation in the variable thermal conductivity parameter (λ) and time (t) for asymmetric and

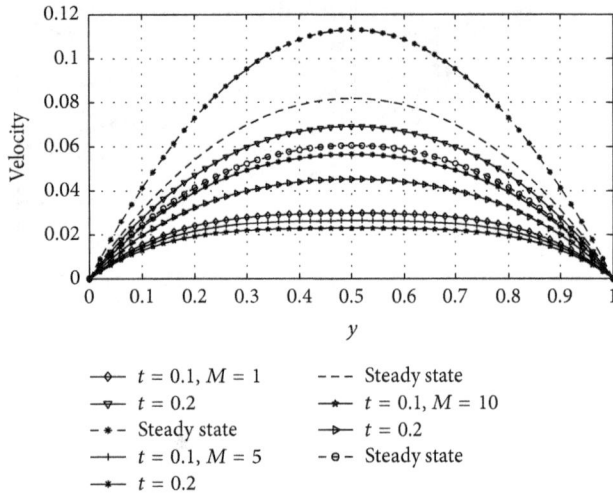

FIGURE 6: Unsteady and steady-state solutions for velocity profiles $R = 1$, $\text{Pr} = 1$, and $\lambda = 0.1$.

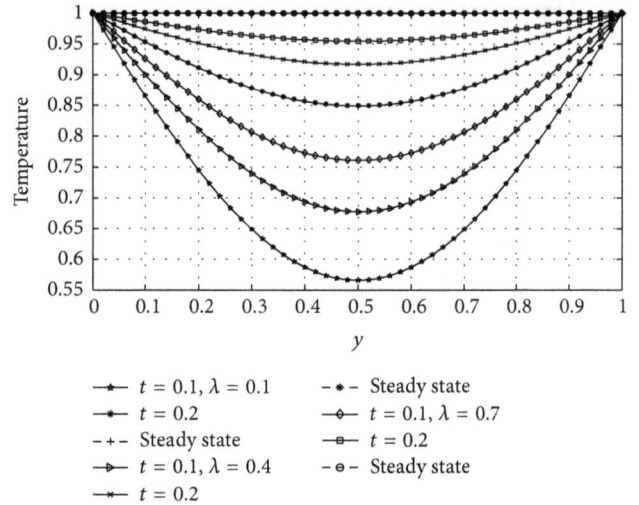

FIGURE 8: Unsteady and steady-state solutions for temperature profiles $R = 1$, $\text{Pr} = 1$, and $M = 1$.

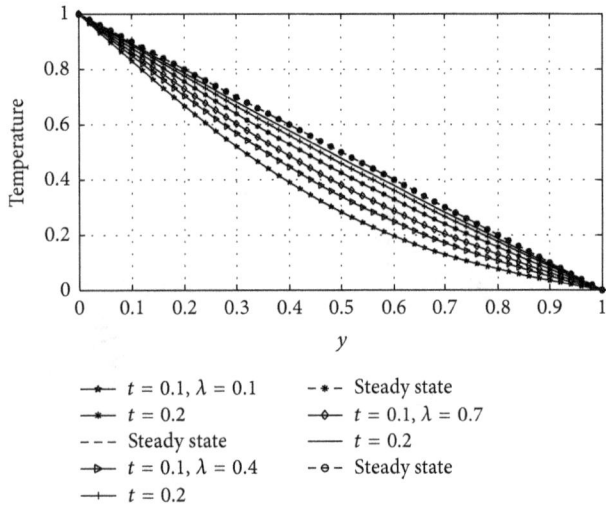

FIGURE 7: Unsteady and steady-state solutions for temperature profiles $R = 0$, $\text{Pr} = 1$, and $M = 1$.

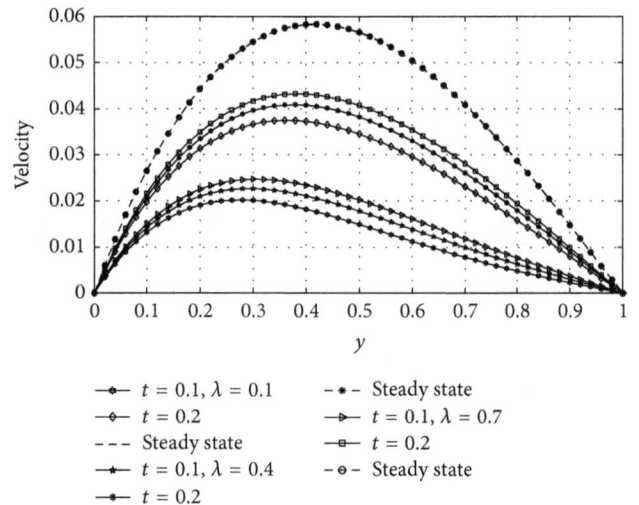

FIGURE 9: Unsteady and steady-state solutions for velocity profiles $R = 0$, $\text{Pr} = 1$, and $M = 1$.

symmetric case. The effect of variable thermal conductivity (λ) and time (t) on velocity profiles for asymmetric and symmetric case is shown in Figures 9 and 10, respectively. From Figures 7 to 10, it is observed that both temperature and velocity of the fluid increase with increasing λ and t until a steady-state condition is attained. This is physically true, since the relation $\lambda = \delta(T_h' - T_m')$ indicates that mounting values of λ increase the temperature difference between outside the plate and outside the boundary layer. As a result, heat is transferred rapidly from plate to fluid within the boundary layer. That is why both velocity and temperature profiles enlarge due to growing λ. It means that the velocity and the thermal boundary layer thickness rise for larger λ. From Figure 9, it is also seen that velocity of the fluid is maximum near the heated wall ($y = 0$) and then progressively decreases as it moves towards the cooled wall ($y = 1$). It is clear from

Figures 8 and 10 that symmetric flow about occurs between the walls for all considered values of λ and t until a steady-state condition is achieved.

The rate of heat transfer (Nusselt number) dependence on λ is illustrated in Figure 11 for symmetric case. Figures 11(a) and 11(b) represent the rate of heat transfer at the walls $y = 0$ and $y = 1$, respectively. Figure 11(b) reveals that the rate of heat transfer increases as λ and t increase until a steady-state condition is reached. A reverse effect is observed at the plate $y = 0$; see Figure 11(a). The wall shear stress (skin friction) dependence on λ for varying values of t is illustrated in Figures 12(a) and 12(b) for symmetric case at the plates $y = 0$ and $y = 1$, respectively. From these figures, it is seen that skin friction increases with increase in λ and t. The skin friction and Nusselt number dependence on λ for varying values of t is displayed in Figures 13(a) and 13(b), respectively,

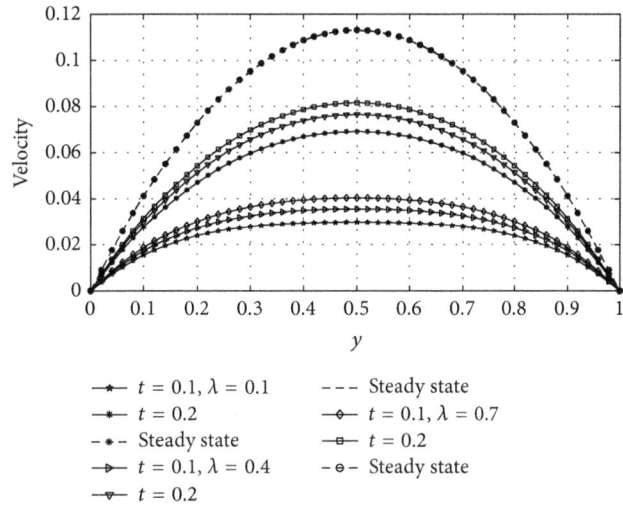

FIGURE 10: Unsteady and steady-state solutions for velocity profiles $R = 1$, $\Pr = 1$, and $M = 1$.

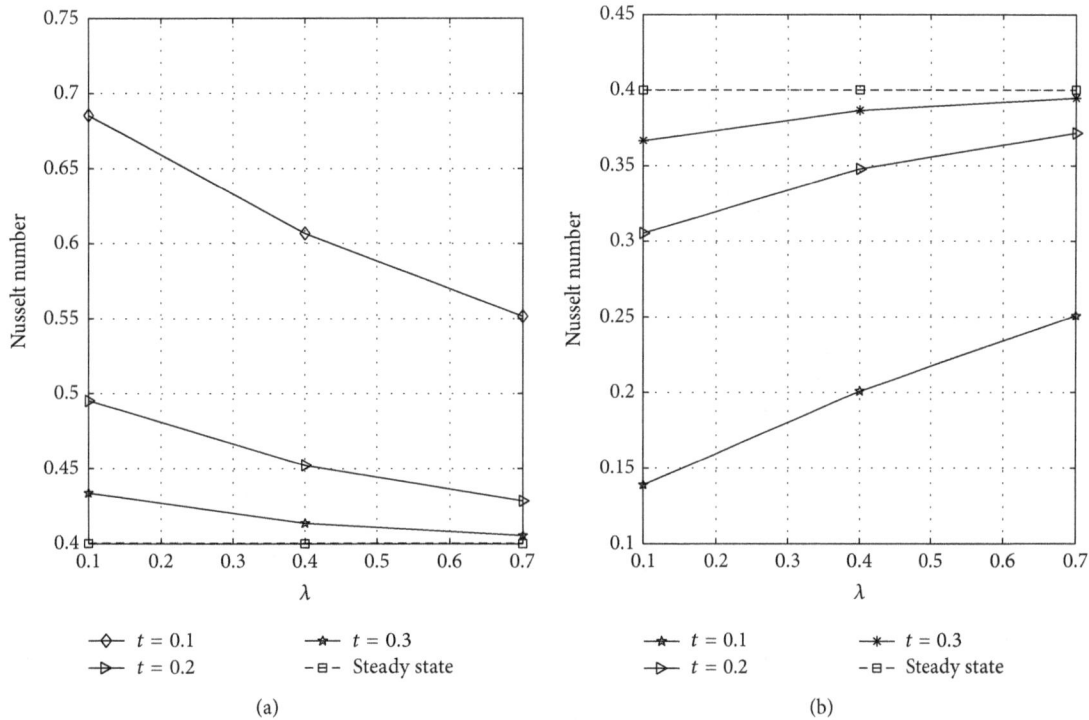

FIGURE 11: Variation of unsteady and steady-state Nusselt number with λ, $R = 0$.

at the plate $y = 0$ for asymmetric case. Figure 13(a) reflected that skin friction increases as λ and t increase until a steady-state condition is attained. Figure 13(b) reveals that Nusselt number decreases with increasing λ and t. It should be noted that the numerical values of skin friction at the plates $y = 0$ and $y = 1$ are the same for $R = 1$ because a symmetric flow occurs for this case. Also, the Nusselt number at the plates $y = 0$ and $y = 1$ is the same for $R = 1$.

The skin friction dependence on M for varying values of t at the plates $y = 0$ and $y = 1$ is displayed in Figures 14(a) and 14(b), respectively, for $R = 0$. These figures reflected that skin friction decreases as M and t increase. Figures 15(a) and

15(b) are plotted to see the effects of M on the skin friction at the plates $y = 0$ and $y = 1$ for $R = 1$. From these figures, it is observed that skin friction decreases as M and t increase. Further, it is noted that Figures 15(a) and 15(b) are exactly the same. This is because a symmetric flow occurs for this case; that is, $R = 1$.

6. Conclusion

The problem of unsteady as well as steady hydromagnetic natural convection flow of a viscous, incompressible, and electrically conducting fluid between two vertical walls having

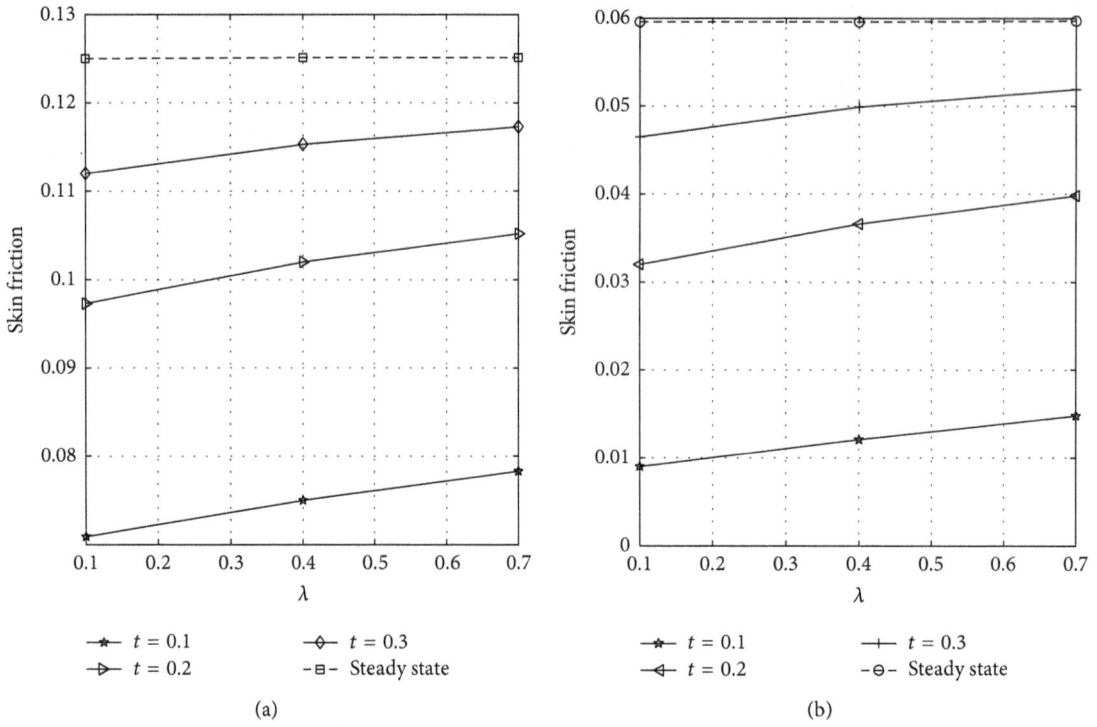

FIGURE 12: Variation of unsteady and steady-state skin friction with λ, $R = 0$.

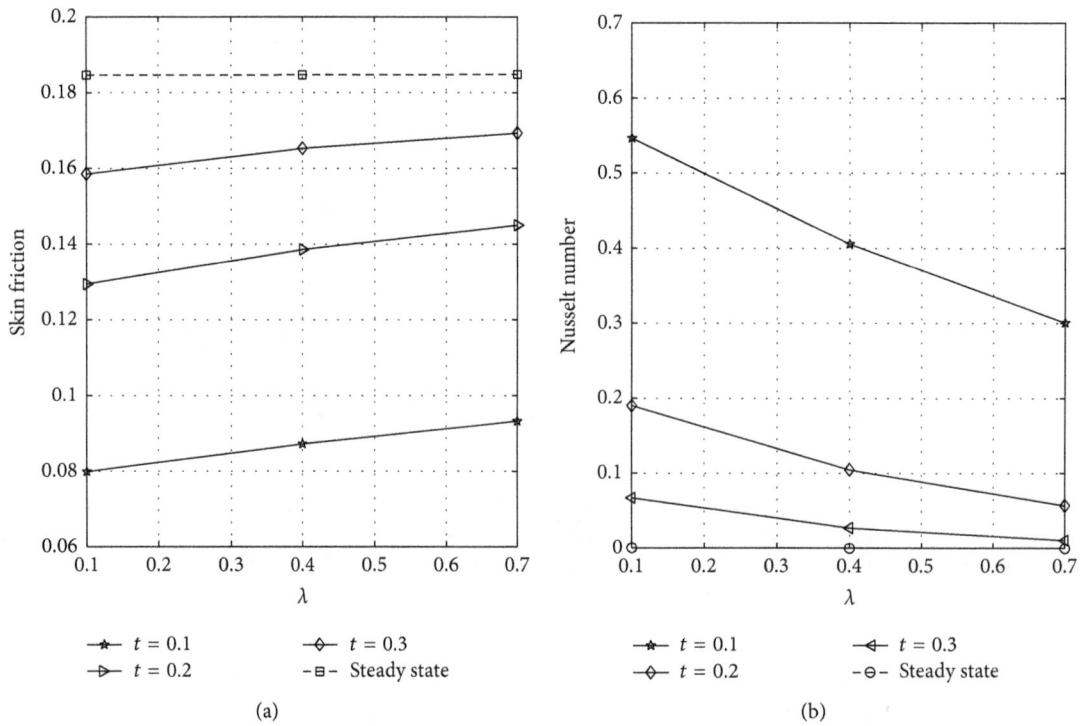

FIGURE 13: Variation of unsteady and steady-state skin friction and Nusselt number with λ, $R = 1$.

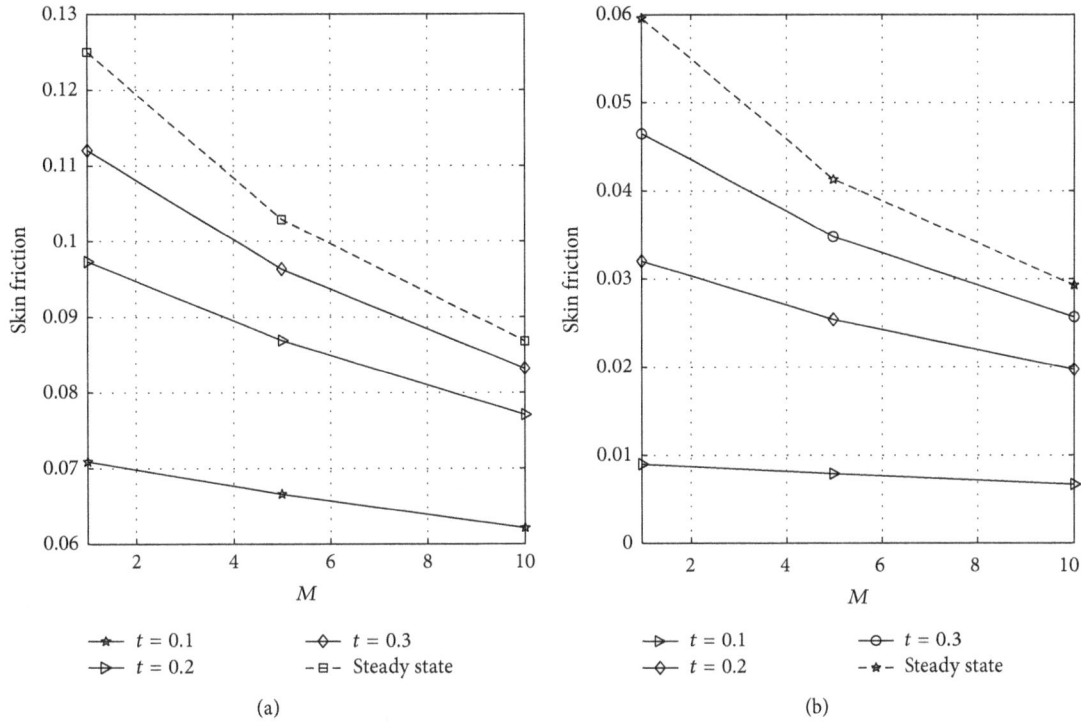

FIGURE 14: Variation of unsteady and steady-state skin friction with M, $R = 0$.

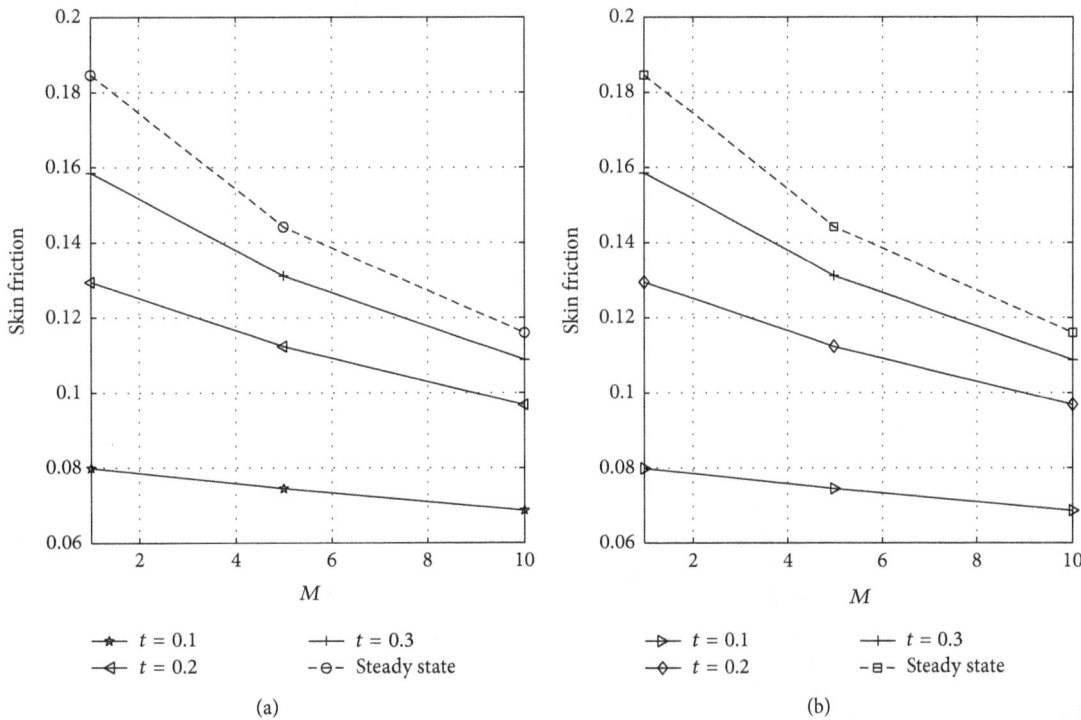

FIGURE 15: Variation of unsteady and steady-state skin friction with M, $R = 1$.

asymmetric/symmetric temperatures on the walls in the presence of variable thermal conductivity and uniform applied magnetic field has been studied by obtaining approximated and numerical solutions of the governing equations, using perturbation series method and semi-implicit finite difference scheme. It is found that the fluid velocity and temperature increase with the increasing of variable thermal conductivity and nondimensional time for both asymmetric and symmetric heating, while magnetic parameter retards the motion of the fluid.

Conflict of Interests

The authors declare that there is no conflict of interests regarding the publication of this paper.

References

[1] J. Hartmann and F. Lazarus, "Hg-dynamics II: 'theory of laminar flow of electrically conductive liquids in a homogeneous magnetic field," *Matematisk-Fysiske Meddelelser*, vol. 15, no. 7, 1937.

[2] R. M. Singer, "Transient magnetohydrodynamic flow and heat transfer," *Zeitschrift für angewandte Mathematik und Physik*, vol. 16, no. 4, pp. 483–494, 1965.

[3] G. S. Seth and R. N. Jana, "Unsteady hydromagnetic flow in a rotating channel with oscillating pressure gradient," *Acta Mechanica*, vol. 37, no. 1-2, pp. 29–41, 1980.

[4] G. S. Seth, R. N. Jana, and M. K. Maiti, "Unsteady hydromagnetic couette flow in a rotating system," *International Journal of Engineering Science*, vol. 20, no. 9, pp. 989–999, 1982.

[5] G. S. Seth and S. K. Ghosh, "Unsteady hydromagnetic flow in a rotating channel in the presence of inclined magnetic field," *International Journal of Engineering Science*, vol. 24, no. 7, pp. 1183–1193, 1986.

[6] P. Chandran, N. C. Sacheti, and A. K. Singh, "Effect of rotation on unsteady hydromagnetic Couette flow," *Astrophysics and Space Science*, vol. 202, no. 1, pp. 1–10, 1993.

[7] A. K. Singh, N. C. Sacheti, and P. Chandran, "Transient effects on magnetohydrodynamic couette flow with rotation: accelerated motion," *International Journal of Engineering Science*, vol. 32, no. 1, pp. 133–139, 1994.

[8] B. K. Jha and C. A. Apere, "Combined effect of hall and ion-slip currents on unsteady MHD couette flows in a rotating system," *Journal of the Physical Society of Japan*, vol. 79, Article ID 104401, 9 pages, 2010.

[9] B. K. Jha and C. A. Apere, "Magnetohydrodynamic free convective Couette flow with suction and injection," *Journal of Heat Transfer*, vol. 133, no. 9, Article ID 092501, 12 pages, 2011.

[10] N. C. Sacheti, P. Chandran, and A. K. Singh, "An exact solution for unsteady magnetohydrodynamic free convection flow with constant heat flux," *International Communications in Heat and Mass Transfer*, vol. 21, no. 1, pp. 131–142, 1994.

[11] M. A. Hossain, K. Khanafer, and K. Vafai, "The effect of radiation on free convection flow of fluid with variable viscosity from a porous vertical plate," *International Journal of Thermal Sciences*, vol. 40, no. 2, pp. 115–124, 2001.

[12] M. Arunachalam and N. R. Rajappa, "Forced convection in liquid metals with variable thermal conductivity and capacity," *Acta Mechanica*, vol. 31, no. 1-2, pp. 25–31, 1978.

[13] T. C. Chiam, "Heat transfer with variable conductivity in a stagnation-point flow towards a stretching sheet," *International Communications in Heat and Mass Transfer*, vol. 23, no. 2, pp. 239–248, 1996.

[14] T. C. Chiam, "Heat transfer in a fluid with variable thermal conductivity over a linearly stretching sheet," *Acta Mechanica*, vol. 129, no. 1-2, pp. 63–72, 1998.

[15] M. A. Seddeek and A. M. Salem, "Laminar mixed convection adjacent to vertical continuously stretching sheets with variable viscosity and variable thermal diffusivity," *Heat and Mass Transfer*, vol. 41, no. 12, pp. 1048–1055, 2005.

[16] M. M. Rahman, A. A. Mamun, M. A. Azim, and M. A. Alim, "Effects of temperature dependent thermal conductivity on magnetohydrodynamic free convection flow along a vertical flat plate with heat conduction," *Nonlinear Analysis: Modelling and Control*, vol. 13, pp. 513–524, 2008.

[17] P. R. Sharma and G. Singh, "Effects of variable thermal conductivity and heat source/sink on MHD flow near a stagnation point on a linearly stretching sheet," *Journal of Applied Fluid Mechanics*, vol. 2, no. 1, pp. 13–21, 2009.

[18] M. A. Alim, M. R. Karim, and M. M. Akand, "Heat generation effects on MHD natural convection flow along a vertical wavy surface with variable thermal conductivity," *American Journal of Computational Mathematics*, vol. 2, no. 1, pp. 42–50, 2012.

[19] M. G. Reddy, "Effects of thermophoresis, viscous dissipation and joule heating on steady MHD flow over an inclined radiative isothermal permeable surface with variable thermal conductivity," *Journal of Applied Fluid Mechanics*, vol. 7, no. 1, pp. 51–61, 2014.

[20] E. M. E. Elbarbary and N. S. Elgazery, "Chebyshev finite difference method for the effects of variable viscosity and variable thermal conductivity on heat transfer from moving surfaces with radiation," *International Journal of Thermal Sciences*, vol. 43, no. 9, pp. 889–899, 2004.

[21] A. K. Singh and T. Paul, "Transient natural convection between two vertical walls heated/cooled asymmetrically," *International Journal of Applied Mechanics and Engineering*, vol. 11, pp. 143–154, 2006.

[22] J. Ettefagh, K. Vafai, and S. J. Kim, "Non-Darcian effects in open-ended cavities filled with a porous medium," *Journal of Heat Transfer*, vol. 113, no. 3, pp. 747–756, 1991.

[23] B. Alazmi and K. Vafai, "Analysis of variants within the porous media transport models," *Journal of Heat Transfer*, vol. 122, no. 2, pp. 303–326, 2000.

12

Hydraulic Analysis of Water Distribution Network Using Shuffled Complex Evolution

Naser Moosavian[1] and Mohammad Reza Jaefarzadeh[2]

[1] *Civil Engineering Department, University of Torbat-e-Heydarieh, Torbat-e-Heydarieh, Iran*
[2] *Civil Engineering Department, Ferdowsi University of Mashhad, Mashhad, Iran*

Correspondence should be addressed to Naser Moosavian; naser.moosavian@yahoo.com

Academic Editor: Prabir Daripa

Hydraulic analysis of water distribution networks is an important problem in civil engineering. A widely used approach in steady-state analysis of water distribution networks is the global gradient algorithm (GGA). However, when the GGA is applied to solve these networks, zero flows cause a computation failure. On the other hand, there are different mathematical formulations for hydraulic analysis under pressure-driven demand and leakage simulation. This paper introduces an optimization model for the hydraulic analysis of water distribution networks using a metaheuristic method called shuffled complex evolution (SCE) algorithm. In this method, applying if-then rules in the optimization model is a simple way in handling pressure-driven demand and leakage simulation, and there is no need for an initial solution vector which must be chosen carefully in many other procedures if numerical convergence is to be achieved. The overall results indicate that the proposed method has the capability of handling various pipe networks problems without changing in model or mathematical formulation. Application of SCE in optimization model can lead to accurate solutions in pipes with zero flows. Finally, it can be concluded that the proposed method is a suitable alternative optimizer challenging other methods especially in terms of accuracy.

1. Introduction

A water distribution network is composed of an edge set consisting of pumps, pipes, valves, and a node set consisting of reservoirs and pipe intersections [1]. The equations governing the flows and heads in a water distribution system are nonlinear, and often a Newton iterative solution algorithm is used in which a linearized set of equations is solved at each iteration [2]. The Newton-based global gradient algorithm (GGA) is a popular method used in solving the water distribution System (WDS) equations [3]. Given the non-linearity of the system of equations, the Newton-based computation of the solution involves an iterative two-step process. The first step includes computing the state variable update, which requires the solution of linear system derived from the Jacobian of the WDS equations. The second step deals with updating estimates of the state variables. The first step is typically the most computationally expensive process within

the GGA [4]. Furthermore, some of the pipes in a network, in which the head losses are modeled by the Hazen-Williams formula, have zero flows. In that case, a key matrix in the method becomes singular and the matrix to be inverted becomes ill conditioned [2]. As a result a failure occurs in the computation. On the other hand, there are no options for pressure-driven demand and leakage simulation in the EPANET program. Meanwhile, there is still a chance to develop a new method for water distribution network analysis in these conditions. In this paper an optimization model is introduced for hydraulic analysis of water distribution networks using a metaheuristic algorithm called shuffled complex evolution (SCE) algorithm.

The analysis of hydraulic networks should be treated as an optimization problem, as shown by Arora [5], Hall [6], and Collins et al. [1]. Arora considered a simple two-piped loop whereas Collins et al. build the basis of their approach on rigorous theoretical background and developed nonlinear

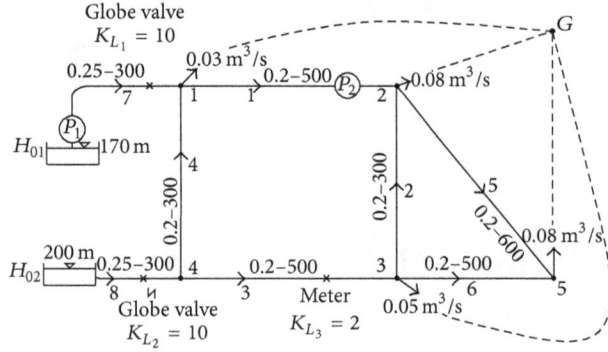

FIGURE 1: Schematic representation of the looped pipe network with 5 unknown nodal heads.

optimization models, whose solutions yielded the hydraulic network analysis [7]. In this paper the Collins model is minimized through the application of shuffled complex evolution algorithm. There is no need to solve linear systems of equations in this methodology, and handling of pressure-driven demand and leakage simulation can be done in a simple way, so an initial solution vector, which is sometimes critical to the convergence, is not required. Furthermore, the proposed model does not entail any complicated mathematical expression and operation. The Collins model is described in the following section.

2. Cocontent Model Approach

Arora [5] is the first researcher who suggested an approach based on the principle of conservation of energy. According to the principle, "Flow in the pipes of a hydraulic network adjusts so that to minimize the expenditure of the system energy." Then, Collins et al. [1] proposed a model termed the cocontent model, which is based on equations having the unknown nodal heads as the basic unknowns, that is, based on H equations. The unknown pipe flows are expressed in terms of the nodal heads and the known pipe resistances, so that the energy loss in pipe x (E_x) is given by [7]

$$E_x = Q_x h_x = \frac{\left[H_i - H_j\right]^{(1/n)+1}}{R_x^{(1/n)}}, \tag{1}$$

in which h_x is head loss in pipe x, $H_i = H_{oi}$ and $H_j = H_{oj}$, for source nodes. R_k is the characteristic parameter of the pipe resistance which depends on roughness length, diameter, and unit of measurement. For example, if the Hazen-Williams equation is used, values n and R in SI units are defined as $R = 10.67L/C^{1.852}D^{4.87}$, $n = 1.852$. C, D, and L are the Hazen-Williams coefficient (depending on the pipe material), the diameter, and its length, respectively.

Now consider the network of Figure 1, with the known and unknown parameters as shown therein. Let the unknown nodal heads at nodes 1, 2, 3, 4, and 5 be H_1, H_2, H_3, H_4, and H_5, respectively. Herein also consider a ground node G with fixed known level H_{0G}, as shown in Figure 1. The nodes 1, 2, 3, and 5 are connected to the ground node G with pseudopipes,

carrying the known nodal outflows q_1, q_2, q_3, and q_5 as shown in Figure 1.

The cocontent optimization model is expressed as

$$\text{Min } C(H) = \frac{\left|H_{01} + h_{p1} - h_{L1} - H_1\right|^{(1/n)+1}}{R_7^{1/n}}$$

$$+ \frac{\left|H_4 - H_1\right|^{(1/n)+1}}{R_4^{1/n}} + \frac{\left|H_1 + h_{p2} - H_2\right|^{(1/n)+1}}{R_1^{1/n}}$$

$$+ \frac{\left|H_2 - H_5\right|^{(1/n)+1}}{R_5^{1/n}} + \frac{\left|H_3 - H_2\right|^{(1/n)+1}}{R_2^{1/n}}$$

$$+ \frac{\left|H_3 - H_5\right|^{(1/n)+1}}{R_6^{1/n}} + \frac{\left|H_4 - h_{L3} - H_3\right|^{(1/n)+1}}{R_3^{1/n}}$$

$$+ \frac{\left|H_{02} - h_{L2} - H_4\right|^{(1/n)+1}}{R_8^{1/n}}$$

$$+ \left(\frac{1}{n} + 1\right) q_1 (H_1 - H_{0G})$$

$$+ \left(\frac{1}{n} + 1\right) q_2 (H_2 - H_{0G})$$

$$+ \left(\frac{1}{n} + 1\right) q_3 (H_3 - H_{0G})$$

$$+ \left(\frac{1}{n} + 1\right) q_5 (H_5 - H_{0G}), \tag{2}$$

where h_L is local loss for valve and h_p is pump head.

The first eight terms of the objective function represent the energy loss in real pipes $1, \ldots, 8$ of the network, respectively, and the last four terms show $(1/n + 1)$ times the energy loss in the pseudopipes [7]. It should be noted that there are no constraints and therefore an unconstrained model in four decision variables is made. For minimization of optimization model, which is partially differentiating in unknown heads, the node-flow continuity equations are created. Therefore, the solution of the cocontent model gives the values of the unknown heads such that the node-flow continuity relationships are satisfied [7].

By partially differentiating (2) with respect to H_1, H_2, H_3, H_4, and H_5, we get

$$\frac{\partial C}{\partial H_1} = -\frac{\left|H_{01} + h_{p1} - h_{L1} - H_1\right|^{(1/n)}}{R_7^{1/n}} - \frac{\left|H_4 - H_1\right|^{(1/n)}}{R_4^{1/n}}$$

$$+ \frac{\left|H_1 + h_{p2} - H_2\right|^{(1/n)}}{R_1^{1/n}} + q_1 = 0,$$

$$\frac{\partial C}{\partial H_2} = -\frac{\left|H_1 + h_{p2} - H_2\right|^{(1/n)}}{R_1^{1/n}} + \frac{\left|H_2 - H_5\right|^{(1/n)}}{R_5^{1/n}}$$

$$- \frac{\left|H_3 - H_2\right|^{(1/n)}}{R_2^{1/n}} + q_2 = 0,$$

$$\frac{\partial C}{\partial H_3} = + \frac{|H_3 - H_2|^{(1/n)}}{R_2^{1/n}} + \frac{|H_3 - H_5|^{(1/n)}}{R_6^{1/n}}$$
$$- \frac{|H_4 - h_{L3} - H_3|^{(1/n)}}{R_3^{1/n}} + q_3 = 0,$$

$$\frac{\partial C}{\partial H_4} = \frac{|H_4 - H_1|^{(1/n)}}{R_4^{1/n}} + \frac{|H_4 - h_{L3} - H_3|^{(1/n)}}{R_3^{1/n}}$$
$$- \frac{|H_{02} - h_{L2} - H_4|^{(1/n)}}{R_8^{1/n}} = 0,$$

$$\frac{\partial C}{\partial H_5} = - \frac{|H_2 - H_5|^{(1/n)}}{R_5^{1/n}} - \frac{|H_3 - H_5|^{(1/n)}}{R_6^{1/n}} + q_5 = 0.$$

$$(3)$$

That is the purely head-based formulations of the network equations. So cocontent model not only minimizes the energy of flow but also preserves water balance in network. For simplicity, H_{0G} can be taken as zero. The general cocontent model can be expressed as

$$\text{Min } C(H) = \sum_x \frac{|H_i + h_{px} - h_{Lx} - H_j|^{(1/n)+1}}{R_x^{1/n}}$$
$$+ \left(\frac{1}{n} + 1\right) \sum_j q_j(H_j). \quad (4)$$

Collins et al. [1] suggested the solution of the NLP optimization of the model. The methods they used were (1) the Frank-Wolfe method, (2) a piecewise linear approximation, and (3) the convex simplex method, which are highly dependent on initial guesses and in some cases converge to an incorrect solution [1].

3. Head Dependent Analysis

In the common approaches, it is presumed that the nodal demands are always satisfied at all demand nodes, irrespective of the available HGL values at demand nodes [7]. But in practice, when the head at a node is insufficient, a reduction in the water flowing from the tap is expected and, at worst, the discharge that can be drafted will be zero, regardless of the actual demand [8]. There are several solutions for these conditions, in the literature. Wagner et al. [9] and Chandapillai [10] suggested a parabolic relationship between required nodal head and minimum head. Their relationships are

$$q_j = \begin{cases} 0, & H_j < H_{\min}, \\ q_j\left(\frac{H_j - H_{\min}}{H^* - H_{\min}}\right)^{1/p}, & H_{\min} \le H_j < H^*, \\ q_j, & H^* \le H_j. \end{cases} \quad (5)$$

H^* is the required nodal head. In this situation, applying if-then rules in the optimization model is a simple way in handling the above formulation.

4. Leakage Simulation

Water losses via leakages constitute a major challenge to the effective operation of municipal WDS since they represent not only diminished revenue for utilities but also undermined service quality [11] and wasted energy resources [12]. In order to conduct a more accurate analysis of a WDS, such as a better estimate of flow through the network (with respect to both satisfied demand and losses through leakage), a hydraulic analysis capable of accounting for pressure-driven (also known as head-driven) demand and leakage flow at the pipe level should prove invaluable. To reach this goal, a leakage model is expressed as follows [13]:

$$q_{k\text{-leak}} = \begin{cases} \beta_k l_k (P_k)^{\alpha_k}, & \text{if } P_k > 0, \\ 0, & \text{if } P_k \le 0, \end{cases} \quad (6)$$

where P_k is average pressure in the pipe computed as the mean of the pressure values at the end nodes i and j of the kth pipe; and l_k is length of that pipe. Variables α_k and β_k are two leakage model parameters [14]. The allocation of leakage to the two end nodes can be performed in a number of ways [15]. Here the nodal leakage flow $q_{j\text{-leak}}$ is computed as the sum of $q_{k\text{-leak}}$ flows of all pipes connected to node j as follows:

$$q_{j\text{-leak}} = \sum_k \frac{1}{2} q_{k\text{-leak}} = \sum_k \begin{cases} \frac{1}{2}\beta_k l_k (P_k)^{\alpha_k}, & \text{if } P_k > 0, \\ 0, & \text{if } P_k \le 0, \end{cases} \quad (7)$$

where $P_k = (P_i + P_j)/2$. This formulation also easily applies to cocontent model without any mathematical complexity. Numerical example 4 demonstrates hydraulic analysis of a real pipe network in this situations.

5. Application of Shuffled Complex Evolution Algorithm for Minimizing Cocontent Model

This study introduces the shuffled complex evolution (SCE) algorithm for the hydraulic analysis. Since the algorithm was originally developed to solve optimization problems, the hydraulic network analysis was introduced into an optimization problem (cocontent model). One advantage of the SCE algorithm is that it does not need an initial solution vector which must be chosen carefully in many other procedures if numerical convergence is to be achieved. Furthermore, application of SCE algorithm in cocontent model does not require any complicated mathematical expression and operation. In this model, pressure-driven demand and leakage can be simulated easily and there is no failure in computation in zero flow conditions.

5.1. Shuffled Complex Evolution (SCE). Shuffled complex evolution (SCE) is a simple powerful and population-based stochastic optimization algorithm that outperforms many metaheuristic algorithms in numerical single-objective optimization problems. This method is based on a synthesis of four concepts: (1) combination of deterministic and probabilistic approaches; (2) systematic evolution of a "complex" of points spanning the parameter space, in the direction of

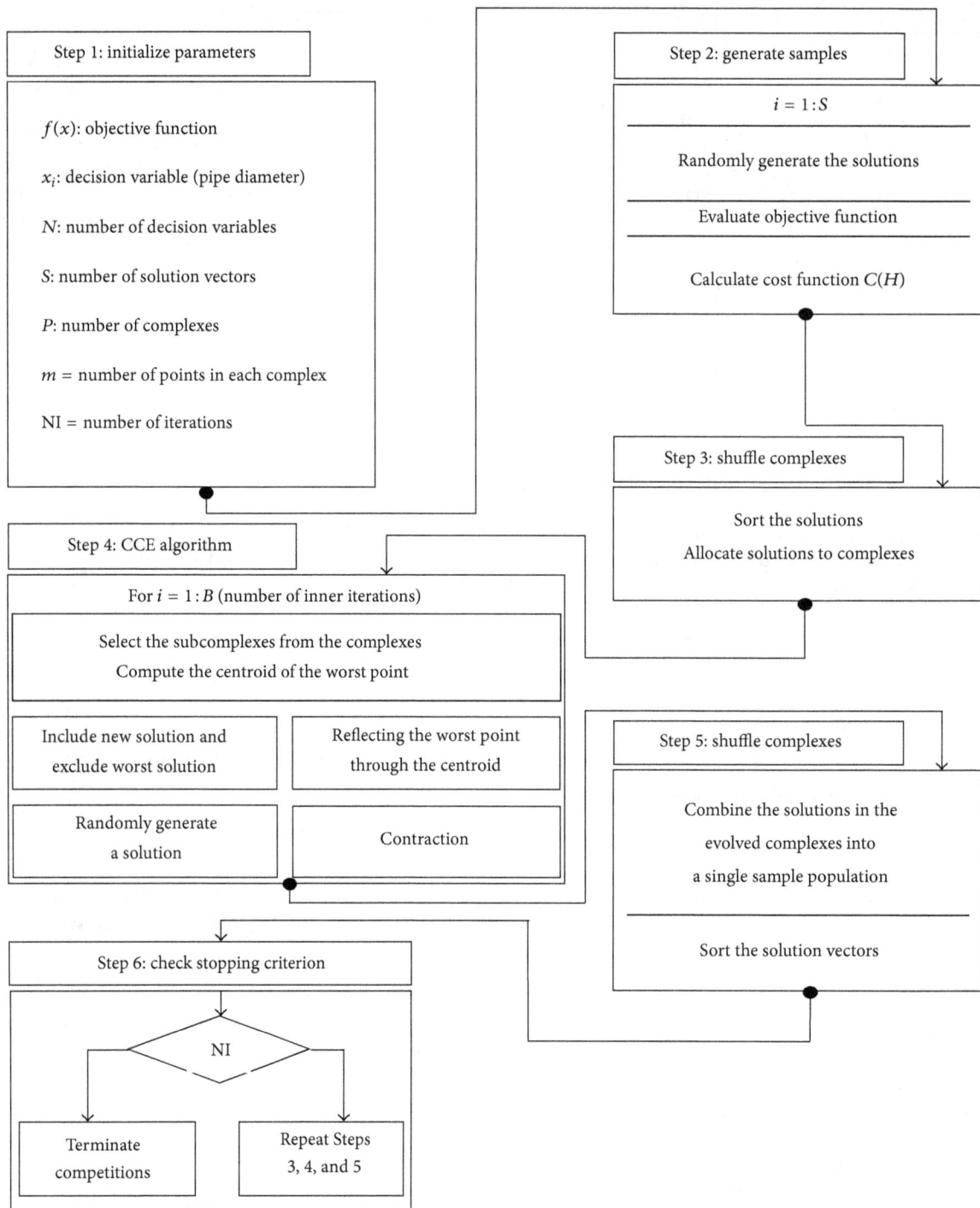

FIGURE 2: SCE procedure for minimization of cocontent model.

global improvement; (3) competitive evolution; (4) complex shuffling [16]. The "complex" is similar to the genetic pool in the GA. The synthesis of these operators makes the SCE method effective and robust and also flexible and efficient [17].

In SCE method, each individual represents a feasible solution for the problem. The search within the feasible region is conducted by first dividing the set of current feasible trial solutions into several complexes, each containing equal number of trial solutions. Each complex represents a local area of the whole domain. Concurrent and independent searches within each complex are conducted until each converges to its local optimal value. Each of the complexes, which are now defined by new trial solutions, is collected into a common pool, shuffled by ranking according to their objective function value, and then further divided into new complexes. The procedure is terminated when none of the local optima found among the complexes can improve on the best current local optimum. The SCE method used the downhill simplex method to accomplish local searches. So, shuffled complex evolution tries to balance between a wide scan of a large solution space and deep search of promising locations. It depends mainly on partitioning the solution space into local communities and performing local search within these communities. Then, it shuffles these local communities to perform global search.

The steps of the procedure of SCE, as shown in Figure 2, include the following.

Step 1: initialize problem and algorithm parameters.

Step 2: samples generation.

Step 3: rank solutions.

Step 4: partition into complexes.

Step 5: start Competitive Complex Evolution (CCE).

Step 6: shuffle complexes.

Step 7: check the stopping criterion.

5.1.1. Step 1: Initialize the Problem and Algorithm Parameters. In Step 1, the optimization problem is specified as follows:

$$
\text{Min } C(H) = \sum_x \frac{\left| H_i + h_{px} - h_{Lx} - H_j \right|^{(1/n)+1}}{R_x^{1/n}} \\
+ \left(\frac{1}{n} + 1 \right) \sum_j q_{oj} \left(H_j \right), \tag{8}
$$

where $C(H)$ is an objective function; H is the set of each decision variable. In this paper, the objective function is the cocontent model; the unknown heads are the decision variables.

5.1.2. Step 2: Samples Generation. The initial population for the DE is created arbitrarily by the following formula:

$$
H(i, j) = H_{\min}(i, j) + \tau \left(H_{\min}(i, j) - H_{\max}(i, j) \right), \tag{9}
$$

where τ denotes a uniformly distributed random value within the range $[0, 1]$. $H_{\min}(i, j)$ and $H_{\max}(i, j)$ are maximum and minimum limits of variable j and node i. Then the fitness values $C(H)$ of all the individuals of population are calculated. The position matrix of the population of generation G can be represented as

$$
P^{(G)} = \begin{bmatrix} C_1 \\ C_2 \\ \cdot \\ \cdot \\ \cdot \\ \cdot \\ C_{nPop} \end{bmatrix} = \begin{bmatrix} H_1^1 & H_2^1 & \cdot & \cdot & \cdot & \cdot & H_N^1 \\ H_1^2 & H_2^2 & \cdot & \cdot & \cdot & \cdot & H_N^2 \\ \cdot & \cdot & \cdot & \cdot & \cdot & \cdot & \cdot \\ \cdot & \cdot & \cdot & \cdot & \cdot & \cdot & \cdot \\ \cdot & \cdot & \cdot & \cdot & \cdot & \cdot & \cdot \\ \cdot & \cdot & \cdot & \cdot & \cdot & \cdot & \cdot \\ H_1^{nPop} & \cdot & \cdot & \cdot & \cdot & \cdot & H_N^{nPop} \end{bmatrix}, \tag{10}
$$

in which N is the number of unknown nodes.

5.1.3. Step 3: Rank Solutions. In this step, the s solutions are sorted in order of increasing criterion value, so that the first vector represents the smallest value of the objective function and the last vector indicates the largest value.

5.1.4. Step 4: Partition into Complexes. The s solutions are partitioned into p complexes, each containing m points. The complexes are partitioned such that the first complex contains every $p(k-1) + 1$ ranked point, the second complex contains every $p(k-1) + 2$ ranked point, and so on, where $k = 1, 2, \ldots, m$ [16].

5.1.5. Step 5: Start Competitive Complex Evolution (CCE). CCE algorithm is based on the simplex downhill search scheme and is one key component of SCE algorithm. This algorithm is presented as follows.

(1) A subcomplex by randomly selecting q solutions from the complex according to a trapezoidal probability distribution is constructed. The probability distribution is specified such that the best solution has the highest chance of being selected to form the subcomplex and the worst point has the least chance.

(2) The worst solution of the subcomplex is identified and the centroid of the subcomplex without including the worst solution is computed as follows:

$$
\text{CR}(i) = (1/(m-1)) \sum_{j=1}^{m-1} s(i, j), \quad i = 1, \ldots, P. \tag{11}
$$

(3) In this step, reflection operator is used, by reflecting the worst point through the centroid according to the following formula:

$$
H^{\text{new}}(i, j) = 2 * \text{CR}(i) - H(i, \text{worst}). \tag{12}
$$

If the newly generated solution is within the feasible space, go to (4); otherwise, randomly generate a point within the feasible space by Equation (9) and go to (6).

(4) If the newly generated solution is better than the worst solution, then it is replaced by the new solution. Otherwise go to (5).

(5) In this step, contraction operator is applied, by computing a solution halfway between the centroid and the worst point:

$$
H^{\text{new}}(i, j) = \frac{(\text{CR}(i) - H(i, \text{worst}))}{2}. \tag{13}
$$

If the contraction solution is better than the worst solution, then it is replaced by the contraction solution. Otherwise, go to (6). This step is imported from competitive complex evolution (CCE).

(6) A solution within the feasible space is generated randomly and the worst solution is replaced by the randomly generated solution.

(7) Steps (2)–(6) are repeated α times and steps (1)–(7) are repeated β times.

5.1.6. Step 6: Shuffle Complexes. The solutions in the evolved complexes into a single sample population are combined and the sample population is sorted in order of increasing criterion value and is shuffled into p complexes.

5.1.7. Step 7: Check the Stopping Criterion. In this section, Steps 3, 4, and 5 are repeated until the termination criterion is satisfied.

It should be noted that the competitive complex evolution (CCE) algorithm is required for the evolution of each complex. Each point of a complex is a potential "parent" with the ability to participate in the process of reproducing offspring. A subcomplex functions like a pair of parents. Use of a stochastic scheme to construct subcomplexes allows the parameter space to be searched more thoroughly. The idea of competitiveness is introduced in forming subcomplexes where the stronger survives better and breeds healthier offspring than the weaker. Inclusion of the competitive measure expedites the search towards promising regions.

A more detailed presentation of the SCE algorithm has been given by Duan et al. [17].

6. Numerical Examples

In this section, the hydraulic analyses for several conditions in some water distribution networks are performed. All computations were executed in MATLAB programming language environment with an Intel(R) Core(TM) 2Duo CPU P8700 @ 2.53 GHz and 4.00 GB RAM. In order to demonstrate the effectiveness of SCE compared with other methods, this study proposes the use of mass balance and energy balance in the network. The average of mass and energy balance is shown by δ and it is calculated by the following formula:

$$\delta = \text{mean}\left(\text{abs}\left(\sum_{j=1}^{N}\sum_{i=x}^{y}\frac{\left|H_i + h_{pk} - h_{Lk} - H_j\right|^{(1/n)}}{R_k^{1/n}} - q_j\right)\right). \tag{14}$$

For Figure 1 and (2) δ is calculated as follows:

$$\delta = \text{mean}\left(\left|\frac{\partial C}{\partial H_1}\right| + \left|\frac{\partial C}{\partial H_2}\right| + \left|\frac{\partial C}{\partial H_3}\right| + \left|\frac{\partial C}{\partial H_4}\right| + \left|\frac{\partial C}{\partial H_5}\right|\right). \tag{15}$$

To check the performance of the SCE for the minimization of cocontent model, in all examples, ten optimization runs were performed using different random initial solutions.

6.1. Numerical Example 1. In this part, the verification of the above mentioned model was conducted via numerical simulation based on an extremely simplified network scheme (5 nodes and 7 pipes) schematically shown in Figure 3 [18]. The pipes resistances were $R_1 = 1.5625$, $R_2 = 50$, $R_3 = 100$, $R_4 = 12.5$, $R_5 = 75$, $R_6 = 200$, and $R_7 = 100$, as reported in Todini [18] for this network.

The SCE technique is applied to solve this problem according to three cases. The bound variables were set between 90 and 100 m. The problem is also solved using the

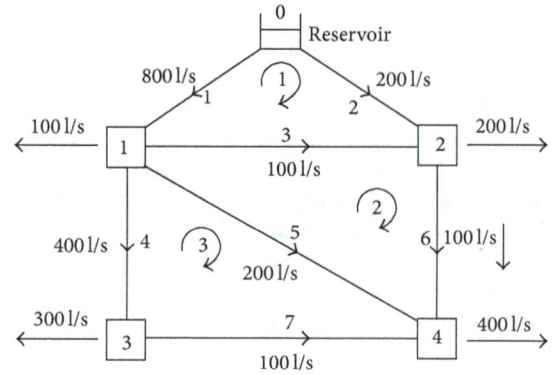

FIGURE 3: Schematic representation of the looped pipe network used in numerical example 1.

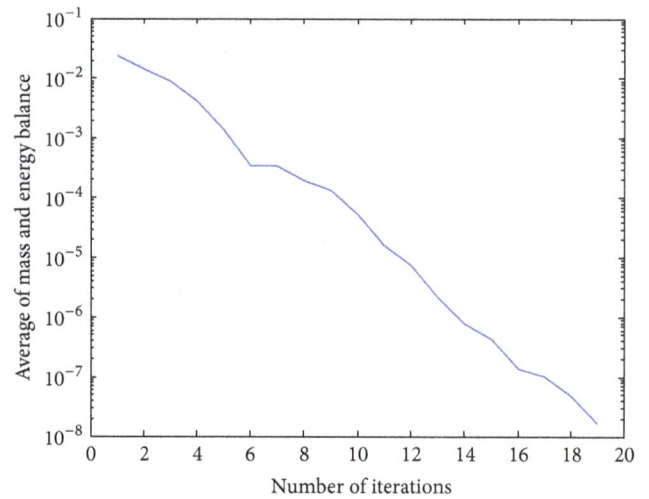

FIGURE 4: Convergence history of numerical example 1 (form 1).

global gradient algorithm (GGA) and the results are compared with those obtained by the SCE. The best, worst, and average solutions of SCE algorithm in three cases are shown in Table 1. As it can be seen in Table 1, in all cases SCE that found the optimal solution more accurately than GGA method. The average number of function evaluation is about 2000 in case 1 and about 100000 in case 3. This shows SCE can converge to global optimum rapidly but reaching high accuracy needs more operations. The convergence process of SCE algorithm has been shown in two forms in Figures 4 and 5. The absolute value of δ is calculated for each iteration in Figure 4 and the amount of objective function $C(H)$ is calculated for each iteration in Figure 5.

6.2. Numerical Example 2. Example 2 considers the symmetric network shown in Figure 6. It has 11 pipes, seven junctions at which the head is unknown, and one fixed head node reservoir at 40 m elevation and all other nodes are at zero elevation. All pipes have diameters, D, of 250 mm and lengths, L, of 1,000 m. Node 8 has a demand of 80 l/s, and all other nodes have zero demands. In the steady state, this network has zero flows in pipes 2, 6, and 9 because of

TABLE 1: Average of mass and energy balance for numerical example 1.

SCE		δ				Average number of function evaluations
		Best	Worst	Mean	Std	
Number of complexes 4 Number of iterations in inner loop 4		$4.60E - 08$	$7.06E - 07$	$3.04E - 07$	$2.30E - 07$	$2.03E + 03$
Number of complexes 9 Number of iterations in inner loop 9		$1.19E - 08$	$3.01E - 08$	$2.04E - 08$	$6.40E - 09$	$1.00E + 05$
Number of complexes 10 Number of iterations in inner loop 15		$1.04E - 08$	$4.19E - 08$	$2.52E - 08$	$8.28E - 09$	$1.00E + 05$
Global gradient algorithm	Maximum accuracy			$6.40E - 06$		

TABLE 2: Head and parameter δ in numerical example 2.

Node number	H (m) [2]	H (m)	δ [2]	δ (SCE)
1	40	40	0	0
2	36.6813	36.6853	**0**	$1.8E - 07$
3	36.6813	36.6853	**0**	$1.8E - 07$
4	33.3626	33.3706	$6.5E - 07$	$\mathbf{8.4E - 08}$
5	33.3626	33.3706	$6.5E - 07$	$\mathbf{8.7E - 08}$
6	30.044	30.0559	$6.5E - 07$	$\mathbf{6.7E - 08}$
7	30.044	30.0559	$6.5E - 07$	$\mathbf{6.6E - 08}$
8	26.7253	26.7411	$5.2E - 05$	$\mathbf{1.2E - 10}$

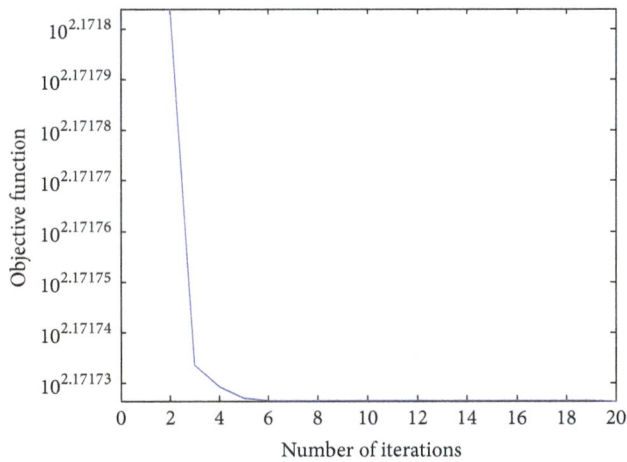

FIGURE 5: Convergence history of numerical example 1 (form 2).

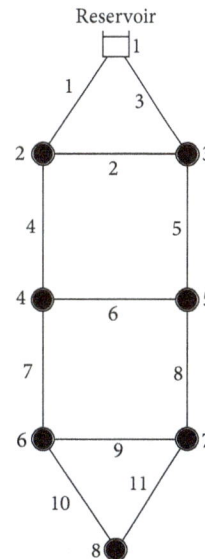

FIGURE 6: Schematic representation of the looped pipe network used in numerical example 2.

symmetry. The head loss is modeled by the Hazen-Williams equation, and each pipe has a Hazen-Williams coefficient $C = 120$ [2]. When the GGA is used in this network, the iterates trend toward the solution and the flows in pipes 2, 6, and 9 approach zero. As this happens, the Jacobian matrix becomes more and more badly conditioned, and the solution computed becomes ill conditioned [2]. Elhay and Simpson [2] proposed a regularization procedure for the GGA which prevents failure of the solution process provided that a flow in the network is ultimately zero or near zero.

The SCE parameters are set as follows: number of decision variables = 7; number of points in each complex = 15; number of complexes for case 1 = 4, case 2 = 9, and case 3 = 10; number of iterations in inner loop for case 1 = 4, case 2 = 9, and

case 3 = 15. The bound variables were set between 25 and 40 m. The previous best solution for this network, when it is simulated using the Elhay algorithm, and the average solution of SCE algorithm are shown in the second and third columns of Table 2, respectively. As can be observed in Table 2, mass and energy balance (δ) in SCE are more accurate than the Elhay algorithm. Table 3 compares the results of applying the SCE algorithm in three cases. The convergence process of SCE algorithm has been shown in two forms in Figures 7 and 8.

TABLE 3: Average of mass and energy balance for numerical example 2.

SCE			δ			Average number of function evaluations
		Best	Worst	Mean	Std	
Number of complexes	4	$2.21E - 06$	$7.24E - 06$	$4.65E - 06$	$1.61E - 06$	$4.19E + 03$
Number of iterations in inner loop	4					
Number of complexes	9	$1.05E - 06$	$5.35E - 06$	$3.46E - 06$	$1.51E - 06$	$9.29E + 03$
Number of iterations in inner loop	9					
Number of complexes	10	$6.36E - 08$	$1.78E - 07$	$1.08E - 07$	$3.89E - 08$	$1.72E + 04$
Number of iterations in inner loop	15					
Global gradient algorithm		Maximum accuracy		Fail		
Elhay algorithm		Maximum accuracy		$7.42E - 06$		

TABLE 4: Head and parameter δ in numerical example 3.

Node number	Z (m)	SCE H (m)	3 steps H (m)	SCE $H - Z$	3 steps $H - Z$	SCE δ	3 steps δ
1	140	140	140	0	0	0	0
2	80	129.304	130.07	49.304	50.07	**$1.49E - 07$**	$3.10E - 04$
3	90	132.288	132.76	42.288	42.76	**$1.26E - 07$**	0.0041
4	70	109.587	110.96	39.587	40.96	**$2.03E - 07$**	0.0022
5	80	80.000	88.54	0.000	8.54	0.058	**$1.51E - 04$**
6	90	90.000	91.45	0.000	1.45	0.0069	**$6.02E - 04$**
7	90	90.000	90.00	0.000	0.00	**0.080**	0.1421
8	100	88.922	90.43	−11.078	−9.57	**$5.84E - 08$**	0.0439

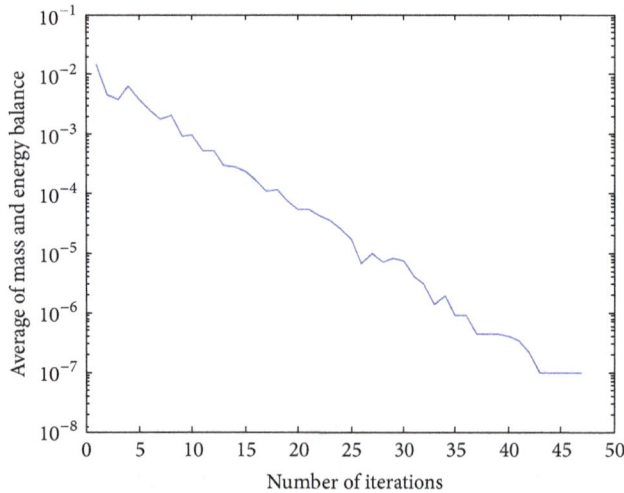

FIGURE 7: Convergence history of numerical example 2 (form 1).

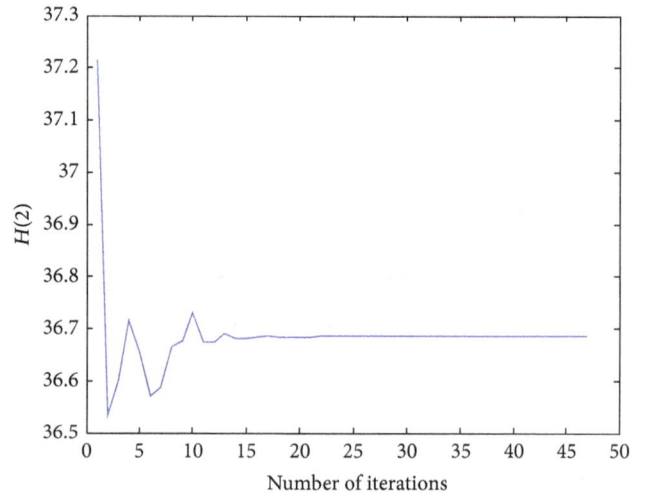

FIGURE 8: Convergence history of head pressure (m) in node 2 for numerical example 2 (form 2).

The absolute value of δ is calculated for each iteration in Figure 7 and the value of head pressure in node 2, $H(2)$, is calculated for each iteration in Figure 8.

6.3. *Numerical Example 3.* The simplified water distribution network shown in Figure 9 was used in order to demonstrate the advantages of the proposed model in pressure-driven demand condition. For the sake of simplicity, the same

Hazen-Williams roughness coefficient $C = 130$ was assumed for all the 14 pipes of identical length of 1000 m, while no local losses have been added. The following diameters have been used in the example: 500 mm (P-2); 400 mm (P-1); 300 mm (P-4, P-7); 250 mm (P-10); 200 mm (P-3, P-5, P-6, and P-13); 150 mm (P-8, P-9, P-11, P-12, and P-14). The nodal demands are listed in the following tables together with the ground elevation Z_i. Without loss of generality, in this example, the

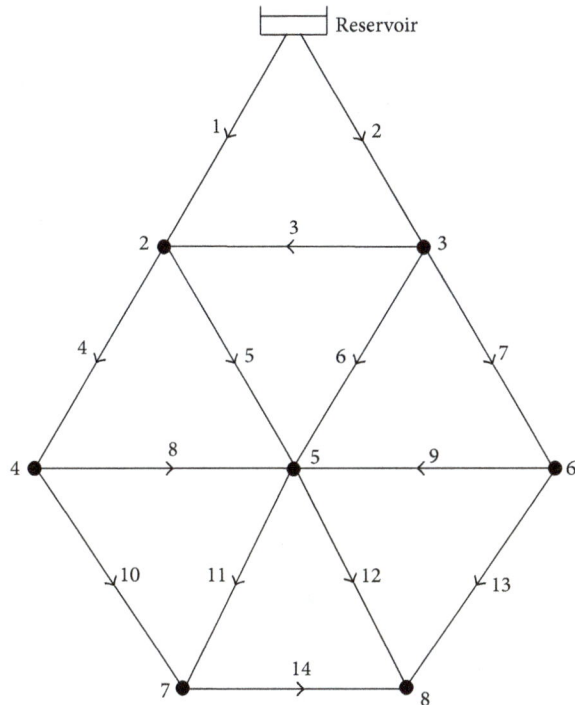

FIGURE 9: Schematic representation of the looped pipe network used in numerical example 3.

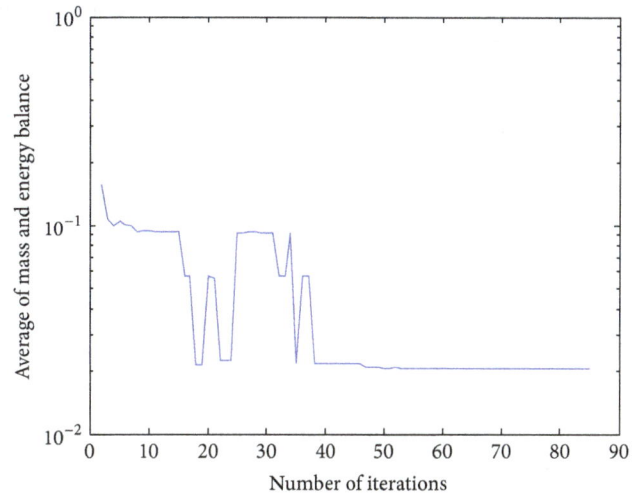

FIGURE 10: Convergence history of numerical example 3 (form 1).

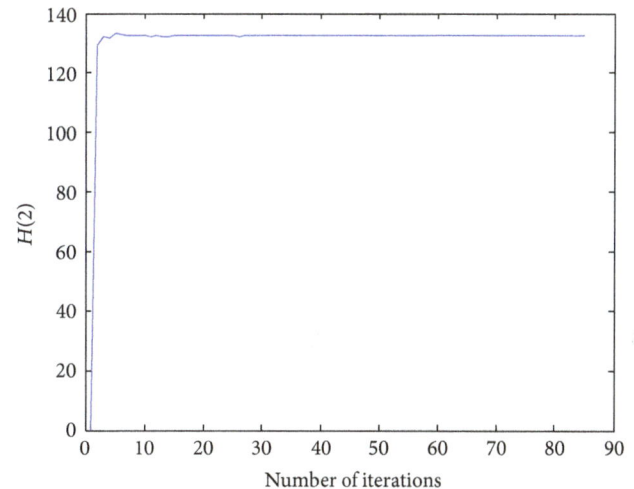

FIGURE 11: Convergence history of head pressure (m) in node 2 for numerical example 3 (form 2).

minimum head requirement H_i^* has been assumed to be equal to the ground elevation Z_i [8]. So the relationship between the required nodal head and minimum head is

$$q_j = \begin{cases} 0, & H_j < Z_j, \\ q_j, & Z_j \leq H_j. \end{cases} \tag{16}$$

Todini [8] proposed a three-step approach for solving this network and its solution is reported in the 4th column of Table 4. In the proposed method pressure-driven model can be applied in hydraulic analysis without any mathematical formulation. In this situation, an if-then rule is added to cocontent model and optimization process is conducted. The number of decision variables in SCE algorithm is 7; the bound variables were set between 50 and 140 m. ten optimization runs were performed using different random initial solutions for all the cases and the results are illustrated in Table 5. Results confirm that SCE is more accurate compared with Todini algorithm in case 2 and case 3. In Table 4, the best result is shown in bold, and it is considered that the method of SCE has calculated the best value of δ at 5 nodes while the Todini method has done it at 2 nodes. The convergence process of SCE algorithm has been shown in two forms in Figures 10 and 11. The absolute value of δ is calculated for each iteration in Figure 10 and the value of head pressure in node 2, $H(2)$, is calculated for each iteration in Figure 11.

6.4. Numerical Example 4. The fourth considered network is a real planned network designed for an industrial area in Apulian Town (Southern Italy). The network layout is

illustrated in Figure 12 and the corresponding data are provided in Table 6. With respect to the leakages, they have been assumed to be pressure-driven (see (6)) given that they are implemented in the pressure driven network simulation model as described above [14]. The parameter $\beta = 1.0632 \times 10^{-7}$ and $\alpha = 1.2$, as reported in Giustolisi et al. [14] for this network. Giustolisi et al. [14] proposed a hydraulic simulation model, which fully integrates a classic hydraulic simulation algorithm, such as that of Todini and Pilati [3] found in EPANET 2, with a pressure-driven model that entails a more realistic representation of the leakage. They applied their model in this network. The results are demonstrated in Table 7. In this table, the best result is shown in bold, and it is considered that the method of SCE has calculated the best value of δ at 14 nodes while the Gistulishi method has done it at 9 nodes. In the proposed method, there is no need to modify the mathematical formulation for hydraulic

TABLE 5: Average of mass and energy balance for numerical example 3.

SCE		δ				Average number of function evaluations
		Best	Worst	Mean	Std	
Number of complexes	5	$3.32E - 02$	$7.97E - 02$	$5.95E - 02$	$1.62E - 02$	$4.30E + 04$
Number of iterations in inner loop	5					
Number of complexes	9	$2.07E - 02$	$5.44E - 02$	$2.52E - 02$	$1.09E - 02$	$6.27E + 04$
Number of iterations in inner loop	9					
Number of complexes	10	$2.07E - 02$	$2.07E - 02$	$2.07E - 02$	$2.17E - 08$	$6.04E + 04$
Number of iterations in inner loop	15					
Three-step approach [8]		Maximum accuracy		$2.76E - 02$		

TABLE 6: Hydraulic data relevant to numerical example 4.

Pipe number	L (m)	D (mm)
1	348.5	327
2	955.7	290
3	483	100
4	400.7	290
5	791.9	100
6	404.4	368
7	390.6	327
8	482.3	100
9	934.4	100
10	431.3	184
11	513.1	100
12	428.4	184
13	419	100
14	1023.1	100
15	455.1	164
16	182.6	290
17	221.3	290
18	583.9	164
19	452	229
20	794.7	100
21	717.7	100
22	655.6	258
23	165.5	100
24	252.1	100
25	331.5	100
26	500	204
27	579.9	164
28	842.8	100
29	792.6	100
30	846.3	184
31	164	258
32	427.9	100
33	379.2	100
34	158.2	368

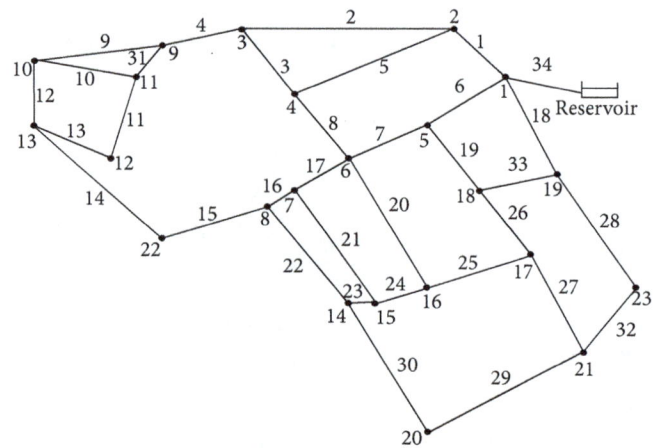

FIGURE 12: Schematic representation of the looped pipe network used in the numerical example 4.

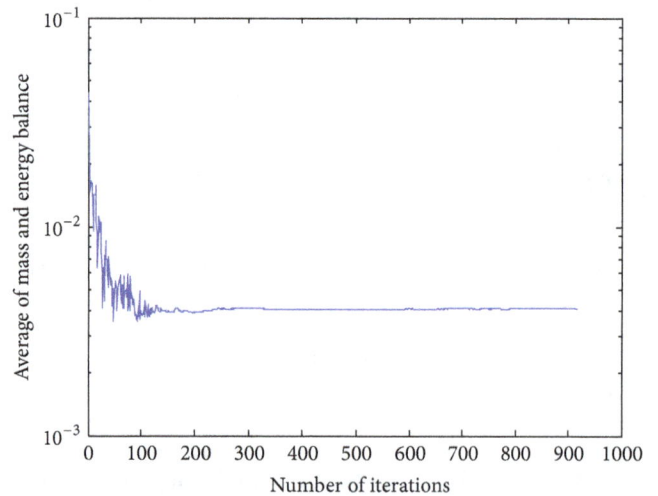

FIGURE 13: Convergence history of numerical example 4 (form 1).

analysis. An if-then rule is added to cocontent model and the optimization process is performed easily. As you can see in Table 8, SCE found the optimal solution more accurately than the Giustolisi algorithm. The convergence process of SCE algorithm has been shown in two forms in Figures 4 and 5. The absolute value of δ is calculated for each iteration in Figure 13 and the amount of head pressure in node 20, $H(20)$, is calculated for each iteration in Figure 14.

TABLE 7: Head and parameter δ in numerical example 4.

Node number	q (l/s)	H (m) (Giustolisi algorithm)	H (m)	δ (Giustolisi algorithm)	δ (SCE)
1	10.863	26.9	33.29	0.154743	**0.004738**
2	17.034	24.81	31.83	0.02131	**0.006532**
3	14.947	21.3	27.39	0.059002	**0.005364**
4	14.28	17.22	25.34	**0.001257**	0.005033
5	10.133	23.54	30.89	0.026184	**0.004018**
6	15.35	20.1	29.02	0.030898	**0.005399**
7	9.114	18.91	27.94	0.017147	**0.003087**
8	10.51	17.9	27.34	**0.00227**	0.003545
9	12.182	17.85	26.35	**0.002936**	0.003867
10	14.579	12.66	23.24	0.008277	**0.004362**
11	9.007	16.23	25.95	0.031554	**0.002723**
12	7.575	10.12	22.05	0.002732	**0.002138**
13	15.2	10.03	22.45	0.0126	**0.004328**
14	13.55	15.41	25.95	**0.001318**	0.004352
15	9.226	14	24.17	**0.002199**	0.002934
16	11.2	14.36	24.05	0.007089	**0.003586**
17	11.469	15.3	25.42	**0.000103**	0.00361
18	10.818	18.83	28.38	0.011889	**0.003908**
19	14.675	19.35	28.39	**5.43E − 05**	0.005097
20	13.318	10.01	23.79	0.013059	**0.003974**
21	14.631	11.48	22.35	**0.003276**	0.004141
22	12.012	14	25.46	0.003901	**0.003677**
23	10.326	10.45	20.11	**0.002551**	0.002979
24	—	36.45	36.45	—	—

TABLE 8: Average of mass and energy balance for numerical example 4.

SCE		δ				Average number of function evaluations
		Best	Worst	Mean	Std	
Number of complexes	5	4.00E − 03	4.50E − 02	4.10E − 03	2.75E − 04	6.59E + 04
Number of iterations in inner loop	5					
Number of complexes	9	3.90E − 03	4.15E − 03	4.00E − 03	3.96E − 05	1.15E + 05
Number of iterations in inner loop	9					
Number of complexes	10	3.70E − 03	4.10E − 03	3.90E − 03	2.53E − 05	1.19E + 05
Number of iterations in inner loop	15					
Giustolisi algorithm		Maximum accuracy		1.81E − 02		

In general, 4 different pipe networks were considered in this paper and different mathematical formulations were used for the hydraulic analysis of these networks. However, the overall results indicate that the proposed method has the capability of handling various pipe networks problems with no change in the model or mathematical formulation. Application of SCE in cocontent model can result in finding accurate solutions in pipes with zero flows and the pressure-driven demand and leakage simulation can be solved through applying if-then rules in cocontent model. As a result, it can be concluded that the proposed method is a suitable alternative optimizer, challenging other methods especially in terms of accuracy.

7. Conclusions

The objective of the present paper was to provide an innovative approach in the analysis of the water distribution networks based on the optimization model. The cocontent model is minimized using shuffled complex evolution (SCE) algorithm. The methodology is illustrated here using four networks with different layouts. The results reveal that the proposed method has the capability to handle various pipe networks problems without changing in model or mathematical formulation. The advantage of the proposed method lies in the fact that there is no need to solve linear systems of equations, pressure-driven demand and

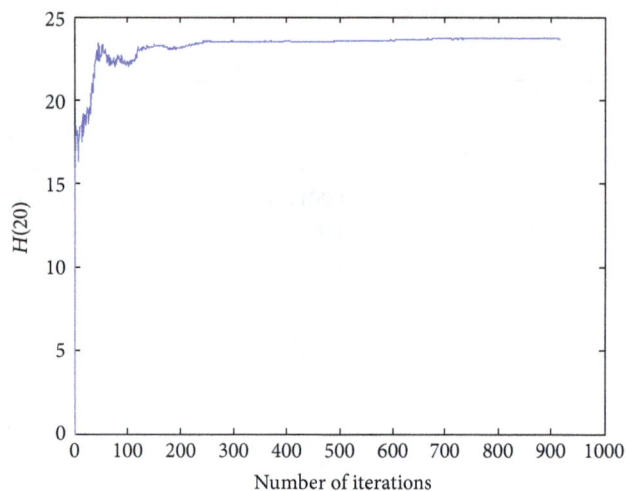

FIGURE 14: Convergence history of head-pressure (m) in node 20 for example, 4 (form 2).

leakage simulation are handled in a simple way, accurate solutions can be found in pipes with zero flows, and it does not need an initial solution vector which must be chosen carefully in many other procedures if numerical convergence is to be achieved. Furthermore, the proposed model does not require any complicated mathematical expression and operation. Finally, it can be concluded that the proposed method is a viable alternative optimizer that challenges other methods particularly in view of accuracy.

Conflict of Interests

The authors declare that there is no conflict of interests regarding the publication of this paper.

References

[1] M. Collins, L. Cooper, R. Helgason, J. Kenningston, and L. LeBlanc, "Solving the pipe network analysis problem using optimization techniques," *Management Science*, vol. 24, no. 7, pp. 747–760, 1978.

[2] S. Elhay and A. R. Simpson, "Dealing with zero flows in solving the nonlinear equations for water distribution systems," *Journal of Hydraulic Engineering*, vol. 137, no. 10, pp. 1216–1224, 2011.

[3] E. Todini and S. Pilati, "A gradient algorithm for the analysis of pipe networks," in *Computer Applications in Water Supply*, vol. 1 of *System Analysis and Simulation*, pp. 1–20, John Wiley & Sons, London, UK, 1988.

[4] A. C. Zecchin, P. Thum, A. R. Simpson, and C. Tischendorf, "Steady-state behaviour of large water distribution systems: the algebraic multigrid method for the fast solution of the linear step," *Journal of Water Resources Planning and Management*, vol. 138, no. 6, pp. 639–650, 2012.

[5] M. L. Arora, "Flow split in closed loops expending least energy," *Journal of the Hydraulics Division*, vol. 102, pp. 455–458, 1976.

[6] M. A. Hall, "Hydraulic network analysis using (generalized) geometric programming," *Networks*, vol. 6, no. 2, pp. 105–130, 1976.

[7] P. R. Bhave and R. Gupta, "Analysis of water distribution networks," in *Alpha Science International, Technology & Engineering*, 2006.

[8] E. Todini, *A More Realistic Approach to the "Extended Period Simulation" of Water Distribution Networks*, Advances in Water Supply Management, chapter 19, Taylor & Francis, London, UK, 2003.

[9] J. M. Wagner, U. Shamir, and D. H. Marks, "Water distribution reliability: analytical methods," *Journal of Water Resources Planning and Management*, vol. 114, no. 3, pp. 253–275, 1988.

[10] J. Chandapillai, "Realistic simulation of water distribution system," *Journal of Transportation Engineering*, vol. 117, no. 2, pp. 258–263, 1991.

[11] J. Almandoz, E. Cabrera, F. Arregui, E. Cabrera Jr., and R. Cobacho, "Leakage assessment through water distribution network simulation," *Journal of Water Resources Planning and Management*, vol. 131, no. 6, pp. 458–466, 2005.

[12] A. F. Colombo and B. W. Karney, "Energy and costs of leaky pipes toward comprehensive picture," *Journal of Water Resources Planning and Management*, vol. 128, no. 6, pp. 441–450, 2002.

[13] G. Germanopoulos, "A technical note on the inclusion of pressure dependent demand and leakage terms in water supply network models," *Civil Engineering Systems*, vol. 2, no. 3, pp. 171–179, 1985.

[14] O. Giustolisi, D. Savic, and Z. Kapelan, "Pressure-driven demand and leakage simulation for water distribution networks," *Journal of Hydraulic Engineering*, vol. 134, no. 5, pp. 626–635, 2008.

[15] L. Ainola, T. Koppel, K. Tiiter, and A. Vassiljev, "Water network model calibration based on grouping pipes with similar leakage and roughness estimates," in *Proceedings of the Joint Conference on Water Resource Engineering and Water Resources Planning and Management (EWRI '00)*, pp. 1–9, Minneapolis, Minn, USA, August 2000.

[16] Q. Duan, S. Sorooshian, and V. K. Gupta, "Optimal use of the SCE-UA global optimization method for calibrating watershed models," *Journal of Hydrology*, vol. 158, no. 3-4, pp. 265–284, 1994.

[17] Q. Y. Duan, S. Sorooshian, and V. K. Gupta, "Shuffled complex evolution approach for effective and efficient global minimization," *Journal of Optimization Theory and Applications*, vol. 76, no. 3, pp. 501–521, 1993.

[18] E. Todini, "On the convergence properties of the different pipe network algorithms," in *Proceedings of the Annual Water Distribution Systems Analysis Symposium (ASCE '06)*, pp. 1–16, August 2006.

13

Hall Effect on Bénard Convection of Compressible Viscoelastic Fluid through Porous Medium

Mahinder Singh[1] and Chander Bhan Mehta[2]

[1] Department of Mathematics, Government Post Graduate College Seema (Rohru), Shimla District, Himachal Pradesh 171207, India
[2] Department of Mathematics, Centre of Excellence, Government Degree College Sanjauli, Shimla District, Himachal Pradesh 171006, India

Correspondence should be addressed to Mahinder Singh; mahinder_singh91@rediffmail.in

Academic Editor: Amy Shen

An investigation made on the effect of Hall currents on thermal instability of a compressible Walter's B' elasticoviscous fluid through porous medium is considered. The analysis is carried out within the framework of linear stability theory and normal mode technique. For the case of stationary convection, Hall currents and compressibility have postponed the onset of convection through porous medium. Moreover, medium permeability hasten postpone the onset of convection, and magnetic field has duel character on the onset of convection. The critical Rayleigh numbers and the wave numbers of the associated disturbances for the onset of instability as stationary convection have been obtained and the behavior of various parameters on critical thermal Rayleigh numbers has been depicted graphically. The magnetic field, Hall currents found to introduce oscillatory modes, in the absence of these effects the principle of exchange of stabilities is valid.

1. Introduction

The theoretical and experimental results of the onset of thermal instability (Bénard convection) under varying assumptions of hydrodynamic and hydromantic stability have been discussed in a treatise by Chandrasekhar [1] in his celebrated monograph. If an electric field is applied at right angles to the magnetic field, the whole current will not flow along the electric field. This tendency of the electric current of flow across an electric field in the presence of a magnetic fluid is called Hall current effect. The Hall effect is likely to be important in many geophysical and astrophysical situations as well as in flows of laboratory plasma. The use of the Boussinesq approximation has been made throughout, which states that the variations of density in the equations of motion can safely be ignored everywhere except in its association with the external force. It has been shown by Sato [2] and Tani [3] that inclusion of Hall currents gave rise to a cross flow, that is, a flow at right angle to the primary flow through a channel in the presence of a transverse magnetic field. In particular, Tani [3] has found that Hall effect produces a cross-flow of double-swirl pattern in incompressible flow through a straight channel with arbitrary cross-section. This breakdown of the primary flow and formation of secondary flow may be presumably attributed to the inherent instability of the primary flow in the presence of Hall current. Sato [2] has pointed out that even if the distribution of the primary flow velocity is stable to external disturbances, the whole layer may become turbulent if the distribution of the cross flow is unstable. Sherman and Sutton [4] have considered the effect of Hall current on the efficiency of a magnetofluid generator. The effect of Hall current on the thermal instability of a horizontal layer of electrically conducting fluid has been studied by Gupta [5].

Hall currents are effects whereby a conductor carrying an electric current perpendicular to an applied magnetic field develops a voltage gradient which is transverse to both the current and the magnetic field. It was discovered by Hall in 1879, while he was working on his doctoral degree at Johns Hopkins University at Baltimore, Maryland. The Hall effect has again become an active area of research with the discovery of the quantized Hall effect by Klaus von Klitzing for which he was bestowed with Nobel prize of physics in 1985. In ionized gases (plasmas), where the magnetic field is very strong and effects the electrical conductivity, cannot be Hall currents.

In the aforementioned studies, the medium has been considered to be nonporous. The development of geothermal power resources has increased general interest in the properties of convection in porous media. The effect of a magnetic field on the stability of such a flow is of interest in geophysics particularly in the study of Earth's core where the Earth's mantle which consists of conducting fluid behaves like a porous medium which can become convectively unstable as a result of differential diffusion. The other application of the results of a magnetic field is in the study of the stability of a convective flow in the geothermal region.

When the fluids are compressible, the equations governing the system become quite complicated to simplify. Boussinesq tried to justify the approximation for compressible fluids when the density variations arise principally from thermal effects. Spiegel and Veronis [6] have simplified the set of equations governing the flow of compressible fluids under the following assumptions.

(a) The depth of the fluid layer is much less than the scale height, as defined by them.

(b) The fluctuations in temperature, density, and pressure, introduced due to motion, do not exceed their total static variations.

Under the previous approximations, the flow equations are the same as those for incompressible fluids, except that the static temperature gradient is replaced by its excess over the adiabatic one and C_v is replaced by C_p.

Chandra [7] observed a contradiction between the theory and experiment for the onset of convection in fluids heated from below. He performed the experiment in an air layer and found that the instability depended on the depth of the layer. Scanlon and Segel [8] have considered the effects of suspended particles on the onset of Bénard convection and found that the critical Rayleigh number is reduced because of the heat capacity of the particles. The suspended particles were thus found to destabilize the layer. The fluids have been considered to be Newtonian, and the medium has been considered to be nonporous in all the previous studies.

One class of elastico-viscous fluids is Walters fluid (model B'), which is not characterized by Maxwell's/Oldroyd's, constitutive relation. When the fluid permeates a porous material, the gross effect is represented by Darcy's law. As a result of this macroscopic law, the usual viscous and viscoelastic terms in the equation of Walters' fluid (model B') motion are replaced by the resistance terms $[-(1/k_1)(\mu - \mu'(\partial/\partial t))\vec{q}]$, where μ and μ' are the viscosity and viscoelasticity of Walters' fluid (model B'), k_1 is the medium permeability and $\vec{q}$ is the Darcian filter velocity of the fluid.

The flow through porous media is of considerable interest for petroleum engineers and geophysical fluid dynamicists. A great number of applications in geophysics may be found in the books by Phillips [9], Ingham and Pop [10], and Nield and Bejan [11]. The scientific importance of the field has also increased because hydrothermal circulation is the dominant heat transfer mechanism in young oceanic crust (Lister [12]).

Generally it is accepted that comets consisting of a dusty "snowball" of a mixture of frozen gasses, which is in the process of their journey, change from solid to gas and vice versa. The physical properties of comets that meteoroids, and interplanetary dust strongly suggest the importance of porosity in astrophysical context have been studied by McDonnell [13].

The stability of two superposed conducting Walters' B' elastico-viscous fluids in hydromagnetics has been studied by Sharma and Kumar [14] and whereas the instability of streaming Walters' viscoelastic fluid B' in porous medium has been considered by Sharma [15], Sunil and Chand [16]. Sunil et al. [17, 18] studied the Hall effect on thermosolutal instability of Rivlin-Ericksen and Walters' (model B') fluid in porous medium. In one study of Singh [19], Hall current effect on thermosolutal instability in a viscoelastic fluid flowing through porous medium and magnetic field stable solute gradient are found to have stabilizing effects on the system, whereas Hall current and medium permeability have a destabilizing effect on the system. The sufficient conditions for the nonexistence of overstability have also obtained. In the one another study, Singh and Kumar [20], hydrodynamic and hydromagnetic stability of two stratified Walter's B' elastico-viscous superposed fluids, where system is stable for stable stratification and unstable for unstable stratification and in case of horizontal magnetic field system having stabilizing effect for unstable stratification, is in contrast to the stability of two superposed Newtonian fluids, where the system is stable for stable stratifications. Gupta et al. [21] have studied thermal convection of dusty compressible Rivlin-Ericksen viscoelastic fluid with Hall currents and found that compressibility and magnetic field postpone the onset of convection, whereas Hall current and suspended particles hasten the onset of convection.

During the survey it has been noticed that Hall effects are completely neglected from the studies of compressible elastico-viscous fluid through porous medium. Keeping in mind the importance of Hall currents, porous medium and compressibility, in elastico-viscous fluid, motivated us to go on detailed study of Walter's B' fluid heated from below through porous medium. We have already studied earlier some problems on Hall current effect with porous as well as nonporous medium and suspended particles found the useful and interesting results, so compressible thermal instability problem of Walter's B' fluid with Hall currents effects through porous medium studied by us here.

2. Mathematical Formulation of the Problem

We have considered an infinite, horizontal, and compressible electrically conducting Walter's B' fluid permeated with porous medium in Hall current effect, bounded by the planes $z = 0$ and $z = d$, as shown in Figure 1. This layer is heated from below so that temperature at bottom (at $z = 0$) and the upper layer (at $z = d$) is T_0 and T_d, respectively, and a uniform temperature gradient $\beta \,(=|dT/dz|)$ is maintained. A uniform vertical magnetic field intensity $\vec{H}(0, 0, H)$ and gravity force $\vec{g}(0, 0, -g)$ pervade the system.

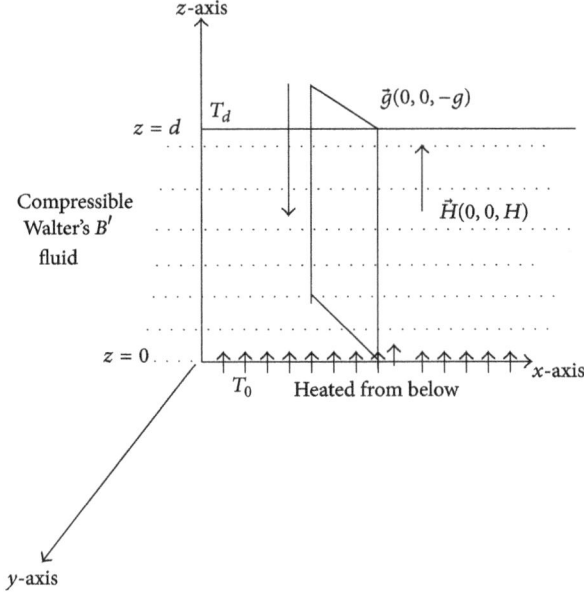

FIGURE 1: Geometrical configuration.

Let p, ρ, T, α, g, η, μ_e, and $\vec{q}(u, v, w)$ denote, respectively, the fluid pressure, density, temperature, thermal coefficient of expansion, gravitational acceleration, resistivity, magnetic permeability, and fluid velocity. Using Spiegel and Veronis' [6] assumptions, the flow equations for compressible fluids are found to be the same as those for incompressible fluids except that in the equation of heat conduction, the temperature gradient β is replaced by its excess over the adiabatic; that is, $(\beta - g/c_p)$. The equations expressing the conservation of momentum, mass, temperature, and equation of state of Walters' (Model B') fluid are

$$\frac{1}{\epsilon}\left[\frac{\partial \vec{q}}{\partial t} + \frac{1}{\epsilon}(\vec{q} \cdot \nabla)\vec{q}\right]$$
$$= -\frac{1}{\rho_m}\nabla p + \vec{g}\left(1 + \frac{\delta\rho}{\rho_m}\right) - \frac{1}{k_1}\left(v - v'\frac{\partial}{\partial t}\right)\vec{q}$$
$$+ \frac{\mu_e}{4\pi\rho_m}\left(\nabla \times \vec{H}\right) \times \vec{H},$$

$$\nabla \cdot \vec{q} = 0,$$

$$E\frac{\partial T}{\partial t} + (\vec{q} \cdot \nabla)T = \left(\beta - \frac{g}{c_p}\right)w + \kappa\nabla^2 T,$$

$$\rho = \rho_m\left[1 - \alpha(T - T_0)\right]. \tag{1}$$

The magnetic permeability μ_e, the kinematic viscosity v, the kinematic viscoelasticity v', and the thermal diffusivity κ are all assumed to be constants. Maxwell's equations relevant to the problems are

$$\epsilon\frac{d\vec{H}}{dt} = \left(\vec{H} \cdot \nabla\right)\vec{q} + \epsilon\eta\nabla^2\vec{H} - \frac{c\epsilon}{4\pi Ne}\nabla \times \left[\left(\nabla \times \vec{H} \times \vec{H}\right)\right],$$

$$\nabla \cdot \vec{H} = 0, \tag{2}$$

where $d/dt = \partial/\partial t + \epsilon^{-1}\,\vec{q} \cdot \nabla$ stands for convective derivative.

Here $E = \epsilon + (1 - \epsilon)(\rho_s c_s/\rho_m c_f)$ is a constant, and E' is a constant analogous to E but corresponding to solute rather than heat. ρ_s, c_s and ρ_m, c_f stand for density and heat capacity of solid (porous matrix) material and fluid, respectively. The steady state solution is

$$\vec{q} = (0, 0, 0), \quad T = -\beta z + T_0,$$
$$\rho = \rho_m\left(1 + \alpha\beta z - \alpha'\beta'z\right). \tag{3}$$

Let $\delta\rho$, δp, θ, $\vec{h}(h_x, h_y, h_z)$, and $\vec{q}(u, v, w)$ denote, respectively, the perturbations in density ρ, pressure p, temperature T, magnetic field $\vec{H}(0, 0, H)$, and filter velocity (zero initially). Then the linearized hydromagnetic perturbation equations through porous medium (Joseph [22], Walter's [23], Sherman and Sutton [4], and Spiegel and Veronis [6]), relevant to the problem, are

$$\left(\frac{1}{\epsilon}\right)\frac{\partial \vec{q}}{\partial t} = -\nabla\left(\frac{\delta p}{\rho_m}\right) + \vec{g}\frac{\delta\rho}{\rho_m} - \frac{1}{k_1}\left(v - v'\frac{\partial}{\partial t}\right)\vec{q}$$
$$+ \frac{\mu_e}{4\pi\rho_m}\left(\nabla \times \vec{h}\right) \times \vec{H},$$

$$\nabla \cdot \vec{q} = 0,$$

$$\epsilon\frac{\partial \vec{h}}{\partial t} = \left(\vec{H} \cdot \nabla\right)\vec{q} + \epsilon\eta\nabla^2\vec{h} - \frac{c\epsilon}{4\pi Ne\eta}\nabla \times \left[\left(\nabla \times \vec{h}\right) \times \vec{H}\right],$$

$$\nabla \cdot \vec{h} = 0,$$

$$E\frac{\partial \theta}{\partial t} = \left(\beta - \frac{g}{c_p}\right)w + \kappa\nabla^2\theta. \tag{4}$$

And change in density $\delta\rho$ caused by perturbation θ in temperature is given by

$$\delta\rho = -\alpha\rho_m\theta, \tag{5}$$

where α is the coefficient of thermal expansion.

Writing (4) in scalar form, using (5), and eliminating u, v, h_x, h_y, and δp between them, we obtain

$$\frac{1}{\epsilon}\frac{\partial}{\partial t}\nabla^2 w = g\alpha\left(\frac{\partial^2}{\partial x^2} + \frac{\partial^2}{\partial y^2}\right)\theta - \frac{1}{k_1}\left(v - v'\frac{\partial}{\partial t}\right)\nabla^2 w$$
$$+ \frac{\mu_e H}{4\pi\rho_m}\nabla^2\frac{\partial h_z}{\partial z},$$

$$\epsilon\frac{\partial h_z}{\partial t} = H\frac{\partial w}{\partial z} + \epsilon\eta\nabla^2 h_z - \frac{cH\epsilon}{4\pi Ne}\frac{\partial\xi}{\partial z},$$

$$\epsilon\frac{\partial\xi}{\partial t} = H\frac{\partial\varsigma}{\partial z} + \epsilon\eta\nabla^2\xi + \frac{cH\epsilon}{4\pi Ne}\nabla^2\frac{\partial h_z}{\partial z},$$

$$\frac{1}{\epsilon}\frac{\partial\varsigma}{\partial t} = -\frac{1}{k_1}\left(v - v'\frac{\partial}{\partial t}\right)\varsigma + \frac{\mu_e H}{4\pi\rho_m}\frac{\partial\xi}{\partial z},$$

$$E\frac{\partial\theta}{\partial t} = \left(\beta - \frac{g}{c_p}\right)w + \kappa\nabla^2\theta, \tag{6}$$

where $\nabla^2 = \partial^2/\partial x^2 + \partial^2/\partial y^2 + \partial^2/\partial z^2$. $\varsigma = \partial v/\partial x - \partial u/\partial y$ and $\xi = (\partial/\partial x)h_y - (\partial/\partial y)h_x$ denote the z-components of vorticity and current density, respectively.

Consider the case in which both the boundaries are free, the medium adjoining the fluid is perfectly conducting and the temperatures at the boundaries are kept fixed. The case of two free boundaries is a little artificial, except in stellar atmospheres (Spiegel [24]) and in certain geophysical situations where it is most appropriate, but it allows us to have an analytical solution. It has been shown by Spiegel that the assumption of free boundary conditions is not a serious one, so in free boundary conditions, the vertical velocity temperature fluctuation horizontal stress and all vanish on the boundaries. The boundary conditions, appropriate to the problem, are (Chandrasekhar [1])

$$w = 0, \quad \frac{\partial^2 w}{\partial z^2} = 0, \quad \theta = 0, \quad \frac{\partial \varsigma}{\partial z} = 0, \tag{7}$$

$$h_z = 0 \quad \text{at } z = 0, z = d.$$

3. The Dispersion Relation

Analyzing the disturbances into normal modes, we seek solutions whose dependence on x, y, and t is given by

$$[w, \theta, h_z, \zeta, \xi] = [W(z), \Theta(z), K(z), Z(z), X(z)]$$
$$\times \exp\left(ik_x x + ik_y y + nt\right), \tag{8}$$

where k_x, k_y are horizontal wave numbers, k $(=\sqrt{k_x^2 + k_y^2})$ is the resultant wave number, and n is a complex constant. Using the dimensionless variables $a = kd$, $\sigma = nd^2/\nu$, $p_1 = \nu/\kappa$, $p_2 = \nu/\eta$, $P_l = k_1/d^2$, $F = \nu'/d^2$, $x^* = x/d$, $y^* = y/d$, $z^* = z/d$, and $D = d/dz^*$ and removing the stars for convenience, (5)-(6) with the help of (8) become

$$\left[\frac{\sigma}{\epsilon} + \frac{1 - \sigma F}{P_l}\right]\left(D^2 - a^2\right)W + \frac{g\alpha d^2 a^2}{\nu}\Theta$$

$$- \frac{\mu_e H d}{4\pi\rho_m \nu}\left(D^2 - a^2\right)DK = 0,$$

$$\left(\frac{\sigma}{\epsilon} + \frac{1 - \sigma F}{P_l}\right)Z = \frac{\mu_e H d}{4\pi\rho_m \nu}DX,$$

$$\left(D^2 - a^2 - p_2\sigma\right)K = -\left(\frac{Hd}{\epsilon\eta}\right)DW + \frac{cHd}{4\pi Ne\eta}DX,$$

$$\left(D^2 - a^2 - p_2\sigma\right)X$$

$$= -\left(\frac{Hd}{\epsilon\eta}\right)DZ - \frac{cH}{4\pi Ne\eta d}\left(D^2 - a^2\right)DK,$$

$$\left(D^2 - a^2 - Ep_1\sigma\right)\Theta = -\left(\frac{G-1}{G}\right)\frac{\beta d^2}{\kappa}W. \tag{9}$$

From boundary conditions (7), using expression (8), we have

$$W = D^2 W = 0, \quad \Theta = 0, \quad DZ = 0, \quad K = 0 \tag{10}$$
$$\text{at } z = 0, z = 1.$$

Eliminating Θ, K, Z, and X from (9), we obtain

$$\left(\frac{\sigma}{\epsilon} + \frac{1 - \sigma F}{P_l}\right)\left(D^2 - a^2\right)\left(D^2 - a^2 - Ep_1\sigma\right)$$

$$\times \left[\left(\frac{\sigma}{\epsilon} + \frac{1 - \sigma F}{P_l}\right)\left(D^2 - a^2 - p_2\sigma\right)^2\right.$$

$$+ \frac{Q}{\epsilon}\left(D^2 - a^2 - p_2\sigma\right)D^2$$

$$\left. - M\left(\frac{\sigma}{\epsilon} + \frac{1 - \sigma F}{P_l}\right)\left(D^2 - a^2\right)D^2\right]W$$

$$- Ra^2\left(\frac{G-1}{G}\right)$$

$$\times \left[\left(\frac{\sigma}{\epsilon} + \frac{1 - \sigma F}{P_l}\right)\left(D^2 - a^2 - p_2\sigma\right)^2\right.$$

$$+ \frac{Q}{\epsilon}\left(D^2 - a^2 - p_2\sigma\right)D^2$$

$$\left. - M\left(\frac{\sigma}{\epsilon} + \frac{1 - \sigma F}{P_l}\right)\left(D^2 - a^2\right)D^2\right]W$$

$$+ \frac{Q}{\epsilon}\left(D^2 - a^2\right)\left(D^2 - a^2 - Ep_1\sigma\right)$$

$$\times \left[\left(\frac{\sigma}{\epsilon} + \frac{1 - \sigma F}{P_l}\right)\left(D^2 - a^2 - p_2\sigma\right) + \frac{Q}{\epsilon}D^2\right]D^2W = 0. \tag{11}$$

Here $R = g\alpha\beta d^4/\nu\kappa$ is thermal Rayleigh number, $Q = \mu_e H^2 d^2/4\pi\rho_m \nu\eta$ is Chandrasekhar number, and $M = (cH/4\pi Ne\eta)^2$ is nondimensional number according to Hall currents.

It can be shown with the help of (9) and boundary conditions (10) that all the even order derivatives of W vanish at the boundaries and hence the proper solution of (10) characterizing the lowest mode is

$$W = W_o \sin \pi z, \tag{12}$$

where W_o is a constant. Substituting (12) in (11) and letting $x = a^2/\pi^2$, $R_1 = R/\pi^4$, $Q_1 = Q/\pi^2$, $i\sigma_1 = \sigma/\pi^2$, and $P = \pi^2 P_l$, we obtain the dispersion relation

$$R_1 x = \left(\frac{G}{G-1}\right)$$

$$\left\{\frac{1}{P} + i\sigma_1\left(\frac{1}{\epsilon} - \frac{\pi^2 F}{P}\right)\right\}(1+x)(1+x+Eip_1\sigma_1)$$

$$\times \left[\left\{\frac{1}{P} + i\sigma_1\left(\frac{1}{\epsilon} - \frac{\pi^2 F}{P}\right)\right\}(1+x+ip_2\sigma_1)^2 + \frac{Q_1}{\epsilon}(1+x+ip_2\sigma_1) - M\left\{\frac{1}{P} + i\sigma_1\left(\frac{1}{\epsilon} - \frac{\pi^2 F}{P}\right)\right\}(1+x)\right] \qquad (13)$$

$$\times \frac{+\frac{Q}{\epsilon}(1+x)(1+x+iEp_1\sigma_1)\left[(1+x+ip_2\sigma_1)\times\left\{\frac{1}{P}+i\sigma_1\left(\frac{1}{\epsilon}-\frac{\pi^2 F}{P}\right)\right\} + \frac{Q_1}{\epsilon}\right]}{(1+x+ip_2\sigma_1)^2\left\{\frac{1}{P}+i\sigma_1\left(\frac{1}{\epsilon}-\frac{\pi^2 F}{P}\right)\right\} + \frac{Q_1}{\epsilon}(1+x+ip_2\sigma_1) - M(1+x)\left\{\frac{1}{P}+i\sigma_1\left(\frac{1}{\epsilon}-\frac{\pi^2 F}{P}\right)\right\}}.$$

4. The Stationary Convection

For the case of stationary convection $\sigma = 0$, and (13) reduces to

$$R_1 = \left(\frac{G}{G-1}\right)\left(\frac{1+x}{x}\right)$$

$$\times \frac{\left((1+x)/P + Q_1/\epsilon\right)^2 - M(1+x)/P^2}{(1+x)/P + Q_1/\epsilon - M/P}. \qquad (14)$$

In order to investigate the effects of Hall current, medium permeability, and magnetic field, we examine the behaviour of dR_1/dM, dR_1/dP, and dR_1/dQ_1 analytically. Equation (14) yields

$$\frac{dR_1}{dM} = \frac{G}{G-1}\frac{Q_1(1+x)}{\epsilon x P}\frac{((1+x)/P + Q_1/\epsilon)}{((1+x)/P + Q_1/\epsilon - M/P)^2}, \qquad (15)$$

which is positive. The Hall current, therefore, had postpone the onset of thermal convection through porous medium for $G > 1$. It is evident from (14) that

$$\frac{dR_1}{dP} = -\frac{G}{G-1}\frac{(1+x)}{xP^2}\frac{\left((1+x)/P^2\right)(1+x-M)^2 + 2Q_1(1+x)(1+x+M)/\epsilon P + (Q_1/\epsilon)^2(1+x+M)}{((1+x-M)/P + Q_1/\epsilon)^2},$$

$$\frac{dR_1}{dQ_1} = \frac{G}{G-1}\frac{(1+x)}{\epsilon x}\frac{((1+x)/P + Q_1/\epsilon)((1+x)/P + Q_1/\epsilon - 2M/P) + \left(M/P^2\right)(1+x)}{((1+x-M)/P + Q_1/\epsilon)^2}, \qquad (16)$$

which imply that for $G > 1$, medium permeability hasten postpone the onset of convection, where as magnetic field has postponed the onset of convection in Walters' B' elastico-viscous fluid through porous medium for $Q_1 > (\epsilon/P)[2M - (1 + x)]$ and hasten postpone the onset of convection, if $Q_1 < (\epsilon/P)[2M - (1 + x)]$. Therefore, magnetic field has duel character in presence of Hall currents through porous medium. For fixed P, Q_1, and M, let G (accounting for the compressibility effects) also, be kept fixed in (14). Then we find that

$$\overline{R}_c = \left(\frac{G}{G-1}\right)R_c, \qquad (17)$$

where $\overline{R}_c$ and R_c denote, respectively, the critical Rayleigh numbers in the presence and absence of compressibility. Thus, the effect of compressibility is to postpone the onset of thermal instability. The cases $G < 1$ and $G = 1$ correspond to negative and infinite values of Rayleigh number which are not relevant in the present study. $G > 1$ is relevant here.

The compressibility, therefore, has postponed the onset of convection.

5. Graphical Results and Discussion

The dispersion relation (14), in case of stationary convection, has been computed by concerning mathematical software. The results have been displayed graphically for various parameters of interest. The effects of these parameters especially Hall parameter, medium permeability, magnetic field, Rayleigh number with wave number have been studied. In Figure 2, Rayleigh number R_1 is plotted against wave number x (=10–80), for different values of Hall parameter M (= 10–40) and fixed values of medium permeability parameter $P = 3$, $G = 10$, magnetic field parameter $Q_1 = 100$, and $\epsilon = 0.5$. Here, we find that with the increase in the value of Hall current parameter, value of Rayleigh number is increased, showing that the Hall currents parameter has stabilizing effect on the system.

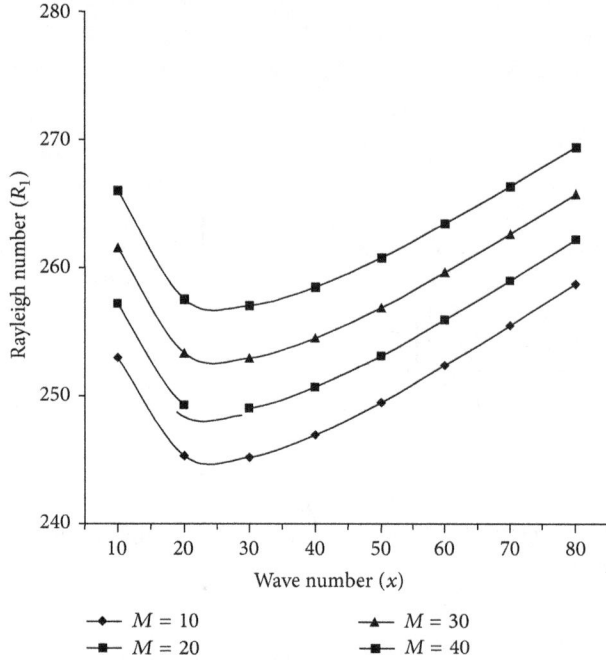

FIGURE 2: Variation of Rayleigh number R_1 against wave number x for $P = 3$, $G = 10$, $Q_1 = 100$, and $\epsilon = 0.5$.

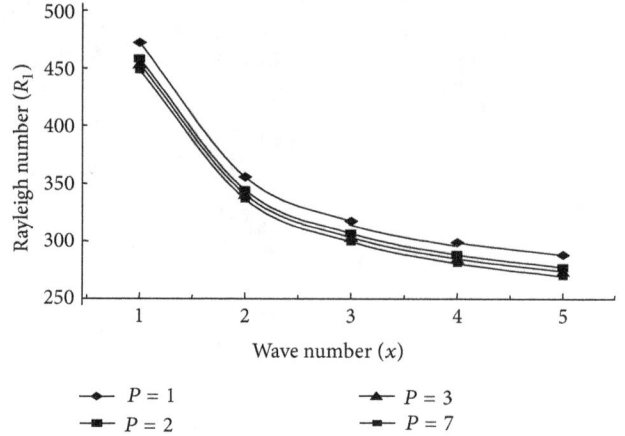

FIGURE 3: Variation of Rayleigh number R_1 against wave number x for $Q_1 = 100$, $M = 10$, $G = 10$, and $\epsilon = 0.5$.

In Figure 3, Rayleigh number R_1 is plotted against wave number x (=1–5), and for different medium permeability parameter P (=1, 2, 3, 7), for fixed magnetic field parameter $Q_1 = 100$, Hall current parameter $M = 10$, $G = 10$, and $\epsilon = 0.5$ are considered. We find that as medium permeability P increases, value of Rayleigh number R_1 decreases, which indicates the destabilizing effect of medium permeability.

In Figure 4, Rayleigh number R_1 is plotted against wave number x (=10–80), and for different values of magnetic field parameter Q_1 (=10–40), for fixed values of medium permeability $P = 3$, Hall current parameter $M = 10$, $G = 10$ and $\epsilon = 0.5$ are considered. It is clear from the graph that with the increase in the value of magnetic field parameter, there is decrease as well as increase in the Rayleigh number R_1, implying the destabilizing as well as stabilizing effect on the system.

6. The Case of Overstability

In the present section, we discuss the possibility as to whether instability may occur as overstability. Since for overstability we wish to determine the critical Rayleigh number for the onset of instability via a state of pure oscillations, it will suffice to find conditions for which (13) will admit of solutions with σ_1 real. Equating real and imaginary parts of (13) and eliminating R_1 between them, we obtain

$$A_3 c_1^3 + A_2 c_1^2 + A_1 c_1 + A_o = 0, \tag{18}$$

where

$$c_1 = \sigma_1^2, \qquad b = 1 + x, \tag{19}$$

$$A_3 = p_2^4 \left(\frac{1}{\epsilon} - \frac{\pi^2 F}{P} \right)^2 \left[\frac{E p_1}{P} + b \left(\frac{1}{\epsilon} - \frac{\pi^2 F}{P} \right) \right], \tag{20}$$

$$
\begin{aligned}
A_o = {} & \frac{1}{P} \left(\frac{1}{\epsilon} - \frac{\pi^2 F}{P} \right) b^5 \\
& + \left[\frac{E p_1}{P} + \frac{2}{P} \left(\frac{Q_1}{\epsilon} - \frac{M}{P} \right) \left(\frac{1}{\epsilon} - \frac{\pi^2 F}{P} \right) \right] b^4 \\
& + \left[\left(\frac{Q_1}{\epsilon} - \frac{M}{P} \right)^2 \left(\frac{1}{\epsilon} - \frac{\pi^2 F}{P} \right) + \frac{2 E p_1}{P^2} \left(\frac{Q_1}{\epsilon} - \frac{M}{P} \right) \right. \\
& \left. + \frac{Q_1}{\epsilon P^2} (E p_1 - p_2) \right] b^3 \\
& + \left[\frac{M Q_1}{\epsilon P^2} (3 E p_1 + p_2) + \left(\frac{Q_1}{\epsilon} \right)^2 \right. \\
& \times \left\{ \frac{2}{P} (E p_1 - p_2) + E p_1 - M \left(\frac{1}{\epsilon} - \frac{\pi^2 F}{P} \right) \right\} \\
& \left. + \frac{E p_1 M^2}{P^3} \right] b^2 \\
& + \left(\frac{Q_1}{\epsilon} \right)^2 \left[\frac{E p_1 M}{P} + \frac{Q_1}{\epsilon} (E p_1 - p_2) \right] b.
\end{aligned}
\tag{21}
$$

The three values of c_1, σ_1 being real, are positive. The product of the roots of (18) is $-A_0/A_3$, and if this is to be positive then

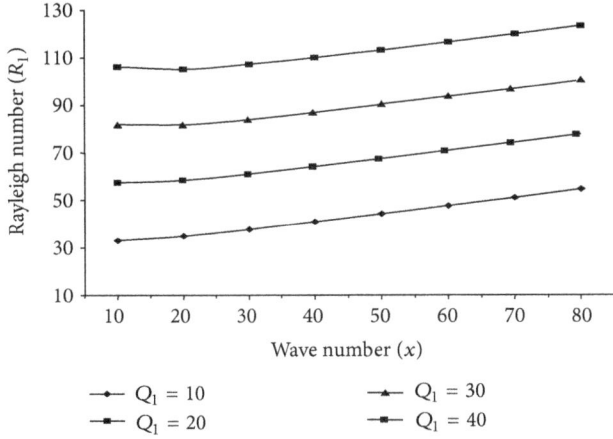

FIGURE 4: Variation of Rayleigh number R_1 against wave number x for $P = 3$, $M = 10$, $G = 100$, and $\epsilon = 0.5$.

$A_0 < 0$, since from (20), $A_3 > 0$ if $1/\epsilon > \pi^2 F/P$. Equation (17) shows that this is clearly impossible if

$$\frac{1}{\epsilon} > \frac{\pi^2 F}{P}, \quad Ep_1 > p_2, \quad Ep_1 > M\left(\frac{1}{\epsilon} - \frac{\pi^2 F}{P}\right), \quad (22)$$

which imply that

$$v' < \frac{k_1}{\epsilon}, \quad E\frac{v}{\kappa} > \max\left[\frac{v}{\eta}, \left(\frac{cH}{4\pi Ne\eta}\right)^2 \frac{k_1 - v'\epsilon}{k_1\epsilon}\right]. \quad (23)$$

Thus, $v' < k_1/\epsilon$ and $E(v/\kappa) > \max[v/\eta, (cH/4\pi Ne\eta)^2((k_1 - v'\epsilon)/k_1\epsilon)]$ are sufficient conditions for the nonexistence of overstability, the violation of which does not necessarily imply the occurrence of overstability.

7. Concluding Remarks

Combined effect of various parameters; that is, magnetic field, compressibility, medium permeability, and hall currents effect has been investigated on thermal instability of a Walter's B' fluid. The principle concluding remarks are as the following.

(i) For the stationary convection. Walter's B' fluid behaves like an ordinary Newtonian fluid due to the vanishing of the viscoelastic parameter.

(ii) The presence of magnetic field (and therefore Hall currents) and medium permeability effects introduce oscillatory modes in the system; in the absence of these effects, the principle of exchange of stabilities is valid.

(iii) The sufficient conditions for the occurrence of overstability are $v' < k_1/\epsilon$ and $E(v/\kappa) > \max[v/\eta, (cH/4\pi Ne\eta)^2((k_1 - v'\epsilon)/k_1\epsilon)]$, violation of which does not necessarily imply the occurrence of overstability.

(iv) From (17), it is clear that effect of compressibility has postponed the onset of convection.

(v) To investigate the effects of medium permeability, magnetic permeability, and Hall currents in compressible Walter's B' viscoelastic fluid, we examined the expressions dR_1/dM, dR_1/dP, and dR_1/dQ_1 analytically. Hall current effect has postponed the onset of convection and medium permeability hastened the onset of convection, where magnetic field has postponed the onset of convection as well as hastened the onset of convection.

Nomenclature

g:	Acceleration due to gravity (ms^{-2})
K:	Stoke's drag coefficient (kg s^{-1})
k:	Wave number (m^{-1})
k_x, k_y:	Horizontal wave-numbers (m^{-1})
k_1:	Medium permeability (m^2)
m:	Mass of single particle (g)
N:	Suspended particle number density (m^{-3})
n:	Growth rate (s^{-1})
p:	Fluid pressure (Pa)
t:	Time (s)
$\vec{u}$:	Fluid velocity (ms^{-1})
$\vec{v}$:	Suspended particle velocity (ms^{-1})
$\vec{H}$:	Magnetic field intensity vector having component $(0, 0, H)$ (G)
$\beta(= \|dT/dz\|)$:	Steady adverse temperature gradient (Km^{-1})
N_{p_1}:	Thermal Prandtl number $(-)$
N_{p_2}:	Magnetic Prandtl number $(-)$
$R = g\alpha\beta d^4/v\kappa$:	thermal Rayleigh number
$Q = \mu_e H^2 d^2/4\pi\rho_m v\eta$:	Chandrasekhar number
$M = (cH/4\pi Ne\eta)^2$:	Nondimensional number according to Hall currents
f:	The mass fraction
ζ:	Z Component of vorticity
ξ:	Z Component of current density
N_R^C and $\overline{N_R^C}$:	Critical Rayleigh numbers in the absence and presence of compressibility.

Greek Letters

ϵ: Medium porosity $(\text{m}^0 \text{ s}^0 \text{ k}^0)$

μ: Dynamic viscosity $(\text{km}^{-1} \text{ s}^{-1})$

μ': Fluid viscoelasticity $(\text{km}^{-1} \text{ s}^{-1})$

v: Kinematic viscosity $(\text{m}^2 \text{ s}^{-1})$

v': Kinematic viscoelasticity $(\text{m}^2 \text{ s}^{-1})$

ρ: Density (kg m^{-3}).

Acknowledgment

The authors are grateful to the referees for their technical comments and valuable suggestions, resulting in a significant improvement of the paper.

References

[1] S. Chandrasekhar, *Hydrodynamic and Hydromagnetic Stability*, Dover Publications, New York, NY, USA, 1981.

[2] H. Sato, "The Hall effect in the viscous flow of ionized gas between parallel plates under transverse magnetic field," *Journal of the Physical Society of Japan*, vol. 16, no. 7, pp. 1427–1433, 1961.

[3] I. Tani, "Steady flow of conducting fluid in channels under transverse magnetic field with consideration of Hall Effect," *Journal of Aerospace Science*, vol. 29, pp. 297–305, 1962.

[4] A. Sherman and G. W. Sutton, *Magnetohydrodynamics*, Northwestern University Press, Evanston, Ill, USA, 1962.

[5] A. S. Gupta, "Hall effects on thermal instability," *Revue Roumaine de Mathématique Pures et Appliquées*, pp. 665–677, 1967.

[6] E. A. Spiegel and G. Veronis', "On the Boussinesq approximation for a compressible fluid," *The Astrophysical Journal*, vol. 131, pp. 442–447, 1960.

[7] K. Chandra, "Instability of fluids heated from below," *Proceedings of the Royal Society A*, vol. 164, pp. 231–242, 1938.

[8] J. W. Scanlon and L. A. Segel, "Some effects of suspended particles on the onset of Bénard convection," *Physics of Fluids*, vol. 16, no. 10, pp. 1573–1578, 1973.

[9] O. M. Phillips, *Flow and Reaction in Permeable Rocks*, Cambridge University Press, Cambridge, UK, 1991.

[10] D. B. Ingham and I. Pop, *Transport Phenomena in Porous Medium*, Pergamon Press, Oxford, UK, 1998.

[11] D. A. Nield and A. Bejan, *Convection in Porous Medium*, Springer, New York, NY, USA, 2nd edition, 1999.

[12] C. R. B. Lister, "On the thermal balance of a mid-ocean ridge," *Geophysics Journal of the Royal Astronomical Society Continues*, vol. 26, pp. 515–535, 1972.

[13] J. A. M. McDonnell, *Cosmic Dust*, John Wiley & Sons, Toronto, Canada, 1978.

[14] R. C. Sharma and P. Kumar, "Rayleigh-Taylor instability of two superposed conducting Walter's B' elastico-viscous fluids in hydromagnetics," *Proceedings of the National Academy of Sciences A*, vol. 68, no. 2, pp. 151–161, 1998.

[15] R. C. Sharma, "MHD instability of rotating superposed fluids through porous medium," *Acta Physica Academiae Scientiarum Hungaricae*, vol. 42, no. 1, pp. 21–28, 1977.

[16] S. Sunil and T. Chand, "Rayleigh-Taylor instability of plasma in presence of a variable magnetic field and suspended particles in porous medium," *Indian Journal of Physics*, vol. 71, no. 1, pp. 95–105, 1997.

[17] S. Sunil, R. C. Sharma, and V. Sharma, "Stability of stratified Walter's B' visco-elastic fluid in stratified porous medium," *Studia Geotechnica et Mechenica*, vol. 261, no. 2, pp. 35–52, 2004.

[18] S. Sunil, R. C. Sharma, and S. Chand, "Hall effect on thermal instability of Rivlin-Ericksen fluid," *Indian Journal of Pure and Applied Mathematics*, vol. 31, no. 1, pp. 49–59, 2000.

[19] M. Singh, "Hall Current effect on thermosolutal instability in a visco-elastic fluid flowing in a porous medium," *International Journal of Applied Mechanics and Engineering*, vol. 16, no. 1, pp. 69–82, 2011.

[20] M. Singh and P. Kumar, "Hydrodynamic and hydromagnetic stability of two stratified Walter's B' elastico-viscous superposed fluids," *International Journal of Applied Mechanics and Engineering*, vol. 16, no. 1, p. 233, 2011.

[21] U. Gupta, P. Aggarwal, and R. K. Wanchoo, "Thermal convection of dusty compressible Rivlin-Ericksen viscoelastic fluid with Hall currents," *Thermal Science*, vol. 16, no. 1, pp. 177–191, 2012.

[22] D. D. Joseph, *Stability of Fluid Motion II*, Springer, New York, NY, USA, 1976.

[23] K. Walter's, "The solution of flow problems in case of materials with memory," *Journal of Mecanique*, vol. 1, pp. 469–479, 1962.

[24] E. A. Spiegel, "Conveive instability in a compressible atmosphere," *Journal of Astrophysics*, vol. 141, pp. 1068–1090, 1965.

14

Enhancement of Impinging Jet Heat Transfer Using Two Parallel Confining Plates Mounted near Rectangular Nozzle Exit

Yoshiaki Haneda,[1] **Akiko Souma,**[1] **Hideo Kurasawa,**[1] **Shouichiro Iio,**[2] **and Toshihiko Ikeda**[2]

[1] *Nagano National College of Technology, Nagano 381-8550, Japan*
[2] *Shinshu University, Nagano 380-8553, Japan*

Correspondence should be addressed to Yoshiaki Haneda; haneda@nagano-nct.ac.jp

Academic Editor: Toshiyuki Gotoh

Impinging jet heat transfer on a target plate was enhanced by using two parallel confining plates mounted between a rectangular nozzle end plate and a jet target plate. The target plate was set equal to 2, 3, 4, and 5 times the jet exit width, h, and the gap ratio of two parallel confining plates, W/h, were changed from 2.7 to 8.0 only by impinging length $H = 5h$ and from 2.7 to 6.7 by $H \neq 5h$. Two confining parallel plates mounted near the jet exit produced swing-type flow under some conditions. As a result, the maximum Nusselt number attained around the stagnation point was augmented by about 50% compared to the one for normal impinging jet without the two parallel plates and then spatial mean Nusselt number was increased by about 40%.

1. Introduction

Impinging jets have been used in several industrial processes, for example, for cooling of steel, glass, electronic components, and gas turbine vanes, for drying of paper, film, and textiles, and for freezing food and tissue in cryosurgery, because the high heat and mass transfer coefficients can be obtained around a jet stagnation region and the heat transfer characteristics can be easily controlled. In the meantime, Deo et al. [1] have examined characteristics of turbulent jets issuing from rectangular nozzles with and without two parallel plates attached as sidewalls to the slot's short sides. They reported that the potential core of the jet without sidewalls was shorter than that with sidewalls and the centerline turbulence intensity of the jet with sidewalls became asymptotical closer to the nozzle exit than that of the jet without sidewalls. The results suggest that a characteristic of impinging jet heat transfer on a target plate with sidewalls is different from that without sidewalls.

By the way, jets often impinge onto a target body mounted in confined spaces for manufacturing processes of frozen food and heat treatment of electronic components, and so forth. San et al. [2] and Lin et al. [3] have studied heat transfer on a surface for mounting an end plate flush with a jet exit. Gao and Ewing [4] have examined static and fluctuating wall pressure and heat transfer on a surface for unconfined jets and confined jets with an end plate. They found that the location of the decrease in the heat transfer and the fluctuating wall pressure in the confined jets shifted laterally outward as a nozzle-to-plate distance increased. Fitzgerald and Garimella [5] have studied characteristics of the flow field of an axisymmetric, confined, and submerged turbulent jet impinging normally on a flat plate experimentally. They mapped the toroidal recirculation pattern in the outflow region characteristic of confined jets with an end plate. Fenot et al. [6] have showed the independence of heat transfer coefficients and effectiveness from a jet injection temperature for confined and unconfined jets. The results showed that the influence of confinement was weak, but it had a great impact on effectiveness. More recently, San and Shiao [7] have examined the effects of jet plate size and plate spacing for mounting removable bars between an end plate mounted flush with a jet exit and a target plate on the heat transfer characteristics for a confined circular air jet

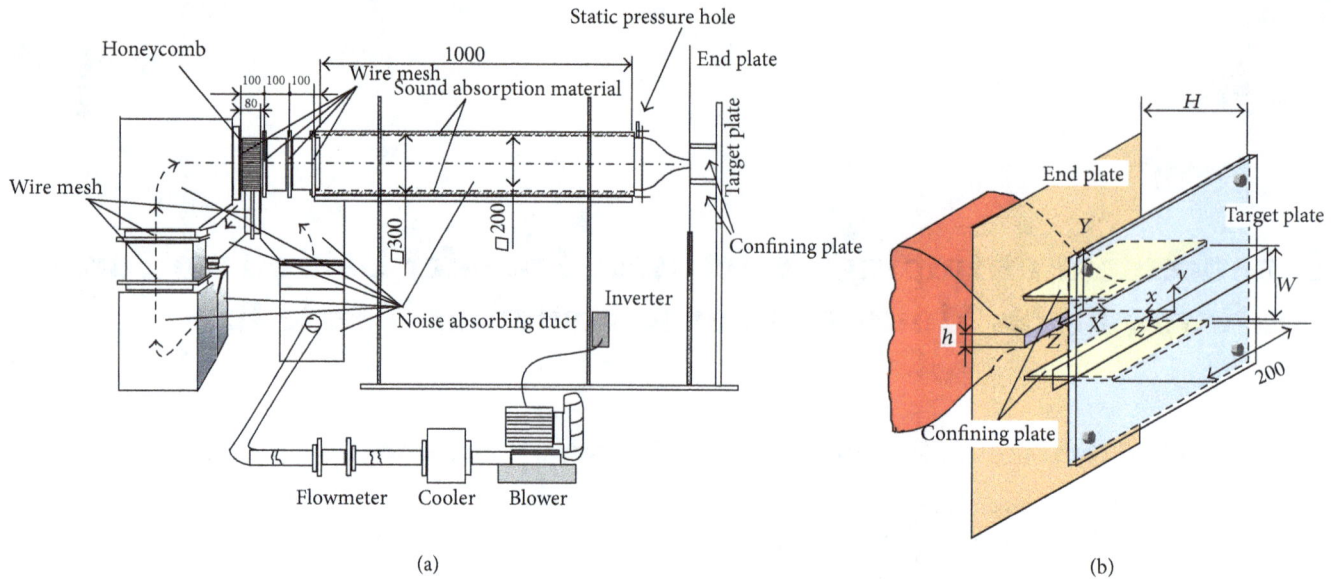

FIGURE 1: Schematic illustrations of experimental setup and the coordinate systems.

vertically impinging on a flat plate. They showed the effects of the jet plate width-to-jet diameter ratio on the stagnation Nusselt number. In addition, San et al. [8] have also studied impingement heat transfer of staggered arrays of air jets confined with an end plate and the removable bars.

Yet, up to date, there have been few investigations concerning heat transfer of a rectangular impinging jet confined with an end plate, target plate, and two confining plates mounted in parallel with long sides of the rectangular nozzle exit. The objective of this work is to examine the flow field and heat transfer influenced by two parallel plates mounted near a rectangular nozzle exit. Measurements of the heat transfer and fluctuating pressure on a target plate were performed for the impinging jet with nozzle-to-target plate distances and with confining plate-to-plate distances. Flow field for the impinging jet was qualitatively visualized using a smoke-wire method and smoke method. The experimental facilities used in this study are presented in the next section. The results of experiments are then presented and discussed. Finally, the conclusions obtained here are presented.

2. Experimental Apparatus and Procedure

Figure 1 shows schematic illustration of experimental setup and the coordinate system. The facility used to produce a jet consisted of a blower, a cooler for maintaining the air temperature within 2°C of an ambient temperature, an orifice flow meter, pipes made from polyvinyl chloride, some noise absorbing ducts, and a contraction. A honeycomb of length 80 mm with hexagon cells of 4 mm and wire meshes were installed in the connection part of the ducts for rectification. The nozzle contraction had an area ratio of about 13.3 : 1 and employs a smooth contraction profile based on sine curve and following the curve had parallel section with the length of 10 mm connecting to the exit to enhance flow uniformity. A rectangular nozzle exit contracted had a cross-section area

with the length of 200 mm and the width, h, of 15 mm, whose outlet rim was arranged flush mount with an end plate. The jet issuing horizontally through the exit impinged onto a target plate perpendicularly mounted in the middle plane of the jet. The nozzle-to-target plate spacing, H, was set equal to $H/h = 2, 3, 4$, and 5, where those places corresponded to a potential core region of the free jet used in the present study. Two parallel confining plates having the length of 200 mm, whose length and width were equal to the length of the nozzle exit and the H value, respectively, were mounted between the end plate and the target plate for each position, H. The spacing between the two confining plates mounted in parallel with the long sides of the exit, W, was set equal to $W/h = 2.7, 3.3, 4, 5.3$, and 6.7 for $H/h = 2, 3$ and 4 and to 2.7, 3.3, 4, 5.3, 6.7, and 8 for $H/h = 5$ only. The intersection between the geometric jet central axis and a target plate was defined as the origin of coordinate, and the z axis was taken across the length of the jet exit from the origin.

For the measurements of pressure and heat transfer on the target plate, the jet exit velocity obtained from flow rate divided by the cross-section area of the nozzle exit was set at 10 m/s, resulting in a Reynolds number, Re, based on the nozzle width, h, and kinematic viscosity, ν, of about 9500. For only flow visualization using a smoke-wire method, the velocity was set at 5 m/s to obtain good quality pictures of the flow field. The adjustment of the flow rate was carried out by changing the rotational frequency of the blower with an inverter. The flow rate was almost constant without the adjustment of its rotational frequency in all cases, even if the confining plate-to-plate spacing and the nozzle exit-to-target plate distance changed.

The target plate for the measurements of fluctuating pressure on the surface consisted of three pieces as shown in Figure 2. The middle sliding plate was manually traversed along the z direction within the region $-8 \leq z/h \leq 8$. A small pressure hole with 0.5 mm in diameter and 5 mm in depth

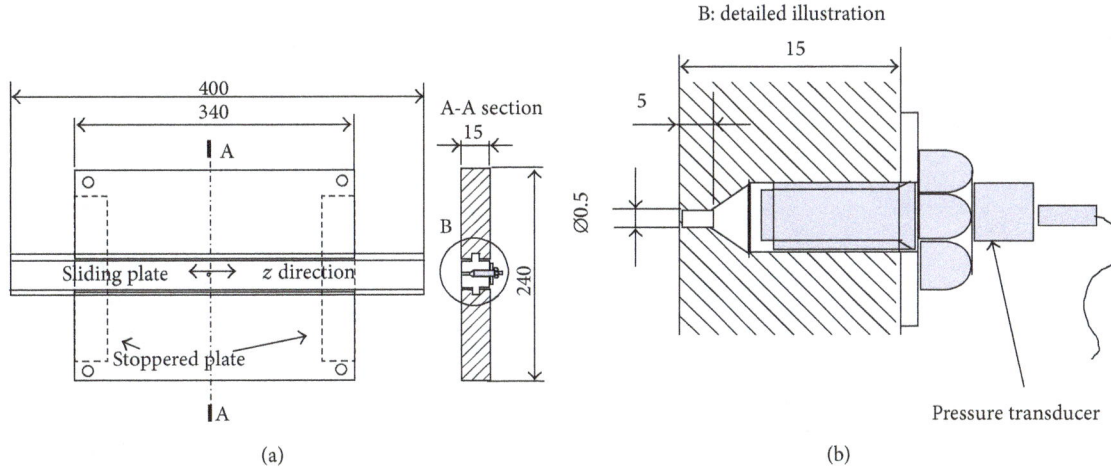

FIGURE 2: Target plate used in the pressure measurement.

was drilled from the front surface at its geometrical center and a differential pressure transducer used in measurement of the pressure was screwed into a blind hole drilled into the wall to the small hole. The transducer output signal was amplified using a strain amplifier module. The voltage signals from the amplifier were recorded onto a data logger in 30,000 data points at a frequency of 5,000 Hz. Time averaged mean pressure and root mean square values of the fluctuating pressure were calculated from the data recorded with a personal computer. The pressure on the target plate was measured at 5 mm interval by traversing the sliding plate within the region $-8 \le z/h \le 8$.

The smoke wire method was used for the visualization of confined flow field. The smoke wire was consisted of two nichrome wires with 50 μm in diameter, which were uniformly twisted together. It was individually placed at the center of the nozzle exit and a little downstream along the y axis. It was coated by paraffin oil before each test and heated by impulse direct current. Pictures of the resultant white streak lines were taken using digital camera and stroboscope light. Additional flow visualization was conducted: oil mist generated from a smoke generator was absorbed to the blower and the oil mist gushed out of the nozzle was observed with a video camera.

For the measurements of local heat transfer, three stainless steel strips 20 μm thick, 8 mm wide, and 260 mm long were glued in parallel with each other along the z-direction. These strips were electrically connected in series and were heated by passing alternating electric current. Distribution of the heated surface temperature was measured with 49 thermocouples allocated to contact with the back surface of the central strip at 5 mm interval so as to cover the region from $z/h = -8$ to $z/h = 8$. Heat flux was basically uniform but correction was carried out for its nonuniformity produced by the conductive loss toward the back surface of the target plate on which glass wool with 20 mm in thickness was mounted. The radiative heat loss to the ambient was neglected as the radiative heat loss was evaluated to be less than 3% of the total

heat flux. Thus, Local Nusselt number to be used hereafter was evaluated by the following equation:

$$\text{Nu} = \frac{q_{\text{net}} h}{\lambda (T_w - T_o)}, \tag{1}$$

where λ is the fluid thermal conductivity, T_w is the local heat transfer surface temperature, T_o is the jet temperature, and q_{net} is the corrected heat flux and evaluated as follows:

$$q_{\text{net}} = \frac{Q}{S} - q_c,$$

$$q_c = \frac{\lambda_p (T_w - T_b)}{\delta}, \tag{2}$$

where Q is the total electric power input, S the total area of the heated strips, δ the thickness of the target plate, λ_p the thermal conductivity of the target plate, and T_b the back surface temperature of the target plate measured with additional 13 thermocouples attached to the back surface of the target plate itself and evaluated from the temperature measured with the 13 thermocouples.

In this experiment, the uncertainty associated with the jet exit velocity and local Nusselt number was estimated to be 4.7% and 7.8% with a 95% confidence level, respectively.

3. Result and Discussion

Preliminary experiment for a free jet issuing from the rectangular exit was performed. The jet velocity for the free jet was measured over the jet central axial range from the exit to $20h$ with a hot-wire anemometer. Figure 3 shows the distributions of centerline mean velocity presented by U_m/U_o and of root mean square value, u', of fluctuating velocity normalized by U_m versus X/h. The results of the previous rectangular jet with aspect ratio 15 (Alnahhal and Panidis [9]) and of the two-dimensional jet (Bradbury [10]) were also shown in Figure 3 to compare them with present results.

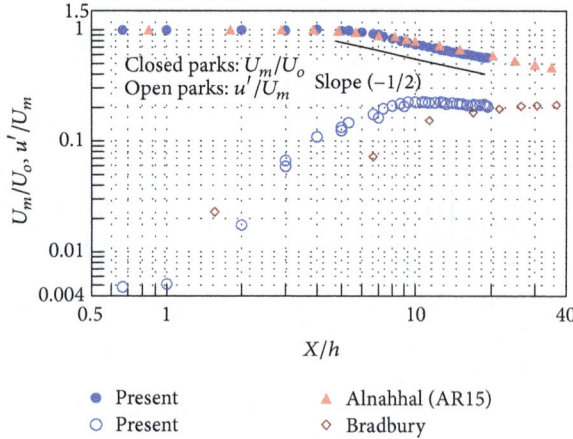

FIGURE 3: Distributions of normalized centerline mean velocity and turbulence intensity.

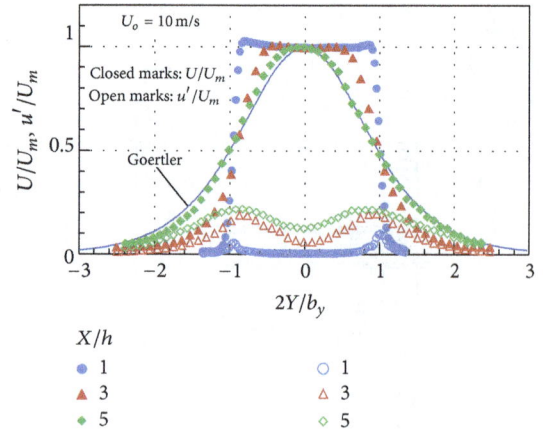

(a) Along the Y axis

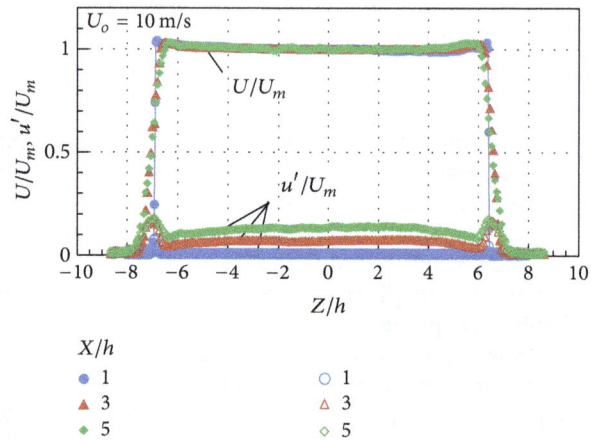

(b) Along the Z axis

FIGURE 4: Distributions of normalized mean velocity and turbulence intensity at different downstream locations.

The jet potential core region where the values of U_m/U_o are equal with approximately one is $X/h \leq 5$. The normalized centerline velocity, U_m/U_o, in the region of $X/h \geq 7$ decays almost proportional to minus one-half power of X/h, that is the jet typically have characteristics of a two-dimensional jet. On the other hand, the normalized turbulence intensity, u'/U_m, near the jet exit is about 0.48% and their values in $X/h \geq 9$ became approximately 0.2, that value is almost the same as the result of a two-dimension jet [10].

Normalized mean streamwise velocity, U/U_m, and turbulence intensity, u'/U_m, along the lateral (Y) and spanwise (Z) directions against $2Y/b_y$ (b_y is the distance where the mean streamwise velocity is half of the centerline velocity at the each X/h location) and Z/h are presented in Figures 4(a)–4(b), respectively. Solid line in Figure 4(a) is the distribution obtained by Goertler's solution for a two-dimensional jet. The mean velocity and turbulence intensity shown in Figure 4(a) are approximately uniform over the region $2Y/b_y \leq |0.6|$ at $X/h = 1$. In contrast, the uniform regions of the mean velocity decrease at $X/h = 3, 5$, and the normalized turbulence intensity, u'/U_m, becomes the saddleback distribution and the u'/U_m value increases with increase of X/h value. The mean velocity along Z axis was almost uniform over the region $-6 \leq Z/h \leq 6$ even at $X/h = 5$ in Figure 4(b).

3.1. Flow Field Characteristics. The mean and fluctuating pressure on the target plate was measured to consider the flow field near its surface and the characteristics of the impingement heat transfer and the association with the fluctuating pressure. Figures 5(a)–5(d) show the distributions of time averaged pressure along z axis for $H/h = 2, 3, 4$, and 5. The P is the time averaged gauge pressure measured on the target plate and the P_{om} is time averaged gauge pressure measured on the target plate at $z/h = 0$ for normal impinging jet without the two confining plates. The open both sides area among the nozzle end plate, confining plates, and plate is less than or equal to the nozzle exit area only for $W/h = 2.7$, 3.3, and $H/h = 2$. The P values measured in the region

$-6 < z/h < 6$ in all cases for $H/h = 2$ and 3 and two cases for $H/h = 4$ and 5 with the two confining plates, however, are larger than that for the normal impinging jet. The shape of the pressure distribution with the two confining plates is probably rather convex for the top around $z/h = 0$ or flat in the region $-6 < z/h < 6$ for $H/h = 2, 3$, and 4. In contrast, for $H/h = 5$, not all the P values measured for the presence of the two confining plates are larger than that for the normal impinging jet and the shape of the pressure distribution is concave for the top around $z/h = 0$ for $W/h = 5.3$. The concave shape may be due to presence of a pair of recirculation regions around the $z/h = 0$. The mean pressure distributions for all cases are approximately symmetrical for $z = 0$. Therefore it is thought that there are not the asymmetry characteristics of the flow on the z axis.

Figures 6(a)–6(d) show the distributions of the fluctuating pressure on the target plate for $H/h = 2, 3, 4$, and 5. The p' shown in Figure 6 is the root mean square value of fluctuating pressure. The p' values increase in most cases owing to mounting the two confining plates. The p' values for each location of H/h vary with the W/h value.

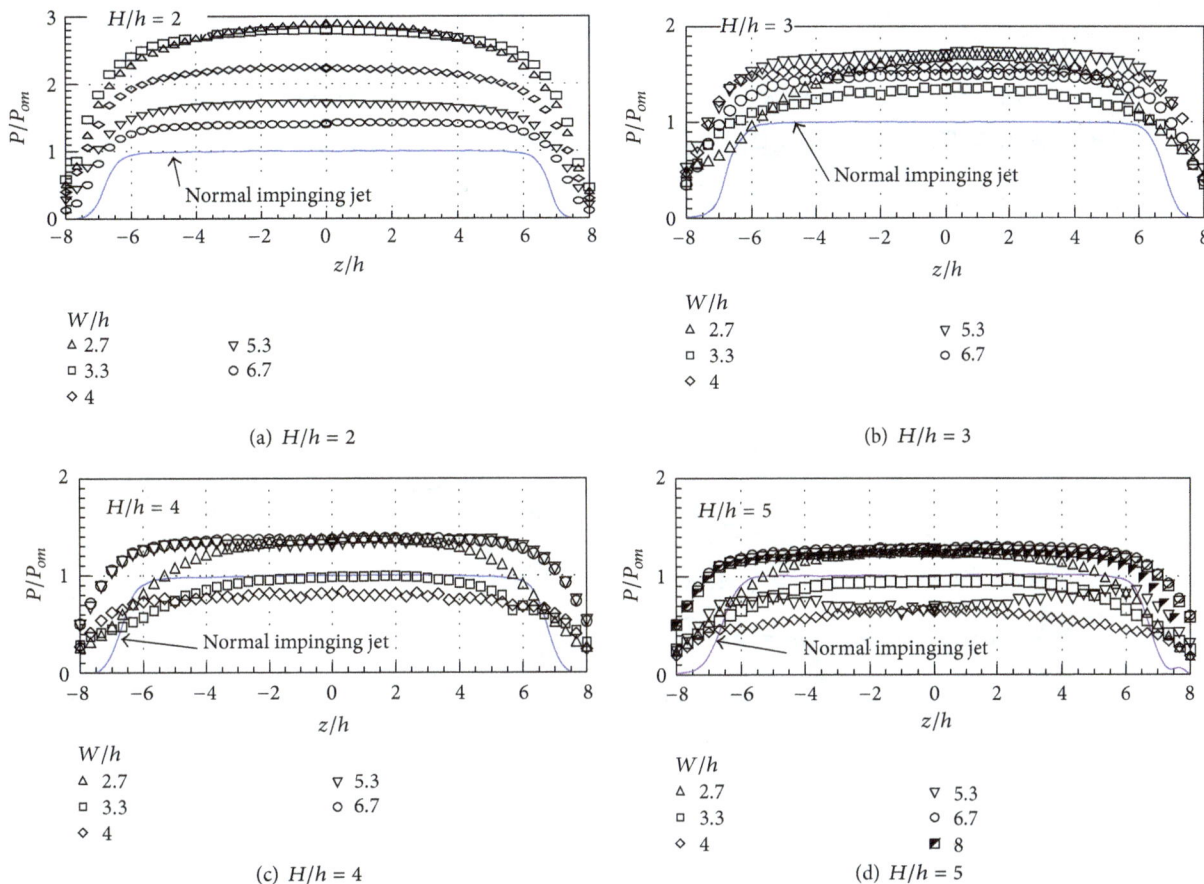

FIGURE 5: Distribution of time averaged pressure on the target plate with confining plates.

TABLE 1: Difference in flow patterns by several conditions.

	$H/h = 2$	$H/h = 3$	$H/h = 4$	$H/h = 5$
$W/h = 2.7$	(b)	(a)	(a)	(a)
$W/h = 3.3$	(b)	(a)	(a)	(a)
$W/h = 4$	(b)	(b)	(b)	(a)
$W/h = 5.3$	(b)	(b)	(b)	(b)
$W/h = 6.7$	(b)	(b)	(b)	(b)
$W/h = 8$				(b)

Therefore, the relation between fluctuating pressure at $z/h = 0$ and the confining plate-to-plate spacing ratio is shown in Figure 7. The p'_{oo} and p'_o are the root mean square value of fluctuating pressure on the target plate at $z/h = 0$ without and with the two confining plates, respectively. The W/h positions of the maximum p'_{oo}/p'_o value exist at each nozzle-to-target plate spacing, H/h, and those positions increase with an increase in the H/h value.

The flow fields of the impinging jet without or with the two confining plates were qualitatively examined by means of smoke wire visualization. Figures 8(a)–8(d) show the instantaneous photographs taken from side view for the absence of the two confining plates with $H/h = 2, 3, 4$, and 5, respectively. The vortices of the jet generated near the long side of the exit were approximately symmetric

for the center axis of the jet. In addition, in the case of each impinging distance, the jet issued out approximately horizontally.

Figures 9(a)-9(b) exemplify the instantaneous photographs taken from side view for presence of the two confining plates with $H/h = 3$ for $W/h = 3.3$ and 4, respectively. In the case of $W/h = 3.3$, the direction of the flow changed into the upper part and flowed toward the bottom after the flow impinged in the upper confining plate, and a recirculation was formed in the bottom of the region surrounded with the plates. In contrast, the flow swung up and down for the jet center surface for $W/h = 4$.

Even in the case of other conditions, we can classify it into two flow patterns such as Figure 10. The flow pattern (a) is the type that the flow is inclined to either upper part or bottom

(a) $H/h = 2$

(b) $H/h = 3$

(c) $H/h = 4$

(d) $H/h = 5$

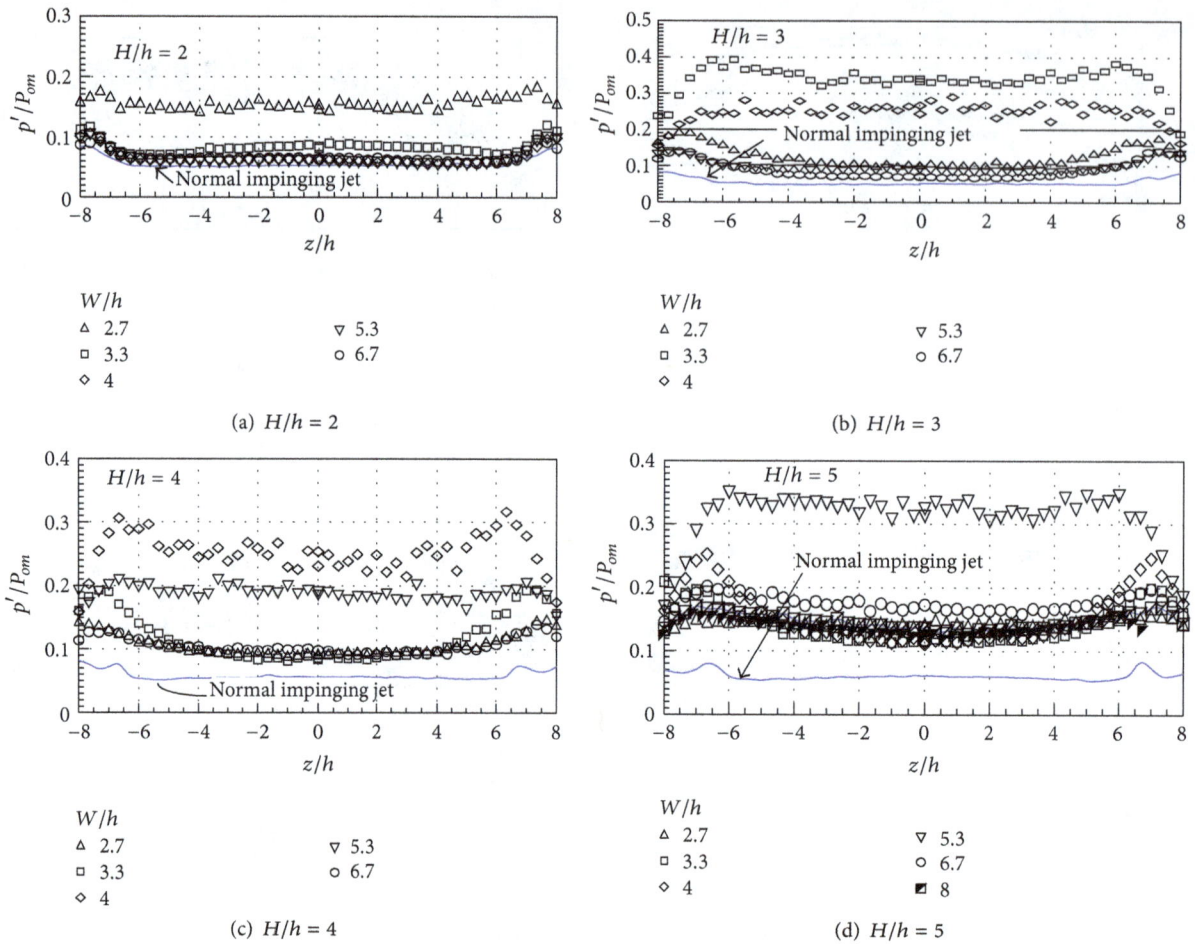

FIGURE 6: Distribution of fluctuating pressure on the target plate with confining plates.

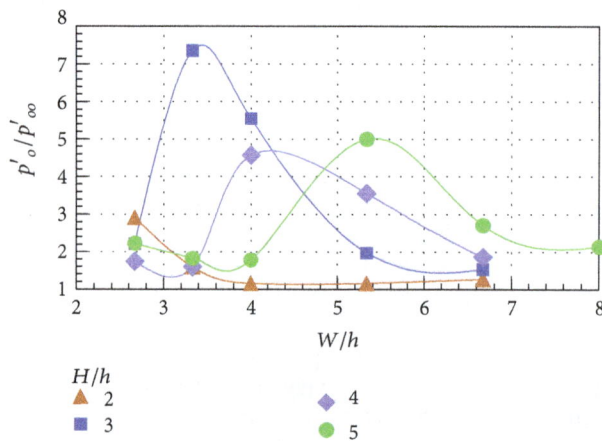

FIGURE 7: Relation between fluctuating pressure and the confining plate-to-plate distance ratio.

and the pattern (b) is the type that the jet flow swings up and down. The presence of the two confining plates mounted near the exit generated the perturbational flow field like that for using a two-dimensional suddenly expanded nozzle [11]. The differences in the flow patterns by the experiment conditions are shown in Table 1. Signs (a) and (b) in Table 1 show two kinds of flow patterns shown in Figure 10. The clear mechanisms to become two kinds of the flow patterns are not clear; however, the conditions are guessed as follows. In the cases of $W < H$, the jet is drawn to a either side of confining plate due to the Coanda effect because of smaller gap between the jet and the confining plates. Disruption of pressure balance between the upper and lower region of the jet for some reason might cause the jet incline to opposite side. In these cases it seems that the vibration of the flow along spanwise also occurs locally.

3.2. *Heat Transfer Characteristics.* Figure 11 shows the distributions of local Nusselt number on the target plate along z axis for normal impinging jet without confining plates. For $H/h = 2, 3,$ and 4 corresponded to a potential core region of the free jet used in the present study, the Nu values over the range $-6 \leq z/h \leq 6$ are closely the same. The Nu values for $H/h = 5$ corresponded to a transition region of the free jet are almost the same as values in the case of $H/h = 2, 3,$ and 4. For all the cases the maximum Nu values obtained near

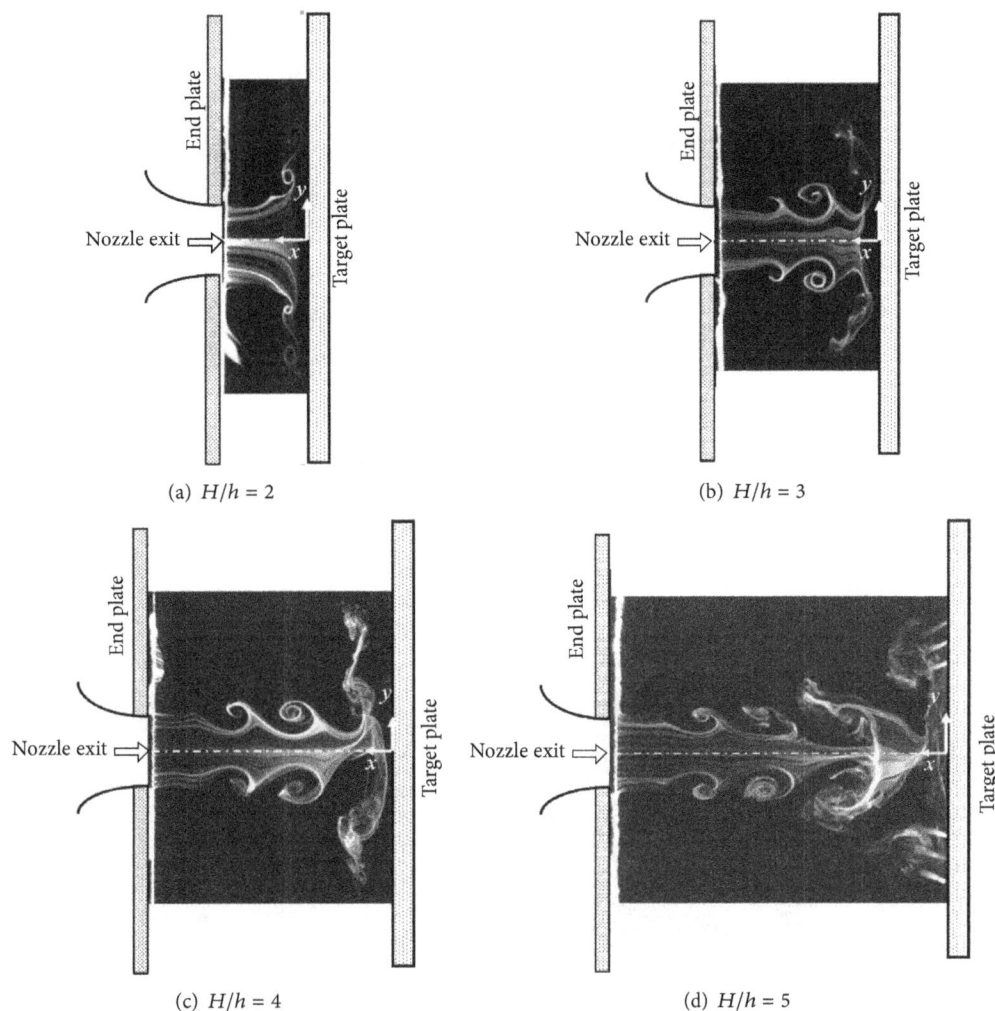

(a) $H/h = 2$

(b) $H/h = 3$

(c) $H/h = 4$

(d) $H/h = 5$

FIGURE 8: Photographs taken from side view without confining plates.

$z/h = \pm 6.7$ resulted from that turbulent intensity measured along the z axis in the preliminary study for the free jet had been maximum near the short sides of the exit. By the way, the Nu values at the geometric stagnation point were large about 11% compared with the results reported by Gardon and Akfirat [12]. This discrepancy may have resulted in the difference in the turbulence intensity near the exit and in the development process of the turbulence along the jet axis reported by them.

Figures 12(a)–12(d) show local Nusselt number, Nu, measured along the z axis for the presence of the two confining plates at $H/h = 2, 3, 4$, and 5. The solid lines shown in the figures are the Nu values for the normal impinging jet shown in Figure 11. As the impinging jet flow passed through both open sides for the presence of the two confining plates, disappearance of the maximum Nu recognized near $z/h = \pm 6.7$ for the normal impinging jet brought almost uniformity of the Nu values over the whole region on the target plate except some conditions. The Nu value over the range $-6 \leq z/h \leq 6$ for the presence of the two confining plates is larger than Nu value for normal impinging jet.

Figure 13 shows the relation between the enhancement ratios of the Nu values obtained at the geometric stagnation point and W/h values. $\overline{\mathrm{Nu}}_{oo}$ and Nu_o are the mean values measured two or three times at $z/h = 0$ for the normal impinging jet and local Nusselt number at $z/h = 0$ for the presence of the two confining plates, respectively. Maximum enhancement ratio of local Nusselt number achieved here is about 50% for $H/h = 3$ and 4. The enhancement of local Nusselt number obtained here probably results from the swing-type flow shown in Figure 10. The difference in heat transfer enhancement ratio seems to depend on the difference in the oscillation frequency of the swing-type flow when we take the visualization result of the flow into consideration. Haneda et al. [13, 14], Fu et al. [15–17], Lin et al. [18], and Chaniotis et al. [19] have reported impinging jet heat transfer enhanced owing to oscillating flow. The W/h values that obtained maxima of $\mathrm{Nu}_o/\overline{\mathrm{Nu}}_{oo}$ for each H/h value are different from the W/h values that obtained maxima of p'_{oo}/p'_o. Narayanan et al. [20] and Gao and Ewing [4] have also reported that peak fluctuating wall pressure disagreed with

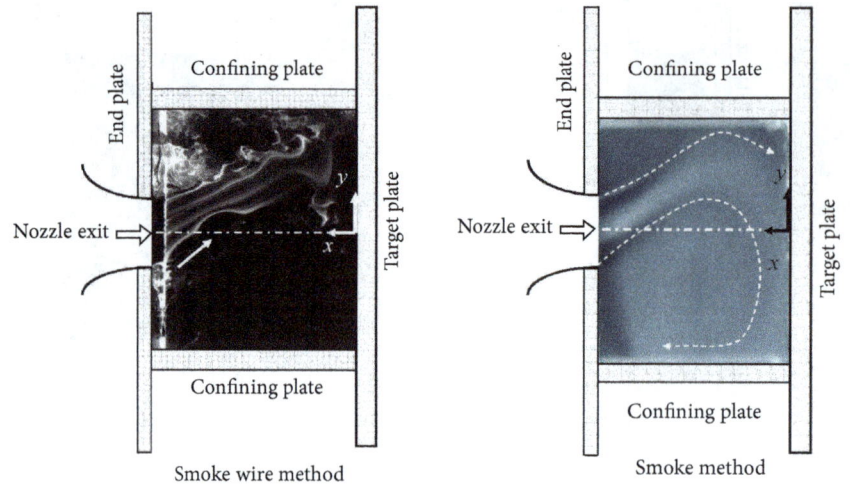

(a) $W/h = 3.3$

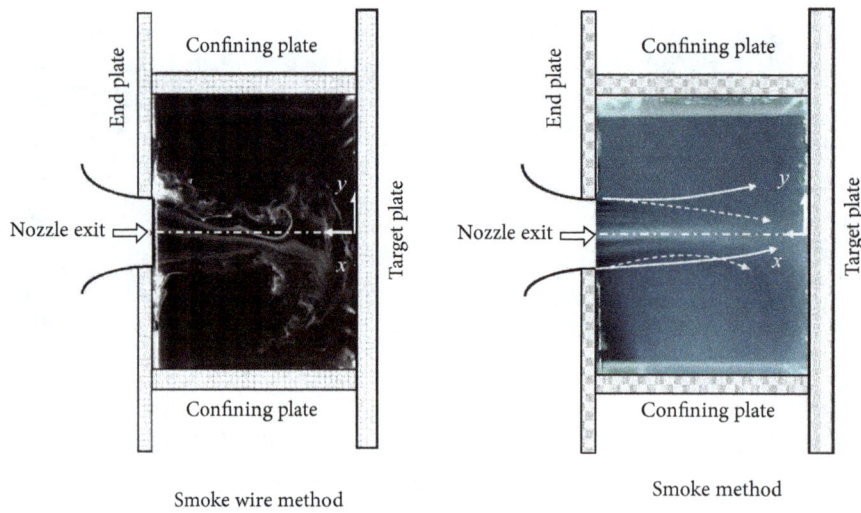

(b) $W/h = 4$

FIGURE 9: Photographs taken from side view for $H/h = 3$ with confining plates.

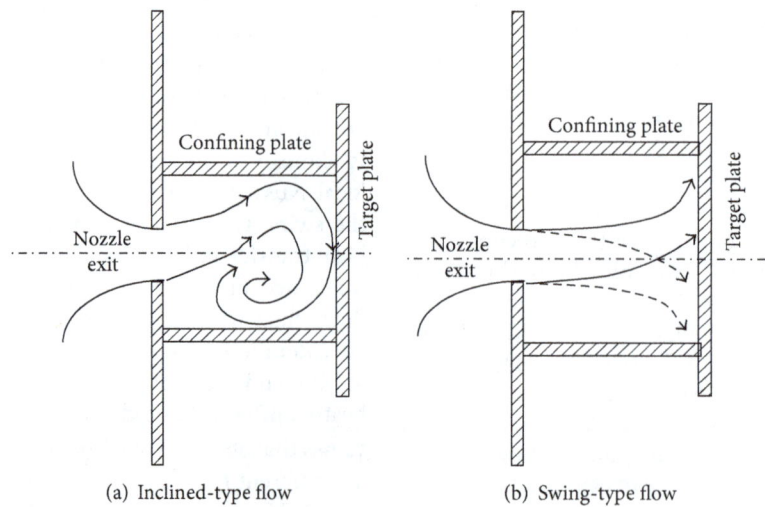

(a) Inclined-type flow

(b) Swing-type flow

FIGURE 10: Flow patterns obtained by the all experimental conditions.

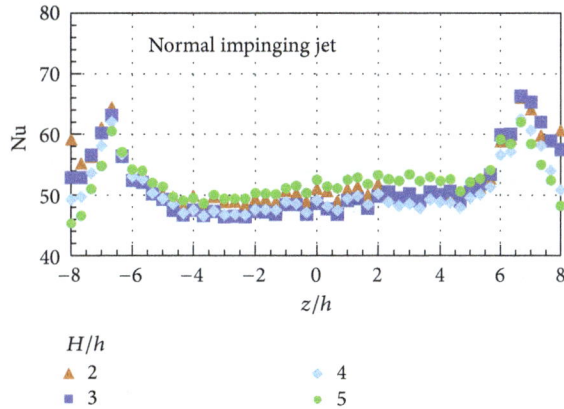

FIGURE 11: Distribution of local Nusselt number for the normal impinging jet.

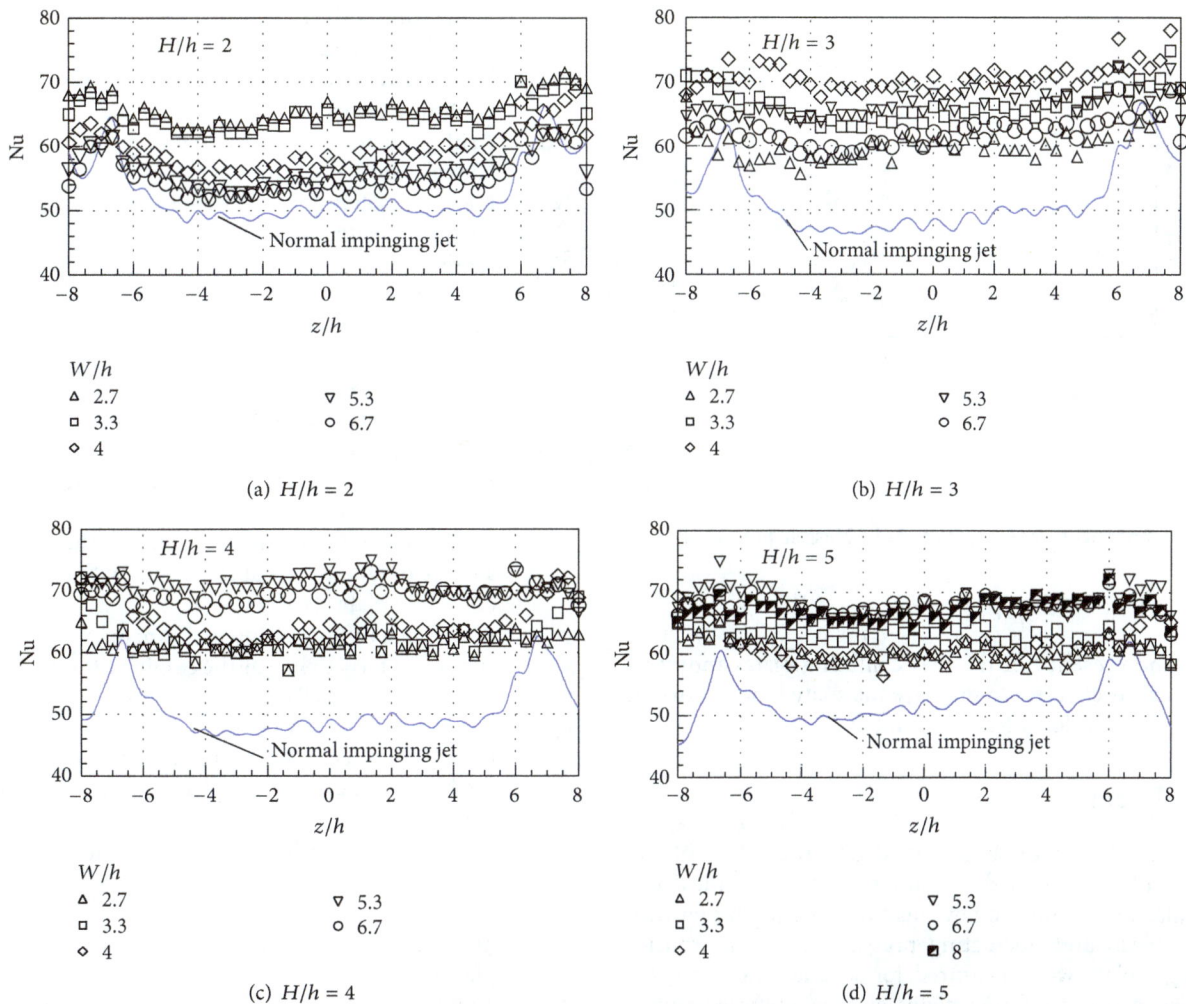

(a) $H/h = 2$

(b) $H/h = 3$

(c) $H/h = 4$

(d) $H/h = 5$

FIGURE 12: Distribution of local Nusselt number for presence of confining plates.

the peak in an impinging jet heat transfer except stagnation region.

Finally, the enhancement of spatial mean Nusselt number averaged over the region $-8 \leq z/h \leq 8$ is shown in Figure 14. $\overline{\mathrm{Nu}}_{mo}$ and Nu_m are the arithmetic mean value of the spatial mean Nusselt number calculated from Nu values

measured two or three times for normal impinging jet and the spatial mean Nusselt number for the presence of the two confining plates, respectively. The spatial mean Nusselt number is enhanced by installing the two confining plates in all cases. The maximum enhancement of mean Nusselt number increased by about 40% for $W/h = 5.3$ and $H/h =$

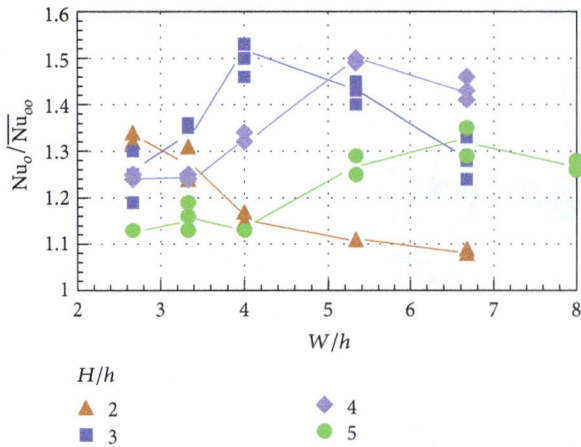

FIGURE 13: Enhancement of local Nusselt number at $Z/h = 0$.

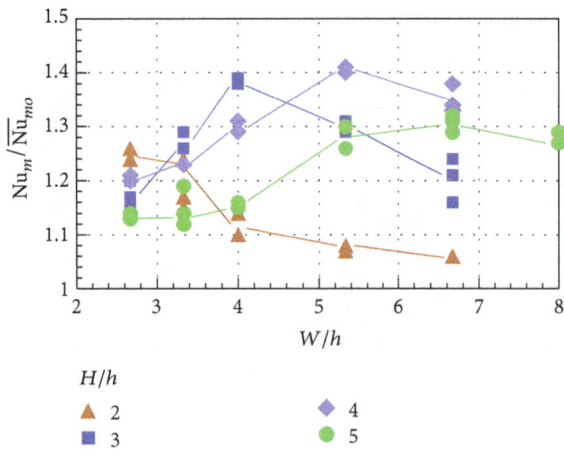

FIGURE 14: Enhancement of spatial mean Nusselt number ($-8 \leq z/h \leq 8$).

4. Thus, the presence of the two confining plates mounted near the exit is effective to improve the spatial mean Nusselt number as well as local Nusselt number.

4. Conclusion

The aim of this study is to investigate the effect of the confining plates mounted in parallel with long sides of a rectangular nozzle on the flow field and impinging jet heat transfer. Mean and fluctuating pressure and heat transfer on a target plate were measured for several kinds of spacing between two confining plates at each impinging plate distance and the flow field was visualized with a smoke wire method and a smoke method. Time averaged pressure and fluctuating pressure on the surface of the target plate varied with the spacing between the two confining plates and became large owing to the presence of the two confining plates.

The result of flow visualization showed two kinds of flow patterns; one was a declined flow type and another was a swing-type flow for the presence of the two confining

plates. This complicated flow enhanced the heat transfer on the target plate and brought almost uniform local Nusselt number over the whole region on the target plate. The enhancement ratio of local Nusselt number and spatial mean Nusselt number for the presence of the two confining plates are about 50% and 40%, respectively.

Nomenclature

b_y: Jet half-velocity width along lateral coordinate (mm)

H: Distance between nozzle end plate and target plate (mm)

h: Jet nozzle width (mm)

Nu: Local Nusselt number on the target plate

Nu_m: Spatial mean Nusselt number of Nu with confining plates

$\overline{Nu_{mo}}$: Arithmetic mean value of spatial mean Nusselt number of Nu

Nu_o: Local Nusselt number at the origin with confining plates

$\overline{Nu_{oo}}$: Arithmetic mean value at the origin without confining plates

P: Time averaged gauge pressure on the surface of target plate (Pa)

P_{om}: Time averaged gauge pressure on the surface of target plate at the origin without confining plates (Pa)

P': RMS value of fluctuating pressure on the surface of target plate (Pa)

p_o': RMS value of fluctuating pressure on the surface of target plate at the origin with confining plates (Pa)

p_{oo}': RMS value of fluctuating pressure on the surface of target plate at the origin without confining plates (Pa)

Q: Total electric power (W)

q_c: Conductive heat loss per unit area toward the back surface of target plate (W/m²)

q_{net}: Corrected heat flux (W/m²)

Re: Reynolds number based on the jet exit velocity, h and ν

S: Area of heated strip (m²)

T_b: Temperature on the rear surface of target plate (K)

T_w: Temperature on the surface of heated strip (K)

U: Jet mean streamwise velocity (m/s)

U_m: Jet mean streamwise velocity on X axis (m/s)

U_o: Exit mean centerline velocity (m/s)

u': RMS value of velocity fluctuation (m/s)

W: Distance between confining plates (mm)

X: Axial coordinate from the center of jet exit (mm)

x: Axial coordinate from geometrical stagnation point on target plate (mm)

Y: Lateral coordinate for free jet (mm)

y: Lateral coordinate on the target plate (mm)

Z: Spanwise coordinate for free jet (mm)

z: Spanwise coordinate on the target plate (mm)

δ: Thickness of target plate (mm)

λ: Thermal conductivity of air (W/mK)

λ_p: Thermal conductivity of target plate (W/mK)

ν: Kinematic viscosity of air (m²/s).

Conflict of Interests

The authors declare that there is no conflict of interests regarding the publication of this paper.

References

[1] R. C. Deo, G. J. Nathan, and J. Mi, "Comparison of turbulent jets issuing from rectangular nozzles with and without sidewalls," *Experimental Thermal and Fluid Science*, vol. 32, no. 2, pp. 596–606, 2007.

[2] J. San, C. Huang, and M. Shu, "Impingement cooling of a confined circular air jet," *International Journal of Heat and Mass Transfer*, vol. 40, no. 6, pp. 1355–1364, 1997.

[3] Z. H. Lin, Y. J. Chou, and Y. H. Hung, "Heat transfer behaviors of a confined slot jet impingement," *International Journal of Heat and Mass Transfer*, vol. 40, no. 5, pp. 1095–1107, 1997.

[4] N. Gao and D. Ewing, "Investigation of the effect of confinement on the heat transfer to round impinging jets exiting a long pipe," *International Journal of Heat and Fluid Flow*, vol. 27, no. 1, pp. 33–41, 2006.

[5] J. A. Fitzgerald and S. V. Garimella, "A study of the flow field of a confined and submerged impinging jet," *International Journal of Heat and Mass Transfer*, vol. 41, no. 8-9, pp. 1025–1034, 1998.

[6] M. Fenot, J.-J. Vullierme, and E. Dorignac, "Local heat transfer due to several configurations of circular air jets impinging on a flat plate with and without semi-confinement," *International Journal of Thermal Sciences*, vol. 44, no. 7, pp. 665–675, 2005.

[7] J. San and W. Shiao, "Effects of jet plate size and plate spacing on the stagnation Nusselt number for a confined circular air jet impinging on a flat surface," *International Journal of Heat and Mass Transfer*, vol. 49, no. 19-20, pp. 3477–3486, 2006.

[8] J. San, Y. Tsou, and Z. Chen, "Impingement heat transfer of staggered arrays of air jets confined in a channel," *International Journal of Heat and Mass Transfer*, vol. 50, no. 19-20, pp. 3718–3727, 2007.

[9] M. Alnahhal and T. Panidis, "The effect of sidewalls on rectangular jets," *Experimental Thermal and Fluid Science*, vol. 33, no. 5, pp. 838–851, 2009.

[10] L. J. S. Bradbury, "The structure of a self-preserving turbulent plane jet," *Journal of Fluid Mechanics*, vol. 23, part 1, pp. 31–64, 1965.

[11] S. Göppert, T. Gürtler, H. Mocikat, and H. Herwig, "Heat transfer under a precessing jet: effects of unsteady jet impingement," *International Journal of Heat and Mass Transfer*, vol. 47, no. 12-13, pp. 2795–2806, 2004.

[12] R. Gardon and J. C. Akfirat, "Heat transfer characteristics of impinging two-dimensional air jet," *ASME: Journal of Heat Transfer*, vol. 88, pp. 101–108, 1966.

[13] Y. Haneda, Y. Tsuchiya, K. Nakabe, and K. Suzuki, "Enhancement of impinging jet heat transfer by making use of mechano-fluid interactive flow oscillation," *International Journal of Heat and Fluid Flow*, vol. 19, no. 2, pp. 115–124, 1998.

[14] Y. Haneda, Y. Tsuchiya, H. Kurasawa, K. Nakabe, and K. Suzuki, "Flow field and heat transfer of a two-dimensional impinging jet disturbed by an elastically suspended circular cylinder," *Heat Transfer—Asian Research*, vol. 30, no. 4, pp. 313–330, 2001.

[15] W. Fu, K. Wang, and W. Ke, "An investigation of block moving back and forth on a heat plate under a slot jet," *International Journal of Heat and Mass Transfer*, vol. 44, no. 14, pp. 2621–2631, 2001.

[16] W.-S. Fu and K.-N. Wang, "An investigation of a block moving back and forth on a heat plate under a slot jet. Part II (the effects of block moving distance and vacant distance)," *International Journal of Heat and Mass Transfer*, vol. 44, no. 24, pp. 4649–4665, 2001.

[17] W. Fu, C. Tseng, C. Huang, and K. Wang, "An experimental investigation of a block moving back and forth on a heat plate under a slot jet," *International Journal of Heat and Mass Transfer*, vol. 50, no. 15-16, pp. 3224–3233, 2007.

[18] Y. Lin, M. Hsu, and C. Hsieh, "Enhancement of the convective heat transfer for a reciprocating impinging jet flow," *International Communications in Heat and Mass Transfer*, vol. 30, no. 6, pp. 825–834, 2003.

[19] A. K. Chaniotis, D. Poulikakos, and Y. Ventikos, "Dual pulsating or steady slot jet cooling of a constant heat flux surface," *ASME: Journal of Heat Transfer*, vol. 125, no. 4, pp. 575–586, 2003.

[20] V. Narayanan, J. Seyed-Yagoobi, and R. H. Page, "An experimental study of fluid mechanics and heat transfer in an impinging slot jet flow," *International Journal of Heat and Mass Transfer*, vol. 47, no. 8-9, pp. 1827–1845, 2004.

15

A Double Diffusive Unsteady MHD Convective Flow Past a Flat Porous Plate Moving through a Binary Mixture with Suction or Injection

D. R. V. S. R. K. Sastry[1] and A. S. N. Murti[2]

[1] *Aditya Engineering College, Surampalem, Andhra Pradesh 533437, India*
[2] *GITAM University, Visakhapatnam, Andhra Pradesh 530045, India*

Correspondence should be addressed to D. R. V. S. R. K. Sastry; sastry_dev@yahoo.co.in

Academic Editor: Andrew W. Cook

The problem of unsteady magnetohydrodynamic convective flow with radiation and chemical reaction past a flat porous plate moving through a binary mixture in an optically thin environment is considered. The governing boundary layer equations are converted to nonlinear ordinary differential equations by similarity transformation and then solved numerically by MATLAB "bvp4c" routine. The velocity, temperature, and concentration profiles are presented graphically for various values of the material parameters. Also a numerical data for the local skin friction coefficient, the local Nusselt number, and local Sherwood number is presented in tabular forms.

1. Introduction

Combined heat and mass transfer problems with chemical reaction are of importance in many processes and have, therefore, received a considerable amount of attention in research. In processes such as drying, evaporation at the surface of a water body, energy transfer in a wet cooling tower, and the flow in a desert cooler, heat and the mass transfer occur simultaneously. Study towards boundary layer flow of a binary mixture of fluids is very important in view of its application in various branches of engineering and technology. A familiar example is an emulsion which is the dispersion of one fluid within another fluid. Typical emulsions are oil dispersed within water or water within oil. Another example where the mixture of fluids plays an important role is in multigrade oils. Some polymeric type fluids are added to the base oil so as to enhance the lubrication properties of mineral oil [1]. Moreover through chemical reaction, all industrial chemical processes are designed to transform cheaper raw materials to high value products. Naturally these transformations occur in reactors. Fluid dynamics plays a pivotal role in establishing relationship between the reactor hardware and reactor performance. Unsteady free convection boundary layer flows with heat and mass transfer encounter an important criterion of species chemical reaction with finite Arrhenius activation energy defined by Makinde [2]. This phenomenon is useful in the areas such as geothermal or oil reservoir engineering where more numbers of experimental works happen. It is very important for theoretical works to predict the effects of the activation energy in these flows. But very few theoretical works are available in the literature as the chemical reaction process involved in the systems are quite complex. This may be simplified by restricting the reaction to binary type chemical reaction. The thermomechanical balance equations for a mixture of general materials were first formulated by Truesdell [3]. Thereafter, Mills [4], Beevers, and Craine [5] have obtained some exact solutions for the boundary layer flow of a binary mixture of incompressible Newtonian fluids. A particular contribution towards the context of binary mixture theory was done by Al-Sharif et al. [6]

and Wang et al. [7]. Recently Kandasamy et al. [8] studied the combined effects of chemical reaction, heat, and mass transfer along a wedge with heat source and concentration in presence of suction or injection. Their result shows that the flow field is influenced appreciably by chemical reaction, heat source, and suction or injection at the wall of the wedge. El-Hakiem [9] studied the unsteady MHD oscillatory flow on free convection-radiation through a porous medium with a vertical infinite surface that absorbs the fluid with a constant velocity. Raptis et al. [10] studied the effect of radiation on two-dimensional steady MHD optically thin gray gas flow along infinite vertical plate taking the induced magnetic field into account. Israel-Cookey et al. [11] discussed the influence of viscous dissipation and radiation on unsteady MHD free convection flow past on infinite heated vertical plate in a porous medium with time-dependent suction. Abd El-Naby et al. [12] employed an implicit finite-difference method to study the effect of radiation on MHD unsteady free convection flow past a semi-infinite vertical porous plate without viscous dissipation. Singh and Dikshift [13] investigated the hydromagntic flow past a continuously moving semi-infinite plate at large suction. Takhar et al. [14] observed the radiation effects on MHD free convection flow past a semi-infinite vertical plate. Kim [15] studied unsteady MHD convective heat transfer past a semi-infinite vertical porous moving plate. He found that an increase in the Prandtl number and magnetic field intensity decreases the fluid velocity. Vajravelu and Hadjinicolaou [16] studied the heat transfer characteristics in the laminar boundary layer of a viscous fluid over a stretching sheet with viscous dissipation or frictional heating and internal heat generation. Alam et al. [17] studied the problem of free convection heat and mass transfer flow past an inclined semi-infinite heated surface of an electrically conducting and steady viscous incompressible fluid in the presence of a magnetic field and heat generation. Chamkha [18] investigated unsteady convective heat and mass transfer past a semi-infinite porous moving plate with heat absorption. Hady et al. [19] studied the problem of free convection flow along a vertical wavy surface embedded in an electrically conducting fluid saturated porous media in the presence of internal heat generation or absorption. More recently Makinde and Olanrewaju [20] have studied the effects of chemical reaction and radiative heat transfer for an unsteady convection flow past porous plate moving through a binary mixture.

In this paper, we have considered the Arrhenius kinetics and thermal radiation in an unsteady MHD convective flow over a moving plate through a binary mixture with suction or injection at the plate surface. The governing equations are converted to ordinary differential equations by applying the similarity transformation. The numerical solution of the similarity equations are then obtained through the MATLAB "bvp4c" routine. The analysis of the results obtained shows that the flow field is influenced by the presence of magnetic field parameter, thermal radiation, chemical reaction, buoyancy force, and suction/injection parameter at surface of the plate. The profiles of velocity, temperature, and concentration are represented graphically. The skin friction, heat transfer, and mass transfer are displayed in tabular forms for different parameters.

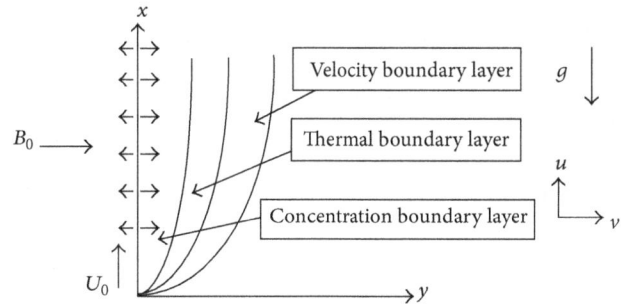

FIGURE 1: Flow configuration and coordinate system.

2. Mathematical Formulation

Consider the unsteady one-dimensional hydromagnetic convective flow with chemical reaction and radiative heat transfer past a vertical porous plate moving through a binary mixture (see Figure 1). Assume that the fluid is electrically conducting and the boundary wall to be of infinite extended so that all quantities are homogeneous in x and hence all derivatives with respect to x are omitted. Further assume the fluid is optically thin with absorption coefficient $\alpha \ll 1$. Let the x-axis be directed in upward direction along the plate and the y-axis is normal to the plate. Let u and v be the velocity components along the x- and y-axes, respectively. A magnetic field B_0 of uniform strength is applied transversely to the direction of the flow. Since the fluid pressure is constant, it is assumed that induced magnetic field is small in comparison to the applied magnetic field; therefore, it is neglected.

Under these assumptions the momentum, energy, and chemical species concentration balance equations which govern the flow may be written as follows:

$$\frac{\partial v}{\partial y} = 0, \tag{1}$$

$$\frac{\partial u}{\partial t} + v \frac{\partial u}{\partial y} = v \frac{\partial^2 u}{\partial y^2} + g\beta_T \left(T - T_\infty\right)$$
$$+ g\beta_C \left(C - C_\infty\right) - \frac{\sigma}{\rho} B_0^2 u, \tag{2}$$

$$\rho C_p \left(\frac{\partial T}{\partial t} + v \frac{\partial T}{\partial y}\right) = k \frac{\partial^2 T}{\partial y^2} + Q - 4\sigma\alpha^2 T^4, \tag{3}$$

$$\frac{\partial C}{\partial t} + v \frac{\partial C}{\partial y} = D_f \frac{\partial^2 C}{\partial y^2} - R_A, \tag{4}$$

where $Q = (-\Delta H)R_A$ is the heat of chemical reaction and is called the activation enthalpy and

$$R_A = k_r e^{-E_A/R_G T} C^n \tag{5}$$

is the Arrhenius type of the nth order irreversible reaction, k_r is the chemical reaction rate, R_G is the universal gas constant, and E_A is the activation energy parameter. The boundary conditions of the above problem are assumed to be

$$u(y,0) = 0, \qquad T(y,0) = T_w, \qquad C(y,0) = C_w,$$

$$u(0,t) = U_0, \qquad T(0,t) = T_w, \qquad C(0,t) = C_w, \quad t > 0,$$

$$u \longrightarrow 0, \quad T \longrightarrow T_\infty, \quad C \longrightarrow C_\infty \text{ as } y \longrightarrow \infty, \ t > 0, \tag{6}$$

where U_0 is the plate characteristic velocity. From the equation of continuity (1), it can be noted that v is either constant or a function of time. Following Makinde [2], we take

$$v = -c\left(\frac{v}{t}\right)^{1/2}, \tag{7}$$

where $c > 0$ is the suction parameter and $c < 0$ is the injection parameter.

We introduce the dimensionless quantities and parameters

$$u = U_0 F(\eta), \qquad (\theta, \theta_w) = \frac{(T, T_w)}{T_\infty},$$

$$(\phi, \phi_w) = \frac{(C, C_w)}{C_\infty}, \qquad \text{Gr} = \frac{4tg\beta_T T_\infty}{U_0},$$

$$\text{Gc} = \frac{4tg\beta_C C_\infty}{U_0}, \qquad \text{Pr} = \frac{v}{\lambda},$$

$$\lambda = \frac{k}{\rho C_p}, \qquad \text{Sc} = \frac{v}{D_f}, \qquad \gamma = \frac{E_A}{R_G T_\infty}, \tag{8}$$

$$\eta = \frac{y}{2\sqrt{vt}}, \qquad h = \frac{(-\Delta H)C_\infty}{\rho C_p T_\infty},$$

$$\text{Ra} = \frac{16\sigma\alpha^2 t T_\infty^3}{\rho C_p}, \qquad \text{Da} = 4tk_0 C_\infty^{n-1},$$

$$k_0 = k_r e^{-E_A/R_G T_\infty}, \qquad M = \frac{4t\sigma B_0^2}{\rho}.$$

With (8) equations in (2), (3), and (4) become

$$F'' + 2(\eta + c)F' - MF = -\text{Gr}(\theta - 1) - \text{Gc}(\phi - 1),$$

$$\frac{1}{\text{Pr}}\theta'' + 2(\eta + c)\theta' = -h\phi^n \exp\left\{\gamma\left(1 - \frac{1}{\theta}\right)\right\} + \text{Ra}\theta^4, \tag{9}$$

$$\frac{1}{\text{Sc}}\phi'' + 2(\eta + c)\phi' = \text{Da}\phi^n \exp\left\{\gamma\left(1 - \frac{1}{\theta}\right)\right\},$$

with boundary conditions

$$F(0) = 1, \quad \theta(0) = \theta_w, \quad \phi(0) = \phi_w,$$

$$F(\infty) = 0, \quad \theta(\infty) = 1, \quad \phi(\infty) = 1, \tag{10}$$

where Da is the Damköhler number, Ra is the radiation parameter, γ is the activation energy parameter, Gr is the thermal Grashof number, Gc is the solutal Grashof number,

Kr is the chemical reaction rate, Pr is the Prandtl number, and M is the magnetic field parameter.

The wall skin-friction

$$\tau_w = -\mu\left(\frac{\partial u}{\partial y}\right)_{y=0} = -\frac{1}{2}\frac{\mu U_0}{\sqrt{vt}}F'(0). \tag{11}$$

Hence, the skin-friction coefficient

$$C_f = \frac{2\tau_w}{\rho U_0^2} = -\frac{\mu}{\rho U_0 \sqrt{vt}}F'(0) = -\frac{1}{\text{Re}}F'(0) \propto -F'(0), \tag{12}$$

where $\text{Re} = \rho U_0 \sqrt{vt}/\mu$ is the Reynolds number.

At the wall, the heat flux (q_w) and the mass flux (m_w) are given by

$$q_w = -k\left(\frac{\partial T}{\partial y}\right)_{y=0},$$

$$m_w = -D_f\left(\frac{\partial C}{\partial y}\right)_{y=0}. \tag{13}$$

The Nusselt number (Nu) and Sherwood number (Sh) are defined as

$$\text{Nu} = \frac{q_w \sqrt{vt}}{k(T_w - T_\infty)} = -\frac{1}{2}\theta'(0) \propto -\theta'(0), \tag{14}$$

$$\text{Sh} = \frac{m_w \sqrt{vt}}{D_f(C_w - C_\infty)} = -\frac{1}{2}\phi'(0) \propto -\phi'(0), \tag{15}$$

where $\sqrt{vt}$ is characteristic length.

The coefficients presented in (12) and (14) are obtained from the procedure of the numerical computations and are sorted for different parameters given in Tables 1–8.

3. Results and Discussion

To get a clear insight of the physical problem, we have assigned various numerical values to the parameters that are incorporated in the problem with which one can discuss the profiles of velocity, temperature, and concentration. In order to compare the present results with previous work [20], we have taken the values of Schmidt number (Sc) for hydrogen 0.22, water vapour 0.62, ammonia 0.78, and propyl benzene 2.62 at temperature 25°C and one atmospheric pressure. The value of Prandtl number is chosen to be Pr = 0.71 which represents air at temperature 25°C and one atmospheric pressure. Moreover the focus is made towards the positive values of the buoyancy parameters; that is, Grashof number Gr > 0, corresponds to the cooling problem, and solutal Grashof number Gc > 0 indicates that the concentration in the free stream region is less than the concentration at the boundary surface. It is worthy to note from Table 1 that increase in chemical reaction parameter (Da) enhances the heat transfer rate and reduces both skin friction and mass transfer rate. From Table 2, it is seen that the increase in the values of Sc leads to enhance both heat, and mass transfer

TABLE 1: Comparison values of $F'(0)$, $\theta'(0)$, and $\phi'(0)$ for different Da (c = Ra = Gr = Gc = γ = θ_w = ϕ_w = 0.1, Sc = 0.22, h = n = 1).

Da	$F'(0)$		$\theta'(0)$		$\phi'(0)$	
	Makinde and Olanrewaju [20]	Present	Makinde and Olanrewaju [20]	Present	Makinde and Olanrewaju [20]	Present
0.1	−1.363220887	−1.363874729	0.940099503	0.939692306	0.45299820	0.450546865
0.2	−1.362524805	−1.363259605	1.017519463	1.019231925	0.40825426	0.403754294
0.3	−1.362049836	−1.362888018	1.085715626	1.088738943	0.36731514	0.361127499

TABLE 2: Comparison values of $F'(0)$, $\theta'(0)$, and $\phi'(0)$ for different Sc (c = Ra = Da = Gr = Gc = γ = θ_w = ϕ_w = 0.1, h = n = 1).

Sc	$F'(0)$		$\theta'(0)$		$\phi'(0)$	
	Makinde and Olanrewaju [20]	Present	Makinde and Olanrewaju [20]	Present	Makinde and Olanrewaju [20]	Present
0.22	−1.363220887	−1.363874729	0.940099503	0.939692306	0.45299820	0.450546865
0.62	−1.347622867	−1.348351348	0.950736367	0.950277754	0.76629786	0.762196009
0.78	−1.344659445	−1.345402697	0.952741674	0.952272964	0.86343256	0.858815122

TABLE 3: Comparison values of $F'(0)$, $\theta'(0)$, and $\phi'(0)$ for different c (Da = Ra = Gr = Gc = γ = θ_w = ϕ_w = 0.1, Sc = 0.22, h = n = 1).

c	$F'(0)$		$\theta'(0)$		$\phi'(0)$	
	Makinde and Olanrewaju [20]	Present	Makinde and Olanrewaju [20]	Present	Makinde and Olanrewaju [20]	Present
0.1	−1.363220887	−1.363874729	0.940099503	0.939692306	0.45299820	0.450546865
1.0	−2.737972223	−2.739021858	1.796428656	1.795804409	0.68928294	0.685864239
−.1	−1.108018112	−1.108575893	0.779055070	0.778698841	0.40619537	0.403951603
−1	−0.311091119	−0.311271144	0.244583695	0.244436402	0.22690545	0.225521578

TABLE 4: Comparison values of $F'(0)$, $\theta'(0)$, and $\phi'(0)$ for different Ra (c = Da = Gr = Gc = γ = θ_w = ϕ_w = 0.1, Sc = 0.22, h = n = 1).

Ra	$F'(0)$		$\theta'(0)$		$\phi'(0)$	
	Makinde and Olanrewaju [20]	Present	Makinde and Olanrewaju [20]	Present	Makinde and Olanrewaju [20]	Present
0.1	−1.363220887	−1.363874729	0.940099503	0.939692306	0.45299820	0.450546865
0.2	−1.370176013	−1.371318069	0.867590872	0.864748108	0.45322418	0.450790024
0.3	−1.375748994	−1.377248017	0.812511169	0.808288321	0.45342377	0.451004257

TABLE 5: Comparison values of $F'(0)$, $\theta'(0)$, and $\phi'(0)$ for different n (c = Da = Ra = Gr = Gc = γ = θ_w = ϕ_w = 0.1, Sc = 0.22, h = 1).

n	$F'(0)$		$\theta'(0)$		$\phi'(0)$	
	Makinde and Olanrewaju [20]	Present	Makinde and Olanrewaju [20]	Present	Makinde and Olanrewaju [20]	Present
1	−1.363220887	−1.363874729	0.940099503	0.939692306	0.45299820	0.450546865
3	−1.363537283	−1.364193050	0.914076370	0.913256158	0.46481660	0.462643274
5	−1.363673665	−1.364329690	0.903949316	0.902860087	0.46999599	0.468013723

TABLE 6: Comparison values of $F'(0)$, $\theta'(0)$, and $\phi'(0)$ for different Gr (c = Da = Ra = Gc = γ = θ_w = ϕ_w = 0.1, Sc = 0.22, h = n = 1).

Gr	$F'(0)$		$\theta'(0)$		$\phi'(0)$	
	Makinde and Olanrewaju [20]	Present	Makinde and Olanrewaju [20]	Present	Makinde and Olanrewaju [20]	Present
0.1	−1.363220887	−1.363874729	0.940099503	0.939692306	0.45299820	0.450546865
1.0	−1.716737895	−1.717731585	0.940099503	0.939692306	0.45299820	0.450546865
5.0	−3.287924599	−3.290428721	0.940099503	0.939692306	0.45299820	0.450546865

TABLE 7: Comparison values of $F'(0)$, $\theta'(0)$, and $\phi'(0)$ for different Gc (c = Da = Ra = Gr = γ = θ_w = ϕ_w = 0.1, Sc = 0.22, $h = n = 1$).

Gc	$F'(0)$		$\theta'(0)$		$\phi'(0)$	
	Makinde and Olanrewaju [20]	Present	Makinde and Olanrewaju [20]	Present	Makinde and Olanrewaju [20]	Present
0.1	−1.363220887	−1.363874729	0.940099503	0.939692306	0.45299820	0.450546865
1.0	−1.950303852	−1.956502423	0.940099503	0.939692306	0.45299820	0.450546865
5.0	−4.559561477	−4.590403287	0.940099503	0.939692306	0.45299820	0.450546865

TABLE 8: Values of $F'(0)$ for different M (Da = Ra = Gr = Gc = γ = θ_w = ϕ_w = 0.1, Sc = 0.22, $h = n = 1$).

M	$F'(0)$ (suction, $c = 0.1$)	$F'(0)$ (injection, $c = -0.1$)
0	−1.363874729	−1.110534369
1	−1.689339475	−1.451374690
2	−1.966674451	−1.737579643
5	−2.635097137	−2.418368678
10	−3.475594470	−3.265944783

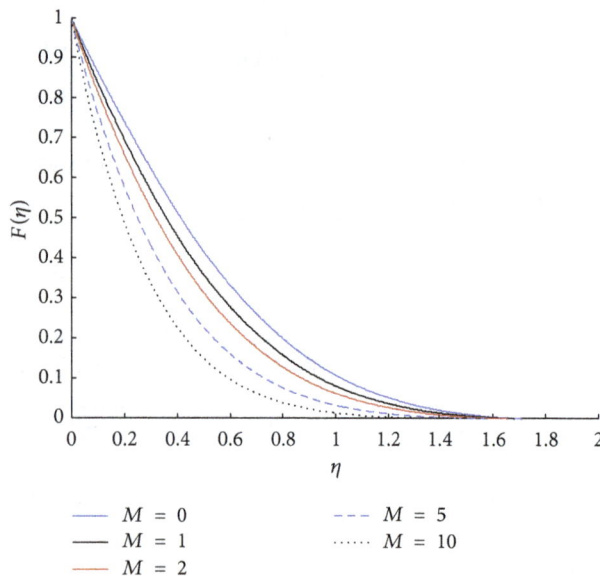

FIGURE 2: Effect of magnetic field parameter on velocity profiles when (Da = c = Ra = Gr = Gc = γ = θ_w = ϕ_w = 0.1, Sc = 0.62, $h = n = 1$).

rates leaving a decrement in the values of skin friction. We note from Table 3 that, in the case of wall suction, increase in the value of c shows an increment in skin friction, heat, and mass transfer rates whereas, in case of injection, increment in c shows a decrement in the respective fluid properties. From Tables 4 and 5 it is seen that the effect of increasing values of Ra and n is to increase both mass transfer rate and skin friction and to decrease the heat transfer rate. From Tables 6 and 7 it is observed that increase in buoyancy parameters enhance the skin friction at the moving plate surface. Also from Table 8 it is observed that skin friction increases with the increase in magnetic field parameter M.

Figures 2–5 display the velocity profiles for different material parameters. Figure 2 illustrates the effect of magnetic field parameter on velocity profiles when the other parameters are fixed. The magnetic field within the boundary layer has produced a resistive type force known as Lorentz force. Due to this force, retardation in the fluid motion along surface is observed. Therefore, it is clear from the same figure that the momentum boundary layer thickness decreases with the increase of magnetic field parameter. Figures 3 and 4 illustrate the effect of buoyancy forces on the horizontal velocity component in the momentum boundary layer. In presence and absence of magnetic field parameter, fluid velocity is highest at the moving plate surface and decreases to free stream zero velocity far away from the plate satisfying the boundary conditions. Also it is observed that in presence of uniform suction at the plate surface, increase of buoyancy forces lead to retardation in the flow and thereby giving rise to a decrease in the velocity profiles; that is, momentum boundary layer thickness decreases with an increase of buoyancy forces. It is observed that a reverse flow occurs within the boundary layer as the intensity of buoyancy forces increases. It is worthy to note that the reverse flow is less in hydromagnetic fluids. Figure 5 depicts the effect of wall suction and injection on the horizontal velocity in momentum boundary layer. It is observed that the momentum boundary layer thickness decreases with increasing the wall suction ($c > 0$) and increases with increasing wall injection ($c < 0$). The same trend is noticed in the magnetohydrodynamic flows.

The effects of various material parameters on the fluid temperature are illustrated in Figures 6–10. It is seen from Figure 6, the variation of temperature profile against similarity variable η for varying values of wall temperature parameter under uniform magnetic field and uniform suction. The temperature increases towards the free stream temperature whenever the surface temperature is lower than the free stream temperature. The temperature decreases toward the free stream temperature whenever the plate temperature is higher than the free stream temperature. Figure 7 shows the effect of chemical reaction rate on the fluid temperature profile within the boundary layer when the wall temperature is lower than the free stream temperature in the presence of uniform suction. It is observed that increasing value of the Damköhler number Da enhances the fluid temperature. This is because of internal heat generation in the fluid due to Arrhenius kinetics. Figure 8 depicts the effect of radiation parameter Ra on temperature profiles in uniform magnetic field. It is noticed from this figure that the fluid temperature starts from a minimum value at the moving plate surface and then increases till it reaches the free stream temperature value

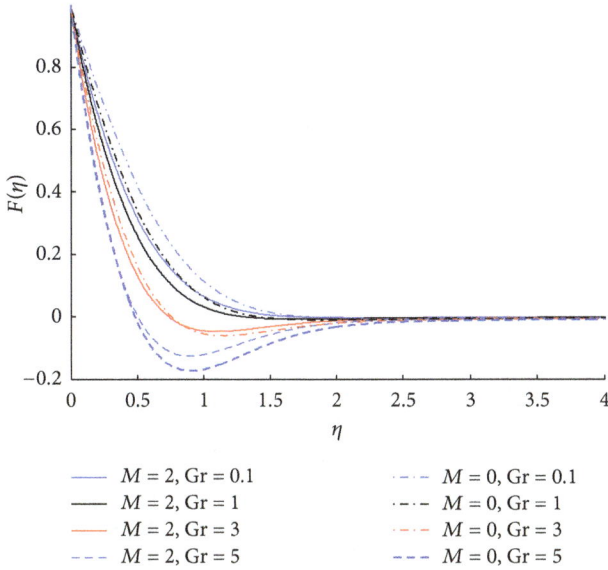

FIGURE 3: Effect of thermal Grashof number on velocity profiles when (Da = c = Ra = Gc = γ = θ_w = ϕ_w = 0.1, Sc = 0.62, h = n = 1).

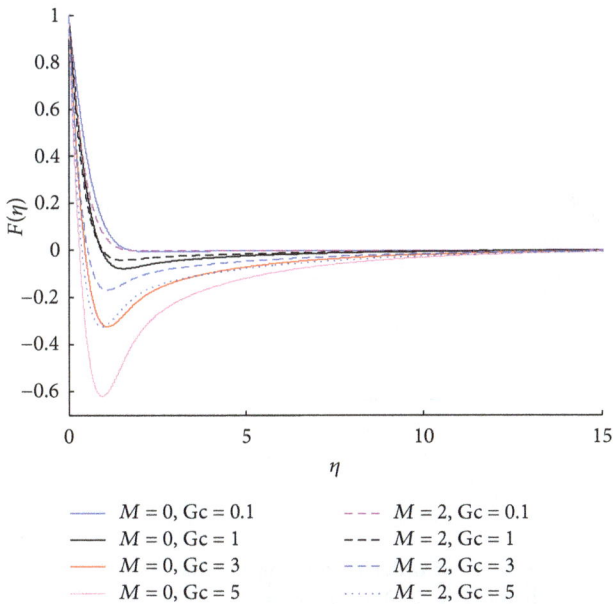

FIGURE 5: Effect of suction/injection parameter on velocity profiles when (Da = Ra = Gr = Gc = γ = θ_w = ϕ_w = 0.1, Sc = 0.62, h = n = 1).

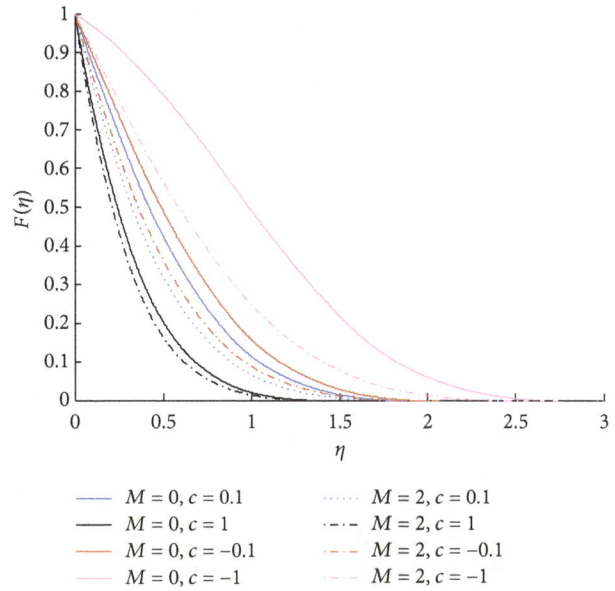

FIGURE 4: Effect of solutal Grashof number on velocity profiles when (Da = c = Ra = Gr = γ = θ_w = ϕ_w = 0.1, Sc = 0.62, h = n = 1).

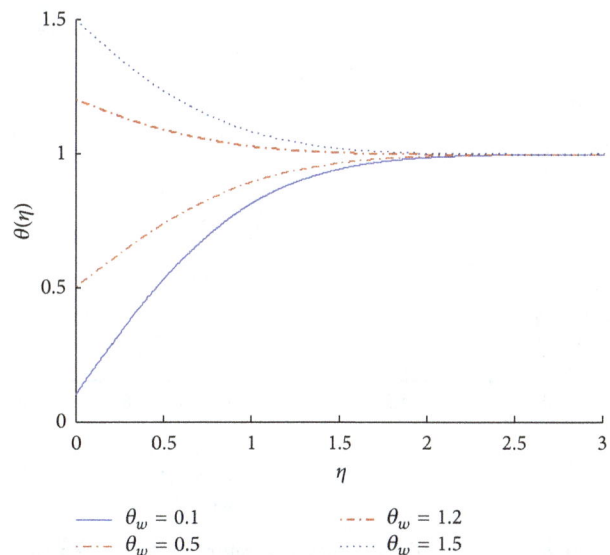

FIGURE 6: Effect of wall temperature on temperature profiles when (M = 2, Da = c = Ra = Gr = Gc = γ = ϕ_w = 0.1, Sc = 0.62, h = n = 1).

at the of the boundary layer for all the values of radiation parameter. It is also observed that temperature of the fluid decreases with increase in the value of Ra. Figure 9 represents the effect of suction and injection parameter on the fluid temperature for different values of magnetic field parameter. It is noticed that the effect of magnetic field parameter on temperature profiles is insignificant. The fluid temperature increases with increase of suction and decreases with increase of injection. The effect of reaction order parameter n, on the temperature profile is represented in Figure 10. The fluid

temperature decreases with increasing order of chemical reaction. Figures 11–14 illustrate the effects of various parameters on concentration profiles. The effect of chemical reaction rate parameter Da on concentration profiles in presence and absence of magnetic field parameter is shown in Figure 11. It is noticed that magnetic field parameter does not affect the concentration profiles. An increase in chemical reaction rate causes a decrease in the concentration of the chemical species in the boundary layer supporting the fact that chemical reaction rate reduces the local concentration.

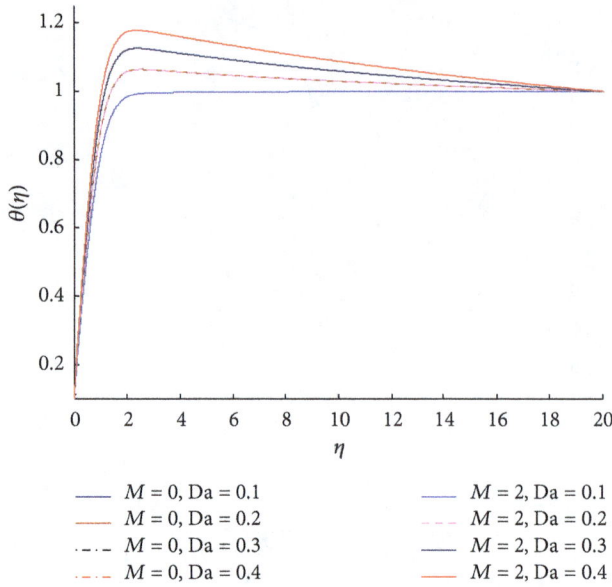

FIGURE 7: Effect of chemical reaction on temperature profiles when ($c = \text{Ra} = \text{Gr} = \text{Gc} = \gamma = \theta_w = \phi_w = 0.1, \text{Sc} = 0.62, h = n = 1$).

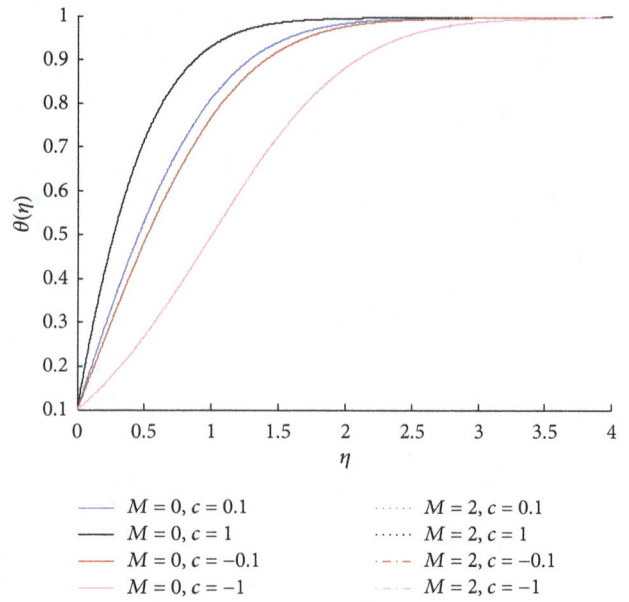

FIGURE 9: Effect of suction/injection parameter on temperature profiles when ($\text{Da} = \text{Ra} = \text{Gr} = \text{Gc} = \gamma = \theta_w = \phi_w = 0.1, \text{Sc} = 0.62, h = n = 1$).

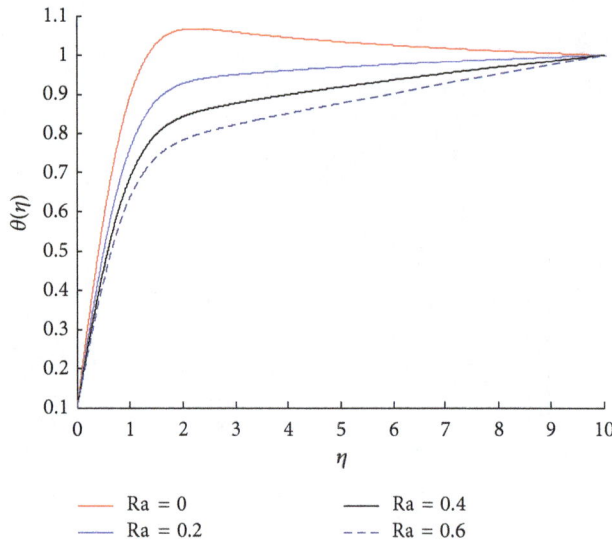

FIGURE 8: Effect of radiation on temperature profiles when ($M = 2, c = \text{Da} = \text{Gr} = \text{Gc} = \gamma = \theta_w = \phi_w = 0.1, \text{Sc} = 0.62, h = n = 1$).

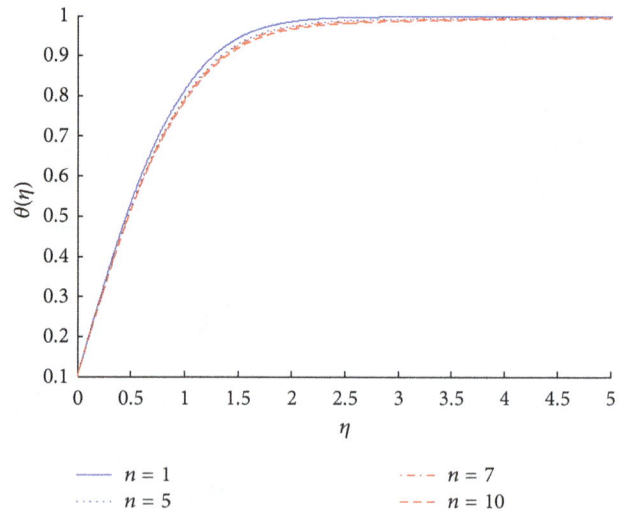

FIGURE 10: Effect of order of reaction parameter on temperature profiles when ($M = 2, \text{Da} = c = \text{Ra} = \text{Gr} = \text{Gc} = \gamma = \theta_w = \phi_w = 0.1, \text{Sc} = 0.62, h = 1$).

Figure 12 shows the effect of Schmidt number Sc on concentration profiles. From this figure we observed that chemical species concentration within the boundary layer increases with an increase in Sc. From Figure 13 it is noted that increasing the order of chemical reaction enhances the species concentration within the boundary layer. Figure 14 shows the effect of suction and injection parameter on the concentration profiles. The species concentration is higher for suction and lower for injection. Figure 15 shows the variation of concentration profiles against similarity variable η for varying values of wall concentration parameter under uniform magnetic field and uniform suction. The concentration

increases towards the free stream concentration whenever the species concentration at surface is lower than the free stream concentration. The reverse trend is observed when surface concentration is higher than the free stream concentration.

4. Conclusion

The effects of magnetism, thermal radiation, suction/injection, buoyancy forces, nth order Arrhenius chemical reaction, and Damköhler number on unsteady convection of viscous incompressible fluid past a vertical porous plate are studied. A set of nonlinear coupled differential equations governing the

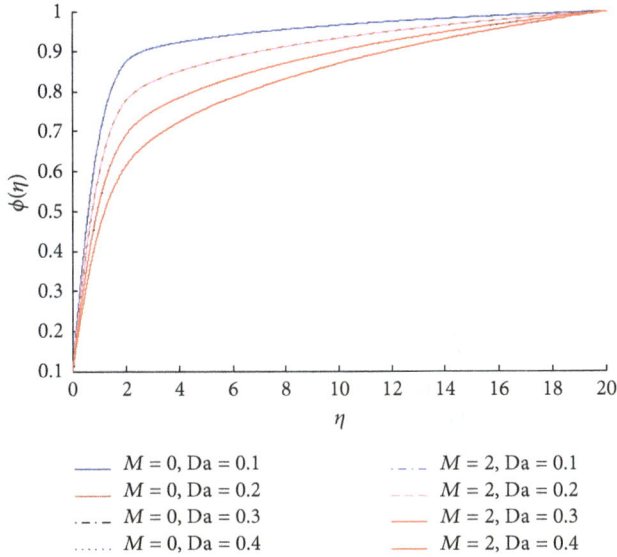

FIGURE 11: Effect of chemical reaction on concentration profiles when (c = Ra = Gr = Gc = γ = θ_w = ϕ_w = 0.1, Sc = 0.62, h = n = 1).

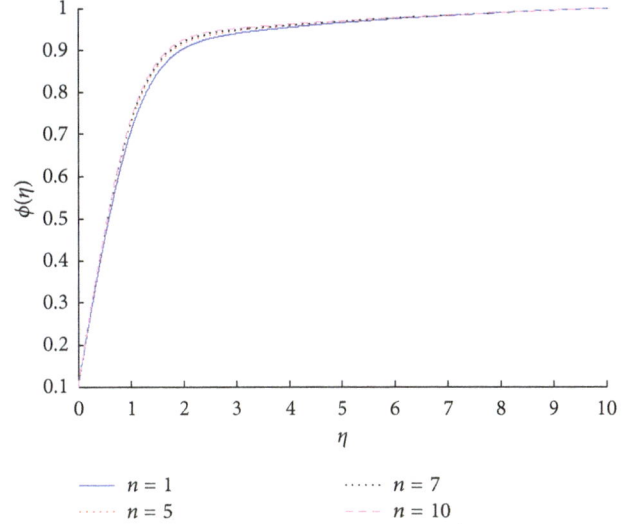

FIGURE 13: Effect of order of reaction on concentration profiles when (M = 2, Da = c = Ra = Gr = Gc = γ = θ_w = ϕ_w = 0.1, Sc = 0.62, h = 1).

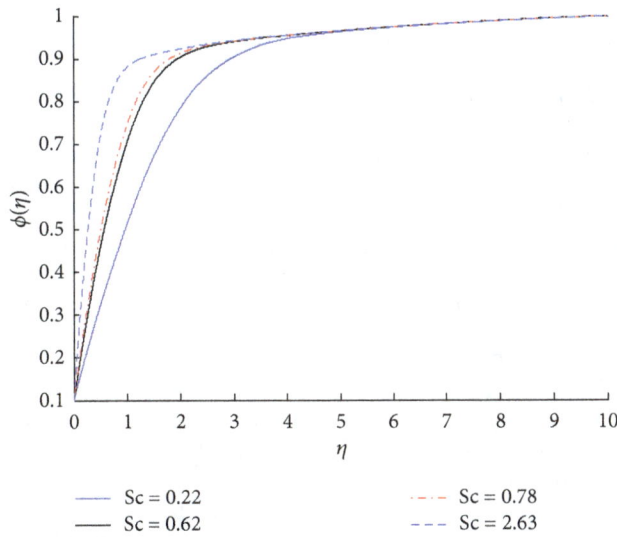

FIGURE 12: Effect of Schmidt number on concentration profiles when (M = 2, Da = c = Ra = Gr = Gc = γ = θ_w = ϕ_w = 0.1, h = n = 1).

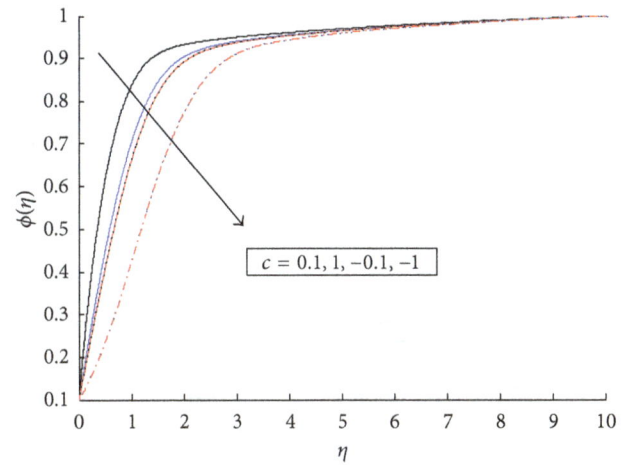

FIGURE 14: Effect of suction/injection on concentration profiles when (M = 2, Da = Ra = Gr = Gc = γ = θ_w = ϕ_w = 0.1, Sc = 0.62, h = n = 1).

fluid velocity, temperature, and chemical species concentration are solved numerically for various material parameters. Results for the velocity, temperature, and concentration are presented and discussed graphically. In the present study we noticed that fluid velocity within the boundary layer decreases with the increasing values of magnetic field parameter and buoyancy forces. Also it is observed that within the boundary layer fluid velocity decreases with increasing values of wall suction and increases with wall injection. The surface temperature decreases in presence of radiation and increases with increasing rate of exothermic chemical reaction Da and reaction order n. Also it is observed that the chemical species concentration within the boundary layer

decreases with increasing values of Da and wall injection. We also noticed that the skin friction increases with increase in magnetic field parameter under uniform suction and injection.

List of Symbols

(x, y) : Cartesian coordinates

(u, v): Velocity components along x, y directions, respectively

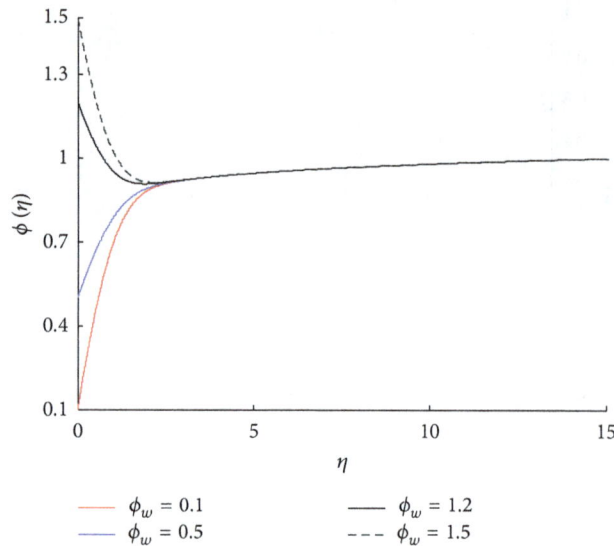

FIGURE 15: Effect of wall concentration on concentration profiles when ($M = 2, \mathrm{Da} = c = \mathrm{Ra} = \mathrm{Gr} = \mathrm{Gc} = \gamma = \theta_w = 0.1, \mathrm{Sc} = 0.62, h = n = 1$).

t:	Time
g:	Acceleration due to gravity
T:	Temperature of the fluid
C:	Concentration of the fluid
T_w:	Surface temperature
C_w:	Surface concentration
T_∞:	Free stream temperature
C_∞:	Free stream concentration
Q:	Activation enthalpy
D_f:	Diffusion coefficient
c:	Suction/injection parameter
E_A:	Activation energy
h:	Heat generation parameter
k:	Thermal conductivity
C_p:	Specific heat at constant pressure
U_0:	Uniform velocity of the plate
R_G:	Universal gas constant
n:	Order of the chemical reaction
Sc:	Schmidt number
ΔH:	Enthalpy change
M:	Magnetic field parameter.

Greek Symbols

θ:	Nondimensional fluid temperature
ϕ:	Nondimensional fluid concentration
θ_w:	Dimensionless wall temperature
ϕ_w:	Dimensionless wall concentration
η:	Similarity variable
γ:	Activation energy parameter
β_T:	Volumetric thermal-expansion coefficient
β_C:	Volumetric solutal-expansion coefficient
σ:	Stefan-Boltzmann constant
α:	Absorption coefficient

ρ:	Density of the fluid
υ:	Kinematic viscosity
μ:	Fluid viscosity
$'$:	Prime represents the derivative with respect to.

References

[1] F. Dai and M. M. Khonsari, "Theory of hydrodynamic lubrication involving the mixture of two fluids," *Journal of Applied Mechanics*, vol. 61, no. 3, pp. 634–641, 1994.

[2] O. D. Makinde, "Free convection flow with thermal radiation and mass transfer past a moving vertical porous plate," *International Communications in Heat and Mass Transfer*, vol. 32, no. 10, pp. 1411–1419, 2005.

[3] C. Truesdell, "Sulle basi della thermomeccanica," *Rendiconti Lincei*, vol. 22, no. 8, pp. 33–38, 1957.

[4] N. Mills, "Incompressible mixtures of newtonian fluids," *International Journal of Engineering Science*, vol. 4, no. 2, pp. 97–112, 1966.

[5] C. E. Beevers and R. E. Craine, "On the determination of response functions for a binary mixture of incompressible newtonian fluids," *International Journal of Engineering Science*, vol. 20, no. 6, pp. 737–745, 1982.

[6] A. Al-Sharif, K. Chamniprasart, K. R. Rajagopal, and A. Z. Szeri, "Lubrication with binary mixtures: liquid-liquid emulsion," *Journal of Tribology*, vol. 115, no. 1, pp. 46–55, 1993.

[7] S. H. Wang, A. Al-Sharif, K. R. Rajagopal, and A. Z. Szeri, "Lubrication with binary mixtures: liquid-liquid emulsion in an EHL conjunction," *Journal of Tribology*, vol. 115, no. 3, pp. 515–522, 1993.

[8] R. Kandasamy, K. Periasamy, and K. K. S. Prabhu, "Effects of chemical reaction, heat and mass transfer along a wedge with heat source and concentration in the presence of suction or injection," *International Journal of Heat and Mass Transfer*, vol. 48, no. 7, pp. 1388–1394, 2005.

[9] M. A. El-Hakiem, "MHD oscillatory flow on free convection-radiation through a porous medium with constant suction velocity," *Journal of Magnetism and Magnetic Materials*, vol. 220, no. 2, pp. 271–276, 2000.

[10] A. Raptis, C. Perdikis, and A. Leontitsis, "Effects of radiation in an optically thin gray gas flowing past a vertical infinite plate in the presence of a magnetic field," *Heat and Mass Transfer*, vol. 39, no. 8-9, pp. 771–773, 2003.

[11] C. Israel-Cookey, A. Ogulu, and V. B. Omubo-Pepple, "Influence of viscous dissipation and radiation on unsteady MHD free-convection flow past an infinite heated vertical plate in a porous medium with time-dependent suction," *International Journal of Heat and Mass Transfer*, vol. 46, no. 13, pp. 2305–2311, 2003.

[12] M. A. Abd El-Naby, E. M. E. Elbarbary, and N. Y. AbdElazem, "Finite difference solution of radiation effects on MHD unsteady free-convection flow over vertical porous plate," *Applied Mathematics and Computation*, vol. 151, no. 2, pp. 327–346, 2004.

[13] A. K. Singh and C. K. Dikshit, "Hydromagnetic flow past a continuously moving semi-infinite plate for large suction," *Astrophysics and Space Science*, vol. 148, no. 2, pp. 249–256, 1988.

[14] H. S. Takhar, R. S. R. Gorla, and V. M. Soundalgekar, "Radiation effects on MHD free convection flow of a gas past a semi-infinite

vertical plate," *International Journal of Numerical Methods for Heat and Fluid Flow*, vol. 6, no. 2, pp. 77–83, 1996.

[15] Y. J. Kim, "Unsteady MHD convection flow of polar fluids past a vertical moving porous plate in a porous medium," *International Journal of Heat and Mass Transfer*, vol. 44, no. 15, pp. 2791–2799, 2001.

[16] K. Vajravelu and A. Hadjinicolaou, "Heat transfer in a viscous fluid over a stretching sheet with viscous dissipation and internal heat generation," *International Communications in Heat and Mass Transfer*, vol. 20, no. 3, pp. 417–430, 1993.

[17] M. S. Alam, M. M. Rahman, and M. A. Sattar, "MHD free convective heat and mass transfer flow past an inclined surface with heat generation," *Thammasat International Journal of Science and Technology*, vol. 11, pp. 1–8, 2006.

[18] A. J. Chamkha, "Unsteady MHD convective heat and mass transfer past a semi-infinite vertical permeable moving plate with heat absorption," *International Journal of Engineering Science*, vol. 42, no. 2, pp. 217–230, 2004.

[19] F. M. Hady, R. A. Mohamed, and A. Mahdy, "MHD free convection flow along a vertical wavy surface with heat generation or absorption effect," *International Communications in Heat and Mass Transfer*, vol. 33, no. 10, pp. 1253–1263, 2006.

[20] O. D. Makinde and P. O. Olanrewaju, "Unsteady mixed convection with Soret and Dufour effects past a porous plate moving through a binary mixture of chemically reacting fluid," *Chemical Engineering Communications*, vol. 22, no. 7, pp. 65–78, 2011.

16

Numerical Simulation of Water Jet Flow Using Diffusion Flux Mixture Model

Zhi Shang, Jing Lou, and Hongying Li

Institute of High Performance Computing (IHPC), Agency for Science, Technology and Research (A*STAR), 1 Fusionopolis Way, No. 16-16 Connexis, Singapore 138632

Correspondence should be addressed to Zhi Shang; shangzhi@tsinghua.org.cn

Academic Editor: Yanzhong Li

A multidimensional diffusion flux mixture model was developed to simulate water jet two-phase flows. Through the modification of the gravity using the gradients of the mixture velocity, the centrifugal force on the water droplets was able to be considered. The slip velocities between the continuous phase (gas) and the dispersed phase (water droplets) were able to be calculated through multidimensional diffusion flux velocities based on the modified multidimensional drift flux model. Through the numerical simulations, comparing with the experiments and the simulations of traditional algebraic slip mixture model on the water mist spray, the model was validated.

1. Introduction

Liquid spray systems are widely used in many chemical, petrochemical, and biochemical industries, such as absorption, oxidation, hydrogenation, coal liquefaction, and aerobic fermentation. The operation of these systems is preferred because of the simple construction, ease of maintenance, and low operating costs. When the spray is injected from nozzle, it causes a turbulent stream to enable an optimum phase exchange. It is built in numerous forms of construction. The mixing is done by the liquid droplets and it requires less energy than mechanical stirring.

A good understanding of the liquid droplet dynamics of the spray will help the engineers to design the high efficient facilities under optimized operating parameters. Although the operation of spray system is simple, due to the complexity of the two-phase turbulent flow, the actual physical flow phenomena are still lacking complete understanding of the fluid dynamics [1].

Many experimental facilities and methods were introduced to study the multiphase flows in spray systems. Ruck and Makiola [2] used a laser Doppler anemometer (LDA) to study the gas-oil droplet passing over a backward facing step. Ferrand et al. [3] used a phase Doppler and laser induced fluorescence technique to study the gas-droplet turbulent velocity and two-phase interaction through a jet with partly responsive droplets. Esposito et al. [4] used a monochrome charge-coupled device CCD camera to study the growth of the droplets. The experimental methods can provide very useful information about the liquid droplets at certain measurement points, but it is difficult for them to show the details of the flow fields and parameters inside the spray.

Following the development of computer technology, it is already allowed to use the numerical method to do the researches in the recent decades [5]. Therefore, many researchers employ the numerical method, called computational fluid dynamics (CFD), to study the details of the flows. Griffiths and Boysan [6] employed the coupled particle Lagrangian model with Eulerian continuous fluid flow model to simulate a cyclone sampler and compared the numerical results with the empirical models. Barton [7] used the stochastic Monte Carlo scheme coupled with k-ε turbulence model to study the particle trajectories in turbulent flow over a backward facing step. Husted [1] used the Eulerian-Lagrangian model provided by FDS open source software to simulate a water mist spray system for the full spray spreading and compared it with experiments. Through the former studies of CFD method, it can be seen that a good mathematical model will not only help to obtain the agreeable

TABLE 1: Constants of standard k-ε turbulence model.

Variable	C_μ	σ_k	σ_ε	C_1	C_2
Constant	0.09	1.0	1.3	1.44	1.92

simulation results but also help to obtain simple, efficient, and accurate ones.

A multidimensional diffusion flux mixture model (DFMM) was developed in this paper. This model was based on the idea of Yang et al. [8] and Shang [9]. It employed a mixture model to describe the multiphase flows based on Eulerian model. The slip velocity can be developed based on the extended 3D drift flux model. Owing to the extension, the effects of gravity and centrifugal force were considered. Through comparisons with experiments and simulations of traditional algebraic slip mixture model (ASMM) on the water mist spray, this model was validated.

2. Mathematical Modeling

Considering a problem of turbulent multicomponent multiphase flow with one continuous phase and several dispersed phases, the time average conservation equations of mass, momentum, and energy for the mixture model, the turbulent kinetic energy equation, and the turbulent kinetic energy transport equation can be written as follows:

$$\frac{\partial \rho_m}{\partial t} + \nabla \cdot (\rho_m U_m) = 0, \tag{1}$$

$$\frac{\partial (\rho_m U_m)}{\partial t} + \nabla \cdot (\rho_m U_m U_m)$$
$$= -\nabla p + \rho_m g \tag{2}$$
$$+ \nabla \cdot \left[(\mu_m + \mu_t)\left(\nabla U_m + \nabla U_m^T\right) \right]$$
$$- \nabla \cdot \sum \alpha_k \rho_k U_{km} U_{km},$$

$$\frac{\partial (\rho_m h_m)}{\partial t} + \nabla \cdot (\rho_m U_m h_m)$$
$$= q + \nabla \cdot \left[\left(\frac{\mu_m}{\mathrm{Pr}} + \frac{\mu_t}{\mathrm{Pr}_t}\right)\nabla h_m \right] - \nabla \cdot \sum \alpha_k \rho_k h_k U_{km}, \tag{3}$$

$$\frac{\partial (\rho_m k)}{\partial t} + \nabla \cdot (\rho_m U_m k)$$
$$= \nabla \cdot \left[\left(\mu_m + \frac{\mu_t}{\sigma_k}\right)\nabla k \right] + G - \rho_m \varepsilon, \tag{4}$$

$$\frac{\partial (\rho_m \varepsilon)}{\partial t} + \nabla \cdot (\rho_m U_m \varepsilon)$$
$$= \nabla \cdot \left[\left(\mu_m + \frac{\mu_t}{\sigma_\varepsilon}\right)\nabla \varepsilon \right] + \frac{\varepsilon}{k}(C_1 G - C_2 \rho_m \varepsilon), \tag{5}$$

in which

$$\rho_m = \sum \alpha_k \rho_k, \tag{6}$$

$$\mu_m = \sum \alpha_k \mu_k, \tag{7}$$

$$\rho_m U_m = \sum \alpha_k \rho_k U_k, \tag{8}$$

$$U_{km} = U_k - U_m, \tag{9}$$

$$G = \frac{1}{2}\mu_t \left[\nabla U_m + (\nabla U_m)^T\right]^2, \tag{10}$$

$$\mu_t = C_\mu \rho_m \frac{k^2}{\varepsilon}, \tag{11}$$

where ρ is the density, U is the velocity vector, α is the volumetric fraction, p is pressure, g is the gravitational acceleration vector, U_{km} is the diffusion velocity vector of k dispersed phase relative to the averaged mixture flow, h is enthalpy, q is heat input, μ is viscosity, μ_t is turbulent viscosity, Pr is molecular Prandtl number, Pr_t is turbulent Prandtl number, and G is stress production. $C_\mu, \sigma_k, \sigma_\varepsilon, C_1, C_2$ are constants for standard $k - \varepsilon$ turbulence model [10], shown in Table 1. The subscript m stands for the averaged mixture flow, and k stands for k dispersed phase.

Additional to the above equations, the following conservation equation for each dispersed phase is also necessary.

Consider the following:

$$\frac{\partial (\alpha_k \rho_k)}{\partial t} + \nabla \cdot (\alpha_k \rho_k U_m) = \Gamma_k - \nabla \cdot (\alpha_k \rho_k U_{km}), \tag{12}$$

where Γ_k is the generation rate of k-phase.

In order to closure the governing equations (1)~(12), it is necessary to determine the diffusion velocities U_{km}. The following equation is employed to covert the diffusion velocities to slip velocities that can be presented as $U_{kc} = U_k - U_c$, where U_c is the velocity of the continuous phase.

Consider the following:

$$U_{km} = U_{kc} - \sum \frac{\alpha_k \rho_k}{\rho_m} U_{kc}. \tag{13}$$

Actually the above equation can be developed from the definition of the mixture density equation (6), the definition of mixture mass flux equation (8), and the diffusion velocity equation (9). Once the slip velocities are obtained, the whole governing equations will be closured.

Because the slip velocities present the difference of the movement between the dispersed phase, for instance, the liquid droplet, and the continuous phase, for instance, the gas, in gas liquid-droplet two-phase flow, the slip velocity can be modeled through the drift flux model [8], shown in (14):

$$U_{lg} = -1.53 \left[\frac{\sigma (\rho_l - \rho_g) g'}{\rho_g^2} \right]^{1/4} \frac{g'}{|g'|}, \tag{14}$$

$$g' = g - \frac{dU_f}{dt} = g - \left(\frac{\partial U_f}{\partial t} + U_f \cdot \nabla U_f \right), \tag{15}$$

$$U_f = \alpha_g U_g + \alpha_l U_l, \tag{16}$$

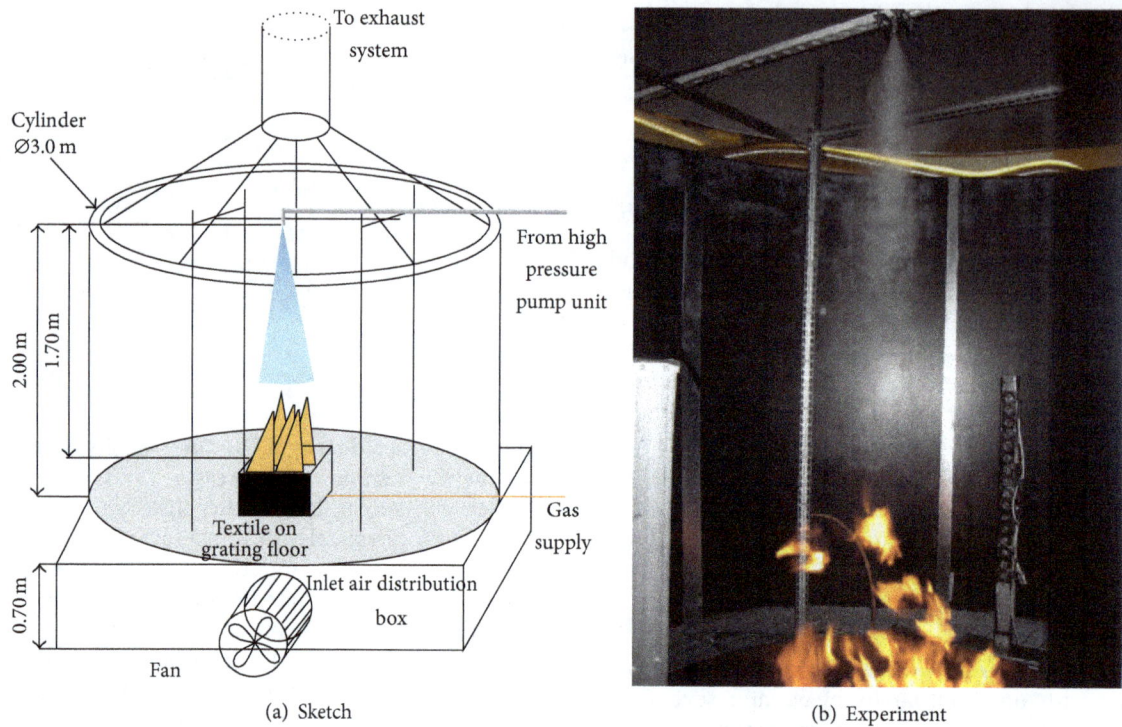

(a) Sketch

(b) Experiment

FIGURE 1: Diagrammatic sketch of the experiment facility [1].

where U_{lg} is the slip velocity between liquid droplet and gas, σ is surface tension, g is gravity, and U_f is drift velocity.

In (14), the drift flux model is different from the traditional 1D drift flux model in Zuber and Findlay [11] and in Hibiki and Ishii [12]. It adopts the form as Yang et al. [8] and Shang [9]. In Yang et al. [8], the centrifugal force was induced by the mixture volumetric flux to revise the gravity considering the natural curve movement of liquid droplets. Owing to the concerns, the traditional 1D drift flux model [11, 12] was extended to 3D. Because of being extended, the revised 3D drift flux model is able to describe complex flow conditions and to be adopted by CFD. Further, in this paper, the gradient of mixture velocity, which is used to calculate the centrifugal force, is innovatively induced to revise and update Yang et al.'s [8] 3D drift flux model. After the updating, the new terminal velocity model, shown in (14), (15), and (16), is able to suit the gas liquid-droplet two-phase flows. Since the slip velocity is determined, the whole equations are closured to be solved.

3. Numerical Procedures

In the simulations, the CFD technique was based on ANSYS FLUENT 13.0. In ANSYS FLUENT, during the numerical computing, all differential governing equations are solved by applying a finite volumes method (FVM). For the fluids, the spatial discretization was performed by upwind scheme of second order for all conservation equations and phase coupled SIMPLE scheme was used for coupling between the pressure and velocity. The overall spatial discretization was of the second-order accuracy. The first order implicit scheme

was used for the time discretization. During the numerical simulations, the pressure-based solver was employed because all the gas and liquid phases were considered as incompressible fluids. The fixed time stepping method was employed to run the transient simulations.

The terminal velocity model was accomplished using the concept of user defined functions (UDF). This numerical solution was implemented as a subroutine and linked to the ANSYS FLUENT solver via a set of the original UDFs. The decision whether the UDFs of the coefficients should be used for the interfacial forces at the given calculation step was made automatically by the solver.

4. Numerical Simulations

Husted [1] did the experiment on a water mist spray system using a particle image velocimetry (PIV), a phase Doppler anemometer (PDA), and a high-speed camera measurement technology. Figure 1 shows the diagrammatic sketch of facility of the experiment. In Husted's [1] experiment, the water mist spray sprayed down from the top nozzle is used for fire extinguishing. The fire can be set at the extruded square at the bottom. If there is no fire, the extruded square can be treated as a solid wall. The purpose of the experiment was to fill the lack of guidelines available for dimensioning water mist systems. The experiment data were good to be used to measure the modeling and numerical simulations. Therefore this experiment was chosen in this paper as the validation tool to validate the drift flux mixture model.

According to the experiment, Figure 2 shows the simplified geometry and the mesh generation for CFD applications.

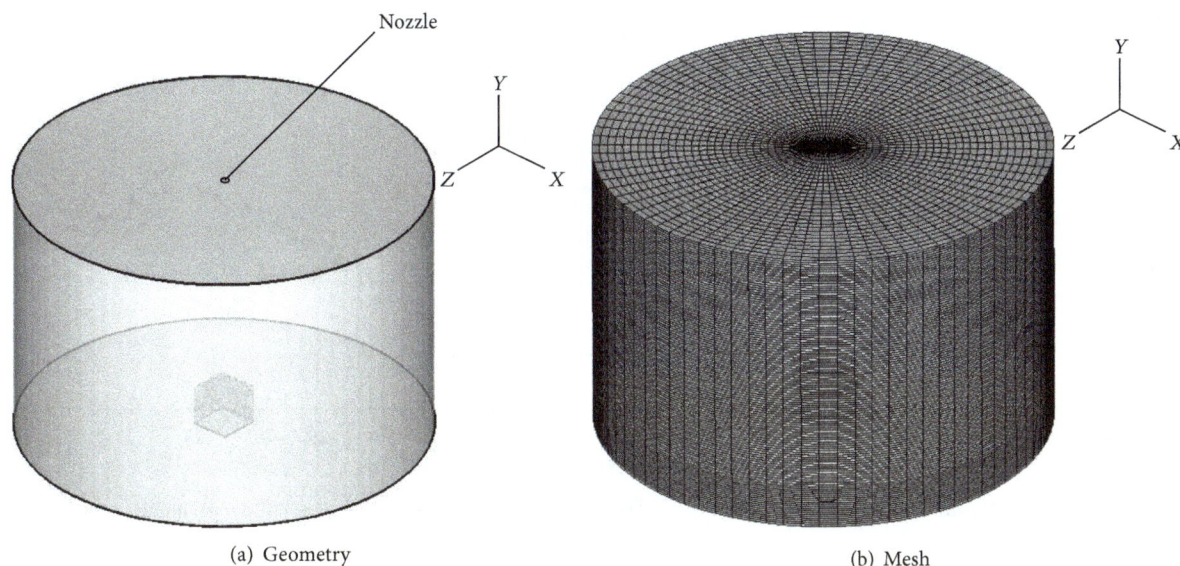

(a) Geometry　　　　　　　　　　　　　　(b) Mesh

FIGURE 2: Simplified geometry and mesh generation.

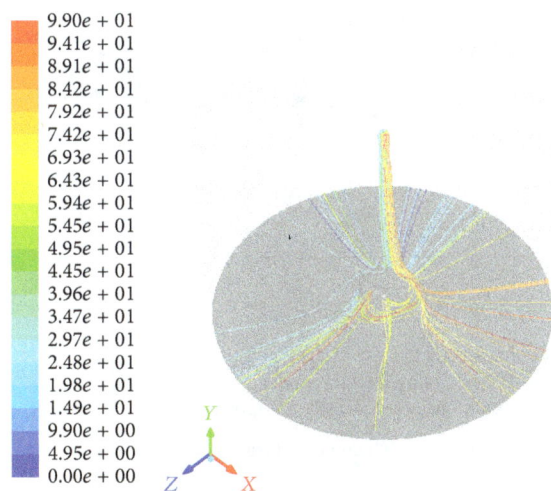

9.90e + 01
9.41e + 01
8.91e + 01
8.42e + 01
7.92e + 01
7.42e + 01
6.93e + 01
6.43e + 01
5.94e + 01
5.45e + 01
4.95e + 01
4.45e + 01
3.96e + 01
3.47e + 01
2.97e + 01
2.48e + 01
1.98e + 01
1.49e + 01
9.90e + 00
4.95e + 00
0.00e + 00

FIGURE 3: Tracked path lines of the water droplets.

The water nozzle is just a simple round hole at the top of the geometry. Due to the fact that the experiment [1] has only the data of the distributions of velocities without considering the heated situation, the simulations are only for the velocity same as the experiment without considering the thermal equation (3) [1].

During the simulations, the numerical exercise was performed in a three-dimensional (3D) environment. All the parameters used were similar to those by Husted [1]. The water droplets are sprayed from the top nozzle with flow rate of 0.38 liter per minute. The bottom is solid wall and the other sides are opened to the environment. Before the formal simulations, the mesh independency studies were performed and it was found that the mesh size of 328000 was enough for the water spray case for guaranteeing the simulation results being mesh-independent. During the simulations, the time

step was set as a constant of 0.01 seconds which is quick enough for the simulations because the simulation time for every case lasted long enough until its steady state.

Figure 3 shows the tracked path lines of the water droplets. From Figure 3, it can be seen that the droplets flow down in the middle and gradually spread like a cone shape down from the nozzle. According to the studies of Husted [1], the stages of the development of the cone spray can be separated as initial conical zone, inflow zone, transition zone, turbulent zone, and full cone zone. When the distance is far from 500 mm down from the nozzle, the spray is fully mixed and drop distribution is uniform. The drop velocity becomes flat.

Figure 4 shows the comparisons of the water mist droplet velocity of the simulations with experiments along different positions down from the nozzle. In Figure 4, the reference numerical simulations were carried out by the traditional algebraic slip mixture model in FLUENT. The Schiller-Naumann drag force model [13] was adopted during the simulations for both of the diffusion flux model and the traditional algebraic slip mixture model [14] due to the fact that the turbulent dispersion force was considered.

From the comparisons, it can be seen that the diffusion flux mixture model is able to capture the velocity profile the same as experiments. The numerical predictions not only have the peak value quite close to the experiment but also have the position approaching to the experiment. The results are much better than the predictions by traditional algebraic slip mixture model in FLUENT.

5. Conclusion

The novel multidimensional diffusion flux mixture model is developed based on the mixture multiphase flow model. The diffusion velocity between the dispersed phase and the mixture is closured through the slip velocity. The slip velocity

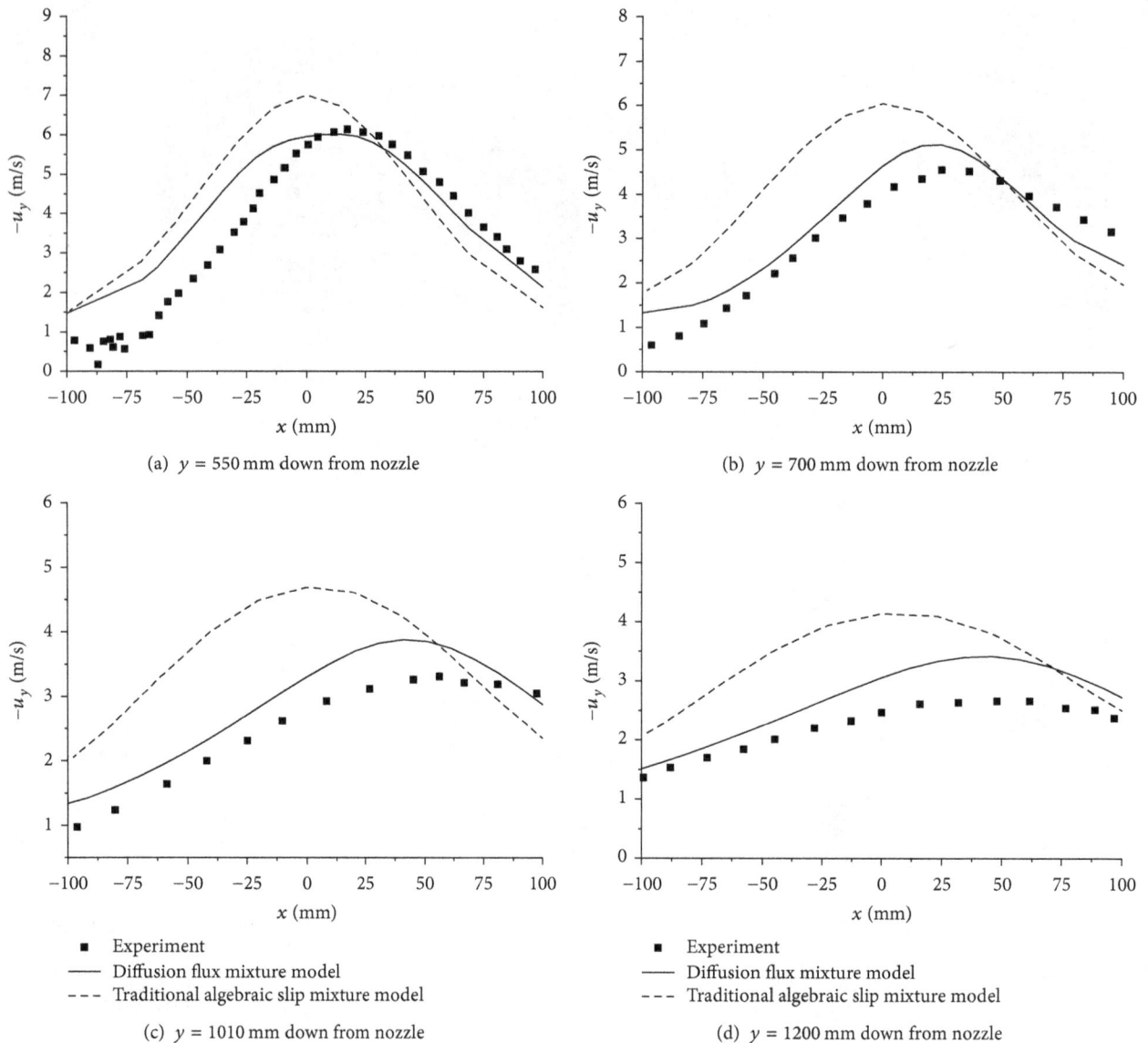

(a) $y = 550\,$mm down from nozzle

(b) $y = 700\,$mm down from nozzle

(c) $y = 1010\,$mm down from nozzle

(d) $y = 1200\,$mm down from nozzle

■ Experiment
— Diffusion flux mixture model
- - - Traditional algebraic slip mixture model

FIGURE 4: Comparisons of water mist droplet velocity under different positions down from nozzle.

is developed through the drift flux model. Accordingly, the multidimensional diffusion flux mixture model is different from the traditional algebraic slip mixture model. Through the comparisons of the numerical simulations between the multidimensional diffusion flux mixture model and experiments, the model is validated. Through comparisons of the numerical simulations between the multidimensional diffusion flux mixture model and the numerical simulations of the traditional algebraic slip mixture model, the efficiency and accuracy of the model are confirmed.

Conflict of Interests

The authors declare that there is no conflict of interests regarding the publication of this paper.

Acknowledgment

The authors would like to acknowledge the support provided by Multiphase Flow for Deep-Sea Oil & Gas Down-Hole Applications - SERC TSRP Programme of Agency for Science, Technology and Research (A*STAR) in Singapore (Reference no. 102 164 0075).

References

[1] B. P. Husted, *Experimental measurements of water mist systems and implications for modeling in CFD [Ph.D. thesis]*, Lund University, Lund, Sweden, 2007.

[2] B. Ruck and B. Makiola, "Particle dispersion in a single-sided backward-facing step flow," *International Journal of Multiphase Flow*, vol. 14, no. 6, pp. 787–800, 1988.

[3] V. Ferrand, R. Bazile, J. Borée, and G. Charnay, "Gas-droplet turbulent velocity correlations and two-phase interaction in

an axisymmetric jet laden with partly responsive droplets," *International Journal of Multiphase Flow*, vol. 29, no. 2, pp. 195–217, 2003.

[4] A. Esposito, A. D. Montello, Y. G. Guezennec, and C. Pianese, "Experimental investigation of water droplet-air flow interaction in a non-reacting PEM fuel cell channel," *Journal of Power Sources*, vol. 195, no. 9, pp. 2691–2699, 2010.

[5] Z. Shang, Y. Yao, and S. Chen, "Numerical investigation of system pressure effect on heat transfer of supercritical water flows in a horizontal round tube," *Chemical Engineering Science*, vol. 63, no. 16, pp. 4150–4158, 2008.

[6] W. D. Griffiths and F. Boysan, "Computational fluid dynamics (CFD) and empirical modelling of the performance of a number of cyclone samplers," *Journal of Aerosol Science*, vol. 27, no. 2, pp. 281–304, 1996.

[7] I. Barton, "Simulation of particle trajectories in turbulent flow over a backward-facing step," *R & D Journal*, vol. 15, no. 3, pp. 65–78, 1999.

[8] R. Yang, R. Zheng, and Y. Wang, "Analysis of two-dimensional two-phase flow in horizontal heated tube bundles using drift flux model," *Heat and Mass Transfer*, vol. 35, no. 1, pp. 81–88, 1999.

[9] Z. Shang, "CFD of turbulent transport of particles behind a backward-facing step using a new model—k-ε-Sp," *Applied Mathematical Modelling*, vol. 29, no. 9, pp. 885–901, 2005.

[10] B. E. Launder and D. B. Spalding, "The numerical computation of turbulent flows," *Computer Methods in Applied Mechanics and Engineering*, vol. 3, no. 2, pp. 269–289, 1974.

[11] N. Zuber and J. A. Findlay, "The effects of non-uniform flow and concentration distributions and the effect of the local relative velocity on the average volumetric concentration in two-phase flow," Tech. Rep. GEAP-4592, 1964.

[12] T. Hibiki and M. Ishii, "Distribution parameter and drift velocity of drift-flux model in bubbly flow," *International Journal of Heat and Mass Transfer*, vol. 45, no. 4, pp. 707–721, 2002.

[13] L. Schiller and Z. Z. Naumann, "Über die grundlegenden Berechungen bei der schwerkraftaufbereitung," *Zeitschrift des Vereines Deutscher Ingenieure*, vol. 77, pp. 318–325, 1935.

[14] O. Simonin and P. L. Viollet, "Modeling of turbulent two-phase jets loaded with discrete particles," in *Phenomena in Multiphase Flows*, pp. 259–269, 1990.

17

Suction/Injection Effects on the Swirling Flow of a Reiner-Rivlin Fluid near a Rough Surface

Bikash Sahoo,[1] **Sébastien Poncet,**[2,3] **and Fotini Labropulu**[4]

[1]*Department of Mathematics, National Institute of Technology Rourkela, Rourkela 769008, India*
[2]*Faculté de Génie, Université de Sherbrooke, Sherbrooke, QC, Canada J1K 2R1*
[3]*Aix-Marseille Université, CNRS, École Centrale, Laboratoire M2P2 UMR 7340, 13451 Marseille, France*
[4]*Department of Mathematics, Luther College, University of Regina, Regina, SK, Canada S4S 0A2*

Correspondence should be addressed to Sébastien Poncet; sebastien.poncet@univ-amu.fr

Academic Editor: Robert M. Kerr

The similarity equations for the Bödewadt flow of a non-Newtonian Reiner-Rivlin fluid, subject to uniform suction/injection, are solved numerically. The conventional no-slip boundary conditions are replaced by corresponding partial slip boundary conditions, owing to the roughness of the infinite stationary disk. The combined effects of surface slip (λ), suction/injection velocity (W), and cross-viscous parameter (L) on the momentum boundary layer are studied in detail. It is interesting to find that suction dominates the oscillations in the velocity profiles and decreases the boundary layer thickness significantly. On the other hand, injection has opposite effects on the velocity profiles and the boundary layer thickness.

1. Introduction

The problem of Newtonian and non-Newtonian swirling flows near a rotating or stationary disk has occupied a central position in the field of fluid mechanics due firstly to the fact that similarity solutions to the Navier-Stokes equations may be found in some idealized infinite configurations and secondly to its industrial and technical applications in rotating machinery (centrifugal pumps, turbines, or computer storage devices), chemical engineering (spinning disk reactors, crystal growth processes, or rheometers), or oceanography among other things.

Recently, Sahoo [1] and Sahoo and Poncet [2] have obtained numerical solution to similarity equations arising due to steady revolving flow (known as Bödewadt flow [3]) of a non-Newtonian Reiner-Rivlin fluid near an infinite rough stationary disk. In this short note, the flow problem studied by Sahoo and Poncet [2] has been reconsidered, including uniform suction/injection at the surface of the stationary disk. Knowledge of the flow structure close to a porous disk is of practical significance with regard to problems of lubrication of porous bearings or gaseous diffusion among other

things. Suction or injection at the surface of a porous disk is also commonly used in chemical engineering to increase the electrochemical reaction time during electrolytic processes [4] or for control purpose as, under given conditions, it delays the transition to turbulence [5]. There are only few attempts in the literature to consider suction/injection at the disk surface. Kelson and Desseaux [6] revisited the von Kármán flow problem over a rotating disk including mass transfer through the disk. Attia [7] extended their work by considering the unsteady flow over an infinite rotating disk with uniform suction and injection and heat transfer effects. Ashraf et al. [8] considered the flow of a micropolar fluid between two stationary disks with constant injection velocity at the surface of one disk. Domairry and Aziz [9] investigated by a homotopy perturbation method the MHD flow between two parallel stationary disks with suction or injection through one of the two disks.

None of these previous works considered the Bödewadt flow of a non-Newtonian fluid over a stationary rough disk with mass transfer through it. This paper is then an endeavour to fill this gap. A second-order finite difference method has been adopted here to solve the resulting system of fully coupled and highly nonlinear similarity equations arising due to Reiner-Rivlin swirling flow over an infinite stationary porous disk. The objective is to check if suction or injection is an effective way to reduce the chances of separation of the boundary layer.

2. Formulation of the Problem

One considers an incompressible non-Newtonian Reiner-Rivlin fluid, whose constitutive equation is given by

$$\mathbf{T} = -p\mathbf{I} + \phi_1 \mathbf{D} + \phi_2 \mathbf{D}^2, \tag{1}$$

where $\mathbf{D} = (1/2)[\nabla \mathbf{v} + (\nabla \mathbf{v})^T]$ [10]. The response functions ϕ_1 and ϕ_2 are functions of the scalar invariants $(\operatorname{tr} \mathbf{D})^2$, $\operatorname{tr}(\mathbf{D}^2)$, and $\det \mathbf{D}$ [11]. The fluid occupies the space $z > 0$ over an infinite stationary disk, which coincides with $z = 0$. The motion is due to the rotation of the fluid like a rigid body with constant angular velocity Ω at large distance from the stationary disk. Let $\mathbf{v} = (u, v, w)$ be the fluid velocity vector in a (r, θ, z) stationary reference frame (see [2] for a schematic view of the flow configuration). Using the von Kármán transformations [12],

$$u = r\Omega F(\zeta), \qquad v = r\Omega G(\zeta), \qquad w = \sqrt{\Omega \nu} H(\zeta),$$

$$z = \sqrt{\frac{\nu}{\Omega}} \zeta, \qquad \frac{p}{\rho} = -\nu \Omega P(\zeta) + \frac{1}{2}\Omega^2 r^2, \tag{2}$$

where ν is the fluid kinematic viscosity. By considering the usual boundary layer approximations, the equations of continuity and motion take the following forms [1, 2]:

$$\frac{dH}{d\zeta} + 2F = 0, \tag{3}$$

$$\frac{d^2 F}{d\zeta^2} - H\frac{dF}{d\zeta} - F^2 + G^2$$
$$- \frac{1}{2}L\left[\left(\frac{dF}{d\zeta}\right)^2 - 3\left(\frac{dG}{d\zeta}\right)^2 - 2F\frac{d^2 F}{d\zeta^2}\right] = 1, \tag{4}$$

$$\frac{d^2 G}{d\zeta^2} - H\frac{dG}{d\zeta} - 2FG + L\left(\frac{dF}{d\zeta}\frac{dG}{d\zeta} + F\frac{d^2 G}{d\zeta^2}\right) = 0, \tag{5}$$

$$\frac{d^2 H}{d\zeta^2} - H\frac{dH}{d\zeta} - \frac{7}{2}L\frac{dH}{d\zeta}\frac{d^2 H}{d\zeta^2} + \frac{dP}{d\zeta} = 0, \tag{6}$$

where $L = \phi_2 \Omega / \phi_1$ corresponds to the non-Newtonian cross-viscous parameter. The above system of equations has to be solved subject to following partial slip boundary conditions:

$$F(0) = \lambda\left[F'(0) - LF(0)F'(0)\right],$$

$$G(0) = \eta\left[G'(0) - 2LF(0)G'(0)\right], \tag{7}$$

$$H(0) = W,$$

$$F(\infty) \longrightarrow 0, \quad G(\infty) \longrightarrow 1,$$

where λ and η are nondimensional slip coefficients and $W = W_0/\sqrt{\Omega \nu}$ is the uniform suction ($W < 0$) or injection ($W > 0$) velocity.

The expression of the nondimensional moment coefficient C_m is given by

$$C_m = \frac{-\pi\left[G'(0) - 2LF(0)G'(0)\right]}{\sqrt{\mathbb{R}}}, \tag{8}$$

where $\mathbb{R} = R^2\Omega/\nu$ is the Reynolds number based on the disk radius and the fluid velocity far from the disk surface. C_m represents the torque required to maintain the disk at rest.

3. Numerical Solution of the Problem

In this section, we will present briefly the finite difference method that has been used to solve the system of coupled, nonlinear equations (3)–(5) subject to slip boundary conditions (7). It is customary to mention that similar scheme has been used by Sahoo et al. [13] to solve the Bödewadt flow problem for a viscous fluid with Navier's slip boundary conditions. In this problem, as $H_0 \neq 0$, a slightly modified scheme has been used in order to get diagonally dominant matrix while using generalized Gauss-Seidel method. The semi-infinite integration domain $[0, \infty)$ is replaced by a finite domain $[0, \zeta_\infty)$. In practice, ζ_∞ should be chosen to be sufficiently large so that the numerical solution closely approximates the terminal boundary conditions and takes into account the asymptotical behavior far from the disk (see Appendix 1 in [14] for the asymptotical behavior of the solutions for large ζ). One approximates the functions and their derivatives by their finite difference counterparts and eventually solves a sequence of linear systems as explained below.

(1) One first solves

$$\left[1 + LF^{(k)}\right]F'' - H^{(k)}F'$$
$$= \left(F^{(k)}\right)^2 - \left(G^{(k)}\right)^2 + \frac{1}{2}L\left[\left(F'^{(k)}\right)^2 - 3\left(G'^{(k)}\right)^2\right] + 1 \tag{9}$$

using mixed boundary conditions (7) and calls the solution of (9) as $\widetilde{F}^{(k+1)}$, with k being the iteration index. To obtain convergence, one defines $F^{(k+1)}$ by the smoothing formula:

$$F^{(k+1)} = \alpha_1 \widetilde{F}^{(k+1)} + (1 - \alpha_1)\widetilde{F}^{(k)}, \quad 0 \leq \alpha_1 \leq 1. \tag{10}$$

(2) Then, one solves

$$\left[1 + LF^{(k+1)}\right]G'' + \left[LF'^{(k+1)} - H^{(k)}\right]G' = 2F^{(k+1)}G^{(k)} \tag{11}$$

using derivative boundary conditions (7) and calls the solution of (11) as $\widetilde{G}^{(k+1)}$. To obtain convergence, one defines $G^{(k+1)}$ by the following smoothing formula:

$$G^{(k+1)} = \alpha_2 \widetilde{G}^{(k+1)} + (1 - \alpha_2)\widetilde{G}^{(k)}, \quad 0 \leq \alpha_2 \leq 1. \tag{12}$$

(3) In this step, one solves

$$H' = -2F^{(k+1)} \qquad (13)$$

and calls the solution as $\widetilde{H}^{(k+1)}$. To obtain convergence, one defines $H^{(k+1)}$ by the smoothing formula:

$$H^{(k+1)} = \alpha_3 \widetilde{H}^{(k+1)} + (1 - \alpha_3) \widetilde{H}^{(k)}, \quad 0 \le \alpha_3 \le 1. \qquad (14)$$

(4) The iterations start with suitable initial guesses $F^{(0)}, G^{(0)}, H^{(0)}, F'^{(0)},$ and $G'^{(0)}$, borrowed from the work by Sahoo and Poncet [2]. If $F^{(k+1)}, F^{(k)}, G^{(k+1)}, G^{(k)},$ and $H^{(k+1)}, H^{(k)}$ are close enough to each other, iterations are stopped; otherwise one sets $k = k + 1$ and goes to step (1).

In order to solve the above system of equations by finite difference method, a uniform grid in $0 \le \zeta \le \zeta_\infty$ is introduced by dividing it into n equal parts with a mesh size h equal to 0.01. One approximates the derivatives by their finite difference counterparts using second-order schemes as follows:

$$F'(\zeta_i) = \frac{F_{i+1} - F_{i-1}}{2h},$$

$$F''(\zeta_i) = \frac{F_{i+1} - 2F_i + F_{i-1}}{h^2}, \qquad (15)$$

$$i = 1, 2, \ldots, n - 1.$$

In order to obtain a diagonally dominant linear algebraic system for (9) and (11), F' and G' are discretized by backward difference approximations as $H_i^{(k+1)} > 0$ for Bödewadt flow. Finally, (13) is discretized by the central difference approximation. The above algebraic system of equations is solved by generalized Gauss-Seidel method [15]. The convergence of the generalized Gauss-Seidel method for the above diagonally dominant system of equations is quite fast. About 19–21 iterations are necessary to achieve an accuracy of 10^{-6}. The Fortran 90 code was compiled and run using one of the NIT Rourkela high-end Linux servers. The typical time per iteration for a given mesh distribution ($\zeta_\infty = 20$ and $h = 0.01$) and $L = \lambda = 1$ is 32.9 seconds for $W = 0$ and increases up to 45.4 seconds for $W = 1$.

4. Results and Discussions

The effects of slip (λ) and cross-viscous parameter (L) on the momentum boundary layer have been already precisely discussed in [2]. This short communication focuses only on the effects of the suction/injection velocity on the momentum boundary layer for fixed values of $L = 1$ and $\lambda = \eta = 1$.

The velocity profiles for the Bödewadt problem exhibit oscillations unlike von Kármán flow. The oscillations occurring in the boundary layer when the fluid rotates near a stationary disk can be explained in the following manner. The radial inflow, induced in the vicinity of the stationary disk, tends to conserve angular momentum and thus to increase

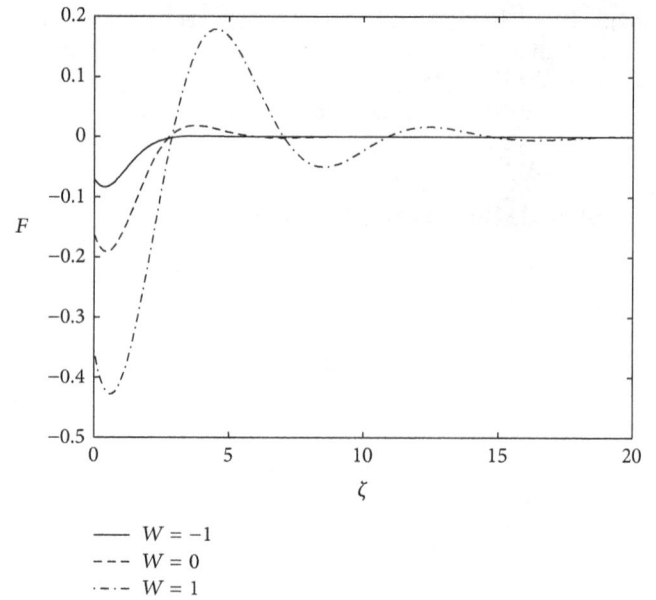

FIGURE 1: Variation of F with W.

the tangential velocity when the local radius is decreased. For an overshoot, radial convection of the angular momentum near the disk must be strong enough to more than balance the dissipation of angular momentum caused by the wall shear. This inward radial convection of surplus angular momentum is possible as long as the distribution of circulation in the outer flow increases with increasing radius. A local overshoot in the tangential velocity increases the centrifugal force locally, which tends to induce a radial outflow. This radial outflow convects an angular momentum defect to force an undershoot in the tangential velocity profile, and the above process is repeated to yield oscillatory approach to infinity. It is interesting to observe that the oscillations in the three velocity components reduce as W changes its sign from positive (injection) to negative (suction), as seen in Figures 1 to 3. The boundary layer thickness decreases significantly when suction is applied. In fact, there is no oscillation in the velocity profiles for $W = -1$. From Figure 1, it is clear that the intensity of the back flow near the disk surface decreases as W changes sign from positive to negative. Suction decreases the amplitude of the oscillations, whereas injection enhances them. To summarize, suction has a stabilizing effect on the velocity profiles, while injection destabilizes the flow. The influence of W on the components F and G confirms the previous results of Attia [7], in the case of a rotating disk in a porous medium. On the contrary, the effect of W on H has an opposite behavior. The axial velocity component H is relatively constant whatever the distance from the disk is when suction is applied. For future comparisons, the variations of H_∞, $F(0)$, and $G(0)$ for different combinations of the flow parameters are provided in Table 1. The velocity gradients close to the disk are more interesting. Suction tends to diminish them, which means that the shear stresses may be reduced by decreasing W. From Figure 2, it can be seen that the boundary layer thickness is also reduced when W decreases. With injection $W = 1$, the fluid is pushed towards

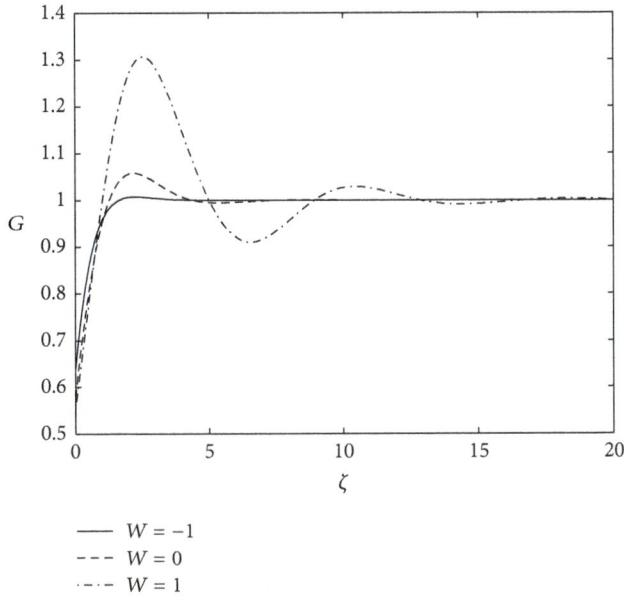

FIGURE 2: Variation of G with W.

TABLE 1: Variations of H_∞, $F(0)$, and $G(0)$ with different flow parameters.

L	$\lambda(=\eta)$	W	H_∞	$F(0)$	$G(0)$
0.0			2.070589	−0.393258	0.510032
1.0	1.0	1.0	1.899414	−0.261875	0.409946
2.0			1.800522	−0.195825	0.345059
	1.0		1.899414	−0.261875	0.409946
1.0	2.0	1.0	1.522734	−0.091550	0.311701
	3.0		1.372901	−0.048644	0.239897
		−1.0	−0.760271	−0.066407	0.593022
1.0	1.0	0.0	0.561328	−0.140015	0.497071
		1.0	1.899414	−0.261876	0.409946

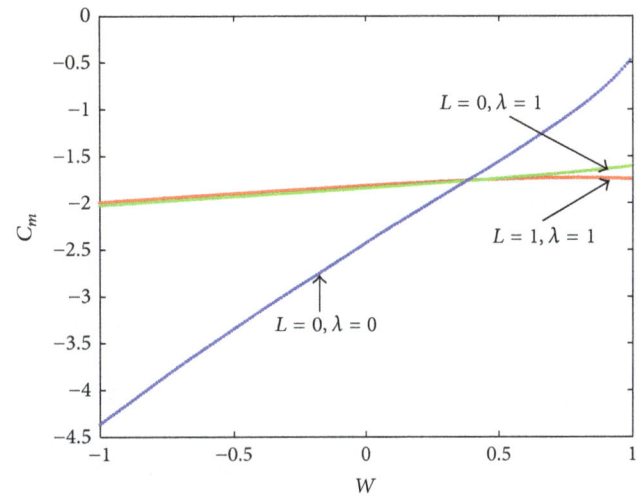

FIGURE 4: Variation of C_m with W.

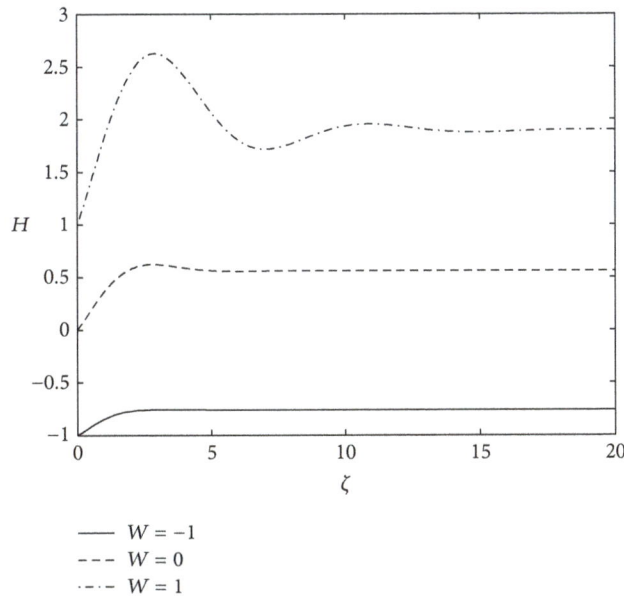

FIGURE 3: Variation of H with W.

($\lambda = 0$) boundary conditions. In this case, the torque required to maintain the disk at rest is high when there is suction at the disk surface and it decreases in magnitude as W changes its sign from negative to positive. The rate of decrease of C_m as W changes sign from negative to positive for the other two cases ($L = 0, \lambda = 1; L = 1, \lambda = 1$) is not significant. It is also observed that, for the Reiner-Rivlin fluid ($L \neq 0$), C_m decreases in magnitude when W varies in the range $[-1, 1]$ up to a critical value W^*. For $L = 1$ and $\lambda = 1$, $W^* \simeq 0.75$.

5. Conclusions

In this short communication, we have investigated the effects of suction and injection on the momentum boundary layer arising due to the swirling flow of a non-Newtonian Reiner-Rivlin fluid over an infinite rough stationary disk. A second-order finite difference method has been adopted to solve the resulting system of fully coupled and highly nonlinear similarity equations. It is observed that suction suppresses the oscillations in the velocity profiles, whereas injection enhances it. The boundary layer thickness decreases as the suction increases. Injection has an opposite effect on it.

a larger distance from the disk, which induces a thickening of the boundary layer. The injection velocity W may be then used to adjust both the shear stress and the boundary layer thickness (and as a consequence the velocity profiles) in given engineering applications.

Finally, Figure 4 shows the variation of the moment coefficient C_m with the suction/injection velocity W for three different sets of values of L and λ. Note that the Reynolds number has been fixed to $\mathbb{R} = 1$. The moment coefficient C_m remains negative for all values of W. It is interesting to observe that W has a significant effect on the moment coefficient for a viscous ($L = 0$) Bödewadt flow with no-slip

Conflict of Interests

The authors declare that there is no conflict of interests regarding the publication of this paper.

References

[1] B. Sahoo, "Effects of slip on steady Bödewadt flow and heat transfer of an electrically conducting non-Newtonian fluid," *Communications in Nonlinear Science and Numerical Simulation*, vol. 16, no. 11, pp. 4284–4295, 2011.

[2] B. Sahoo and S. Poncet, "Effects of slip on steady Bödewadt flow of a non-Newtonian fluid," *Communications in Nonlinear Science and Numerical Simulation*, vol. 17, no. 11, pp. 4181–4191, 2012.

[3] U. T. Bödewadt, "Die Drehströmung über festem Grunde," *ZAMM-Zeitschrift für Angewandte Mathematik und Mechanik*, vol. 20, no. 5, pp. 241–253, 1940.

[4] R. T. Bonnecaze, N. Mano, B. Nam, and A. Heller, "On the behavior of the porous rotating disk electrode," *Journal of the Electrochemical Society*, vol. 154, no. 2, pp. F44–F47, 2007.

[5] R. J. Lingwood, "On the effects of suction and injection on the absolute instability of the rotating-disk boundary layer," *Physics of Fluids*, vol. 9, no. 5, pp. 1317–1328, 1997.

[6] N. Kelson and A. Desseaux, "Note on porous rotating disk flow," *ANZIAM Journal*, vol. 42, pp. C837–C855, 2000.

[7] H. A. Attia, "On the effectiveness of uniform suction and injection on unsteady rotating disk flow in porous medium with heat transfer," *Computational Materials Science*, vol. 38, no. 2, pp. 240–244, 2006.

[8] M. Ashraf, M. A. Kamal, and K. S. Syed, "Numerical simulation of flow of a micropolar fluid between a porous disk and a nonporous disk," *Applied Mathematical Modelling*, vol. 33, no. 4, pp. 1933–1943, 2009.

[9] G. Domairry and A. Aziz, "Approximate analysis of MHD dqueeze flow between two parallel disks with suction or injection by homotopy perturbation method," *Mathematical Problems in Engineering*, vol. 2009, Article ID 603916, 19 pages, 2009.

[10] W. R. Schowalter, *Mechanics of Non-Newtonian Fluids*, Pergamon Press, Oxford, UK, 1978.

[11] D. R. Smith, *An Introduction to Continuum Mechanics—after Truesdell and Noll*, vol. 22 of *Solid Mechanics and its Applications*, Kluwer Academic, Dordrecht, The Netherlands, 1993.

[12] T. von Kármán, "Über laminare und turbulente Reibung," *Zeitschrift für Angewandte Mathematik und Mechanik*, vol. 1, no. 4, pp. 233–252, 1921.

[13] B. Sahoo, S. Abbasbandy, and S. Poncet, "A brief note on the computation of the Bödewadt flow with Navier slip boundary conditions," *Computers and Fluids*, vol. 90, pp. 133–137, 2014.

[14] A. I. van de Vooren, E. F. F. Botta, and J. Stout, "The boundary layer on a disk at rest in a rotating fluid," *Quarterly Journal of Mechanics and Applied Mathematics*, vol. 40, no. 1, pp. 15–32, 1987.

[15] D. K. Salkuyeh, "Generalized Jacobi and Gauss-Seidel methods for solving linear system of equations," *Numerical Mathematics A: Journal of Chinese Universities*, vol. 16, no. 2, pp. 164–170, 2007.

18

Modeling the Uniformity of Manifold with Various Configurations

Jafar M. Hassan,[1] Thamer A. Mohamed,[2] Wahid S. Mohammed,[1] and Wissam H. Alawee[1]

[1] *Department of Mechanical Engineering, University of Technology, Baghdad, Iraq*
[2] *Department of Civil Engineering, Faculty of Engineering, Universiti Putra Malaysia (UPM), 43400 Serdang, Selangor, Malaysia*

Correspondence should be addressed to Jafar M. Hassan; jafarmehdi1951@yahoo.com

Academic Editor: Mohy S. Mansour

The flow distribution in manifolds is highly dependent on inlet pressure, configuration, and total inlet flow to the manifold. The flow from a manifold has many applications and in various fields of engineering such as civil, mechanical, and chemical engineering. In this study, physical and numerical models were employed to study the uniformity of the flow distribution from manifold with various configurations. The physical model consists of main manifold with uniform longitudinal section having diameter of 10.16 cm (4 in), five laterals with diameter of 5.08 cm (2 in), and spacing of 22 cm. Different inlet flows were tested and the values of these flows are 500, 750, and 1000 L/min. A manifold with tapered longitudinal section having inlet diameters of 10.16 cm (4 in) and dead end diameter of 5.08 cm (2 in) with the same above later specifications and flow rates was tested for its uniformity too. The percentage of absolute mean deviation for manifold with uniform diameter was found to be 34% while its value for the manifold with nonuniform diameter was found to be 14%. This result confirms the efficiency of the nonuniform distribution of fluids.

1. Introduction

Flow in manifold is of great importance in many industrial processes when it is necessary to distribute a large fluid stream into several smaller streams and then to collect them into one discharge stream. Manifolds can usually be categorized into one of the following types [1]: dividing, combining, parallel, and reverse flow manifolds as shown in Figure 1. Parallel and reverse flow manifolds are those which combine dividing and combining flow manifolds and are most commonly used in plate heat exchangers. In a parallel flow manifold, the flow directions in dividing and combining flow headers are the same which is generally referred to as a Z-manifold. In a reverse flow manifold, the flow directions are opposite and it is referred to as a U-manifold. A uniform flow distribution requirement is a common issue in many engineering circumstances such as plate-type heat exchangers, piping system, heat sinks for cooling of electronic devices, fuel cells, chemical reactors, solar thermal collectors, flow distribution systems in treatment plant, and the piping system of pumping stations. Therefore, for most applications, the goal of manifold design is to achieve a uniform flow distribution through all of the lateral exit ports. A great number of experimental, analytical, and numerical studies deal with flow in manifold.

The flow in distribution manifold has been studied by several investigators [2–6]. For instance, Bajura [2] developed the general theoretical model for investigation of the performance of single-phase flow distribution for both intake and exhaust manifolds. Bajura and Jones Jr. [3] extended the previous model and the prediction for the flow rates and the pressures in the headers of dividing, combining, reverse, and parallel manifold configurations. Majumdar [4] developed a mathematical model with one-dimensional elliptic solution procedure for predicting flows in dividing and combining flow manifolds. Bassiouny and Martin [5, 6] presented an analytical solution for the prediction of flow and pressure distribution in both intake and exhaust conduits of heat exchanger for both types flow (U-type and Z-type). A great number of experimental and numerical studies covered the effect of design parameters on flow distribution in manifold. Choi et al. [7, 8] studied numerically the effect of Reynolds number and the width ratio on the flow distribution in manifolds of a liquid cooling module for electronic packaging. Kim et al. [9] investigated numerically the effects of the header shapes and Reynolds number on the flow distribution in a parallel flow manifold of a liquid cooling module for electronic packaging, for three different header geometries (i.e., rectangular, triangular, and trapezoidal) with the Z-type flow direction. Jiao et al. [10] investigated experimentally the effect of the inlet pipe diameter, the first header's diameter of equivalent area, and the second header's diameter of equivalent area on the flow maldistribution in plate-fin heat exchanger. Wen et al. [11] investigated flow characteristics in the entrance region of plate-fin heat exchanger by means of particle image velocimetry (PIV). Tong et al. [12] investigated numerically the strategies capable of perfecting manifold design to achieve the same rate of mass outflow through each of the exit ports of a distribution manifold. Minqiang et al. [13] performed a three-dimensional computational fluid dynamics (CFD) model to calculate the velocity distribution among multiple parallel microchannels with triangle manifolds. The effect of channel width and channel spacing on flow distribution among microchannels with U-shape rectangular manifolds has been investigated by Mathew et al. [14]. Chen and Sparrow [15] present a method to investigate the effect of geometric shape of the exit ports on mass flow rate uniformity effusing from a distribution manifold; three candidate exit-port geometries were considered: (a) an array of discrete slots, (b) an array of discrete circular apertures, and (c) a single continuous longitudinal rectangular slot. In order to have a valid comparison of the impacts of these individual geometries, the total exit areas were made identical. Dharaiya et al. [16] studied numerically the effect of tapered header configuration to reduce flow maldistribution in minichannels and microchannels. Tong et al. [17] applied a logic-based systematic method of designing manifold systems to achieve flow rate uniformity among the channels that interconnect a distribution manifold and a collection manifold. The method was based on tailoring the flow resistance of the individual channels to achieve equal pressure drops for all the channels. The tailoring of the flow resistance was accomplished by the use of gate-valve-like obstructions. Huang and Wang [18] examined an inverse design problem to determine the optimum variables for a three-dimensional Z-type compact parallel flow heat exchanger with the Levenberg-Marquardt method (LMM) [19]. To obtain the uniform tube flow rates, five different optimization design problems were examined to demonstrate the validity of the study. Wang et al. [20] investigated experimentally and numerically the single-phase flow into parallel flow heat exchangers with inlet and outlet rectangular headers having square cross-section and 9 circular tubes. Wang et al. [21] presented experimentally the results of liquid flow distribution in compact parallel flow heat exchanger through a rectangular and 5 modified inlet headers (i.e., 1 trapezoidal, one multistep, 2 baffle plates, and 1 baffle tubes header). Zeng et al. [22] performed a three-dimensional computational fluid dynamics (CFD) model to calculate the velocity distribution among microchannels with two different manifold structures. A similar performance improvement with a more uniform flow distribution in methanol steam reformers was reported by Jang et al. [23]. Such findings affirm the influence of flow distribution uniformity on

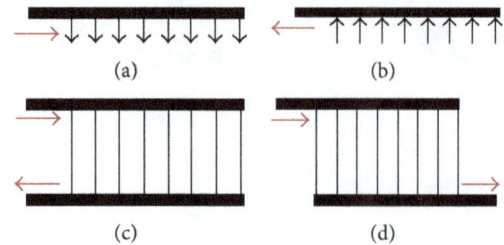

FIGURE 1: Different types of flow manifolds: (a) dividing, (b) combining, (c) U-manifold, and (d) Z-manifold.

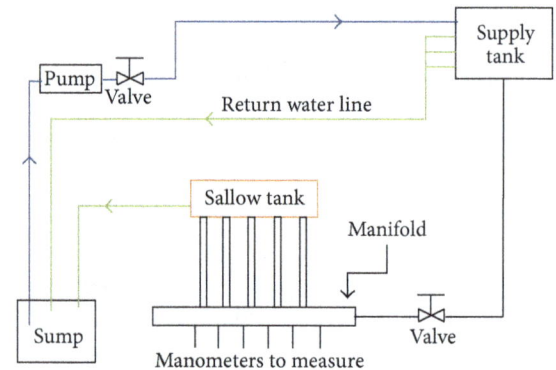

FIGURE 2: Experimental setup.

the performance of microreactor devices and reflect the importance of efficient manifold design. Tuo and Hrnjak [24] investigated experimentally and numerically the flow maldistribution caused by the pressure drop in headers and its impact on the performance of a microchannel evaporator with horizontal headers and vertically oriented tubes. Kim and Byun [25] studied experimentally the effect of inlet configuration on upward branching of two-phase refrigerant in a parallel flow heat exchanger. Three different inlet orientations (parallel, normal, and vertical) were investigated.

In general, all previous studies for manifolds with different applications had shown that typical manifold design does not give a uniform flow distribution at outlets. Therefore, the objective of the study was to predict the flow distribution through each outlet for circular cross-section header and to develop an optimized tapered cross-section header design having a better flow distribution through outlets.

2. Methodology

2.1. Experimental Setup. The nonuniformity of flow distribution through parallel outlets is found to be more severe in models with constant cross-sectional area headers [16]. Hence, the objectives of the study are to predict the flow distribution through each outlet of manifold with uniform longitudinal section and to develop an optimized manifold with longitudinal section design having a better flow distribution through outlets. The schematic diagram of the experimental setup is shown in Figure 2. The experimental setup consists of water tank with over flow, steel support,

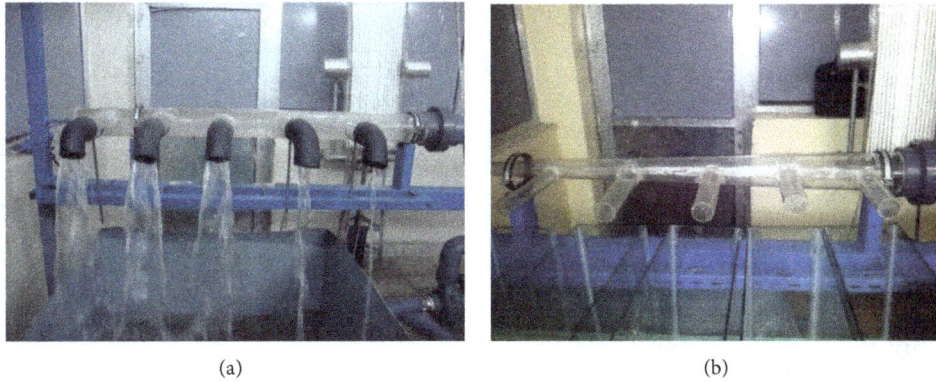

(a)   (b)

FIGURE 3: Experimental setup of two manifold configurations: (a) manifold with uniform longitudinal section and (b) manifold with tapered longitudinal section.

pump, sump, and number of valves to set the required flow rate through two dividing manifolds. The first manifold is with uniform longitudinal section while the second manifold is with tapered longitudinal section (optimal taper shape from numerical section).

The rig was assembled at a selected site in fluid laboratory of Machines and Equipment Engineering Department, University of Technology, Iraq. The water tank is rested on 3 m high steel elevated frame. At the outlet of each branch pipe, a shallow tank with cross-section 150 cm × 150 cm is used to collect the water flowing from the branch pipes as shown in Figure 3. The water from the branch pipes is measured using 50 liter capacity rectangular tank. A constant head was ensured during the experiments and, as a result, constant flow rate from branch pipes was obtained. Six uniformly spaced piezometers were installed along the pipe to monitor the pressure head at the branch pipes. The spacing was 25 cm.

Dimensions of two configuration manifolds are shown in Figure 4. The manifolds have been fabricated with acrylic material to ensure the developed flow and the good visibility of flow pattern. The branch pipes junctions are at right angles with header. The difference between two models only lies in the header configuration.

2.2. CFD Model.
In the CFD analysis, a model of the manifold with uniform longitudinal section was prepared. The configuration used in the analysis is as shown in Figure 1(a). Later, the simulation was performed to develop a manifold design to achieve nearly uniform flow distribution through the outlets. The geometry of manifold with tapered longitudinal section is shown in Figure 1(b). The manifold diameter ratio (D_1/D_2) is varied parametrically to estimate the optimal tapered ratio and uniform flow distribution.

In the present problem, the fluid flow is three-dimensional; that is, all three possible velocity components (x, y, and z) exist and all three components depend on the three coordinates of cartesian geometry. The statements of the governing parietal equations are

$$\frac{\partial u}{\partial x} + \frac{\partial v}{\partial y} + \frac{\partial w}{\partial z} = 0. \tag{1}$$

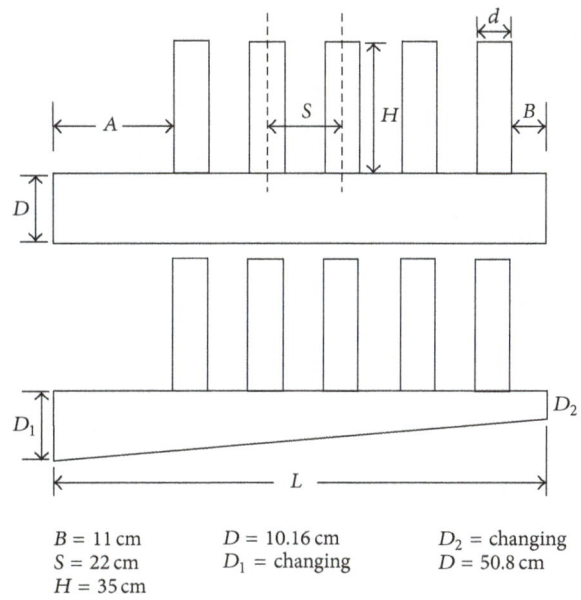

$B = 11$ cm	$D = 10.16$ cm	D_2 = changing
$S = 22$ cm	D_1 = changing	$D = 50.8$ cm
$H = 35$ cm		

FIGURE 4: Manifolds used for conducting experiment.

x-momentum is

$$P\left[\frac{\partial}{\partial x}\left(u^2\right) + \frac{\partial}{\partial y}\left(uv\right) + \frac{\partial}{\partial z}\left(uw\right)\right]$$
$$= -\frac{\partial p}{\partial x} + \frac{\partial}{\partial x}\left(\mu_{\text{eff}}\frac{\partial u}{\partial x}\right) + \frac{\partial}{\partial y}\left(\mu_{\text{eff}}\frac{\partial u}{\partial y}\right) \tag{2}$$
$$+ \frac{\partial}{\partial z}\left(\mu_{\text{eff}}\frac{\partial u}{\partial z}\right),$$

y-momentum is

$$P\left[\frac{\partial}{\partial x}\left(vu\right) + \frac{\partial}{\partial y}\left(v^2\right) + \frac{\partial}{\partial z}\left(vw\right)\right]$$
$$= -\frac{\partial p}{\partial y} + \frac{\partial}{\partial x}\left(\mu_{\text{eff}}\frac{\partial v}{\partial x}\right) + \frac{\partial}{\partial y}\left(\mu_{\text{eff}}\frac{\partial v}{\partial y}\right) \tag{3}$$
$$+ \frac{\partial}{\partial z}\left(\mu_{\text{eff}}\frac{\partial v}{\partial z}\right),$$

and z-momentum is

$$P\left[\frac{\partial}{\partial x}(wu) + \frac{\partial}{\partial y}(wv) + \frac{\partial}{\partial z}(w^2)\right]$$

$$= -\frac{\partial p}{\partial z} + \frac{\partial}{\partial x}\left(\mu_{\text{eff}}\frac{\partial w}{\partial x}\right) + \frac{\partial}{\partial y}\left(\mu_{\text{eff}}\frac{\partial w}{\partial y}\right) \quad (4)$$

$$+ \frac{\partial}{\partial z}\left(\mu_{\text{eff}}\frac{\partial w}{\partial z}\right),$$

where u, v, and w are the velocity components in three dimensions, respectively. ρ is the fluid density, and the effective viscosity, μ_{eff}, is defined as $\mu_{\text{eff}} = \mu + \mu_t$. The turbulent viscosity depends on the selected turbulence model as well as on the specific application. In the present study, the realizable k–e model was chosen for application here [15, 25].

The simulation of the two geometries was conducted using a commercial CFD software FLUENT. The design, meshing, and boundary definition of the geometries were done using the presolver software, GAMBIT. Tet/Hybrid T-grid scheme was used for the mesh generation [16]. The numbers of elements in each geometrical model were approximately 1,000,000. Grid independence test was carried out to determine the best mesh spacing for the geometrical model. The solutions are considered to be converged when all of the residuals for the continuity and momentum equations are less than or equal to 10^{-6}.

2.2.1. Boundary Condition. The boundary condition used for the simulation is shown in Table 1.

3. Result and Discussions

3.1. Numerical Result. A numerical model was prepared in this study to

(1) determine the flow distribution and pressure drop at the parallel pipes and to validate the result with the data obtained from experimental setup,

(2) determine the optimum design of the tapered manifold that can give uniform water distribution through changing the diameter ratio (D_1/D_2) parametrically.

CFD simulation was first performed on manifold with uniform longitudinal section having circular diameter of 10.16 cm (4 in) and straight flow with outlets of constant cross-sectional areas. The axial momentum would progressively decrease. This would give rise to the static pressure from the entrance to the manifold dead end. Such an increase in static pressure should favour a higher efflux through the downstream outflows. Figure 5 represents the static pressure contour for circular cross-section manifold ($D = 10.16$ cm) with Reynolds number (Re = 150,000). It can be clearly seen from Figure 5 that the pressure along the manifold is increasing which results in nonuniformity flow.

To study the flow distribution among the parallel tubes, the dimensionless parameters, Φ and β_i, are used to evaluate

TABLE 1: Boundary condition for two manifolds.

Boundary condition	Test, 1	Test, 2	Test, 3
Reynolds number	10×10^4	15×10^4	20×10^4
Inlet volume rate L/m	500	750	1000
Inlet water temperature °C	20	20	20
Outlet gage pressure	Zero	Zero	Zero

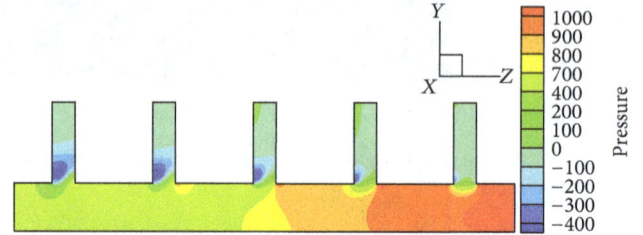

FIGURE 5: Pressure contour for flow in manifold with uniform longitudinal section.

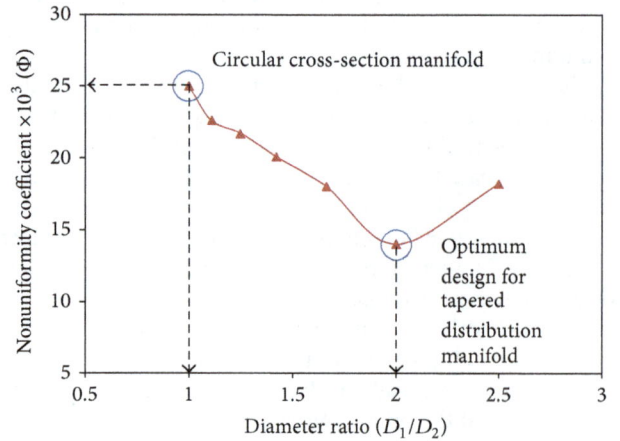

FIGURE 6: The nonuniformity coefficient (Φ) for different diameter ratio.

the flow distribution. Their definitions are given as follows [20]:

$$\Phi = \sqrt{\frac{\sum_{i=1}^n (\beta_i - \overline{\beta})^2}{N}} \quad (5)$$

$$\overline{\beta} = \frac{Q_i}{Q}, \quad (6)$$

where Φ is the nonuniformity, β_i denotes the flow ratio for ith pipe, Q_i represents volume flow rate for ith pipe (m^3/s), Q is total flow rate (m^3/s), N is the number of parallel pipes in the manifold, and $\overline{\beta}$ is the average flow ratio for the total tubes which is defined as $\overline{\beta} = (\sum_{i=1}^n \beta_i)/N$. The large value of Φ indicates high nonuniformity. For this reason, the minimum value of nonuniformity coefficient will give the optimum configuration for the tapered manifold.

From Figure 6, the values of nonuniformities (Φ) of tapered manifold are 0.025, 0.0226, 0.0222, 0.020, 0.019, 0.014,

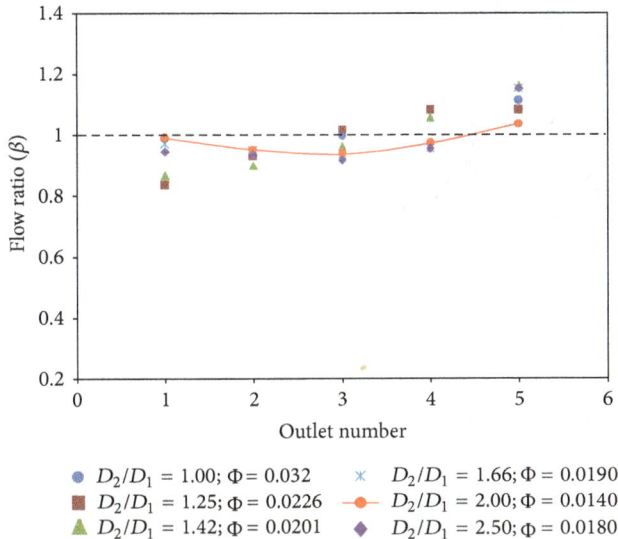

FIGURE 7: The flow ratio (β) for different diameter ratio.

Legend:
- $D_2/D_1 = 1.00$; $\Phi = 0.032$
- $D_2/D_1 = 1.25$; $\Phi = 0.0226$
- $D_2/D_1 = 1.42$; $\Phi = 0.0201$
- $D_2/D_1 = 1.66$; $\Phi = 0.0190$
- $D_2/D_1 = 2.00$; $\Phi = 0.0140$
- $D_2/D_1 = 2.50$; $\Phi = 0.0180$

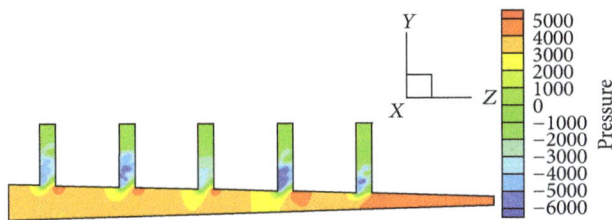

FIGURE 9: Flow distribution plot for manifold with uniform longitudinal section (Re = 100,000, 150,000, and 200,000).

- Re = 200,000
- Re = 150,000
- Re = 100,000

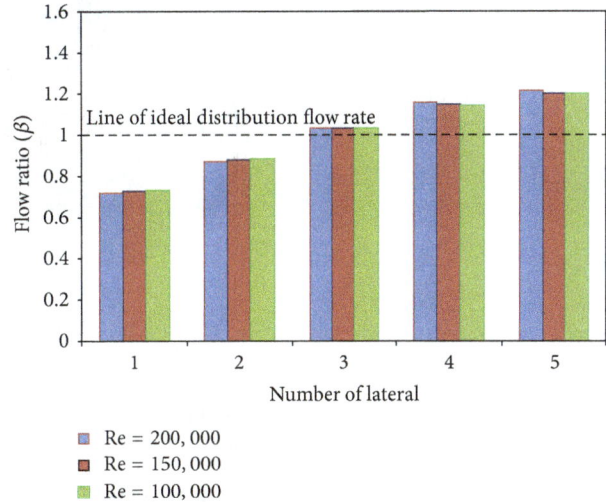

FIGURE 8: Pressure contour for flow in manifold with tapered longitudinal section.

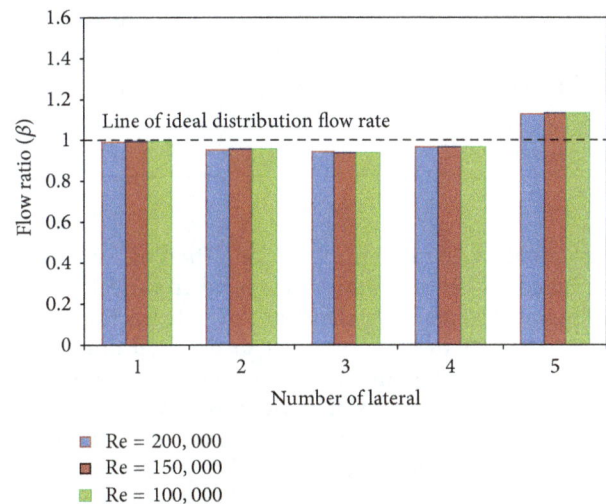

FIGURE 10: Flow distribution plot for manifold with tapered longitudinal section (Re = 100,000, 150,000, and 200,000).

- Re = 200,000
- Re = 150,000
- Re = 100,000

and 0.0182 at diameter ratio of 1, 1.1, 1.25, 1.42, 1.66, 2, and 2.5, respectively. The (Φ) values are lower for the manifold with circular cross-section and the corresponding value is 0.0345 as shown in Figure 6. Then the dead end diameter is reduced gradually from 10.16 cm (4 in) to 5.08 cm (2 in); the flow distribution is generally improved. The nonuniformity (Φ) decreases until it reaches a minimum value (optimal design); then it starts to increase although the diameter ratio was increased too as shown in Figure 6.

From the results shown in Figures 6 and 7, the optimum configuration of distribution manifold can be determined using diameter ratio (D_1/D_2) which is equal to 2. Figure 8 shows the pressure contour for tapered distribution manifold. The pressure along the manifold was found to be nearly uniform which resulted in a better flow distribution through to outlets.

3.2. Experimental Results. Figure 9 shows the flow distribution plots for manifold with uniform longitudinal section (diameter 10.16 cm) for three values of Reynolds number (100,000, 150,000, and 200,000). The flow through the first outlet was found to be very small compared with the last outlet as shown by pressure contours (Figure 5).

Uniform flow distribution through the manifold with tapered longitudinal section can be achieved using the design

obtained from the numerical model. Figure 10 represents the flow distribution from manifold with tapered longitudinal section having inlet diameters of 10.16 cm (4 inch) and dead diameter of 5.08 cm (2 in). The improvement of flow distribution through the outlets is compared to that obtained from circular cross-section manifold as shown in Figures 9 and 10.

The nonuniformity flow coefficient (Φ) was taken as a parameter to quantify the uniformity in flow distribution through the manifold outlet. The Φ can be defined using (5). Flow distribution through the outlets is better at lower (Φ) values. Table 2 shows the nonuniformity coefficient for circular and tapered manifold cross-sections from the values of Reynolds number (100,000, 150,000, and 200,000). It can be seen that the flow distribution was severe in case

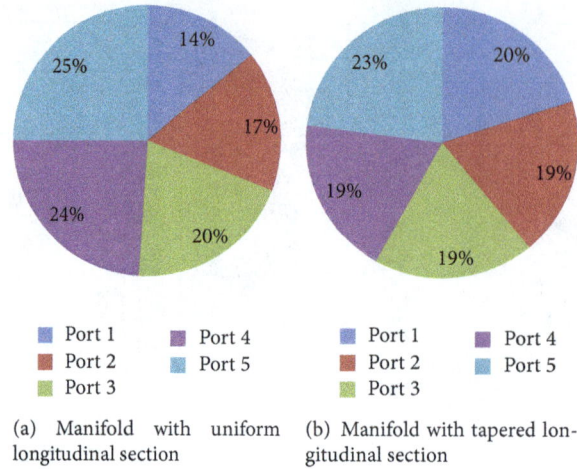

(a) Manifold with uniform longitudinal section

(b) Manifold with tapered longitudinal section

FIGURE 11: Flow rate fraction percentage for two manifold configurations (Re = 150,000, Q = 750 L/min).

TABLE 2: The nonuniformity flow coefficient (Φ) for circular and tapered cross-section manifold.

Manifold cross-section	The nonuniformity flow coefficient (Φ)	
	Circular	Tapered
Re = 100,000	0.0367	0.0142
Re = 150,000	0.0345	0.0140
Re = 200,000	0.0340	0.0139

of manifold of circular cross-section. The flow was evenly distributed for the manifold with tapered cross-section.

Figures 11(a) and 11(b) show the flow rate fraction of each outlet (which is the rate of outlet to the total flow rate in the manifold). For nonuniform flow, results show that the smallest flow rate occurred in outlet closest to manifold inlet while highest flow rate occurred in the last manifold outlet. Let the respective outlets be numbered as (1) which is the first outlet while the last is outlet (5). The discharge from outlet (1) is lower by 44% than outlet (5), while, for the tapered cross-section manifold, the percentage is reduced from 44% to 13%.

4. Model Validation Using Experimental Data

Experimental tests for flow distribution from two manifolds with different configurations have been conducted. The numerical simulation results obtain by using FLUENT[@] CFD package. The experimental test was conducted to measure the flow rate at the 5 outlets. The accuracy of the solution from the FLUENT[@] CFD package in flow field calculation of the manifolds system is used to determine the optimal design. If the solution from the FLUENT[@] CFD code cannot reproduce the actual performance of the manifold, this means that configuration for the taper distribution manifold is not optimum. The first task is thus to demonstrate the accuracy of the numerical solution. The computed and experimental flow rate distribution per outlet for Q_{total} = 750 liter/minute (Re = 1500000) are shown in Figures 12 and 13, respectively. It can be clearly seen that the differences in flow rates between

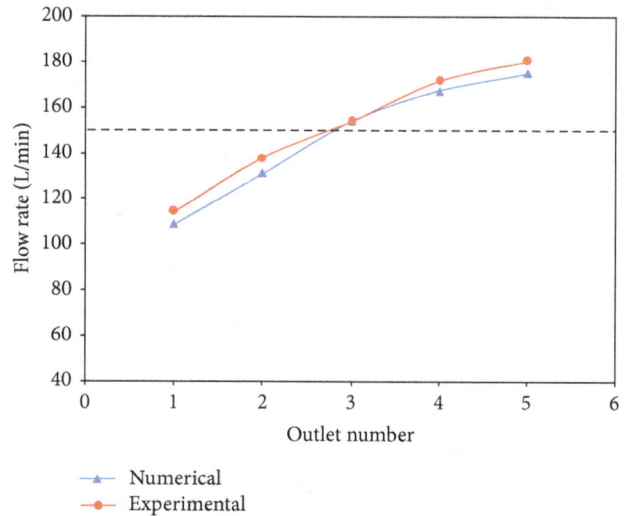

FIGURE 12: Flow distribution per outlet for manifold with uniform longitudinal section (Re = 150,000, Q = 750 L/m).

computed and measured are acceptable and therefore the validity of present numerical solution is evident.

5. Conclusions

The goal of this investigation is to evaluate the hydraulic parameter of manifold so that same rate of mass outflow can be obtained from outlet of the manifold. The CFD simulation and experimental data at different outlets and configurations, namely, circular and tapered cross-section, were carried out. Severe maldistribution was found at the outlet of the manifold with circular cross-section whereas the flow through the manifold with tapered cross-section was nearly uniform. A numerical model was used to predict the flow across each lateral for three different Reynolds numbers (i.e., 100,000, 150,000, and 200,000) and the results were found to have the same trend compared with experimental data. The flow

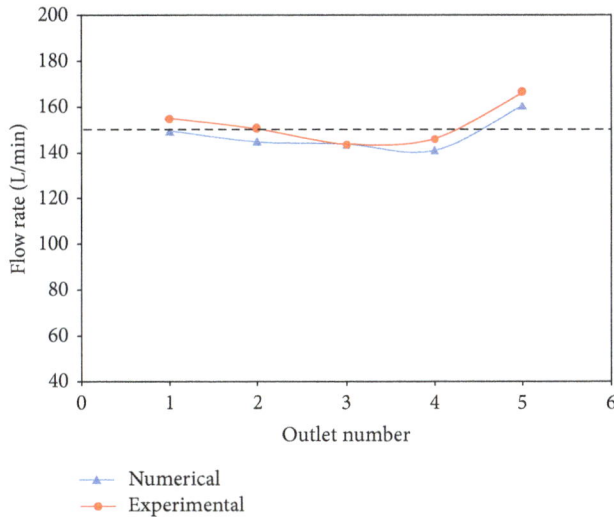

FIGURE 13: Flow distribution per outlet for manifold with tapered longitudinal section ($Re = 150{,}000$, $Q = 750$ L/min).

distribution in manifolds is independent of Reynolds number because Reynolds number was found to have slight effect on the uniformity of the mass effusion from the outlets.

Conflict of Interests

The authors declare that there is no conflict of interests regarding the publication of this paper.

References

[1] M. S. Gandhi, A. A. Ganguli, J. B. Joshi, and P. K. Vijayan, "CFD simulation for steam distribution in header and tube assemblies," *Chemical Engineering Research and Design*, vol. 90, no. 4, pp. 487–506, 2012.

[2] R. A. Bajura, "A model for flow distribution in manifolds," *Journal of Engineering for Gas Turbines and Power*, vol. 93, no. 1, pp. 7–12, 1971.

[3] R. A. Bajura and E. H. Jones Jr., "Flow distribution manifolds," *Journal of Fluids Engineering, Transactions of the ASME*, vol. 98, no. 4, pp. 654–666, 1976.

[4] A. K. Majumdar, "Mathematical modeling of flows in dividing and combining flow manifold," *Applied Mathematical Modelling*, vol. 4, no. 6, pp. 424–432, 1980.

[5] M. K. Bassiouny and H. Martin, "Flow distribution and pressure drop in plate heat exchangers-I U-type arrangement," *Chemical Engineering Science*, vol. 39, no. 4, pp. 693–700, 1984.

[6] M. K. Bassiouny and H. Martin, "Flow distribution and pressure drop in plate heat exchangers-II Z-type arrangement," *Chemical Engineering Science*, vol. 39, no. 4, pp. 701–704, 1984.

[7] S. H. Choi, S. Shin, and Y. I. Cho, "The effect of area ratio on the flow distribution in liquid cooling module manifolds for electronic packaging," *International Communications in Heat and Mass Transfer*, vol. 20, no. 2, pp. 221–234, 1993.

[8] S. H. Choi, S. Shin, and Y. I. Cho, "The effects of the Reynolds number and width ratio on the flow distribution in manifolds of liquid cooling modules for electronic packaging," *International Communications in Heat and Mass Transfer*, vol. 20, no. 5, pp. 607–617, 1993.

[9] S. Kim, E. Choi, and Y. I. Cho, "The effect of header shapes on the flow distribution in a manifold for electronic packaging applications," *International Communications in Heat and Mass Transfer*, vol. 22, no. 3, pp. 329–341, 1995.

[10] A. Jiao, R. Zhang, and S. Jeong, "Experimental investigation of header configuration on flow maldistribution in plate-fin heat exchanger," *Applied Thermal Engineering*, vol. 23, no. 10, pp. 1235–1246, 2003.

[11] J. Wen, Y. Li, A. Zhou, and Y. Ma, "PIV investigations of flow patterns in the entrance configuration of plate-fin heat exchanger," *Chinese Journal of Chemical Engineering*, vol. 14, no. 1, pp. 15–23, 2006.

[12] J. C. K. Tong, E. M. Sparrow, and J. P. Abraham, "Geometric strategies for attainment of identical outflows through all of the exit ports of a distribution manifold in a manifold system," *Applied Thermal Engineering*, vol. 29, no. 17-18, pp. 3552–3560, 2009.

[13] P. Minqiang, Z. Dehuai, T. Yong, and C. Dongqing, "CFD-based study of velocity distribution among multiple parallel microchannels," *Journal of Computers*, vol. 4, no. 11, pp. 1133–1138, 2009.

[14] B. Mathew, T. J. John, and H. Hegab, "Effect of manifold design on flow distribution in multichanneled microfluidic devices," in *Proceedings of the ASME Fluids Engineering Division Summer Conference (FEDSM '09)*, pp. 543–548, August 2009.

[15] A. W. Chen and E. M. Sparrow, "Effect of exit-port geometry on the performance of a flow distribution manifold," *Applied Thermal Engineering*, vol. 29, no. 13, pp. 2689–2692, 2009.

[16] V. V. Dharaiya, A. Radhakrishnan, and S. G. Kandlikar, "Evaluation of a tapered header configuration to reduce flow maldistribution in minichannels and microchannels," in *Proceedings of the ASME 7th International Conference on Nanochannels, Microchannels, and Minichannels (ICNMM '09)*, June 2009.

[17] J. C. K. Tong, E. M. Sparrow, and J. P. Abraham, "Attainment of flowrate uniformity in the channels that link a distribution manifold to a collection manifold," *Journal of Fluids Engineering*, vol. 129, no. 9, pp. 1186–1192, 2007.

[18] C. Huang and C. Wang, "The design of uniform tube flow rates for Z-type compact parallel flow heat exchangers," *International Journal of Heat and Mass Transfer*, vol. 57, no. 2, pp. 608–622, 2013.

[19] D. W. Marquardt, "An algorithm for least-squares estimation of nonlinear parameters," *Journal of Society for Industrial and Applied Mathematics*, vol. 11, pp. 431–441, 1963.

[20] C.-C. Wang, K.-S. Yang, J.-S. Tsai, and I. Y. Chen, "Characteristics of flow distribution in compact parallel flow heat exchangers, part I: typical inlet header," *Applied Thermal Engineering*, vol. 31, no. 16, pp. 3226–3234, 2011.

[21] C. C. Wang, K. S. Yang, J. S. Tsai, and I. Y. Chen, "Characteristics of flow distribution in compact parallel flow heat exchangers, part II: modified inlet header," *Applied Thermal Engineering*, vol. 31, no. 16, pp. 3235–3242, 2011.

[22] D. Zeng, M. Pan, and Y. Tang, "Qualitative investigation on effects of manifold shape on methanol steam reforming for hydrogen production," *Renewable Energy*, vol. 39, no. 1, pp. 313–322, 2012.

[23] J.-Y. Jang, Y.-X. Huang, and C.-H. Cheng, "The effects of geometric and operating conditions on the hydrogen production performance of a micro-methanol steam reformer," *Chemical Engineering Science*, vol. 65, no. 20, pp. 5495–5506, 2010.

[24] H. Tuo and P. Hrnjak, "Effect of the header pressure drop induced flow maldistribution on the microchannel evaporator performance," *International Journal of Refrigeration*, vol. 36, pp. 2176–2186, 2013.

[25] N. Kim and H. Byun, "Effect of inlet configuration on upward branching of two-phase refrigerant in a parallel flow heat exchanger," *International Journal of Refrigeration*, vol. 36, no. 3, pp. 1062–1077, 2013.

19

Numerical and Experimental Analysis of the Growth of Gravitational Interfacial Instability Generated by Two Viscous Fluids of Different Densities

Snehamoy Majumder, Debajit Saha, and Partha Mishra

Department of Mechanical Engineering, Jadavpur University, Kolkata 700 032, West Bengal, India

Correspondence should be addressed to Debajit Saha; debajit.saha1986@gmail.com

Academic Editor: Andrew W. Cook

In the geophysical context, there are a wide variety of mechanisms which may lead to the formation of unstable density stratification, leading in turn to the development of the Rayleigh-Taylor instability and, more generally, interfacial gravity-driven instabilities, which involves moving boundaries and interfaces. The purpose of this work is to study the level set method and to apply the process to study the Rayleigh-Taylor instability experimentally and numerically. With the help of a simple, inexpensive experimental arrangement, the R-T instability has been visualized with moderate accuracy for real fluids. The same physical phenomenon has been investigated numerically to track the interface of two fluids of different densities to observe the gravitational instability with the application of level set method coupled with volume of fraction replacing the Heaviside function. Good agreement between theory and experimental results was found and growth of instability for both of the methods has been plotted.

1. Introduction

The Rayleigh-Taylor instability is instability of an interface of two fluids of different densities which occurs when the interface between the two fluids is subjected to a normal pressure gradient with direction such that the pressure is higher in the light fluid than in the dense fluid. This is the case with an interstellar cloud and shock system. A similar situation occurs when gravity is acting on two fluids of different density—with the denser fluid above a fluid of lesser density—such as water balancing on light oil. Considering two completely plane-parallel layers of immiscible fluid, the heavier on top of the light one and both subject to the Earth's gravity, the equilibrium here is unstable to certain perturbations or disturbances. An unstable disturbance will grow and direct to a release of potential energy, as the heavier material moves down under the gravitational field and the lighter material is displaced upwards. Such instability can be observed in many situations including technological applications as laser implosion of deuterium-tritium fusion targets, electromagnetic implosion of a metal liner and natural phenomena as overturn of the outer portion of

the collapsed core of a massive star, and the formation of high luminosity twin-exhaust jets in rotating gas clouds in an external gravitational potential.

Various numerical and experimental works have been done by many researchers concentrating on the growth of single wavelength perturbations as well as considering different wavelength modes. Sharp [1] presented some of the critical issues concerning Rayleigh-Taylor instability. The importance to carry out the three-dimensional study of Taylor instability and the role of statistically distributed heterogeneities on the growth of instability have been analyzed in his work. Read [2] experimentally investigated the turbulent mixing by Rayleigh-Taylor instability and the results showed that if the instability arises from small random perturbations, the width of the mixed region grows in proportion to t^2. The same investigation has been done numerically by Youngs [3] to simulate the growth of perturbations at an interface between two fluids of different density. If the mixing process evolves from small perturbations then the growth of instability is controlled by the non-linear interaction between bubbles of different sizes. Dalziel [4] investigated the Rayleigh-Taylor instability experimentally using a simple apparatus of novel design where the initial nonlinear perturbations to the flow have been introduced by the removal of the barrier separating the two fluid layers and a good agreement between the results of this work and a previous one has been achieved. Velocity measurements have been done by particle tracking using the method of particle image velocimetry. Voropayev et al. [5] experimentally analyzed the evolution of gravitational instability of an overturned, initially stable stratified fluid. In the present analysis, the instability is initiated by overturning the experimental setup such that the heavy fluid lies over the lighter one. The present study is mainly concerned about the propagating interface between the two fluids and its formation and growth rate. A propagating interface is a closed surface in some space that is moving under a function of local, global, and independent properties. A variety of numerical algorithms are available to track propagating interfaces, and in the present numerical simulation level set method coupled with volume of fraction has been used. Level set method is a computational technique for tracking moving interfaces which rely on an implicit representation of the interface whose equation of motion is numerically approximated using schemes built from those for hyperbolic-conservation laws. The consequential techniques are able to handle problems in which the speed of the evolving interface may sensitively depend on local properties such as curvature and normal direction, as well as complex physics of the front and internal jump and boundary conditions determined by the interface location.

The volume of fluid (VOF) technique has been presented by Hirt and Nichols [6] as a simple and efficient means for numerically handling free boundaries in a calculation mesh of Eulerian or arbitrary Lagrangian-Eulerian cells. It works extremely well for a wide range of complicated problems and this process is very much conservative in nature, but the appropriate tracking of the interface is not possible by this method.

Sethian [7] presented a case of the evolution of a front propagating along its normal vector field with curvature-dependent speed. Numerical methods based on finite difference schemes for marker particles along the front are shown to be unstable in regions where the curvature builds rapidly. And then the front tracking based on volume of fluid techniques has been used together with the entropy condition.

Various numerical methods were developed to study the propagating interfaces. Osher and Sethian [8] devised new numerical algorithms, called PSC algorithms, for fronts propagating with curvature-dependent speed. Merriman et al. [9] extended the Hamilton Jacobi formulation of Osher and Sethian and proposed a level set method for the motion of multiple junctions where the diffusion equation was shown to generate curvature-dependent motion. Zhu and Sethian [10] considered hydrodynamic problems with cold flame propagation by merging a second-order projection method for viscous Navier stokes equations with modern techniques for computing the motion of interfaces propagating with curvature-dependent speed. A new method was presented by Unverdi and Tryggvasan [11] to simulate unsteady multifluid

flows in which a sharp interface or a front separates incompressible fluids of different density and viscosity. Chopp and Sethian [12] studied hyper surfaces moving under flow that depends on the mean curvature. The approach was based on a numerical technique that embeds the evolving hypersurface as the zero Level Set of a family of evolving surfaces. Sussman et al. [13] combined a level set method with a variable density projection method for capturing the interface between two fluids to allow for computation of two-phase flow where the interface can merge or break considering a high Reynolds number. Chang et al. [14] presented a level set formulation for incompressible, immiscible multi fluid flow separated by a free surface and the interface was identified as the zero Level Set of a smooth function.

Theory and algorithms of level set method were reviewed by Sethian [15] for the evaluation of the complex interfaces. Topological changes, corner and cusp development, and accurate determination of geometric properties such as curvature and normal direction were obtained by the method. Few years later, Sethian [16] summarized the development and interconnection between narrow band level set method and fast marching method, which provides efficient techniques for tracking moving fronts. At another paper, Sethian [17] reviewed past works on fast marching method and level set method for tracking propagating interfaces in two or three space dimensions.

Kaliakatos and Tsangaris [18] studied the motion of deformable drops in pipes and channels using a level set approach in order to capture the interface of two fluids. The shape of the drop, the velocity field, and the additional pressure loss due to the presence of the drop, the relative size of the drop to the size of the pipe or channel cross-section, the ratio of the drop viscosity to the viscosity of the suspending fluid, and the relative magnitude of viscous forces to the surface tension forces were computed. Son and Hur [19] combined a level set method with the volume of fluid method to calculate an interfacial curvature accurately as well as to achieve mass conservation. They developed a complete and efficient interface reconstruction algorithm which was based on the explicit relationship between the interface configuration and the fluid volume function.

Sethian and Smereka [20] provided an overview of level set methods, introduced by Osher and Sethian [8], for computing the solution to fluid-interface problems. They discussed the essential ideas behind the computational techniques that rely on an implicit formulation of the interface and the coupling of these techniques to finite-difference methods for incompressible and compressible flow. Majumder and Chakraborty [21] developed a novel physically based mass conservation model in the skeleton of a level set method, as a substitute to the Heaviside function based formulation. The transient evolution of a rounded bubble in a developing shear flow and rising bubbles in a static fluid, the Cox angle, and the deformation parameter characterizing the bubble evolution were critically examined. Carlès et al. [22] used a magnetic field gradient to draw down a low density paramagnetic fluid below a more dense fluid in a Hele-Shaw cell. An extended level set method for classical shape and topology optimization was proposed based on the popular radial basis functions by Wang et al. [23]. The implicit level set function was approximated by using the RBF implicit modeling with multiquadric splines. Sun and Tao [24] presented a coupled volume of fluid and level set (VOSET) method for computing incompressible two-phase flows. VOF method was used to conserve the mass and level set method was used to get the accuracy of curvature and smoothness of discontinuous physical quantities near interfaces. Sussman and Puckett [25] presented a coupled level set/volume-of-fluid (CLSVOF) method for computing 3D and axisymmetric incompressible two-phase flows and Sussman [26] presented a coupled level set and volume of fluid method for computing growth and collapse of vapor bubbles. A level set method was combined with the volume of fluid method by Son [27] for computing incompressible two-phase flows in three dimensions where the interface configurations were much more diverse and complicated. A passive scalar transport model has been studied by Wang et al. [28] to study the 3D Rayleigh-Taylor instability. The characteristic behavior and the principle of the interfacial motion from both sinusoidal and random perturbations have been achieved. Youngs [29] numerically simulated three-dimensional turbulent mixing of miscible fluids of RT instability which concluded that significant dissipation of turbulent fluctuations and kinetic energy occurs via the cascade to high wave numbers. The chaotic stage of Rayleigh-Taylor instability is characterized by the evolution of bubbles of the light fluid and spikes of the heavy fluid. Gardner et al. [30] proposed a statistical model to analyze the growth of bubbles in a Rayleigh-Taylor unstable interface. The model using numerical solutions based on the front tracking method has been compared to the solutions of the full Euler equations for compressible two-phase flow. Later, Glimm et al. [31] numerically studied the dynamics of the bubbles in chaotic environment and their interactions with each other as well as the acceleration of the bubble envelope.

The Rayleigh-Taylor instability is a gravity driven instability of a contact surface and this growth of this instability is sensitive to numerical or physical mass diffusion. Li et al. [32] addressed this problem using a second-order TVD finite difference scheme with artificial compression. They numerically simulated the 3D Rayleigh-Taylor instability using this scheme. A new model was proposed by Chen et al. [33, 34] for the momentum coupling between the two phases. The Rayleigh-Taylor instability of an interface separating fluids of distinct density is driven by acceleration across the interface. Two-phase turbulent mixing data were analyzed, which have been obtained from direct numerical simulation of the two-fluid Euler equations by the front tracking method. Direct numerical simulation of three-dimensional Rayleigh-Taylor instability (RTI) between two incompressible, miscible fluids has been presented by Cook and Dimotakis [35]. Mixing was found to be even more sensitive to initial conditions than growth rates. The flow structure and energy budget for Rayleigh-Taylor instability using the results of a high resolution direct numerical simulation have been examined by Cook and Zhou [36]. Later Cook et al. [37] described large eddy simulation for computing RT instability. A relation has been obtained between the rate of growth of the mixing layer

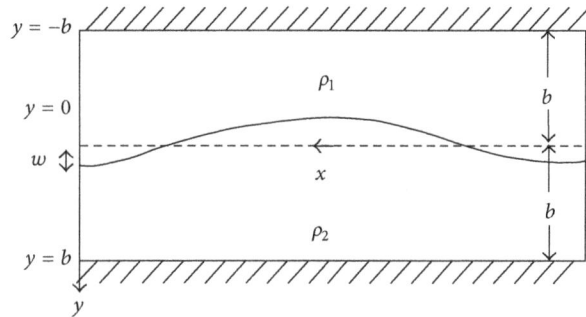

FIGURE 1: Geometrical presentation of analysis of Rayleigh-Taylor instability of a dense fluid overlying a lighter fluid.

and the net mass flux through the plane associated with the initial location of the interface.

In this present work, the level set methodology has been applied to visualize theoretically the RT instability using a triangular distribution of initial disturbance. The fraction of volume in the interface control volumes has been successfully incorporated for identifying the interface very accurately. The topological changes with time have been captured accurately and this has been matched effectively with the experimental results. The instability growth rate which is predicted by the theory is confirmed by the experimentation with the initial incipience of linear distribution of disturbances as already stated. This is a positive contribution along with the theoretical topological visualization of the RT effects.

On the other hand, the merging and consequent breaking up of the interfaces has been captured while the RT instability growth takes place. These results are important as they provides the probable trapping, merging, and consequent breaking of the oil and natural gas pools trapped between the formation of salt domes and overlying sedimentary rocks. These effects of the geothermal RT instabilities and deformation of the rocks above the salt domes are important as they provide the possibility of exploration of oil and gas pools, thus coagulated and subsequent fragmented in huge mass under the earth for million of years. These results are encouraging and can bridge our knowledge of RT to apply to the oil and gas industry.

2. Geometrical Description

The geometry of the problem is shown in Figure 1. A fluid layer with a thickness b and density ρ_1 overlies a second layer of thickness b and density ρ_2. The upper boundary and lower boundary are assumed to be rigid surfaces. Here, ρ_1 is greater than ρ_2. The undisturbed interface between fluid layers is taken to be at $y = 0$. Due to gravitational instability, the interface between the fluids distorts and motions occurs in the fluid layers. The displacement of the disturbed fluid layers is denoted by w. When the heavy fluid lies above the light fluid, the configuration becomes unstable. The time to grow the instability depends on the viscosity of the fluid and the density difference of the fluids. When the viscosity of the fluid is high and the density difference is smaller, the instability takes longer to grow.

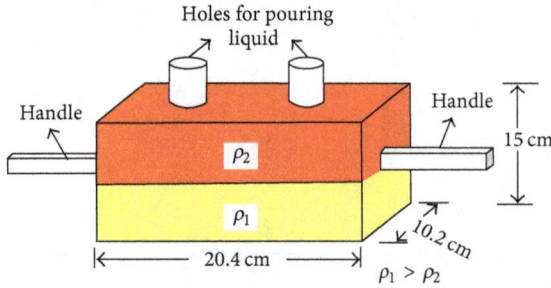

FIGURE 2: Diagram of the experimental setup.

3. Experimental Setup

The experimental setup consists of a closed rectangular box made of Perspex of 20.4 cm × 10.2 cm × 15 cm dimension. There are two openings at the top surface with valve arrangement for the purpose of filling the box with the required liquids. The two side handles are provided for convenience turning of the setup to upside down or vice versa in quick time. The setup is placed on a preleveled surface and lower half of the box is filled with glucose solution and upper half is filled with colored refined soya bean oil, with the help of funnels. The viscosities of both the liquids were measured in the laboratory at room temperature by "Falling Sphere method" and density of the fluids was measured by simply measuring their mass and volume (see Figure 2).

The viscosity and specific gravity of the liquids have been measured as follows:

Viscosity of glucose syrup = 350 Pa-S,

Viscosity of oil = 0.0791 Pa-S,

Specific Gravity of Glucose syrup = 1.4,

Specific Gravity of oil = 0.92.

4. Experimental Technique

In the experiment, first the setup rests at position 3 where the light fluid lies over the heavy one. In this position, it is totally balanced and stable. Then the setup is turned upside down quickly so that heavy liquid lies in the upper half and thus instability is initiated. The instability can also be initiated by keeping the setup at position 2 where the heavy and light liquids stand vertically side by side in an unbalanced and unstable condition. Naturally all these configurations want to return to position 3 to minimize the potential energy and to gain a stable and balanced position. The whole process is captured to track the moving interface and to study the growth rate of the instability with time (see Figure 3).

5. Formulation of Two-Phase Flow with Surface Tension

The term two-phase flow refers to the motion of two different interacting fluids or with fluids that are in different phases. In the present analysis, only two immiscible incompressible fluids have been considered and a low enough Reynolds

number is assumed so that the flow can be considered as laminar flow. Level set method may be applied to track the interface efficiently in case of incompressible, immiscible fluids in which steep gradient in viscosity and density existed across the interface. In these problems, the role of surface tension is crucial and formed an important part of the algorithm.

6. Numerical Modeling

For mathematical analysis, we assume a system of two-fluid phases constituting a two-dimensional domain. The individual fluid phases are assumed to be incompressible but deformable in shape on account of shear stresses prevailing between various fluid layers as well as fluid-solid interfaces. We assume the flow field to be two dimensional and laminar.

Navier-Stokes equation is given as

$$u_t + (u \cdot \nabla) u = F + \frac{1}{\rho} \left(-\nabla P + \mu \nabla^2 u + \text{ST} \right). \quad (1)$$

Assume a sharp fluid interface between two fluids with different densities, and also the flow is incompressible, and thus

$$\nabla \cdot u = 0. \quad (2)$$

The surface tension term acts normal to the fluid interface and is proportional to the curvature, due to balance of force argument between the pressure on each side of the interface. This leads to the relation,

$$\text{ST} = \sigma \kappa \delta (d) n. \quad (3)$$

Thus, surface tension acts as an additionally forcing term in the direction normal to the fluid interface.

Now replacing normal n by $\nabla \phi / |\nabla \phi|$ and when distance d is approximated by $\nabla \phi / |\nabla \phi|$, we have

$$\sigma \kappa \delta (d) n = \sigma \kappa (\phi) \delta (\phi) \nabla \phi. \quad (4)$$

The curvature $k(\phi)$ can be expressed by ϕ and its derivatives as follows,

$$k (\phi) = -\frac{\phi_y^2 \phi_{xx} - 2\phi_x \phi_y \phi_{xy} + \phi_x^2 \phi_{yy}}{\left(\phi_x^2 + \phi_y^2 \right)^{3/2}}. \quad (5)$$

As in [14], regularized delta function $\delta(\phi)$ can be defined as

$$\delta (\phi) \equiv \begin{cases} \dfrac{1/2 \left(1 + \cos (\pi x / \varepsilon) \right)}{\varepsilon} & \text{if } |x| < \varepsilon, \\ 0 & \text{Otherwise.} \end{cases} \quad (6)$$

This recasts the surface tension in the level set framework. If ϕ is always reinitialized to the distance function, the Dirac delta function itself can be smoothed.

Thus the equation of motion become

$$u_t + (u \cdot \nabla) u = F + \frac{1}{\rho} \left(-\nabla P + \mu \nabla^2 u + \sigma \kappa (\phi) \delta (\phi) \nabla \phi \right),$$

$$\nabla \cdot u = 0.$$

$$(7)$$

The governing equations can be written as follows.

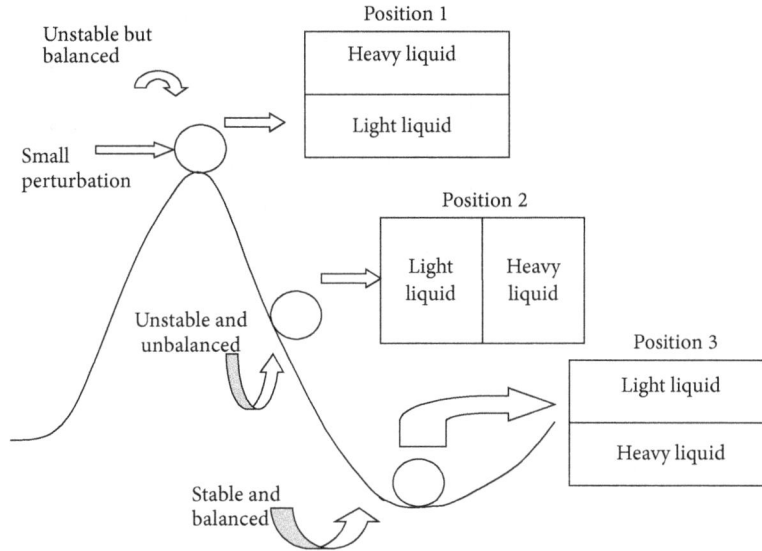

FIGURE 3: Illustration of different types of stability and experimental procedure.

6.1. Continuity. Consider

$$\frac{\partial \rho}{\partial t} + \frac{\partial (\rho u_j)}{\partial x_j} = 0. \tag{8}$$

6.2. Momentum. Consider

$$\rho \frac{\partial u_i}{\partial t} + \rho u_j \frac{\rho u_i}{\partial x_j} = \frac{\partial}{\partial x_j} \left(\mu \frac{\partial u_i}{\partial x_j} \right) - \frac{\partial p}{\partial x_i} + \rho g_i$$
$$+ \sigma \kappa (\phi) \nabla \phi \delta (\phi) \tag{9}$$
$$(i, j = 1, 2).$$

A scalar variable, level set function is used to identify the interface between two fluids and also acts as a distance function. The equation transporting the interface can be written as

$$\frac{\partial \phi}{\partial t} + u_j \frac{\partial \phi}{\partial x_j} = 0, \tag{10}$$

where $\phi(x_{j,t})$ is the level set function prescribing position of the interface at any specified time instant. If the value of the ϕ at the interface is taken as zero, it effectively becomes a distance function satisfying

$$|\nabla \phi| = 1. \tag{11}$$

But at all instant of times ϕ must remain a distance function, to ensure that another scalar variable needs to be introduced and solved. This variable (ψ) must be constrained to constitute a distance function having the same interface value as ϕ. This can be achieved by obtaining a pseudo-steady-state solution for the following transient transport equation of ψ:

$$\frac{\partial \psi}{\partial \bar{t}} = \text{sign} (\psi) (1 - |\nabla \psi|), \tag{12}$$

where

$$|\nabla \psi| = \sqrt{(\psi_x^2 + \psi_y^2)} \tag{13}$$

with $\bar{t}$ being a pseudo-time step.

Equation (12) is subjected to the following initial condition

$$\psi (X, 0) = \phi (X, t + \Delta t). \tag{14}$$

The reinitialization process is iteration of (12) with a pseudo-time step, and within a few iterations it comes to a steady state solution. Then the reinitialization procedure ends leading to reassignment of the level set value.

It is evident that pseudo-steady-state value of ψ is the value of ϕ at the time instant $(t + \Delta t)$. Success of the mass correction is affected by (12) which depends on the accuracy of the interpolation of physical properties such as density, and viscosity across the interface. This can be achieved by calculating a property ξ within a control volume as

$$\xi = [1 - H (\phi)] \xi_1 + H (\phi) \xi_2, \tag{15}$$

where $H(\phi)$ is called Heaviside function.

The equation for the one-dimensional volume fraction is given by

$$H = 0.5 + \left(\frac{\phi}{\Delta X} \right), \tag{16}$$

and for two-dimensional volume fraction the concept has been taken from [21].

At the solid boundary, the Neumann boundary condition for the level set function has been utilized.

7. Solution Procedure

The governing differential equations, coupled with appropriate boundary conditions, are solved using a pressure based finite volume method, as per the SIMPLER algorithm [38].

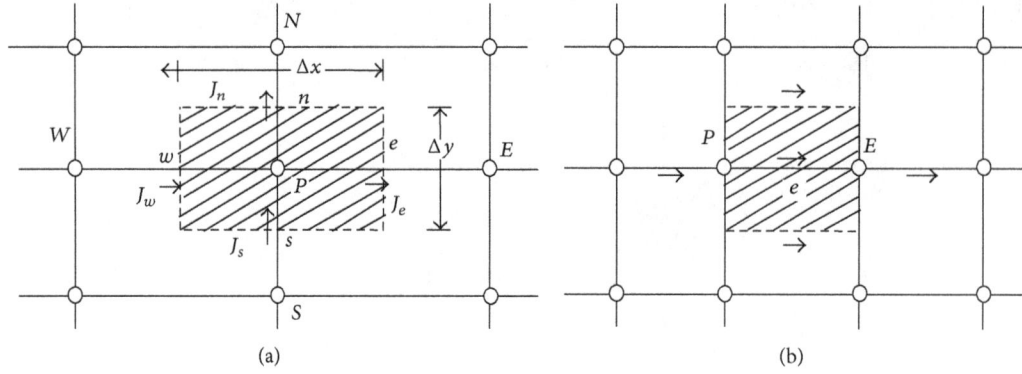

FIGURE 4: (a) Control volume for the two-dimensional situation. (b) Control volume of u.

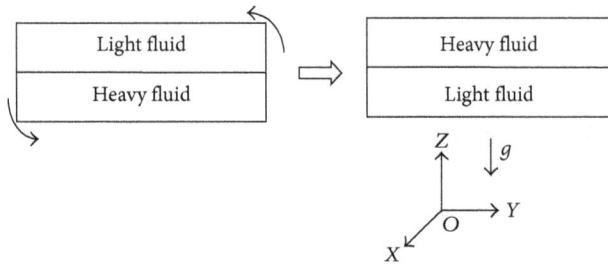

FIGURE 5: Illustration of the experimental technique.

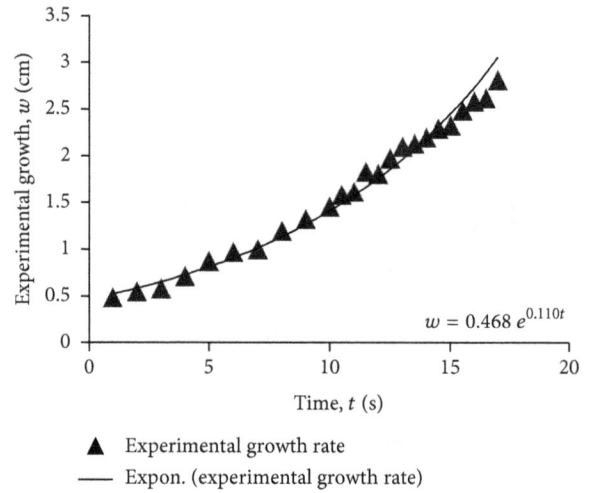

$$w = 0.468\, e^{0.110t}$$

▲ Experimental growth rate
— Expon. (experimental growth rate)

FIGURE 6: Development of growth of instability.

Convection-diffusion terms in the conservation equations are discretized using the power law scheme [38].

The location of the interface at time $t = 0$ has been specified and then the normal distance for all nodes from the interface is calculated. The properties at all nodes have been specified using (15). The continuity and momentum conservation equations at time instant $(t + \Delta t)$ are solved. Then using the velocities obtained in a previous step, using (10), ϕ has been solved. Next, using the values of ϕ from preceding step as initial values, the pseudo-steady-state ϕ (12) has been solved. Setting $\phi(x, t + \Delta t) = \psi(x)$, the procedure is going on until the desired convergence is achieved.

The temporal term of the momentum equation has been discretized as follows. Equation (9) in two-dimensional form is discretized to get algebraic linear simultaneous equations as follows:

$$\frac{\partial}{\partial t}(\rho\phi) + \frac{\partial J_x}{\partial x} + \frac{\partial J_y}{\partial y} = S, \qquad (17)$$

where ϕ represents general variables and J_x and J_y are the total (convection plus diffusion) fluxes defined by

$$J_x = \rho u\phi - \Gamma\frac{\partial\phi}{\partial x}, \qquad J_y = \rho v\phi - \Gamma\frac{\partial\phi}{\partial y}, \qquad (18)$$

where u and v denote the velocity components in the x and y directions, S is the source term, and Γ represents the diffusion coefficient. The integration of (17) over the control volume (Figure 4(a)) gives

$$\frac{\left(\rho_P\phi_P - \rho_P^0\phi_P^0\right)\Delta x\Delta y}{\Delta t} + J_e - J_w + J_n - J_s \qquad (19)$$

$$= \left(S_C + S_P\phi_P\right)\Delta x\Delta y.$$

The source term is linearized in the usual manner anticipating negative slope while the unsteady terms ρ_P and ϕ_P are assumed to prevail over the whole control volume. In a similar fashion, the continuity equation is also linearized.

Similarly the pressure gradient term is discretized considering the staggered control volume as:

$$u_e = \frac{\sum a_{nb}u_{nb} + b}{a_e} + \left(P_P - P_E\right)d_e, \qquad (20)$$

where $d_e = A_e/a_e$.

This is for the u equation as shown in Figure 4(b). The corresponding other equations are discretized in a similar fashion. Finally, guessing the velocity field, the pressure equation is solved, and consequently by correcting the velocity field the variables are solved. This method has an essence physically possible solution by removing unrealistic checker board results.

8. Results and Discussions

8.1. Gravitational Instability due to Density Difference with Initially Horizontal Layers of Fluids. If the box is rotated in the *YZ* plane quickly so that the heavy liquid occupies the upper portion, then instability will initiate at once in

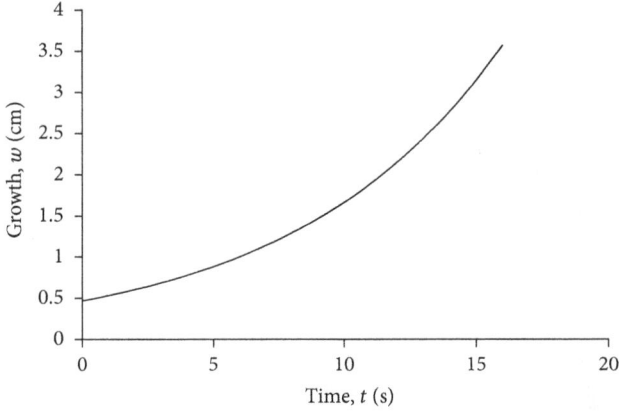

FIGURE 7: Development of growth of instability as found by theoretical modeling.

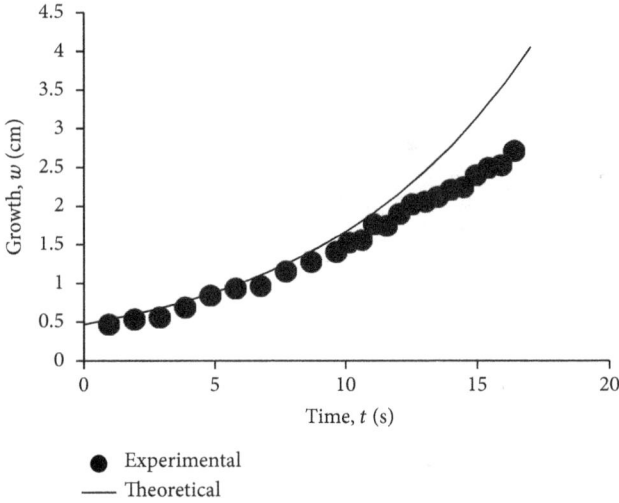

FIGURE 9: Comparison of the growth Rate of the instability.

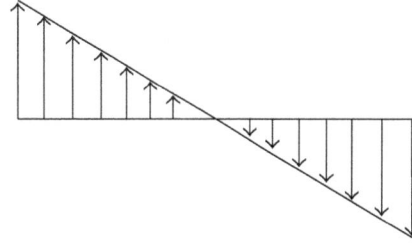

FIGURE 8: Comparison of the development of the growth of instability.

FIGURE 10: Distribution of initial instability triggered for the numerical analysis.

the presence of sufficient unavoidable perturbations and the interface starts moving. The position of the interface at different times, especially in the initial stage of growing instability, has been analyzed in the present investigation (see Figure 5).

8.2. Theoretical Growth of Instability. From theoretical analysis of the problem, the growth rate is given by

$$\frac{\partial w}{\partial t} = \frac{(\rho_1 - \rho_2)\, gb}{4\mu}$$

$$\times \left(\left(\left(\frac{\lambda}{2\pi b} \right)^2 \tanh \frac{2\pi b}{\lambda} \right. \right.$$

$$\left. - \frac{1}{\sinh\left(2\pi b/\lambda\right)\cosh\left(2\pi b/\lambda\right)} \right)$$

$$\left. \times \left(\frac{\lambda}{2\pi b} + \frac{1}{\sinh\left(2\pi b/\lambda\right)\cosh\left(2\pi b/\lambda\right)} \right)^{-1} \right) \times w. \tag{21}$$

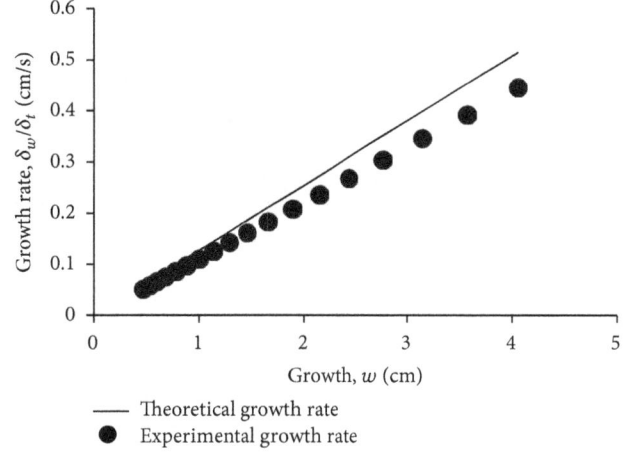

Here it has been assumed that the flow is laminar owing to the fact that the viscosities are of very high order in a two-dimensional, incompressible flow.

The solution of this equation is

$$w = w_0 e^{t/\tau_a}, \tag{22}$$

where w_0 is the initial ($t = 0$) displacement of that point of the interface from the undisturbed interface and τ_a is the growth time.

Now, it can be seen from the growth equation that growth rate varies linearly with displacement of that point at a particular time.

Now, for comparison purpose, a point at a distance of 6.1 cm, that is, approximately $\lambda/4$ distance from the left vertical wall, is considered, where λ is the wavelength of the applied perturbation sine curve which in our case is the length of the box = 20.3 cm.

The displacement of the considered point is measured at different times from the undisturbed interface by proper measurement in the series of snapshots presented in Figure 11 and the graph between growth rate (w) and time (t) has been plotted (see Figure 6).

It can be seen that the best fitted curve is unbounded exponential in nature, which agrees very much with the theory demanding exponential growth of the instability. The curve is of the form $w = 0.468e^{0.110t}$, or we can write $w = 0.468e^{t/9.09}$ where growth time τ_a is 9.09 seconds. We also see

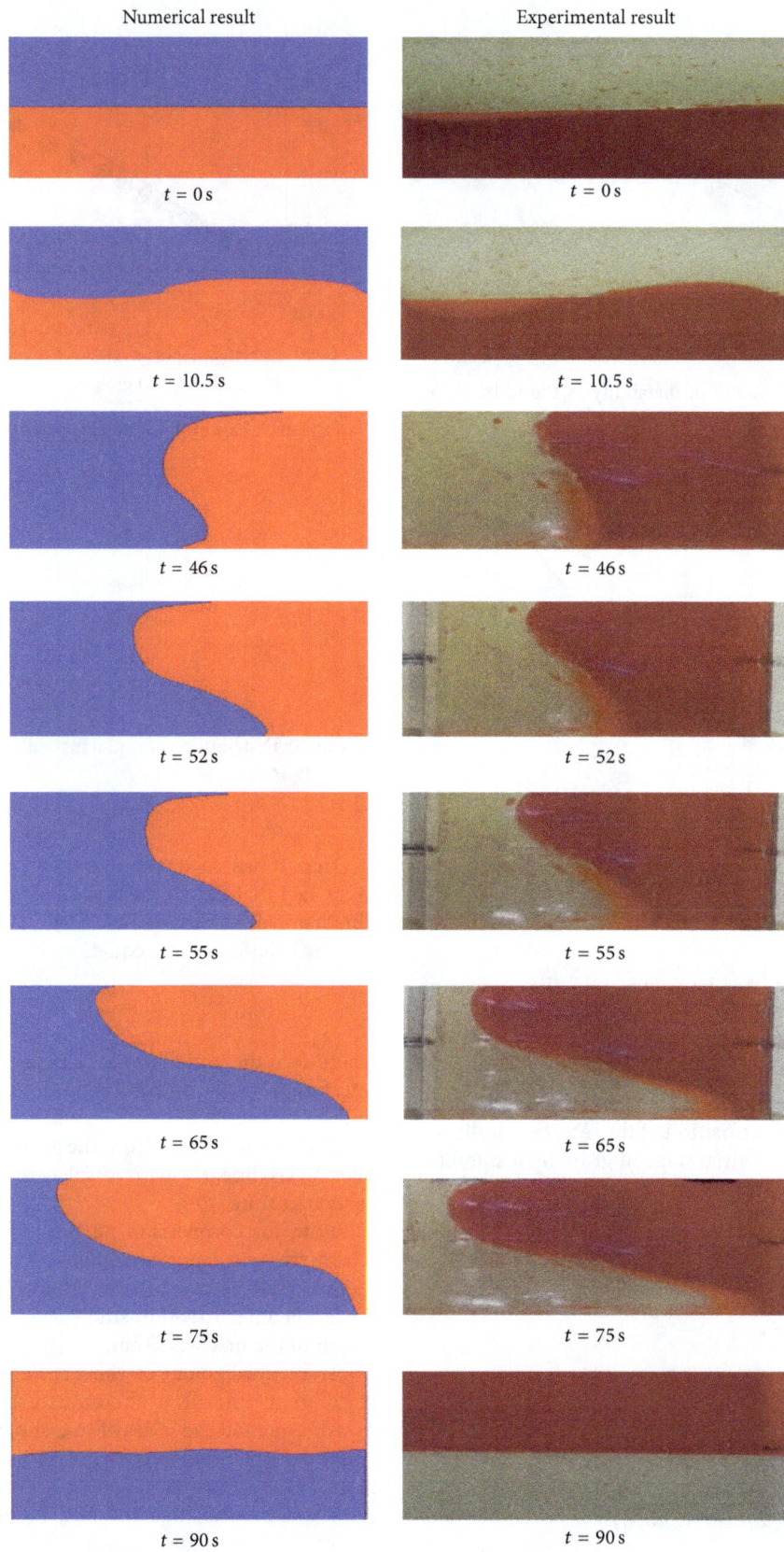

Numerical result   Experimental result

$t = 0\,\mathrm{s}$ $t = 0\,\mathrm{s}$

$t = 10.5\,\mathrm{s}$ $t = 10.5\,\mathrm{s}$

$t = 46\,\mathrm{s}$ $t = 46\,\mathrm{s}$

$t = 52\,\mathrm{s}$ $t = 52\,\mathrm{s}$

$t = 55\,\mathrm{s}$ $t = 55\,\mathrm{s}$

$t = 65\,\mathrm{s}$ $t = 65\,\mathrm{s}$

$t = 75\,\mathrm{s}$ $t = 75\,\mathrm{s}$

$t = 90\,\mathrm{s}$ $t = 90\,\mathrm{s}$

FIGURE 11: Comparison between experimental and numerical results.

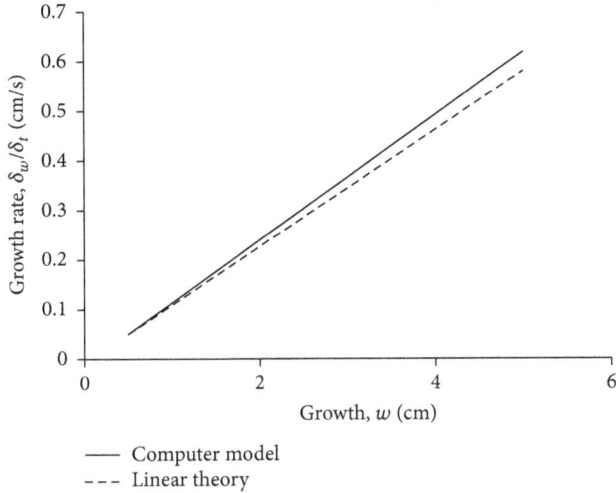

FIGURE 12: Comparison of the growth rate of instability (numerical result versus linear theory).

that the initial ($t = 0$) perturbation of the observed point is 0.468 cm.

Theoretically, the value of τ_a can be calculated as

$$
\tau_a = \frac{4\mu}{(\rho_2 - \rho_1)\,gb}
$$
$$
\times \left(\frac{\lambda}{2\pi b} + \frac{1}{\sinh(2\pi b/\lambda)\cosh(2\pi b/\lambda)} \right)
$$
$$
\times \left(\left(\frac{\lambda}{2\pi b} \right)^2 \tanh\frac{2\pi b}{\lambda} \right. \tag{23}
$$
$$
\left. - \frac{1}{\sinh(2\pi b/\lambda)\cosh(2\pi b/\lambda)} \right)^{-1},
$$

where μ is equivalent coefficient of dynamic viscosity and it can be expressed as

$$
\mu = \frac{(\rho_1 \mu_1 + \rho_2 \mu_2)}{(\rho_1 + \rho_2)}, \tag{24}
$$

where ρ_1 = density of lighter liquid = 0.92×10^3 kg/m^3, ρ_2 = density of heavier liquid = 1.4×10^3 kg/m^3, μ_1 = viscosity of lighter liquid = 0.0791 Pa-s, and μ_2 = viscosity of Heavier liquid = 350 Pa-s. So μ becomes 211.23 Pa-s.

Here, b = height of the upper or lower rigid boundary from the undisturbed, liquid interface = 7.5 cm = 0.075 m, g = acceleration due to gravity = 9.8 m/s^2, and λ = wave length of the perturbation sine curve = 20.3 cm. So, τ_a becomes 7.846 seconds.

Now, from the experimental study, the initial ($t = 0$) perturbation of the specified point is 0.468 cm.

So the theoretical growth equation becomes,

$$
w = 0.468e^{t/7.846} \text{—theoretical growth equation,}
$$
$$
w = 0.468e^{t/9.09} \text{—experimental growth equation.}
$$

It can be observed from the above two expressions that the characteristics of the development of growth of instability are quite similar, with a slight difference in the growth time. The growth time is slightly higher in case of experimental observation than the numerical investigation.

Figure 7 shows the growth of instability with time and Figure 8 shows the comparison between the theoretical and experimental results. Figure 9 depicts the theoretical and experimental comparisons of growth rate of instability. From the figures, it is observed that, at the early stage of growth of instability, the experimental and theoretical results match considerably while the growth rate differs with the increase in time. This may be due to the fact that theoretically the flow has been assumed to be two dimensional, but in case of experimentation the three-dimensional characteristics come into consideration and due to this effect of three dimensionality the theoretical results differ with the experimental result and it increases with increase of time.

The same problem is numerically analyzed considering a rectangular two-dimensional domain. Two arrays of 61 × 21 and 121 × 41 grid points in axial and radial directions, respectively, have been used. It has been observed that the grid independent study has shown 0.001% change and the results are almost unaffected considering both the grid meshes. The grid array of 61 × 21 has been used for all subsequent results reported here with uniform mesh size and time step DT = 0.01 s. A disturbance of the vertical component of velocity having triangular distribution has been introduced as shown in Figure 10.

The variation of the propagating interface with time has been shown. Both the experimental and numerical results are presented here (see Figure 11). In the numerical results, red color represents the lighter fluid and blue color represents the heavy fluid, whereas in the experimental results white solution is the heavy fluid and the red colored fluid is the light fluid.

It can be observed from the above figures that the experimental results are in good agreement with the numerical results. The interface between the two fluids shows similar pattern during the study for both experimental and numerical analysis. However, three-dimensional features are observed to affect the results as seen in the experimental study. Figure 12 shows the comparison of growth rate of instability. The numerical result matches the linear theory at the initial period, whereas with the increase of time it varies with the linear theory.

In Figure 13, two-bubble merging and consequent breaking up have been evaluated with time. The matrix is the heavier fluid. This is the result of numerical experimentation.

9. Conclusions

The nature of the development of instability was experimentally found as a function of sine curve as predicted by theoretical model. A numerical methodology was devised and validated with experimental results so that the methodology can handle any gravitational interfacial instability. It was found that, in the early stages of the growth of instability,

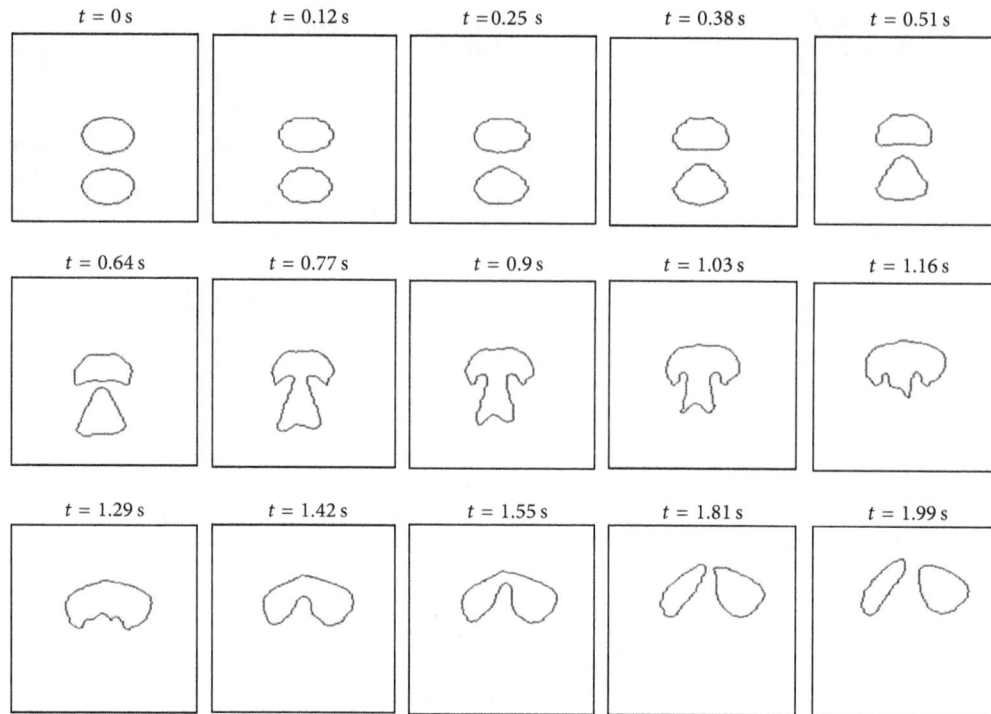

FIGURE 13: Merging and consequent breaking of two bubbles in a heavier matrix.

the growth rate is proportional to the instantaneous growth in a particular position, that is, growth rate varies linearly with growth at that moment at a particular point on the interface. But, at the later stage of development of instability, substantial deviation from the linear theory was observed. The pictorial views of the interface between the two fluids have been studied both theoretically and experimentally and they have matched satisfactorily.

Nomenclature

A: Atwood number (–)
b: Height of one fluid layer (cm)
F: Speed function (m/s)
g: Acceleration due to gravity (m/s^2)
H: Heaviside function (–)
P: Pressure (N/m^2)
t: Time (s)
u_j: Velocity (m/s)
W: Growth at a particular time (cm)
W_0: Initial growth (cm)
Δt: Small time step (s).

Greek Symbols

ϕ: Level set function (m)
ψ: Dummy variable for level set function (m)
δ: Dirac delta function (–)
ξ: Fluid properties such as density and viscosity (–)
σ: Surface tension coefficient (N/m)

μ: Coefficient of viscosity (Pa-s)
κ: Curvature (m^{-1})
ρ: Density of the liquid (kg/m^3)
τ_a: Growth time (s)
λ: Wavelength of the perturbation (cm).

Acknowledgment

This work is supported by the Council of Scientific and Industrial Research (CSIR), Government of India.

References

[1] D. H. Sharp, "An overview of Rayleigh-Taylor instability," *Physica D*, vol. 12, no. 1–3, pp. 3–18, 1984.

[2] K. I. Read, "Experimental investigation of turbulent mixing by Rayleigh-Taylor instability," *Physica D*, vol. 12, no. 1–3, pp. 45–58, 1984.

[3] D. L. Youngs, "Numerical simulation of turbulent mixing by Rayleigh-Taylor instability," *Physica D*, vol. 12, no. 1–3, pp. 32–44, 1984.

[4] S. B. Dalziel, "Rayleigh-Taylor instability: experiments with image analysis," *Dynamics of Atmospheres and Oceans*, vol. 20, no. 1-2, pp. 127–153, 1993.

[5] S. I. Voropayev, Y. D. Afanasyev, and G. J. F. van Heijst, "Experiments on the evolution of gravitational instability of an overturned, initially stably stratified fluid," *Physics of Fluids A*, vol. 5, no. 10, pp. 2461–2466, 1993.

[6] C. W. Hirt and B. D. Nichols, "Volume of fluid (VOF) method for the dynamics of free boundaries," *Journal of Computational Physics*, vol. 39, no. 1, pp. 201–225, 1981.

[7] J. A. Sethian, "Curvature and the evolution of fronts," *Communications in Mathematical Physics*, vol. 101, no. 4, pp. 487–499, 1985.

[8] S. Osher and J. A. Sethian, "Fronts propagating with curvature-dependent speed: algorithms based on Hamilton-Jacobi formulations," *Journal of Computational Physics*, vol. 79, no. 1, pp. 12–49, 1988.

[9] B. Merriman, J. K. Bence, and S. J. Osher, "Motion of multiple junctions: a level set approach," *Journal of Computational Physics*, vol. 112, no. 2, pp. 334–363, 1994.

[10] J. Zhu and J. Sethian, "Projection methods coupled to level set interface techniques," *Journal of Computational Physics*, vol. 102, no. 1, pp. 128–138, 1992.

[11] S. Unverdi and G. Tryggvasan, "A front-tracking method for viscous, incompressible, multifluid flows," *Journal of Computational Physics*, vol. 100, no. 1, pp. 25–37, 1992.

[12] D. L. Chopp and J. A. Sethian, "Flow under curvature: singularity formation, minimal surfaces, and geodesics," *Journal of Experimental Mathematics*, vol. 24, pp. 235–255, 1993.

[13] M. Sussman, P. Smereka, and S. Osher, "A level set approach for computing solutions to incompressible two-phase flow," *Journal of Computational Physics*, vol. 114, no. 1, pp. 146–159, 1994.

[14] Y. C. Chang, T. Y. Hou, B. Merriman, and S. Osher, "A level set formulation of Eulerian interface capturing methods for incompressible fluid flows," *Journal of Computational Physics*, vol. 124, no. 2, pp. 449–464, 1996.

[15] J. A. Sethian, "Theory, algorithms, and applications of level set methods for propagating interfaces," *Acta Numerica*, vol. 5, pp. 309–395, 1996.

[16] J. A. Sethian, "Adaptive fast marching and level set methods for propagating interfaces," *Acta Mathematica Universitatis Comenianae*, vol. 67, no. 1, pp. 3–15, 1998.

[17] J. A. Sethian, *Fast Marching Methods and Level Set Methods for Propagating Interfaces*, The Computational Fluid Dynamics Lecture Series, Von Karman Institute, 1998.

[18] C. Kaliakatos and S. Tsangaris, "Motion of deformable drops in pipes and channels using Navier-Stokes equations," *International Journal of Numerical Methods in Fluids*, vol. 34, no. 7, pp. 609–626, 2000.

[19] G. Son and N. Hur, "A coupled level set and volume-of-fluid method for the buoyancy-driven motion of fluid particles," *Numerical Heat Transfer B*, vol. 42, no. 6, pp. 523–542, 2002.

[20] J. A. Sethian and P. Smereka, "Level set methods for fluid interfaces," *Annual Review of Fluid Mechanics*, vol. 35, pp. 341–372, 2003.

[21] S. Majumder and S. Chakraborty, "New physically based approach of mass conservation correction in level set formulation for incompressible two-phase flows," *Journal of Fluids Engineering*, vol. 127, no. 3, pp. 554–563, 2005.

[22] P. Carlès, Z. Huang, G. Carbone, and C. Rosenblatt, "Rayleigh-Taylor instability for immiscible fluids of arbitrary viscosities: a magnetic levitation investigation and theoretical model," *Physical Review Letters*, vol. 96, no. 10, Article ID 104501, 4 pages, 2006.

[23] S. Y. Wang, K. M. Lim, B. C. Khoo, and M. Y. Wang, "An extended level set method for shape and topology optimization," *Journal of Computational Physics*, vol. 221, no. 1, pp. 395–421, 2007.

[24] D. L. Sun and W. Q. Tao, "A coupled volume-of-fluid and level set (VOSET) method for computing incompressible two-phase flows," *International Journal of Heat and Mass Transfer*, vol. 53, no. 4, pp. 645–655, 2010.

[25] M. Sussman and E. G. Puckett, "A coupled level set and volume of fluid method for computing 3D and axisymmetric incompressible two-phase flows," *Journal of Computational Physics*, vol. 162, no. 2, pp. 301–337, 2000.

[26] M. Sussman, "A second order coupled level set and volume-of-fluid method for computing growth and collapse of vapor bubbles," *Journal of Computational Physics*, vol. 187, no. 1, pp. 110–136, 2003.

[27] G. Son, "Efficient implementation of a coupled level-set and volume-of-fluid method for three-dimensional incompressible two-phase flows," *Numerical Heat Transfer B*, vol. 43, no. 6, pp. 549–565, 2003.

[28] L. Wang, J. Li, and Z. Xie, "Large-eddy-simulation of 3-dimensional Rayleigh-Taylor instability in incompressible fluids," *Science in China, Series A*, vol. 45, no. 1, pp. 95–106, 2002.

[29] D. L. Youngs, "Three-dimensional numerical simulation of turbulent mixing by Rayleigh-Taylor instability," *Physics of Fluids A*, vol. 3, no. 5, pp. 1312–1320, 1991.

[30] C. L. Gardner, J. Glimm, O. McBryan, R. Menikoff, D. H. Sharp, and Q. Zhang, "The dynamics of bubble growth for Rayleigh-Taylor unstable interfaces," *Physics of Fluids*, vol. 31, no. 3, pp. 447–465, 1988.

[31] J. Glimm, X. L. Li, R. Menikoff, D. H. Sharp, and Q. Zhang, "A numerical study of bubble interactions in Rayleigh-Taylor instability for compressible fluids," *Physics of Fluids A*, vol. 2, no. 11, pp. 2046–2054, 1990.

[32] X. L. Li, B. X. Jin, and J. Glimm, "Numerical study for the three-dimensional Rayleigh-Taylor: instability through the TVD/AC scheme and parallel computation," *Journal of Computational Physics*, vol. 126, no. 2, pp. 343–355, 1996.

[33] Y. Chen, J. Glimm, D. H. Sharp, and Q. Zhang, "A two-phase flow model of the Rayleigh-Taylor mixina zone," *Physics of Fluids*, vol. 8, no. 3, pp. 816–825, 1996.

[34] Y. Chen, J. Glimm, D. Saltz, D. H. Sharp, and Q. Zhang, "A two-phase flow formulation for the Rayleigh-Taylor mixing zone and its renormalization group solution," in *Proceedings of the Fifth International Workshop on Compressible Turbulent Mixing*, World Scientific, Singapore, 1996.

[35] A. W. Cook and P. E. Dimotakis, "Transition stages of Rayleigh-Taylor instability between miscible fluids," *Journal of Fluid Mechanics*, vol. 443, pp. 69–99, 2001.

[36] A. W. Cook and Y. Zhou, "Energy transfer in Rayleigh-Taylor instability," *Physical Review E*, vol. 66, no. 2, Article ID 026312, 12 pages, 2002.

[37] A. W. Cook, W. Cabot, and P. L. Miller, "The mixing transition in Rayleigh-Taylor instability," *Journal of Fluid Mechanics*, vol. 511, pp. 333–362, 2004.

[38] S. V. Patankar, *Nuemrical Heat Transfer and Fluid Flow*, McGraw-Hill, New York, NY, USA, 1981.

20

Free Convection Heat and Mass Transfer MHD Flow in a Vertical Porous Channel in the Presence of Chemical Reaction

R. N. Barik,[1] G. C. Dash,[2] and M. Kar[3]

[1] Department of Mathematics, Trident Academy of Technology, Infocity, Bhubaneswar, Odisha 751024, India
[2] Department of Mathematics, S.O.A. University, Bhubaneswar, Odisha 751030, India
[3] Department of Mathematics, Christ College, Cuttack, Odisha, India

Correspondence should be addressed to R. N. Barik; rnbmath22@yahoo.com

Academic Editor: Hideki Tsuge

The objective of the present study is to examine the fully developed free convective MHD flow of an electrically conducting viscous incompressible fluid in a vertical porous channel under influence of asymmetric wall temperature and concentration in the presence of chemical reaction. The heat and mass transfer coupled with diffusion-thermo effect renders the present analysis interesting and curious. The analytical solution by Laplace transform technique of partial differential equations is used to obtain the expressions for the velocity, temperature, and concentration. It is observed that under the influence of dominating mass diffusivity over thermal diffusivity with stronger Lorentz force the velocity is reduced at all points Further, low rate of thermal diffusion delays the attainment of free stream state. Flow of aqueous solution in the presence of heavier species is prone to back flow.

1. Introduction

In many transport processes and industrial applications, transfer of heat and mass occurs simultaneously as a result of combined buoyancy effects of thermal diffusion and diffusion of chemical species. Unsteady natural convection of heat and mass transfer is of great importance in designing control systems for modern free convection heat exchangers. More recently, Jha and Ajibade [1] have studied the heat and mass transfer aspect of the flow of a viscous incompressible fluid in a vertical channel considering the Dufour effect.

Soundalgekar and Akolkar [2] studied the effect of mass transfer and free convection currents on the flow past an impulsively started infinite vertical plate and observed that the presence of foreign gasses in the flow domain leads to reduce the shear stress and rate of mass transfer significantly.

In nature, flow occurs due to density differences caused by temperature as well as chemical composition gradients. Therefore, it warrants the simultaneous consideration of temperature difference as well as concentration difference when heat and mass transfer occurs simultaneously. It has been found that an energy flux can be created not only by temperature gradients but by composition gradients also. This is

called Dufour effect. If, on the other hand, mass fluxes are created by temperature gradients, it is called the Soret effect.

The Soret and the Dufour effects have been found to be useful as the Soret effect is utilized for isotope separation and, in a mixture of gases of light and medium molecular weight, the Dufour effect is found to be of considerable order of magnitude such that it cannot be neglected. In view of importance of the diffusion-thermo effect Kafoussias and Williams [3] have studied the effects of thermal diffusion and diffusion-thermo on mixed free and forced convective and mass transfer boundary layer flow with temperature dependent viscosity. Jha and Singh [4], Kafoussias [5], and Alam et al. [6, 7] have contributed significantly to this field of study. Recently, Beg et al. [8] studied chemically reacting mixed convective heat and mass transfer along inclined and vertical plates considering Soret and Dufour effects. More recently, Dursunkaya and Worek [9] have studied the diffusion-thermo and thermal diffusion effects in transient and steady natural convections from a vertical surface.

MHD flow with thermal diffusion and chemical reaction finds numerous applications in various areas such as thermo nuclear fusion, liquid metal cooling of nuclear reactions, and electromagnetic casting of metals. Abreu et al. [10] have studied the boundary layer flows with Dufour and Soret effects. Osalusi et al. [11] have worked on mixed and free convective heat and mass transfer of an electrically conducting fluid considering Dufour and Soret effects. Anghel et al. [12] and Postelnicu [13] obtained numerical solutions of chemical reacting mixed convective heat and mass transfer along inclined and vertical plates with the Soret and the Dufour effects and concluded that skin friction increases with a positive increase in the concentration-to-thermal-buoyancy ratio parameter.

The analysis of natural convection heat and mass transfer near a moving vertical plate has received much attention in recent times due to its wide application in engineering and technological processes. There are applications of interest in which combined heat and mass transfer by natural convection occurs between a moving material and ambient medium, such as the design and operation of chemical processing equipment, design of heat exchangers, transpiration cooling of a surface, chemical vapour deposition of solid layer, nuclear reactor, and many manufacturing processes like hot rolling, hot extrusion, wire drawing, continuous casting, and fiber drawing. Gebhart and Pera [14] studied the effects of mass transfer on a steady free convection flow past a semi-infinite vertical plate by the similarity method and it was assumed that the concentration level of the diffusing species in fluid medium was very low. This assumption enabled them to neglect the diffusion-thermo and thermo-diffusion effects as well as the interfacial velocity at the wall due to species diffusion. Following this assumption, Das et al. [15] investigated the effects of simultaneous heat and mass transfer on free convection flow past an infinite vertical plate under different physical situations.

Recently, the unsteady MHD heat and mass transfer free convection flow of a polar fluid past a vertical moving porous plate in a porous medium with heat generation and thermal diffusion has been studied by Saxena and Dubey [16]. Raveendra Babu et al. [17] studied diffusion-thermo and radiation effects on MHD free convective heat and mass transfer flow past an infinite vertical plate in the presence of a chemical reaction of first order. Sudhakar et al. [18] discussed chemical reaction effect on an unsteady MHD free convection flow past an infinite vertical accelerated plate with constant heat flux, thermal diffusion, and diffusion thermo.

In the present paper, the flow of an electrically conducting viscous incompressible fluid in a vertical porous channel formed by two vertical parallel porous plates in the presence of a transverse magnetic field is studied. Further, the mass transfer phenomena considered in this problem is associated with chemically reacting species. The objective of the present study is to extend the work of Jha and Ajibade [1] by incorporating the permeability and the magnetic field effect as well as chemical reaction on the flow, heat and mass transfer phenomena.

2. Mathematical Formulation

Let us consider the free convective heat and mass transfer MHD flow of a viscous incompressible fluid in a vertical porous channel formed by two infinite vertical parallel porous

plates in the presence of chemical reaction. A magnetic field of uniform strength B_0 is applied transversely to the plate. The induced magnetic field is neglected as the magnetic Reynolds number of the flow is taken to be very small. The convection current is induced due to both the temperature and concentration differences. The flow is assumed to be in the x'-direction which is taken to be vertically upward along the channel walls and y'-axis is taken to be normal to the plates that are h distance apart.

Under the usual Boussinesq's approximations, the governing equations for flow are given by

$$\frac{\partial u'}{\partial t'} = \nu \frac{\partial^2 u'}{\partial y'^2} + g\beta\left(T' - T_0\right)$$

$$+ g\beta^*\left(C' - C_0\right) - \frac{\sigma B_0^2 u'}{\rho} - \frac{\nu}{K'_p}u',$$

$$\frac{\partial C'}{\partial t'} = D\frac{\partial^2 C'}{\partial y'^2} - K'_c\left(C' - C_0\right),$$

$$\frac{\partial T'}{\partial t'} = \alpha\frac{\partial^2 T'}{\partial y'^2} + D_1\frac{\partial^2 C'}{\partial y'^2},$$

(1)

where ν, g, β, β^*, σ, B_0, ρ, D, K'_c, α, D_1, and K'_p are kinematic viscosity, acceleration due to gravity, coefficient of thermal expansion, coefficient of mass expansion, electrical conductivity of the fluid, uniform magnetic field, density of the fluid, chemical molecular diffusivity, the dimensional chemical reaction parameter in the diffusion equation, thermal diffusivity, the dimensional coefficient of the diffusion-thermo effect, and the permeability of the medium, respectively.

Initial and boundary conditions of the problem are

$$t' \leq 0 : u'\left(y', t'\right) = 0, \quad T'\left(y', t'\right) = T_0, \quad C'\left(y', t'\right) = C_0$$

$$t' > 0 : \begin{array}{l} u'\left(o, t'\right) = 0, \quad T'\left(0, t'\right) = T_w, \quad C'\left(0, t'\right) = C_w \\ u'\left(h, t'\right) = 0, \quad T'\left(h, t'\right) = T_0, \quad C'\left(h, t'\right) = C_0. \end{array}$$

(2)

We now introduce the following dimensionless quantities:

$$y = \frac{y'}{h}, \qquad t = \frac{t'\nu}{h^2}, \qquad u = \frac{u'\nu}{g\beta h^2\left(T_w - T_0\right)},$$

$$P_r = \frac{\nu}{\alpha}, \qquad S_c = \frac{\nu}{D},$$

$$T = \frac{T' - T_0}{T_w - T_0}, \qquad C = \frac{C' - C_0}{C_w - C_0},$$

(3)

$$N = \frac{\beta^*\left(C_w - C_0\right)}{\beta\left(T_w - T_0\right)},$$

$$D^* = \frac{D_1\left(C_w - C_0\right)}{\alpha\left(T_w - T_0\right)}, \qquad M = \frac{\sigma B_0^2 h^2}{\rho\nu},$$

$$K_c = \frac{K_c^* h^2}{\nu}, \qquad K_p = \frac{K'_P}{h^2}.$$

(4)

Here P_r, S_c, N, D^*, M, K_c, and K_p are the Prandtl number, the Schmidt number, the Sustentation parameter, the Dufour number, the magnetic field parameter, the chemical reaction parameter, and the Porosity parameter, respectively.

The nondimensional form of (1) is given by

$$\frac{\partial u}{\partial t} = \frac{\partial^2 u}{\partial y^2} + NC + T - \left(M + \frac{1}{K_p}\right)u,$$

$$\frac{\partial C}{\partial t} = \frac{1}{S_c}\frac{\partial^2 C}{\partial y^2} - K_c C, \tag{5}$$

$$\frac{\partial T}{\partial t} = \frac{1}{P_r}\frac{\partial^2 T}{\partial y^2} + \frac{D^*}{P_r}\frac{\partial^2 C}{\partial y^2}.$$

Subject to boundary conditions,

$$t \le 0 : u(y,t) = 0, \quad T(y,t) = 0, \quad C(y,t) = 0,$$

$$t \le 0 : \begin{array}{lll} u(0,t) = 0, & T(0,t) = 1, & C(0,t) = 1, \\ u(1,t) = 0, & T(1,t) = 0, & C(1,t) = 0. \end{array} \tag{6}$$

3. Solution of the Problem

Applying Laplace transform to (5) with boundary condition (6), we get

$$\frac{d^2\overline{C}}{dy^2} - S_c(s + K_c)\overline{C} = 0,$$

$$\frac{d^2\overline{T}}{dy^2} - sP_r\overline{T} = S_c D^*(s + K_c)\overline{C}, \tag{7}$$

$$s\overline{u} = \frac{d^2\overline{u}}{dy^2} + N\overline{C} + \overline{T} - \left(M + \frac{1}{K_p}\right)\overline{u}. \tag{8}$$

The boundary condition (6) becomes

$$\overline{C}(0,s) = \frac{1}{s}, \qquad \overline{C}(1,s) = 0,$$

$$\overline{T}(0,s) = \frac{1}{s}, \qquad \overline{T}(1,s) = 0, \tag{9}$$

$$\overline{u}(0,s) = 0, \qquad \overline{u}(1,s) = 0.$$

Applying inverse Laplace transform to (7) and (8) subject to boundary condition (9), following Debanath and Bhatta [19], and using the formula

$$L^{-1}\left(\frac{e^{-b\sqrt{s+a}}}{s}\right) = \frac{1}{2}\left\{e^{-b\sqrt{a}}\operatorname{erfc}\left(\frac{b}{2\sqrt{t}} - \sqrt{at}\right)\right.$$

$$\left. + e^{b\sqrt{a}}\operatorname{erfc}\left(\frac{b}{2\sqrt{t}} + \sqrt{at}\right)\right\}. \tag{10}$$

The solutions of (5) subject to the boundary conditions (6) are obtained by Laplace transform technique.

Case 1 $(P_r \ne 1, S_c \ne 1)$. The solutions of (5) subject to boundary conditions (6) are given by

$$C = \sum_{n=0}^{\infty}\left[f\left(K_c, a_n\sqrt{S_c}, 1, 0, t\right) - f\left(K_c, b_n\sqrt{S_c}, 1, 0, t\right)\right]$$

$$T = A_1\sum_{n=0}^{\infty}\left[f\left(0, a_n\sqrt{P_r}, P_r - S_c, S_c K_c, t\right)\right.$$

$$\left. - f\left(0, b_n\sqrt{P_r}, P_r - S_c, S_c K_c, t\right)\right]$$

$$- A_1\sum_{n=0}^{\infty}\left[f\left(K_c, a_n\sqrt{S_c}, P_r - S_c, S_c K_c, t\right)\right.$$

$$\left. - f\left(K_c, b_n\sqrt{S_c}, P_r - S_c, S_c K_c, t\right)\right]$$

$$+ (D^* + 1)\sum_{n=0}^{\infty}\left[f\left(0, a_n\sqrt{P_r}, 1, 0, t\right)\right.$$

$$\left. - f\left(0, b_n\sqrt{P_r}, 1, 0, t\right)\right] - D^*C$$

$$u = A_2\sum_{n=0}^{\infty}\left[f\left(M + \frac{1}{K_p}, a_n, 1 - S_c, S_c K_c - M - \frac{1}{K_p}, t\right)\right.$$

$$\left. - f\left(M + \frac{1}{K_p}, b_n, 1 - S_c, S_c K_c - M - \frac{1}{K_p}, t\right)\right]$$

$$- A_2\sum_{n=0}^{\infty}\left[f\left(K_c, a_n\sqrt{S_c}, 1 - S_c, S_c K_c - M - \frac{1}{K_p}, t\right)\right.$$

$$\left. - f\left(K_c, b_n\sqrt{S_c}, 1 - S_c, S_c K_c - M - \frac{1}{K_p}, t\right)\right]$$

$$+ A_3\sum_{n=0}^{\infty}\left[f\left(M + \frac{1}{K_p}, a_n, 1 - P_r, -M - \frac{1}{K_p}, t\right)\right.$$

$$\left. - f\left(M + \frac{1}{K_p}, b_n, 1 - P_r, -M - \frac{1}{K_p}, t\right)\right]$$

$$- A_3\sum_{n=0}^{\infty}\left[f\left(K_c, a_n\sqrt{P_r}, 1 - P_r, -M - \frac{1}{K_p}, t\right)\right.$$

$$\left. - f\left(K_c, b_n\sqrt{P_r}, 1 - P_r, -M - \frac{1}{K_p}, t\right)\right]$$

$$+ 2A_4\sum_{n=0}^{\infty}\left[f\left(M + \frac{1}{K_p}, a_n, P_r - S_c, S_c K_c, t\right)\right.$$

$$\left. - f\left(M + \frac{1}{K_p}, b_n, P_r - S_c, S_c K_c, t\right)\right]$$

$$- A_4\sum_{n=0}^{\infty}\left[f\left(K_c, a_n\sqrt{S_c}, P_r - S_c, S_c K_c, t\right)\right.$$

$$\left. - f\left(K_c, b_n\sqrt{S_c}, P_r - S_c, S_c K_c, t\right)\right]$$

$$- A_4 \sum_{n=0}^{\infty} \left[f\left(0, a_n \sqrt{P_r}, P_r - S_c, S_c K_c, t \right) \right.$$

$$\left. - f\left(0, b_n \sqrt{S_c}, P_r - S_c, S_c K_c, t \right) \right]$$

$$+ (A_5 - A_6) \sum_{n=0}^{\infty} \left[f\left(M + \frac{1}{K_p}, a_n, 1, 0, t \right) \right.$$

$$\left. - f\left(M + \frac{1}{K_p}, b_n, 1, 0, t \right) \right]$$

$$+ A_6 \sum_{n=0}^{\infty} \left[f\left(0, a_n \sqrt{P_r}, 1, 0, t \right) - f\left(0, b_n \sqrt{P_r}, 1, 0, t \right) \right],$$

$$(11)$$

where $a_n = 2n + y$ and $b_n = 2n + 2 - y$.

The functions f and erfc are given in Appendix A. The constants A_1, A_2, A_3, A_4, A_5, and A_6 are given in Appendix B.

The rate of mass transfer (Sh), the rate of heat transfer (Nu), and the skin function (τ) at the walls of the channel are obtained as follows:

Sherwood Number (Sh). Consider

$$Sh_0 = \left. \frac{\partial C}{\partial y} \right|_{y=0}$$

$$= -\sqrt{S_c} \sum_{n=0}^{\infty} \left[g\left(K_c, 2n\sqrt{S_c}, 1, 0, t \right) \right.$$

$$\left. + g\left(K_c, (2n+2)\sqrt{S_c}, 1, 0, t \right) \right],$$

$$Sh_1 = \left. \frac{\partial C}{\partial y} \right|_{y=1} = 2\sqrt{S_c} \sum_{n=0}^{\infty} \left[g\left(K_c, (2n+1)\sqrt{S_c}, 1, 0, t \right) \right].$$

$$(12)$$

Nusselt Number (Nu). Consider

$$Nu_0 = -\left. \frac{\partial T}{\partial y} \right|_{y=0}$$

$$= A_1 \sqrt{S_c} \sum_{n=0}^{\infty} \left[g\left(K_c, 2n\sqrt{S_c}, P_r - S_c, S_c K_c, t \right) \right.$$

$$\left. + g\left(K_c, (2n+2)\sqrt{S_c}, P_r - S_c, S_c K_c, t \right) \right],$$

$$- A_1 \sqrt{P_r} \sum_{n=0}^{\infty} \left[g\left(0, 2n\sqrt{P_r}, P_r - S_c, S_c K_c, t \right) \right.$$

$$\left. + g\left(0, (2n+2)\sqrt{P_r}, P_r - S_c, S_c K_c, t \right) \right]$$

$$- (D^* + 1) \sqrt{P_r} \sum_{n=0}^{\infty} \left[g\left(0, 2n\sqrt{P_r}, 1, 0, t \right) \right.$$

$$\left. + g\left(0, (2n+2)\sqrt{P_r}, 1, 0, t \right) \right]$$

$$- D^* Sh_0$$

$$Nu_1 = \left. \frac{\partial T}{\partial y} \right|_{y=1}$$

$$= 2A_1 \sqrt{P_r} \sum_{n=0}^{\infty} \left[g\left(0, (2n+1)\sqrt{P_r}, P_r - S_c, S_c K_c, t \right) \right]$$

$$- 2A_1 \sqrt{S_c} \sum_{n=0}^{\infty} \left[g\left(K_c, (2n+1)\sqrt{S_c}, P_r - S_c, S_c K_c, t \right) \right]$$

$$+ 2(D^* + 1) \sqrt{P_r} \sum_{n=0}^{\infty} \left[g\left(0, (2n+1)\sqrt{P_r}, 1, 0, t \right) \right]$$

$$- D^* Sh_1.$$

$$(13)$$

Skin Friction (τ). Consider

$$\tau_0 = -\left. \frac{\partial u}{\partial y} \right|_{y=0}$$

$$= A_2 \sum_{n=0}^{\infty} \left[g\left(M + \frac{1}{K_p}, 2n, 1 - S_c, S_c K_c - M - \frac{1}{K_p}, t \right) \right.$$

$$+ g\left(M + \frac{1}{K_p}, 2n + 2, 1 - S_c, S_c K_c \right.$$

$$\left. \left. -M - \frac{1}{K_p}, t \right) \right]$$

$$- A_2 \sqrt{S_c} \sum_{n=0}^{\infty} \left[g\left(K_c, 2n\sqrt{S_c}, 1 - S_c, S_c K_c - M - \frac{1}{K_p}, t \right) \right.$$

$$+ g\left(K_c, (2n+2)\sqrt{S_c}, 1 - S_c, S_c K_c \right.$$

$$\left. \left. -M - \frac{1}{K_p}, t \right) \right]$$

$$+ A_3 \sum_{n=0}^{\infty} \left[g\left(M + \frac{1}{K_p}, 2n, 1 - P_r, -M - \frac{1}{K_p}, t \right) \right.$$

$$+ g\left(M + \frac{1}{K_p}, 2n + 2, 1 - P_r, \right.$$

$$\left. \left. -M - \frac{1}{K_p}, t \right) \right]$$

$$- A_3 \sqrt{P_r} \sum_{n=0}^{\infty} \left[g\left(K_c, 2n\sqrt{P_r}, 1 - P_r - M - \frac{1}{K_p}, t \right) \right.$$

$$+ g\left(K_c, (2n+2)\sqrt{P_r}, 1 - P_r \right.$$

$$\left. \left. -M - \frac{1}{K_p}, t \right) \right]$$

$$+ 2A_4 \sum_{n=0}^{\infty} \left[g\left(M + \frac{1}{K_p}, 2n, P_r - S_c, S_c K_c, t \right) \right.$$

$$\left. + g\left(M + \frac{1}{K_p}, 2n + 2, P_r - S_c, S_c K_c, t \right) \right]$$

$$- A_4 \sqrt{S_C} \sum_{n=0}^{\infty} \left[g\left(K_c, 2n\sqrt{S_c}, P_r - S_c, S_c K_c, t \right) \right.$$

$$\left. + g\left(K_c, (2n+2)\sqrt{S_c}, P_r - S_c, S_c K_c, t \right) \right]$$

$$- A_4 \sqrt{P_r} \sum_{n=0}^{\infty} \left[g\left(0, 2n\sqrt{P_r}, P_r - S_c, S_c K_c, t \right) \right.$$

$$\left. + g\left(0, (2n+2)\sqrt{P_r}, P_r - S_c, S_c K_c, t \right) \right]$$

$$+ (A_5 - A_6) \sum_{n=0}^{\infty} \left[g\left(M + \frac{1}{K_p}, 2n, 1, 0, t \right) \right.$$

$$\left. + g\left(M + \frac{1}{K_p}, 2n + 2, 1, 0, t \right) \right]$$

$$+ A_5 Sh_0 + A_6 \sqrt{P_r} \sum_{n=0}^{\infty} \left[g\left(0, 2n\sqrt{P_r}, 1, 0, t \right) \right.$$

$$\left. + g\left(0, (2n+2)\sqrt{P_r}, 1, 0, t \right) \right]$$

$$\tau_1 = - \frac{\partial u}{\partial y} \bigg|_{y=1}$$

$$= -2A_2 \sum_{n=0}^{\infty} \left[g\left(M + \frac{1}{K_p}, 2n + 1, 1 - S_c, S_c K_c \right. \right.$$

$$\left. \left. -M - \frac{1}{K_p}, t \right) \right]$$

$$+ 2A_2 \sqrt{S_c} \sum_{n=0}^{\infty} \left[g\left(K_c, (2n+1)\sqrt{S_c}, 1 - S_c, S_c K_c \right. \right.$$

$$\left. \left. -M, t \right) \right]$$

$$- 2A_3 \sum_{n=0}^{\infty} \left[g\left(M + \frac{1}{K_p}, 2n + 1, 1 - P_r, -M - \frac{1}{K_p}, t \right) \right]$$

$$+ 2A_3 \sqrt{P_r} \sum_{n=0}^{\infty} \left[g\left(K_c, (2n+1)\sqrt{P_r}, 1 - P_r, -M \right. \right.$$

$$\left. \left. -\frac{1}{K_p}, t \right) \right]$$

$$- 4A_4 \sum_{n=0}^{\infty} \left[g\left(M + \frac{1}{K_p}, 2n + 1, P_r - S_c, S_c K_c, t \right) \right]$$

$$+ 2A_4 \sqrt{S_c} \sum_{n=0}^{\infty} \left[g\left(K_c, (2n+1)\sqrt{S_c}, P_r - S_c, S_c K_c, t \right) \right]$$

$$+ 2A_4 \sqrt{P_r} \sum_{n=0}^{\infty} \left[g\left(0, (2n+1)\sqrt{P_r}, P_r - S_c, S_c K_c, t \right) \right]$$

$$- 2(A_5 - A_6) \sum_{n=0}^{\infty} \left[g\left(M + \frac{1}{K_p}, 2n + 1, 1, 0, t \right) \right]$$

$$- 2A_6 \sqrt{P_r} \sum_{n=0}^{\infty} \left[g\left(0, (2n+1)\sqrt{P_r}, 1, 0, t \right) \right] + A_5 Sh_1,$$

$$(14)$$

where the function g is given in Appendix A.

The mass flux (volumetric flow rate) for the problem is given by

$$Q = \int_0^1 u \, dy. \tag{15}$$

This is computed by using the trapezoidal rule for numerical integration.

Case 2 $(P_r = S_c = 1)$. The solutions of (5) subject to boundary conditions (6) are given by

$$C = \sum_{n=0}^{\infty} \left[f\left(0, a_n, 1, 0, t \right) - f\left(0, b_n, 1, 0, t \right) \right]$$

$$T = \frac{D^*\left(1 - e^{-K_c t} \right)}{2 K_c t \sqrt{\pi t}} \sum_{n=0}^{\infty} \left[a_n \exp\left(-\frac{a_n^2}{4t} \right) - b_n \exp\left(-\frac{b_n^2}{4t} \right) \right]$$

$$+ (D^* + 1) \sum_{n=0}^{\infty} \left[f\left(0, a_n, 1, 0, t \right) - f\left(0, b_n, 1, 0, t \right) \right] - D^* C$$

$$u = A_8 C + \frac{D^*}{2t\sqrt{\pi t}} \left[\frac{\exp(-K_c t)}{K_c\left(K_c - M - \left(1/K_p \right) \right)} \right.$$

$$+ \frac{1}{\left(M + \left(1/K_p \right) \right) K_c}$$

$$\left. - \frac{\exp(-Mt)}{\left(M + \left(1/K_p \right) \right)\left(K_c - M - \left(1/K_p \right) \right)} \right]$$

$$\times \sum_{n=0}^{\infty} \left[a_n \exp\left(-\frac{a_n^2}{4t} \right) - b_n \exp\left(-\frac{b_n^2}{4t} \right) \right]$$

$$+ A_6 \sum_{n=0}^{\infty} \left[f\left(0, a_n, 1, 0, t \right) - f\left(0, b_n, 1, 0, t \right) \right],$$

$$(16)$$

where A_7 and A_8 are given in Appendix B.

In this case, the rate of mass transfer (Sh), the rate of heat transfer (Nu), and the skin frictions (τ) on the walls of the channel are obtained as follows:

Sherwood Number. Consider

$$Sh_0 = - \frac{\partial C}{\partial y} \bigg|_{y=0}$$

$$= - \sum_{n=0}^{\infty} \left[g\left(K_c, 2n, 1, 0, t \right) - g\left(K_c, 2n + 2, 1, 0, t \right) \right],$$

$$Sh_1 = - \frac{\partial C}{\partial y} \bigg|_{y=1} = 2 \sum_{n=0}^{\infty} \left[g\left(K_c, 2n + 1, 1, 0, t \right) \right].$$

$$(17)$$

Nusselt Number. Consider

$$\text{Nu}_0 = -\left.\frac{\partial T}{\partial y}\right|_{y=0}$$

$$= -\frac{D^* \left[1 - \exp\left(-K_c t\right)\right]}{2 K_c t \sqrt{\pi t}}$$

$$\times \left[\sum_{n=0}^{\infty} \left\{\exp\left(-\frac{n^2}{t}\right) + \exp\left(-\frac{(n+1)^2}{t}\right)\right\}\right.$$

$$-\frac{2}{t} \sum_{n=0}^{\infty} \left\{n^2 \exp\left(-\frac{n^2}{t}\right)\right.$$

$$\left.\left. + (n+1)^2 \exp\left(-\frac{(n+1)^2}{t}\right)\right\}\right]$$

$$- (D^* + 1) \sum_{n=0}^{\infty} \left[g\left(0, 2n, 1, 0, t\right) + g\left(0, 2n+2, 1, 0, t\right)\right]$$

$$+ D^* \text{Sh}_0,$$

$$\text{Nu}_1 = -\left.\frac{\partial T}{\partial y}\right|_{y=0}$$

$$= \frac{D^* \left[1 - \exp\left(-K_c t\right)\right]}{2 K_c t \sqrt{\pi t}}$$

$$\times \left[2 \sum_{n=0}^{\infty} \exp\left(-\frac{(2n+1)^2}{4t}\right)\right.$$

$$\left. -\frac{1}{t} \sum_{n=0}^{\infty} (2n+1)^2 \exp\left(-\frac{(2n+1)^2}{4t}\right)\right]$$

$$+ 2\left(D^* + 1\right) \sum_{n=0}^{\infty} \left[g\left(0, 2n+1, 1, 0, t\right)\right] - D^* \text{Sh}_1.$$

(18)

Skin Friction. Consider

$$\tau_0 = -\left.\frac{\partial u}{\partial y}\right|_{y=0}$$

$$= A_8 \text{Sh}_0 + \frac{D^*}{2t \sqrt{\pi t}}$$

$$\times \left[\frac{\exp\left(-K_c t\right)}{K_c \left(K_c - M - \left(1/K_p\right)\right)} + \frac{1}{K_c \left(M + \left(1/K_p\right)\right)}\right.$$

$$\left. -\frac{\exp\left(-Mt\right)}{\left(M + \left(1/K_p\right)\right)\left(K_c - M - \left(1/K_p\right)\right)}\right]$$

$$\times \left[\sum_{n=0}^{\infty} \left\{\exp\left(-\frac{n^2}{t}\right) + \exp\left(-\frac{(n+1)^2}{t}\right)\right\}\right.$$

$$-\frac{2}{t} \sum_{n=0}^{\infty} \left\{n^2 \exp\left(-\frac{n^2}{t}\right)\right.$$

$$+ (n+1)^2 \exp\left(-\frac{(n+1)^2}{t}\right)\right\}\right]$$

$$+ A_6 \sum_{n=0}^{\infty} \left[g\left(0, 2n, 1, 0, t\right) + g\left(0, 2n+2, 1, 0, t\right)\right]$$

$$- A_7 \sum_{n=0}^{\infty} \left[g\left(M + \frac{1}{K_p}, 2n, 1, 0, t\right)\right.$$

$$\left. + g\left(M + \frac{1}{K_p}, 2n+2, 1, 0, t\right)\right]$$

(19)

$$\tau_1 = -\left.\frac{\partial u}{\partial y}\right|_{y=1}$$

$$= A_8 \text{Sh}_1 + \frac{D^*}{t \sqrt{\pi t}}$$

$$\times \left[\frac{\exp\left(-M - \left(1/K_p\right)\right)t}{\left(M + \left(1/K_p\right)\right)\left(K_c - M - \left(1/K_p\right)\right)}\right.$$

$$\left. -\frac{1}{K_c \left(M + \left(1/K_p\right)\right)} - \frac{\exp\left(-K_c t\right)}{K_c \left(K_c - M - \left(1/K_p\right)\right)}\right]$$

$$\times \sum_{n=0}^{\infty} \left[\exp\left(-\frac{(2n+1)^2}{t}\right)\right.$$

$$\left. -2 \frac{(2n+1)^2}{t} \exp\left(-\frac{(2n+1)^2}{t}\right)\right]$$

$$- 2 A_6 \sum_{n=0}^{\infty} \left[g\left(0, 2n+1, 1, 0, t\right)\right]$$

$$\times 2 A_7 \sum_{n=0}^{\infty} \left[g\left(M + \frac{1}{K_p}, 2n+1, 1, 0, t\right)\right].$$

(20)

Steady State. Setting $\partial u / \partial t = 0$, $\partial T / \partial t = 0$ and $\partial C / \partial t = 0$ in (5), the steady state of the problem is obtained as

$$\frac{d^2 u}{dy^2} + NC + T - \left(M + \frac{1}{K_p}\right) u = 0,$$

$$\frac{d^2 C}{dy^2} - S_c K_c C = 0,$$

(21)

$$\frac{d^2 T}{dy^2} + D^* \frac{d^2 C}{dy^2} = 0.$$

The solutions of (21) subject to boundary conditions (6) are given by

$$C = \frac{\sinh\left(\sqrt{S_c K_c}\left(1 - y\right)\right)}{\sinh\left(\sqrt{S_c K_c}\right)},$$

$$T = \left(D^* + 1\right)\left(1 - y\right) - D^* C,$$

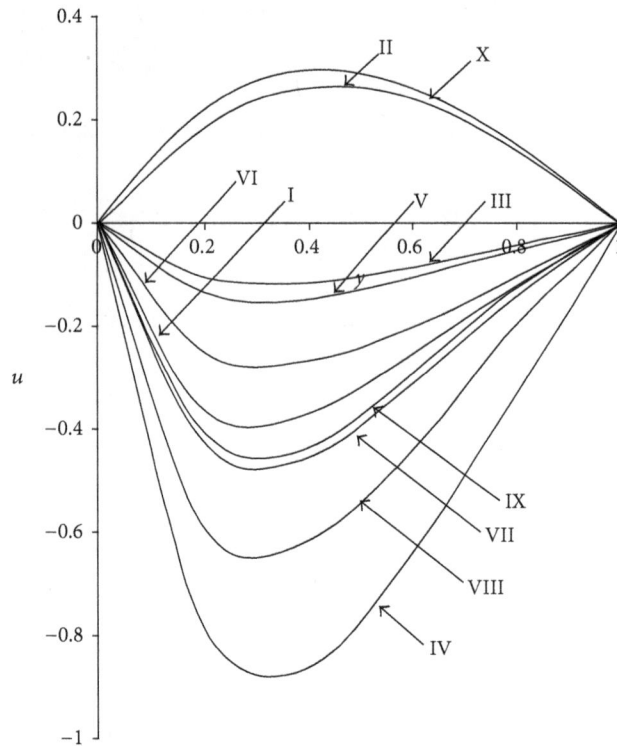

FIGURE 1: Velocity profile.

Curve	t	M	S_c	K_c	N	P_r	D^*	K_p
I	0.4	0.5	0.22	2	2	1.8	2	100
II	0.2	0.5	0.22	2	2	1.8	2	100
III	0.4	2	0.22	2	2	1.8	2	100
IV	0.4	0.5	0.3	2	2	1.8	2	100
V	0.4	0.5	0.22	1	2	1.8	2	100
VI	0.4	0.5	0.22	2	4	1.8	2	100
VII	0.4	0.5	0.22	2	2	2	2	100
VIII	0.4	0.5	0.22	2	2	1.8	4	100
IX	0.4	0.5	0.22	2	2	1.8	2	0.5
X	Steady state	0.5	0.22	2	4	—	2	100

$$u = A_9 \frac{\sinh \sqrt{\left(M + \left(1/K_p\right)\right)}\left(1 - y\right)}{\sinh \sqrt{M + \left(1/K_p\right)}} + A_5 C + A_6 \left(1 - y\right),$$

(22)

where A_9 is given in Appendix B.

In this case, the Sherwood number (Sh), the Nusselt number (Nu), and the skin friction (τ) are given by the following:

Sherwood Number (Sh). Consider

$$\mathrm{Sh}_0 = \sqrt{S_c K_c} \coth \sqrt{S_c K_c},$$

$$\mathrm{Sh}_1 = \sqrt{S_c K_c} \operatorname{cosech} \sqrt{S_c K_c}.$$

(23)

Nusselt Number (Nu). Consider

$$\mathrm{Nu}_0 = D^* + 1 + D^* \mathrm{Sh}_0,$$

$$\mathrm{Nu}_1 = -\left(D^* + 1\right) - D^* \mathrm{Sh}_1.$$

(24)

Skin Friction (τ). Consider

$$\tau_0 = -A_9 \sqrt{M + \frac{1}{K_p}} \coth \sqrt{M + \frac{1}{K_p}} + A_5 \mathrm{Sh}_0 - A_6,$$

$$\tau_1 = \frac{A_9 \sqrt{M + \left(1/K_p\right)}}{\sinh \sqrt{M + \left(1/K_p\right)}} - A_5 \mathrm{Sh}_1 + A_6.$$

(25)

Mass Flux. Consider

$$Q = A_9 \sqrt{M + \frac{1}{K_p}} \frac{\cosh \sqrt{M + \left(1/K_p\right)} - 1}{\sinh \sqrt{M + \left(1/K_p\right)}}$$

$$+ \frac{A_5}{\sqrt{S_c K_c}} \frac{\cosh \sqrt{S_c K_c} - 1}{\sinh \sqrt{S_c K_c}} + \frac{A_6}{2}.$$

(26)

4. Results and Discussion

Computations have been carried out by assigning the values to the pertinent parameters characterising the fluids of practical interest. The flow phenomenon is characterized by magnetic parameter (M), chemical reaction parameter (K_c), sustentation parameter (N), Dufour number (D^*), Schmidt number (S_c), Prandtl number (P_r), and porosity parameter (K_p). The effects of various parameters on velocity, temperature, and concentration profiles are shown graphically and in tabulated form for Sherwood number (Sh), Nusselt number (Nu), and skin friction (τ).

The particular case without magnetic field and porosity ($M = 0$, $K_p \rightarrow \infty$) can be obtained from (8) which is in good agreement with the work reported earlier [1].

Another interesting point to note is that diffusion in aqueous solution, that is, for higher values of S_c gives rise to oscillations in the velocity as well as temperature field in the presence of chemical reaction parameter (K_c).

Figure 1 depicts the velocity profile for various values of pertinent parameters characterising flow fields. The common

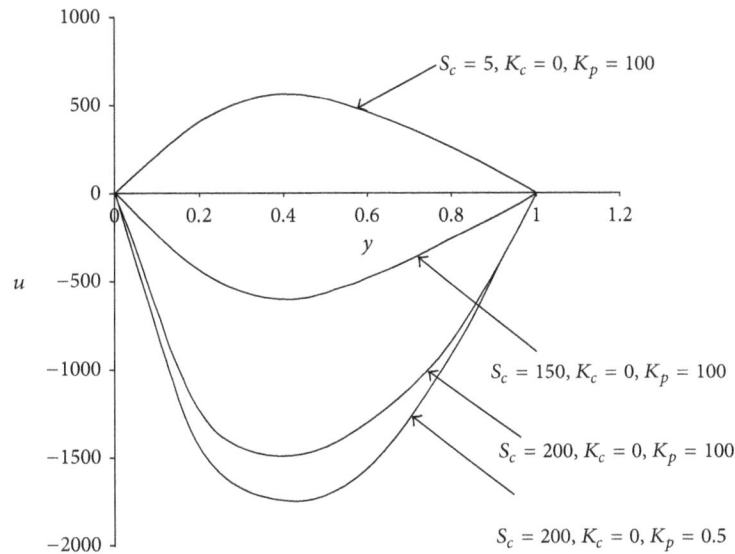

FIGURE 2: Velocity profile for $t = 0.4$, $M = 0.5$, $N = 2$, $P_r = 7$, and $D^* = 2$.

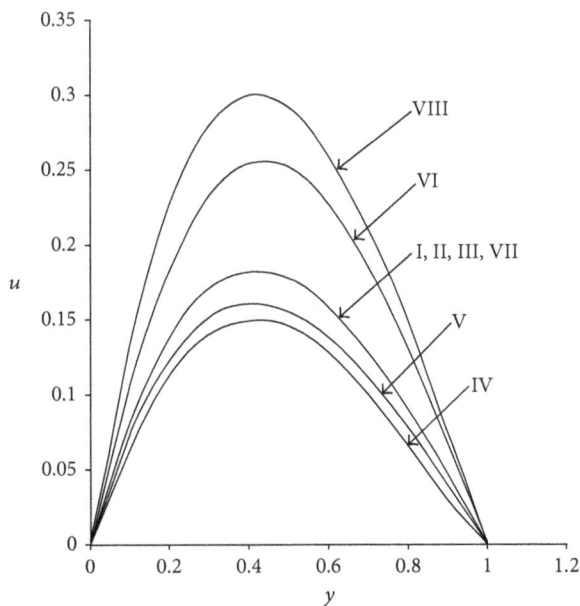

Curve	M	S_c	K_c	N	D^*	K_p
I	0.5	100	1	2	2	100
II	0.5	0.22	2	2	2	100
III	0.5	152	1	2	2	100
IV	0.5	0.22	2	2	2	0.5
V	2	0.22	2	2	2	100
VI	0.5	0.22	2	2	30	100
VII	0.5	0.3	1	2	2	100
VIII	0.5	0.22	2	4	2	100

FIGURE 3: Velocity profile (steady state).

feature of the profiles is parabolic. This is evident from curves II and X ($t = 0.2$, Curve II; the steady state curve X) in case of steady state as well as $t \leq 0.3$; the velocity profiles remain positive throughout the flow field. This is evident from curves II and X ($t = 0.2$, Curve II; the steady state curve X). Thus it is concluded that the time span plays a vital role for engendering back flow (Curve I and Curve II).

Further, magnetic parameter (M), chemical reaction parameter (K_c), sustentation parameter (N), and porosity parameter (K_p) decrease the velocity $|u|$ at all points. In all other cases such as S_c, P_r, and D^*, the reverse effect is observed. If the mass diffusivity becomes greater than the thermal diffusivity (i.e., N, the sustentation parameter >1.0) then the velocity decreases. Moreover, with the stronger

magnetic field Lorentz force also reduces the velocity field which is in conformity with the result of Cramer and Pai [20].

An increase in P_r leads to decrease in the velocity which suggests that low rate of thermal diffusion leads to increase in the velocity boundary layer thickness.

Figure 2 depicts the velocity distribution in the absence of chemical reaction. It is noteworthy to observe that the back flow occurs for higher values of S_c; that is, $S_c = 150$ and $S_c = 200$ and the permeability of the medium representing the diffusion in aqueous solution in the presence of constant magnetic field and Dufour effect. Further, it is to mention that for $S_c < 1.0$ and $K_c = 0.0$, no velocity profile could be graphed/presented. Further, it is to note that heavier species that is, with increasing S_c lead to flow reversal. It is most

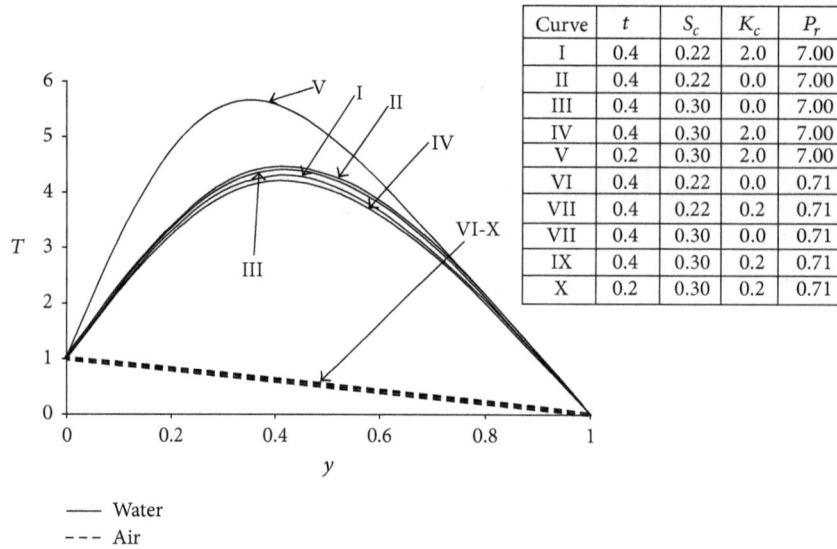

Curve	t	S_c	K_c	P_r
I	0.4	0.22	2.0	7.00
II	0.4	0.22	0.0	7.00
III	0.4	0.30	0.0	7.00
IV	0.4	0.30	2.0	7.00
V	0.2	0.30	2.0	7.00
VI	0.4	0.22	0.0	0.71
VII	0.4	0.22	0.2	0.71
VII	0.4	0.30	0.0	0.71
IX	0.4	0.30	0.2	0.71
X	0.2	0.30	0.2	0.71

FIGURE 4: Temperature profile for $D^* = 2.0$, $P_r = 7.0$ (water), and $P_r = 0.71$ (air).

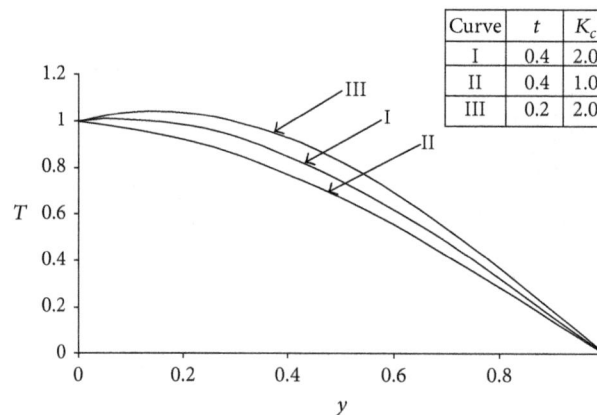

Curve	t	K_c
I	0.4	2.0
II	0.4	1.0
III	0.2	2.0

FIGURE 5: Temperature profile ($P_r = 1$, $S_c = 1$).

interesting to note that in the presence of destructive reaction that is, $K_c > 0$, higher value of S_c leads to oscillatory flow and, in case of diffusion in the aqueous solution, it leads to back flow. Thus, this suggests that the back flow is prevented by diffusing lighter species which is in conformity with the results reported earlier by Pop [21], Hossain and Mohammad [22], and Rath et al. [23].

It is also remarked that when S_c varies from 0 to 5 the back flow is prevented in the absence of chemical reaction. But in the presence of heavier species the sharp fall of velocity is well marked.

Figure 3 exhibits the steady state velocity profiles indicating no back flow irrespective of higher or lower value of mass transfer coefficient and Dufour effect.

On careful observation, it is revealed that for higher values of Dufour number D^* (curve VI), sustentation parameter N (Curve VIII), magnetic parameter M (Curve V), and porosity parameter K_p (Curve IV) steady state velocity is effected significantly whereas variation in reaction parameter

and Schmidt number produces no change. Thus, it is important to note that mass transfer phenomena with chemical reaction does not affect the steady flow significantly but increase in M, N, D^*, and K_p, decrease the velocity, increase the velocity, increase the velocity, and decrease the velocity, respectively. It may be inferred that magnetic parameter, porosity parameter, and Dufour effect produce the same effect irrespective of steady or unsteady state except for sustentation parameter.

Figures 4, 5, and 6 present the temperature distribution. It is seen that for an increase in destructive reaction parameter ($K_c > 0$) temperature increases whereas increase in P_r, t, and S_c reduces the temperature at all points for $P_r = 7.0$ (water) when $y < 0.7$. It is noted that, the slow rate of diffusion leaves the heat energy spread in the fluid mass and it is enhanced for heavier species.

It is observed in case of $P_r = 0.71$ (air) that the temperature falls sharply at all points with almost linear distribution regardless of other parameters. Thus, it may be pointed out

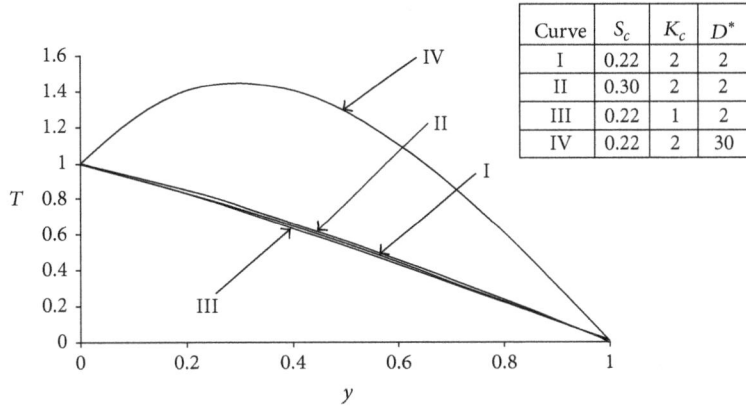

Curve	S_c	K_c	D^*
I	0.22	2	2
II	0.30	2	2
III	0.22	1	2
IV	0.22	2	30

FIGURE 6: Temperature profile (steady state).

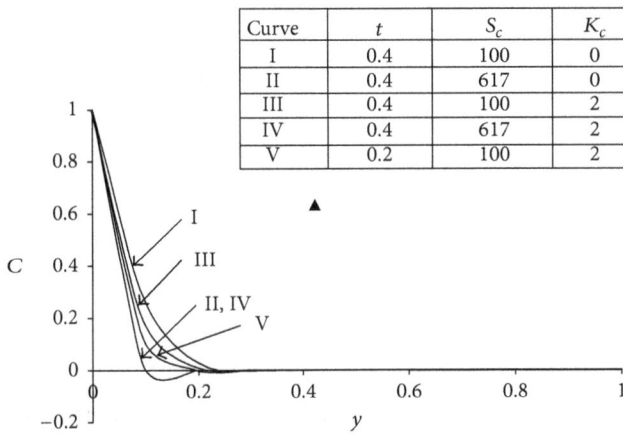

Curve	t	S_c	K_c
I	0.4	100	0
II	0.4	617	0
III	0.4	100	2
IV	0.4	617	2
V	0.2	100	2

FIGURE 7: Concentration profile.

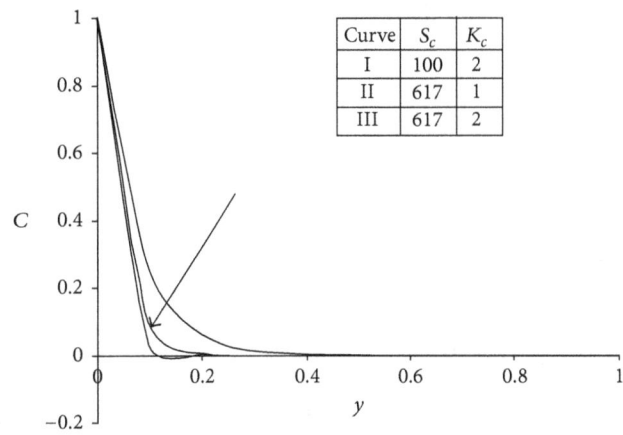

Curve	S_c	K_c
I	100	2
II	617	1
III	617	2

FIGURE 9: Concentration profile (steady state).

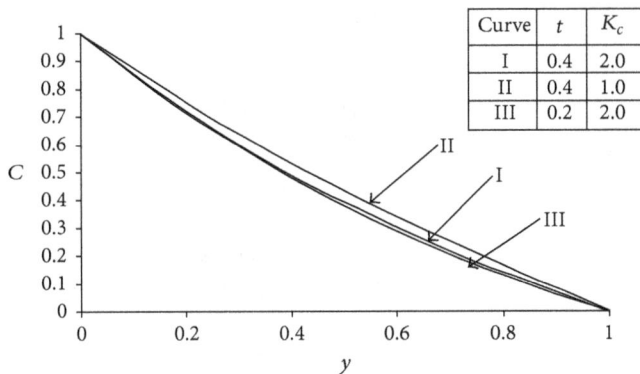

Curve	t	K_c
I	0.4	2.0
II	0.4	1.0
III	0.2	2.0

FIGURE 8: Concentration profile ($S_c = 1$).

that thermal diffusivity property of the fluid plays a vital role in controlling the temperature distribution and hence contributing to thermal boundary layer thickness.

Figure 5 also exhibits the temperature distribution when $P_r = 1$ and $S_c = 1$ which means that thermal diffusivity and mass diffusivity are at par. This contributes to the uniform variation/decrease in temperature. Increase in reaction parameter gives rise to increase in temperature whereas allowing for a large span of time reduces it.

Figure 6 presents the nonlinearity distribution of temperature due to high value of D^*, that is, for large Dufour effect. Thus, sharp rise and fall of temperature is due to low diffusivity and high Dufour effect.

Figures 7, 8 and 9 exhibit the concentration variation in the flow domain. Sharp fall of concentration is indicated in Figures 7 and 9 along with mass absorption near the plate for very high value of S_c (i.e., $S_c = 617$). This means that heavier species give rise to sharp fall of concentration accelerating the process of mass diffusion whereas in case of lighter species the variation is smooth.

Skin friction plays an important role in flow characteristics. It measures the frictional forces encountered at the solid surfaces due to the motion of the fluid. One striking feature of the entries of the skin friction in Table 1, in case of unsteady motion, is that the skin frictions bear same sign in case of lower and upper plate. Further, on careful observation it is revealed that at the lower plate all are negative for $t = 0.4$ except for the case when $t = 0.2$. Thus, it may be concluded that time span plays an important role to modify the frictional drag due to shear stress at the plates.

Moreover, from (18)-(19) it is clear that the Prandtl number (P_r) has no role to play to affect the velocity, temperature, and concentration fields in the steady flow and hence the skin

TABLE 1: Skin friction (unsteady case).

Sl. no.	t	M	S_c	K_c	N	P_r	D^*	K_p	τ_0	τ_1
1	0.4	0.5	0.22	2	2	1.8	2	100	−1.37638	−0.3153
2	0.2	0.5	0.22	2	2	1.8	2	100	0.649408	0.466308
3	0.4	2	0.22	2	2	1.8	2	100	−0.40403	−0.12279
4	0.4	0.5	0.3	2	2	1.8	2	100	−2.90421	−0.9827
5	0.4	0.5	0.22	1	2	1.8	2	100	−0.46631	−0.15838
6	0.4	0.5	0.22	2	4	1.8	2	100	−0.91633	−0.28675
7	0.4	0.5	0.22	2	2	2	2	100	−1.66428	−0.3346
8	0.4	0.5	0.22	2	2	1.8	4	100	−2.24604	−0.34433
9	0.4	0.5	0.22	2	2	1.8	2	0.5	−1.25513	−0.2425

TABLE 2: Skin friction (steady case).

Sl. no.	M	S_c	K_c	N	D^*	K_p	τ_0	τ_1
1	0.5	0.22	2	2	2	100	1.732053	0.932515
2	2	0.22	2	2	2	100	1.66425	0.795436
3	0.5	0.3	2	2	2	100	1.732052	0.932515
4	0.5	0.22	1	2	2	100	1.73205	0.932515
5	0.5	0.22	2	4	2	100	1.664279	0.869287
6	0.5	0.22	2	2	4	100	2.144507	1.376382
7	0.5	0.22	2	2	2	0.5	1.42312	0.524363

TABLE 3: Nusselt number.

Sl. no.	t	S_c	K_c	P_r	D^*	Nu_0	Nu_1
1	0.4	0.22	2	7.00	2	−1.37638192	−1.150368407
2	0.2	0.22	2	7.00	2	−2.674621494	−1.213097004
3	0.4	0.30	2	7.00	2	−1.327044822	−1.110612515
4	0.4	0.22	0	7.00	2	−1.455009029	−1.191753593
5	0.4	0.22	2	0.71	2	0.113935608	−0.113935608
6	0.4	0.22	2	7.00	4	−1.569685558	−1.279941632

friction. In case of unsteady flow, as P_r increases, the shearing stress increases at both the plates.

Considering Tables 1 and 2 for both steady and unsteady cases, it is concluded that an increase in magnetic parameter, porosity parameter, and sustentation parameter leads to reduce the magnitude of frictional drag at both plates. Moreover, an increase in S_c, K_c, and D^* leads to enhance the frictional drag at the plates for steady and unsteady flow.

Physically, we may interpret the above results as follows.

Lorentz force and sustentation parameter, that is, ratios of buoyancy effects due to temperature and concentration differences reduce the frictional drag which agrees well with the established result reported earlier. Moreover, presence of heavier species and exothermic reaction and Dufour effect enhance the magnitude of the shearing stress at both plates.

From Table 3 it is seen that an increase in t, S_c, and K_c leads to decrease the Nusselt number at both plates. One interesting point is that rate of heat transfer remains negative for all the parameters at both the plates for aqueous solution $P_r = 7.0$ but in case of air, that is, $P_r = 0.71$, Nusselt number is positive at the lower plate and negative at the upper plate.

TABLE 4: Sherwood number.

Sl. no.	t	S_c	K_c	Sh_0	Sh_1
1	0.4	100	2	5.9757644	−0.612800788
2	0.2	100	2	6.313751515	0.15838444
3	0.4	617	2	7.026366229	0.470564281
4	0.2	100	0	5.67128182	−0.466307658

Table 4 presents the variation of mass transfer for various values of t, S_c, and K_c. It is seen that an increase in chemical reaction parameter K_c increases the mass transfer at the lower plate and decreases it at the other. Further, it is to note that Sherwood number at both the plates increases as S_c increases, that is, for heavier species mass transfer increases at the plates. The effect of increase in time span is to reduce the rate of mass transfer at both plates.

Tables 5 and 6 show the effects of various parameters on mass flux both for unsteady and steady cases, respectively. From Table 5, it is observed that mass flux increases with the

TABLE 5: Mass flux (unsteady case).

Sl. no.	t	M	S_c	K_c	N	P_r	D^*	K_p	Q
1	0.4	0.5	0.22	2	2	1.8	2	100	0.197836
2	0.2	0.5	0.22	2	2	1.8	2	100	0.176572
3	0.4	2	0.22	2	2	1.8	2	100	0.057655
4	0.4	0.5	0.3	2	2	1.8	2	100	0.568475
5	0.4	0.5	0.22	1	2	1.8	2	100	0.091193
6	0.4	0.5	0.22	2	4	1.8	2	100	0.166667
7	0.4	0.5	0.22	2	2	2	2	100	0.264939
8	0.4	0.5	0.22	2	2	1.8	4	100	0.364583
9	0.4	0.5	0.22	2	2	1.8	2	0.5	0.034241

TABLE 6: Mass flux (steady case).

Sl. no.	M	S_c	K_c	N	D^*	K_p	Q
1	0.5	0.22	2	2	2	100	0.119052
2	2	0.22	2	2	2	100	0.104207
3	0.5	0.3	2	2	2	100	0.119052
4	0.5	0.22	1	2	2	100	0.119052
5	0.5	0.22	2	4	2	100	0.195077
6	0.5	0.22	2	2	4	100	0.122392
7	0.5	0.22	2	2	2	0.5	0.094632

increase in the values in S_c, K_c, P_r, and D^* but the reverse effect is observed in case of t, M, N, and K_p when the motion is unsteady. Hence, it is concluded that in case of unsteady motion more flux fluid was experienced with heavier species having low thermal diffusivity and increasing Dufour effect in the presence of constructive chemical reaction.

Now, if we analyse the steady case, we observe that more flux is measured with higher value of N and D^* whereas the reverse effect is observed in case of M and K_p. In case of S_c and K_c there is no significant effect on mass flux.

Comparing both cases steady and unsteady motion, it is observed that an increase in K_c and D^* gives rise to higher flux whereas magnetic field yields less amount of flux.

5. Conclusion

(i) Under the influence of dominating mass diffusivity over thermal diffusivity with stronger Lorentz force the velocity is reduced at all points of the channel.

(ii) Low rate of thermal diffusion leads to increase the thickness of boundary layer.

(iii) Diffusion of heavier species leads to flow reversal. It is well marked in case of aqueous solution. Flow of aqueous solution in the presence of heavier species is prone to back flow.

(iv) Destructive reaction in the presence of heavier species leads to oscillatory flow.

(v) In case of steady state, no back flow occurs irrespective of high or low value of mass transfer coefficient and Dufour effects.

(vi) Sharp rise and fall of temperature is the outcome of low diffusivity and high Dufour effects.

(vii) Dufour effect and chemical reaction rate in the presence of heavier species enhance the frictional drag.

(viii) Heat transfer bears same sign at both the plates in the presence of aqueous solution but in case of air it is of opposite sign.

(ix) Mass transfer increases at both plates due to the presence of heavier species but it reduces with increasing time span.

(x) In case of unsteady motion more flux of fluid is experienced due to heavier species with low thermal diffusivity. Dufour effect enhances the flux both in steady and unsteady motion.

Appendices

A.

The complementary error function is given by

$$\text{erfc}(x) = \frac{2}{\sqrt{\pi}} \int_x^{\infty} e^{-\lambda^2} d\lambda,$$

$$f(x_1, x_2, x_3, x_4, x_5)$$

$$= \frac{1}{2} \exp\left(\frac{x_4 x_5}{x_3}\right)$$

$$\times \left[\exp\left(-x_2 \sqrt{\frac{x_4}{x_3}} + x_1\right) \right.$$

$$\times \operatorname{erfc}\left(\frac{x_2}{2\sqrt{x_5}} - \sqrt{\left(\frac{x_4}{x_3} + x_1\right)x_5}\right)$$

$$+ \exp\left(x_2\sqrt{\frac{x_4}{x_3} + x_1}\right)$$

$$\times \operatorname{erfc}\left(\frac{x_2}{2\sqrt{x_5}} + \sqrt{\left(\frac{x_4}{x_3} + x_1\right)x_5}\right)\Bigg],$$

$$g(x_1, x_2, x_3, x_4, x_5)$$

$$= \frac{1}{2}\sqrt{\frac{x_4}{x_3} + x_1}\exp\left(\frac{x_4 x_5}{x_3}\right)$$

$$\times \left[\exp\left(x_2\sqrt{\frac{x_4}{x_3} + x_1}\right)\right.$$

$$\times \operatorname{erfc}\left(\frac{x_2}{2\sqrt{x_5}} + \sqrt{\left(\frac{x_4}{x_3} + x_1\right)x_5}\right)$$

$$- \exp\left(-x_2\sqrt{\frac{x_4}{x_3} + x_1}\right)$$

$$\left.\times \operatorname{erfc}\left(\frac{x_2}{2\sqrt{x_5}} - \sqrt{\left(\frac{x_4}{x_3} + x_1\right)x_5}\right)\right]$$

$$- \frac{2}{\sqrt{\pi}}\exp\left(-\left(\frac{x_3^2}{4x_5} + x_1 x_5\right)\right).$$

$$(A.1)$$

B.

$$A_1 = -D^*\left[\frac{S_c}{P_r - S_c} + P_r - S_c\right],$$

$$A_2 = -\frac{1 - S_c}{S_c K_c - M - \left(1/K_p\right)}$$

$$\times \left[N + \frac{S_c D^*\left(K_c - M - \left(1/K_p\right)\right)}{S_c K_c\left(1 - P_r\right) + \left(M + \left(1/K_p\right)\right)\left(P_r - S_c\right)}\right],$$

$$A_3 = \frac{1 - P_r}{M + \left(1/K_p\right)}$$

$$\times \left[1 + \frac{S_c D^*\left(K_c - M - \left(1/K_p\right)\right)}{S_c K_c\left(1 - P_r\right) + \left(M + \left(1/K_p\right)\right)\left(P_r - S_c\right)}\right],$$

$$A_4 = \frac{D^* P_r\left(P_r - S_c\right)}{S_c K_c\left(1 - P_r\right) + \left(M + \left(1/K_p\right)\right)\left(P_r - S_c\right)},$$

$$A_5 = \frac{D^* - N}{S_c K_c - M - \left(1/K_p\right)},$$

$$A_6 = \frac{D^* + 1}{M + \left(1/K_p\right)},$$

$$A_7 = \frac{K_c\left(D^* + 1\right) - \left(M + \left(1/K_p\right)\right)\left(N + 1\right)}{\left(M + \left(1/K_p\right)\right)\left(K_c - M - \left(1/K_p\right)\right)},$$

$$A_8 = \frac{D^* - N}{K_c - M - \left(1/K_p\right)},$$

$$A_9 = \frac{-S_c K_c\left(D^* + 1\right) + \left(M + \left(1/K_p\right)\right)\left(N + 1\right)}{\left(M + \left(1/K_p\right)\right)\left(S_c K_c - M - \left(1/K_p\right)\right)}.$$

References

[1] B. K. Jha and A. O. Ajibade, "Free convection heat and mass transfer flow in a vertical channel with the Dufour effect," *Journal of Process Mechanical Engineering*, vol. 224, no. 2, pp. 91–101, 2010.

[2] V. M. Soundalgekar and S. P. Akolkar, "Effects of free convection currents and mass transfer on flow past a vertical oscillating plate," *Astrophysics and Space Science*, vol. 89, no. 2, pp. 241–254, 1983.

[3] N. G. Kafoussias and E. W. Williams, "Thermal-diffusion and diffusion-thermo effects on mixed free-forced convective and mass transfer boundary layer flow with temperature dependent viscosity," *International Journal of Engineering Science*, vol. 33, no. 9, pp. 1369–1384, 1995.

[4] B. K. Jha and A. K. Singh, "Soret effects on free-convection and mass transfer flow in the stokes problem for a infinite vertical plate," *Astrophysics and Space Science*, vol. 173, no. 2, pp. 251–255, 1990.

[5] N. G. Kafoussias, "MHD thermal-diffusion effects on free-convective and mass-transfer flow over an infinite vertical moving plate," *Astrophysics and Space Science*, vol. 192, no. 1, pp. 11–19, 1992.

[6] M. S. Alam, M. M. Rahman, and M. A. Samad, "Dufour and Soret effects on unsteady MHD free convection and mass transfer flow past a vertical porous plate in a porous medium," *Nonlinear Analysis. Modelling and Control*, vol. 11, no. 3, pp. 217–226, 2006.

[7] M. S. Alam, M. M. Rahman, M. Ferdous, K. Maino, E. Mureithi, and A. Postelnicu, "Diffusion-thermo and thermal-diffusion effects on free convective heat and mass transfer flow in a porous medium with time dependent temperature and concentration," *International Journal of Applied Engineering Research*, vol. 2, no. 1, pp. 81–96, 2007.

[8] O. A. Beg, T. A. Beg, A. Y. Bakier, and V. R. Prasad, "Chemically-reacting mixed convective heat and mass transfer along inclined and vertical plates with soret and Dufour effects: numerical solution," *International Journal of Applied Mathematics and Mechanics*, vol. 5, no. 2, pp. 39–57, 2009.

[9] Z. Dursunkaya and W. M. Worek, "Diffusion-thermo and thermal-diffusion effects in transient and steady natural convection from a vertical surface," *International Journal of Heat and Mass Transfer*, vol. 35, no. 8, pp. 2060–2065, 1992.

[10] C. R. A. Abreu, M. F. Alfradique, and A. S. Telles, "Boundary layer flows with Dufour and Soret effects: I: forced and natural convection," *Chemical Engineering Science*, vol. 61, no. 13, pp. 4282–4289, 2006.

[11] E. Osalusi, J. Side, and R. Harris, "Thermal-diffusion and diffusion-thermo effects on combined heat and mass transfer of a steady MHD convective and slip flow due to a rotating disk with viscous dissipation and Ohmic heating," *International Communications in Heat and Mass Transfer*, vol. 35, no. 8, pp. 908–915, 2008.

[12] M. Anghel, H. S. Takhar, and I. Pop, "Dufour and Soret effects on free convection boundary layer over a vertical surface embedded in a porous medium," *Studia Universitatis Babes-Bolyai. Matematica*, vol. 11, no. 4, pp. 11–21, 2000.

[13] A. Postelnicu, "Influence of a magnetic field on heat and mass transfer by natural convection from vertical surfaces in porous media considering Soret and Dufour effects," *International Journal of Heat and Mass Transfer*, vol. 47, no. 6-7, pp. 1467–1472, 2004.

[14] B. Gebhart and L. Pera, "The nature of vertical natural convection flows resulting from the combined buoyancy effects of thermal and mass diffusion," *International Journal of Heat and Mass Transfer*, vol. 14, no. 12, pp. 2025–2050, 1971.

[15] U. N. Das, R. Deka, and V. M. Soundalgekar, "Effects of mass transfer on flow past an impulsively started infinite vertical plate with constant heat flux and chemical reaction," *Forschung im Ingenieurwesen*, vol. 60, no. 10, pp. 284–287, 1994.

[16] S. S. Saxena and G. K. Dubey, "Unsteady MHD heat and mass transfer free convection flow of a polar fluid past a vertical moving porous plate in a porous medium with heat generation and thermal diffusion," *Advances in Applied Science Research*, vol. 2, no. 4, pp. 259–278, 2011.

[17] K. Raveendra Babu, A. G. Vijaya Kumar, and S. V. K. Varma, "Diffusion-thermo and radiation effects on MHD free convective heat and mass transfer flow past an infinite vertical plate in the presence of a chemical reaction of first order," *Advances in Applied Science Research*, vol. 3, no. 4, pp. 2446–2462, 2012.

[18] K. Sudhakar, R. Srinibasa Raju, and M. Rangamma, "Chemical reaction effect on an unsteady MHD free convection flow past an infinite vertical accelerated plate with constant heat flux, thermal diffusion and diffusion thermo," *International Journal of Modern Engineering Research*, vol. 2, no. 5, pp. 3329–3339, 2012.

[19] L. Debanath and D. Bhatta, *Integral Transforms and Their Applications, Applications of Laplace Transforms*, Chapman and Hall/CRC, Taylor and Francies Group, London, UK, 2007.

[20] K. R. Cramer and S. I. Pai, *Magnetofluid Dynamics for Engineers and Applied Physics*, McGraw-Hill, New York, NY, USA, 1973.

[21] I. Pop, "Effect of Hall current on hydromagnetic flow near an accelerated plate," *International Journal of Physics and Mathematical Sciences*, vol. 5, p. 375, 1971.

[22] M. A. Hossain and K. Mohammad, "Effect of hall current on hydromagnetic free convection flow near an accelerated porous plate," *Japanese Journal of Applied Physics*, vol. 27, no. 8, pp. 1531–1535, 1988.

[23] P. K. Rath, G. C. Dash, and A. K. Patra, "Effect of Hall current and chemical reaction on MHD flow along an exponentially accelerated porous flat plate with internal heat absorption/generation," *Proceedings of the National Academy of Sciences India A*, vol. 80, no. 4, pp. 295–308, 2010.

21

An Exact Analytical Solution of the Strong Shock Wave Problem in Nonideal Magnetogasdynamics

S. D. Ram,[1] **R. Singh,**[2] **and L. P. Singh**[2]

[1] *Department of Mathematics, Mata Sundri College, University of Delhi, Delhi 110002, India*
[2] *Department of Applied Mathematics, Indian Institute of Technology (BHU), Varanasi 221005, India*

Correspondence should be addressed to S. D. Ram; sdram.apm@gmail.com

Academic Editor: Miguel Onorato

We construct the solutions to the strong shock wave problem with generalized geometries in nonideal magnetogasdynamics. Here, it is assumed that the density ahead of the shock front varies according to a power of distance from the source of the disturbance. Also, an analytical expression for the total energy carried by the wave motion in nonideal medium under the influence of magnetic field is derived.

1. Introduction

The propagation of shock waves, generated by a strong explosion in earth's atmosphere is of great interest both from mathematical and physical point of view due to its numerous applications in various fields. They result from a sudden release of a relatively large amount of energy; typical examples are lightening and chemical or nuclear explosions. Assume that we have an explosion, following which there may exist for a while a very small region filled with hot matter at high pressure, which starts to expand outwards with its front headed by a strong shock. The process generally takes place in a very short time after which a forward-moving shock wave develops, which continuously assimilates the ambient air into the blast wave. The study of strong shock wave problems has been of long interest for researchers in fields ranging from condensed matter to fluid dynamics due to its theoretical and practical importance. Practically, it is recognized that strong shock waves are excellent means for generating very high-pressure, high temperature plasma at the center of explosion. Many authors, for example, Arora and Sharma [1], Sakurai [2, 3] and Rogers [4] have presented exact solutions for the problem of strong shock wave with spherical geometry, since the study of spherically symmetric motion is important for the theory of explosion in various gasdynamic regimes. Recently, Singh et al. [5, 6] presented an approximate analytical solution to the system of first order quasilinear partial differential equations that govern a one dimensional unsteady planar and nonplanar motion in ideal and nonideal gases, involving discontinuities.

The study of interaction between gasdynamic motion of an electrically conducting medium and a magnetic field has been of great interest to scientists and engineers due to its application in astrophysics, geophysics, and interstellar gas masses. Taylor [7] obtained exact solution of the equations governing the motion in a gas generated by a point explosion. Singh et al. [8] studied the influence of the magnetic field upon the collapse of the cylindrical shock wave problem. Menon and Sharma [9] investigated the influence of magnetic field on the process of steepening and flattening of the characteristic wave front in a plane and cylindrically symmetric motion of ideal plasma. Oliveri and Speciale [10] used the substitution principle to obtain an exact solution for unsteady equation of perfect gas. Murata [11] obtained the exact solution for the one dimensional blast wave problem with generalized geometry. Singh et al. [12] have used quasisimilar theory to construct an analytical solution for the strong shock wave problem with generalized geometries in a nonideal gas satisfying the equation of state of the Van der Waals type.

The present paper aims to construct the closed form solution of the basic equations governing the one dimensional unsteady flows of a nonideal gas involving strong shock waves under the influence of transverse magnetic field. The basic configuration investigated here is that which arises when a transverse magnetic field is generated by a current of finite, constant strength passing along a straight wire of infinite length and either a shock or detonation wave propagates with uniform speed outwards from the wire into the ambient undisturbed gas at rest. Also, an expression for the total energy carried by the wave motion is obtained.

2. Problem Formulation

The basic equations for unsteady flow of a one dimensional gasdynamic motion may be written as [8, 12–14]

$$\frac{\partial \rho}{\partial t} + u\frac{\partial \rho}{\partial r} + \rho\left(\frac{\partial u}{\partial r} + \frac{m}{r}u\right) = 0, \tag{1}$$

$$\frac{\partial u}{\partial t} + u\frac{\partial u}{\partial r} + \frac{1}{\rho}\left(\frac{\partial p}{\partial r} + \frac{\partial h}{\partial r}\right) = 0, \tag{2}$$

$$\frac{\partial p}{\partial t} + u\frac{\partial p}{\partial r} + a^2\rho\left(\frac{\partial u}{\partial r} + \frac{m}{r}u\right) = 0, \tag{3}$$

$$\frac{\partial h}{\partial t} + u\frac{\partial h}{\partial r} + 2h\left(\frac{\partial u}{\partial r} + \frac{m}{r}u\right) = 0, \tag{4}$$

where ρ is the gas density, u is the velocity along the x-axis, p is the pressure, $h = \mu H^2/2$ is the magnetic pressure with H as the magnetic field strength, μ the magnetic permeability, t is the time, $a = (\gamma p/\rho(1 - b\rho))^{1/2}$ is the speed of sound, r is the single spatial coordinate being either axial in flows with planer ($m = 0$) geometry or radial in cylindrically symmetric ($m = 1$) and spherically symmetric ($m = 2$) flows, and γ is the constant specific heat ratio.

The system of (1)–(4) is supplemented with a Van der Waals equation of state of the form:

$$p(1 - b\rho) = \rho\mathfrak{R}T, \tag{5}$$

where b is the Van der Waals excluded volume which is known in terms of the molecular interaction potential in high temperature gases, T is the absolute temperature, and $\mathfrak{R}$ is the gas constant. It may be noted that the case $b = 0$ corresponds to an ideal gas (ideal in the sense that the particle interactions are absent).

The propagation velocity of the shock front U may be given as

$$\frac{dR}{dt} = U, \tag{6}$$

where R is the position of the shock front which is a function of time t.

3. Rankine-Hugoniot Conditions

The Rankine-Hugoniot relations, given by the principle of conservation of mass, momentum, and energy across the shock front may be expressed as [5, 6, 15]

$$\rho = \frac{\gamma + 1}{(\gamma - 1 + 2\bar{b})}\rho_a, \tag{7a}$$

$$u = \frac{2(1 - \bar{b})}{\gamma + 1}U, \tag{7b}$$

$$p = \left\{\frac{2(1 - \bar{b})}{\gamma + 1} + \frac{2C_0((\gamma - 1)\bar{b} - \gamma)}{(\gamma - 1 + 2\bar{b})^2}\right\}\rho_a U^2, \tag{7c}$$

$$h = \frac{C_0}{2}\left(\frac{(\gamma + 1)}{(\gamma - 1 + 2\bar{b})}\right)^2\rho_a U^2, \tag{7d}$$

where $\bar{b} = b\rho_a$ and $C_0 = 2h_a/\rho_a U^2$ is the cowling number.

In the present problem the density ρ_a is taken to vary according to the power law of radius of the shock front R, given as

$$\rho_a = \rho_0 R^\lambda, \tag{8}$$

where ρ_0 and λ are constants. The constant λ is determined in the subsequent analysis.

4. Solution of the Strong Shock Wave Problem

We construct a relation for the pressure in the flow field satisfying the Rankine-Hugoniot relations (7a)–(7d) as

$$p = \left\{\frac{C_0(\gamma + 1)(\bar{b} + \gamma) - (\gamma - 1 + 2\bar{b})^2}{(\bar{b} - 1)(\gamma - 1 + 2\bar{b})}\right\}\rho u^2. \tag{9}$$

Inserting (9) into (2) and (3) yields

$$\frac{\partial u}{\partial t} + u\frac{\partial u}{\partial r} + \frac{\{C_0(\gamma + 1) + 4(1 - \bar{b})\}(\gamma - 1 + 2\bar{b})}{8(1 - \bar{b})^2} \tag{10}$$

$$\times\left(\frac{u^2}{\rho}\frac{\partial \rho}{\partial r} + 2u\frac{\partial u}{\partial r}\right) = 0,$$

$$\frac{\partial u}{\partial t} + u\frac{\partial u}{\partial r} + \frac{(\gamma - 1)^2 + (3\gamma - 1)\bar{b}}{2(\gamma - 1)(1 - \bar{b})}u\left(\frac{\partial u}{\partial r} + \frac{m}{r}u\right) = 0. \tag{11}$$

Combining (10) and (4) and integrating the resulting equation we get

$$S(t) = \rho u^{2-\alpha}r^{-m\alpha}, \tag{12}$$

where $S(t)$ is a arbitrary function of integration. Also, α is constants given as

$$\alpha = \frac{4(1 - \bar{b})\{(\gamma - 1)^2 + (3\gamma - 1)\bar{b}\}}{(\gamma - 1)\{(C_0(\gamma + 1) + 4(1 - \bar{b}))(\gamma - 1 + 2\bar{b})\}}. \tag{13}$$

Using the solution (12), (1) takes the form

$$\frac{(2-\alpha)}{u}\frac{\partial u}{\partial t} + (1-\alpha)\frac{\partial u}{\partial r} - (1+\alpha)\frac{mu}{r} - \frac{1}{S}\frac{dS}{dt} = 0. \quad (14)$$

Solving (4) and (14), we have

$$u = -\Omega\frac{r}{S}\frac{dS}{dt}, \quad (15)$$

where $\Omega = 1/\{\delta + m(\alpha + \delta)\}$ is a constant and δ given as

$$\delta = 1 + \left(1 - \frac{\alpha}{2}\right)\left(\frac{(\gamma-1)^2 + \bar{b}(3\gamma-1)}{(\gamma-1)(1-\bar{b})}\right). \quad (16)$$

Plugging the value of u from (15) into (14), yields

$$S(t) = S_0 t^{-\eta}. \quad (17)$$

Here, S_0 is arbitrary constant and constant η is given as

$$\eta = \frac{(\alpha-2)}{(\alpha-1)\Omega + (\alpha+1)m\Omega - 1}. \quad (18)$$

Using the Rankine-Hugoniot relations (7a)–(7d) and (15) in (6) yield the analytical expression for the radius of the shock front as

$$R(t) = t^{((\gamma+1)/2)(\Omega\eta/(1-\bar{b}))}. \quad (19)$$

With the help of Rankine-Hugoniot condition (7a)–(7d), we may determine the value of the constant λ as

$$\lambda = \frac{(\gamma+1)(m-1) - 2(1-\bar{b})[(\alpha+1)(m+1)-2]}{(\gamma+1)}. \quad (20)$$

Consequently, the analytical solution of the blast wave problem described in the prior section is given as

$$u = \Omega\eta\frac{r}{t},$$

$$\rho = \frac{1}{(\Omega\eta)^{(2-\alpha)}}S_0 r^{(m+1)\alpha-2}t^{(2-\alpha-\eta)},$$

$$p = \left\{\frac{C_0(\gamma+1)(\bar{b}+\gamma) - (\gamma-1+2\bar{b})^2}{(\bar{b}-1)(\gamma-1+2\bar{b})}\right\}$$

$$\times \frac{1}{(\Omega\eta)^{-\alpha}}S_0 r^{(m+1)\alpha}t^{-(\alpha+\eta)}, \quad (21)$$

$$h = \left\{\frac{(C_0(\gamma+1) + 4(1-\bar{b}))(\gamma-1+2\bar{b})}{4(1-\bar{b})^2}\right.$$

$$\left. - \frac{C_0(\gamma+1)(\bar{b}+\gamma) - (\gamma-1+2\bar{b})^2}{(\bar{b}-1)(\gamma-1+2\bar{b})}\right\}$$

$$\times \frac{1}{(\Omega\eta)^{-\alpha}}S_0 r^{(m+1)\alpha}t^{-(\alpha+\eta)}.$$

For a spherical motion of an ideal gas, the solutions (21) recover the results presented by some authors [2, 4, 7, 11].

5. Behaviour of the Energy

The total energy carried by the wave motion in nonideal medium under the influence of transverse magnetic field is expressed as [15]

$$E = 4\pi\int_0^R \left\{\frac{1}{2}\rho u^2 + \frac{(1-b\rho)}{\gamma-1}p + h\right\}r^m dr, \quad (22)$$

where E is the function of time and represents the sum of the kinetic and internal energy of the gas.

Using solutions (21) into (22), we have

$$E(t) = \Pi t^{-(\alpha+\eta)+((m+1)(\gamma+1)(\alpha+1)\Omega\eta/2(1-\bar{b}))}. \quad (23)$$

The constant Π appearing in the above relation may be expressed as

$$\Pi = \frac{4\pi}{(m+1)(\alpha+1)}$$

$$\times \left\{1 + \frac{C_0(\gamma+1)^3}{4(1-\bar{b})^2}\right.$$

$$+ \left(4\bar{b}^2 - \bar{b}(4 + C_0 + (C_0-4)\gamma)\right.$$

$$\left. + (\gamma-1)^2 - C_0(\gamma+1)\right)\left((\gamma-1+2\bar{b})^2\right)^{-1}\right\}$$

$$\times \frac{S_0}{(\Omega\eta)^{-\alpha}}. \quad (24)$$

6. Result and Discussion

The solution (21), which is an exact solution for nonideal magnetogasdynamic problems, obtained in the form of a power in the distance and time, is same as provided by [6–11] for strong and weak shock waves in ideal gases. However, for a planar and nonplanar nonideal magnetogasdynamic motion, the solutions (21) are a new exact solution in case of strong shock waves for arbitrary values of adiabatic index. From the solution equations (21), it is also clear that in the absence of magnetic field solutions coincides with the solution given by Singh et al. [5].

It may be noted here that the total energy carried by the wave remains constant in planar case of ideal gas, whereas in case of nonideal medium it varies with respect time t and is given by (23). Also, due to increase in magnetic field strength total energy monotonically decreases with time t and this process is slowed down due to the increase in Van der Waals excluded volume.

7. Conclusion

In the present paper, a new exact solution is derived for a strong shock wave problem in nonideal magnetogasdynamics with the density ahead of the shock which is varied as a power of the distance from the origin of the shock wave. The effect of coupling between the nonideal effects and magnetogasdynamics phenomena on the flow field is analyzed. The exact solution presented in (21) becomes the same results, for the absence of magnetic field, presented by Singh et al. [5]. The behavior of the total energy is also presented by (23).

Acknowledgment

R. Singh acknowledges the financial support from the UGC, New Delhi, India, under the SRF scheme.

References

[1] R. Arora and V. D. Sharma, "Convergence of strong shock in a Van der Waals gas," *SIAM Journal on Applied Mathematics*, vol. 66, no. 5, pp. 1825–1837, 2006.

[2] A. Sakurai, "On the propagation and structure of the Blast wave, I," *Journal of the Physical Society of Japan*, vol. 8, no. 5, pp. 662–665, 1953.

[3] A. Sakurai, "On the propagation and structure of a Blast wave, II," *Journal of the Physical Society of Japan*, vol. 9, no. 2, pp. 256–266, 1954.

[4] M. H. Rogers, "Analytic solutions for the Blast-waves problem with an atmosphere of varying density," *Astrophysical Journal*, vol. 125, pp. 478–493, 1957.

[5] L. P. Singh, S. D. Ram, and D. B. Singh, "Analytical solution of the Blast wave problem in a non-ideal gas," *Chinese Physics Letters*, vol. 28, no. 11, Article ID 114303, 2011.

[6] L. P. Singh, S. D. Ram, and D. B. Singh, "Exact solution of planar and nonplanar weak shock wave problem in gasdynamics," *Chaos, Solitons & Fractals*, vol. 44, no. 11, pp. 964–967, 2011.

[7] J. L. Taylor, "An exact solution of the spherical Blast wave problem," *Philosophical Magazine*, vol. 46, pp. 317–320, 1955.

[8] L. P. Singh, S. D. Ram, and D. B. Singh, "The influence of magnetic field upon the collapse of a cylindrical shock," *Meccanica*, vol. 48, no. 4, pp. 841–850, 2013.

[9] V. V. Menon and V. D. Sharma, "Characteristic wave fronts in magnetohydrodynamics," *Journal of Mathematical Analysis and Applications*, vol. 81, no. 1, pp. 189–203, 1981.

[10] F. Oliveri and M. P. Speciale, "Exact solutions to the unsteady equations of perfect gases through Lie group analysis and substitution principles," *International Journal of Non-Linear Mechanics*, vol. 37, no. 2, pp. 257–274, 2002.

[11] S. Murata, "New exact solution of the Blast wave problem in gas dynamics," *Chaos, Solitons and Fractals*, vol. 28, no. 2, pp. 327–330, 2006.

[12] L. P. Singh, S. D. Ram, and D. B. Singh, "Quasi-similar solution of the strong shock wave problem in non-ideal gas dynamics," *Astrophysics and Space Science*, vol. 337, no. 2, pp. 597–604, 2012.

[13] R. Courant and K. O. Friedrichs, *Supersonic Flow and Shock Wave*, Wiley, New York, NY, USA, 1948.

[14] G. B. Whitham, *Linear and Non-Linear Waves*, Wiley, New York, NY, USA, 1974.

[15] C. C. Wu and P. H. Roberts, "Structure and stability of a spherical shock wave in a van der Waals gas," *Quarterly Journal of Mechanics and Applied Mathematics*, vol. 49, no. 4, pp. 501–543, 1996.

22

Thermal Jump Effects on Boundary Layer Flow of a Jeffrey Fluid Near the Stagnation Point on a Stretching/Shrinking Sheet with Variable Thermal Conductivity

M. A. A. Hamad,[1] **S. M. AbdEl-Gaied,**[1] **and W. A. Khan**[2]

[1] *Mathematics Department, Faculty of Science, Assiut University, Assiut 71516, Egypt*
[2] *Department of Engineering Sciences, PN Engineering College, National University of Science, Pakistan*

Correspondence should be addressed to S. M. AbdEl-Gaied; sagaied123@gmail.com

Academic Editor: Boming Yu

A mathematical model will be analyzed in order to study the effects of thermal jump and variable thermal conductivity on flow and heat transfer near the stagnation point on a stretching/shrinking sheet in a Jeffrey fluid. The highly nonlinear partial differential equation of Jeffrey fluid flow along with the energy equation are transformed to an ordinary system using nondimensional transformations. The arising equations are solved for temperature, velocity, shear stress, and heat flux using finite difference method. The effect of the influences parameters is discussed. For nonradiation regular viscous fluid our results are as that by Nazar et al. (2002).

1. Introduction

It is well known that the thermophysical properties of a fluid play an important role in the engineering applications in aerodynamics, geothermal systems, crude oil extractions, ground water pollution, thermal insulation, heat exchanger, storage of nuclear waste, and so forth, convective flows over bodies. The change in the thermal conductivity with temperature is an important property [1–5]. Prasad and Vajravelu [6] investigated the effect of variable thermal conductivity in a nonisothermal sheet stretching through power law fluids while Prasad et al. [4] reported similar studies for viscoelastic fluids. Abel et al. [7] studied combined effects of thermal buoyancy and variable thermal conductivity on a magnetohydrodynamic flow and the associated heat transfer in a power-law fluid past a vertical stretching sheet in the presence of a nonuniform heat source. The general findings of these studies were that the effects of variable thermal conductivity increase the shear stress. The temperature at wall increase with an increase in variable thermal conductivity by Seddeek et al. [8]. Prasad et al. [9] found that the variable thermal conductivity has an impact in enhancing the skin friction coefficient;

hence, fluids with less thermal conductivity may be opted for effective cooling. Abel et al. [10] concluded that the variable thermal conductivity increases the temperature distribution in both prescribed surface temperature and prescribed heat flux cases. Mahanti and Gaur [11] investigated the effects of linearly varying viscosity and thermal conductivity on steady free convective flow of a viscous incompressible fluid along an isothermal vertical plate in the presence of heat sink. Deissler [12] obtained that the effects of second-order terms on the velocity and temperature jumps at a wall are by a physical derivation. The analysis used the concepts of effective mean free paths for momentum and energy transfer; the effective mean free paths are obtained from known viscosities and thermal conductivities. Rahman and Eltayeb [13] studied numerically the convective slip flow of slightly rarefied fluids over a wedge with thermal jump and temperature dependent transport properties such as fluid viscosity and thermal conductivity. Cipolla Jr. [14] studied the temperature jump in polyatomic gas, also Kao [15] and Latyshev and Yushkanov [16] studied the temperature jump. The flow and heat transfer of Jeffry fluid near stagnation point on a stretching/shrinking sheet with parallel external flow was investigated by Turkyilmazoglu and Pop [17]; Akram and Nadeem [18] discussed the peristaltic motion of a two-dimensional Jeffry fluid. Authors in [19–22] studied more properties in Jeffrey fluid. Different non-Newtonian fluids were considered in studies by Pandey and Tripathi [23–25] and Tripathi [26].

Interest in boundary layer flow and heat transfer over a stretching sheet has gained considerable attention because of its application in industry and manufacturing processes. Such applications include polymer extrusion drawing of copper wires, continuous stretching of plastic films and artificial fibers, hot rolling, wire drawing, glass fiber, metal extrusion, and metal spinning. For example, Liu and Andersson [27] studied the heat transfer in a liquid film on an unsteady stretching sheet. The effects of variable fluid properties and thermocapillarity on the flow of a thin film on an unsteady stretching sheet were studied by Dandapat et al. [28]. Hayat et al. [29] investigated the peristaltic mechanism of a Jeffrey fluid in a circular tube. Nadeem et al. [30] analyzed the boundary layer flow of a Jeffrey fluid over an exponentially stretching surface. The effects of thermal radiation are carried out for two cases of heat transfer analysis known as (1) pre-scribed exponential order surface temperature (PEST) and (2) prescribed exponential order heat flux (PEHF). Hamad [31] studied the convective flow and heat transfer of an incompressible viscous nanofluid past a semi-infinite vertical stretching sheet in the presence of a magnetic field. Hamad and Pop [32] studied theoretically the steady boundary layer flow near the stagnation-point flow on a permeable stretching sheet in a porous medium saturated with a nanofluid and in the presence of internal heat generation/absorption.

The objective of the present study is to investigate the dynamics of the thermal boundary layer flow of a viscous incompressible Jeffrey fluid near the stagnation point on a stretching sheet taking into account the thermal jump condition at the surface. Thus, the main focus of the analysis is to investigate how the flow field, temperature field, shear stress, and heat flux vary within the boundary layer with thermal jump at the wall when the thermal conductivity is temperature dependent. The similarity equations are derived and solved numerically with the widely used and robust computer algebra software. Graphs and tables are presented to illustrate and discuss important hydrodynamic and thermal features of the flow.

2. Problem Formulation

Consider a steady two dimensional flow of an incompressible Jeffrey fluid near the stagnation point on a stretching/shrinking sheet. The thermal conductivity is assumed to be functions of temperature. A thermal jump condition is assumed to occur at the wall. We are considering Cartesian coordinate system in such a way that x-axis is taken along the stretching sheet in the direction of the motion and y-axis is normal to it. The plate is stretched in the x-direction with a velocity $u_w = cx$ defined at $y = 0$. The flow and heat transfer characteristics under the boundary layer approximations are governed by the following equations:

$$\frac{\partial u}{\partial x} + \frac{\partial v}{\partial y} = 0, \tag{1}$$

$$u\frac{\partial u}{\partial x} + v\frac{\partial u}{\partial y}$$
$$= u_e\frac{du_e}{dx} + \frac{\nu}{1+\gamma_1}\left[\frac{\partial^2 u}{\partial y^2}\right.$$
$$+ \gamma_2\left(u\frac{\partial^3 u}{\partial x\partial y^2} + v\frac{\partial^3 u}{\partial y^3} - \frac{\partial u}{\partial x}\frac{\partial^2 u}{\partial y^2}\right.$$
$$\left.\left. + \frac{\partial u}{\partial y}\frac{\partial^2 u}{\partial x\partial y}\right)\right], \tag{2}$$

$$\rho C_p\left(u\frac{\partial T}{\partial x} + v\frac{\partial T}{\partial y}\right) = \frac{\partial}{\partial y}\left(\kappa(T)\frac{\partial T}{\partial y}\right), \tag{3}$$

with the boundary conditions (see Rahman and Eltayeb [13]):

$$v = v_w(x), \qquad u = u_w(x) = cx,$$
$$T_{\text{jump}} = T_f - T_w$$
$$= \lambda_1\left(\frac{2}{\sigma_T} - 1\right)\frac{2\gamma}{\gamma+1}\frac{\kappa(T)}{\mu C_p}\frac{\partial T}{\partial y} \quad \text{at } y = 0, \tag{4}$$
$$u = u_e(x) = ax, \qquad \frac{\partial u}{\partial y} = 0, \qquad T = T_\infty$$
$$\text{as } y \longrightarrow \infty.$$

Here x and y are the Cartesian coordinates along the plate and normal to it, respectively, u and v are the velocity components along x- and y-axes, $v_w(x)$ is the mass transfer velocity with $v_w(x) < 0$ for suction and $v_w(x) > 0$ for injection or withdrawal, T is the fluid temperature, α is thermal diffusivity, ν is the kinematic viscosity, γ_1 is the ratio of relaxation and retardation times, γ_2 is the relaxation time, γ is the ratio of specific heats, σ_T is the thermal accommodation coefficient, λ_1 is the mean free path, μ is the dynamic viscosity, and $\kappa(T)$ is the thermal conductivity which can be, following Chiam [1], written as

$$\kappa = \kappa_\infty\left(1 + \varepsilon\frac{T - T_\infty}{T_w - T_\infty}\right), \tag{5}$$

where ε is the thermal conductivity parameter.

We introduce now the following similarity variables:

$$\psi = \sqrt{a\nu x}f(\eta), \qquad \theta(\eta) = \frac{(T - T_\infty)}{(T_w - T_\infty)},$$
$$\eta = \sqrt{\frac{a}{\nu}}y, \tag{6}$$

where ψ is the stream function which is defined in the usual way as $u = \partial\psi/\partial y$ and $v = -\partial\psi/\partial x$. Thus, $v_w(x) = -\sqrt{a\nu}s$, where s is the mass transfer parameter with $s > 0$ for suction and $s < 0$ for injection, respectively. Substituting (5) and

(6) into (2) and (3), the following set of ordinary differential equations results in

$$f''' + (1 + \gamma_1)\left(ff'' - f'^2\right)$$
$$+ \beta\left(f''^2 - ff''''\right) + (1 + \gamma_1) = 0, \qquad (7)$$

$$(1 + \varepsilon\,\theta)\,\theta'' + \varepsilon\theta'^2 + \Pr f\theta' = 0,$$

and the boundary conditions (4) become

$$f(0) = s, \qquad f'(0) = \lambda,$$

$$\theta(0) = 1 + \frac{2\gamma T_s}{\gamma + 1}\Pr\left(1 + \varepsilon\theta(0)\right)\theta'(0), \qquad (8)$$

$$f'(\infty) = 1, \qquad f''(\infty) = 0, \qquad \theta(\infty) = 0,$$

where $\Pr = \mu C_p/\kappa_\infty$ is the Prandtl number, $\lambda = c/a$ is the stretching ($\lambda > 0$) or shrinking ($\lambda < 0$) parameter, $\beta = c\gamma_2$ is the Deborah number, $T_s = \lambda_1(2/(\sigma_T - 1))\sqrt{a/\nu}$ is the slip parameter, and primes denote differentiation with respect to η.

2.1. Particular Case.

It is worth mentioning that for a regular viscous fluid ($\beta = \gamma_1 = 0$), (7) reduce to the steady state equations from the paper by Nazar et al. [33] when we neglect the radiation effect.

2.2. Physical Quantities.

The physical quantities of interest are the skin friction coefficient C_f and the local Nusselt number Nu_x, which are defined as

$$C_f = \frac{\tau_w}{\rho u_e^2(x)}, \qquad \text{Nu}_x = \frac{x q_w}{\kappa\left(T_w - T_\infty\right)}, \qquad (9)$$

where τ_w is the skin friction or shear stress along the stretching surface and q_w is the heat flux from the surface, which are given by

$$\tau_w = \mu\left(\frac{\partial u}{\partial y}\right)_{y=0}, \qquad q_w = -\kappa\left(\frac{\partial T_{\text{jump}}}{\partial y}\right)_{y=0}. \qquad (10)$$

Using (6), we get

$$\text{Re}_x^{1/2}C_f = f''(0),$$

$$\text{Re}_x^{-1/2}\text{Nu}_x \qquad\qquad\qquad (11)$$
$$= -\frac{2\gamma}{\gamma + 1}T_s\Pr\left[(1 + \varepsilon\theta)\,\theta''(0) + \varepsilon\left(\theta'(0)\right)^2\right],$$

where $\text{Re}_x = u_e(x)x/\nu$ is the local Reynolds number.

3. Results and Discussion

The transformed equations (7) with boundary conditions (8) are solved numerically by using a finite difference method. The asymptotic boundary conditions at $\eta = \infty$ are replaced by $\eta = 6$. In Table 1, we have shown the variation of wall temperature and heat transfer rates with the Prandtl numbers for three different values of Deborah numbers. It is observed

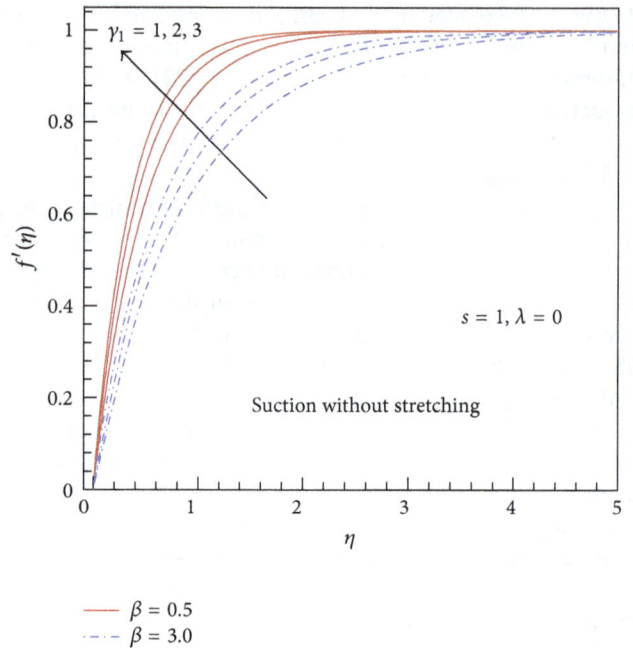

FIGURE 1: Effect of β and γ_1 on the velocity profiles.

that the wall temperature decreases whereas the heat transfer rates increase with an increase in Prandtl number. However, the wall temperature increases slightly and heat transfer values decrease slightly with an increase in Deborah number, whereas both decrease with an increase in the ratio of specific heats. This is shown in Tables 1(a) and 1(b). Tables 1(c) and 1(d) show the effects of slip temperature on the wall temperature and heat transfer rates for the constant thermal conductivity. Table 1(e) shows the effects of thermal slip on the wall temperature and heat transfer rates when the thermal conductivity varies with temperature. It can be seen that both the wall temperature and heat transfer rates decrease with an increase in the Prandtl number due to decrease in the thermal conductivity. The effects of the ratio of relaxation and retardation times, Deborah number, and suction and stretching parameters on the x-component of velocity are shown in Figures 1 and 2. Figure 1 shows the effects in the absence of stretching. It shows that the velocity increases as the ratio of relaxation and retardation times increases but decreases with an increase in Deborah number. The velocity boundary layer converges quickly for small Deborah numbers. In fact, small Deborah numbers correspond to situations where the material has time to relax (and behaves in a viscous manner), while high Deborah numbers correspond to situations where the material behaves rather elastically. Figure 2 shows the effects of stretching parameter on the velocity for different values of Deborah number. It is observed that the velocity becomes constant when $\lambda = 0$, increases when $\lambda < 1$, and decreases when $\lambda > 0$. Accordingly, the velocity decreases or increases with Deborah number when $\lambda < 1$ or $\lambda > 0$.

The effect of thermal conductivity parameter on temperature profiles is shown in Figure 3 for two different Prandtl numbers. It is observed that the thermal boundary layer

TABLE 1: Wall temperature and heat transfer values when one has the following.

(a) $s = 1, \lambda = 0.5, \gamma_1 = 1, \varepsilon = 0, T_s = 0.1, \gamma = 0.5$

Pr	$\beta = 0.5$		$\beta = 1.0$		$\beta = 1.5$	
	$\theta(0)$	$-\theta'(0)$	$\theta(0)$	$-\theta'(0)$	$\theta(0)$	$-\theta'(0)$
1	0.9132	1.3014	0.9139	1.2922	0.9142	1.2865
2	0.7516	1.8627	0.7526	1.8553	0.7532	1.8509
3	0.5883	2.0586	0.5892	2.0540	0.5898	2.0512
4	0.4540	2.0474	0.4547	2.0447	0.4552	2.0431
5	0.3523	1.9430	0.3529	1.9415	0.3532	1.9405
6	0.2772	1.8069	0.2776	1.8060	0.2778	1.8055
7	0.2218	1.6676	0.2221	1.6670	0.2222	1.6667
8	0.1804	1.5368	0.1806	1.5364	0.1807	1.5362
9	0.1490	1.4184	0.1491	1.4181	0.1492	1.4180
10	0.1248	1.3128	0.1249	1.3127	0.1250	1.3126

(b) $s = 1, \lambda = 0.5, \gamma_1 = 1, \varepsilon = 0, T_s = 0.1, \gamma = 1.0$

Pr	$\beta = 0.5$		$\beta = 1.0$		$\beta = 1.5$	
	$\theta(0)$	$-\theta'(0)$	$\theta(0)$	$-\theta'(0)$	$\theta(0)$	$-\theta'(0)$
1	0.8753	1.2473	0.8761	1.2389	0.8766	1.2336
2	0.6686	1.6570	0.6698	1.6511	0.6705	1.6476
3	0.4878	1.7072	0.4888	1.7040	0.4894	1.7021
4	0.3567	1.6084	0.3573	1.6067	0.3577	1.6057
5	0.2661	1.4677	0.2666	1.4668	0.2669	1.4663
6	0.2036	1.3273	0.2039	1.3268	0.2041	1.3265
7	0.1597	1.2005	0.1599	1.2002	0.1600	1.2000
8	0.1279	1.0901	0.1281	1.0899	0.1282	1.0898
9	0.1045	0.9950	0.1046	0.9949	0.1047	0.9948
10	0.0868	0.9132	0.0869	0.9131	0.0869	0.9131

(c) $s = 1, \lambda = 0.5, \gamma_1 = 1, \varepsilon = 0, T_s = 0.5, \gamma = 0.5$

Pr	$\beta = 0.5$		$\beta = 1.0$		$\beta = 1.5$	
	$\theta(0)$	$-\theta'(0)$	$\theta(0)$	$-\theta'(0)$	$\theta(0)$	$-\theta'(0)$
1	0.6780	0.9661	0.6797	0.9611	0.6807	0.9579
2	0.3771	0.9344	0.3783	0.9326	0.3791	0.9314
3	0.2223	0.7778	0.2229	0.7771	0.2233	0.7767
4	0.1426	0.6431	0.1430	0.6428	0.1432	0.6426
5	0.0981	0.5411	0.0983	0.5410	0.0984	0.5409
6	0.0713	0.4644	0.0714	0.4643	0.0714	0.4643
7	0.0539	0.4055	0.0540	0.4054	0.0541	0.4054
8	0.0422	0.3592	0.0422	0.3592	0.0422	0.3592
9	0.0338	0.3221	0.0339	0.3220	0.0339	0.3220
10	0.0277	0.2917	0.0278	0.2917	0.0278	0.2917

(d) $s = 1, \lambda = 0.5, \gamma_1 = 1, \varepsilon = 0, T_s = 0.5, \gamma = 1$

Pr	$\beta = 0.5$		$\beta = 1.0$		$\beta = 1.5$	
	$\theta(0)$	$-\theta'(0)$	$\theta(0)$	$-\theta'(0)$	$\theta(0)$	$-\theta'(0)$
1	0.5839	0.8321	0.5858	0.8284	0.5870	0.8260
2	0.2875	0.7125	0.2886	0.7114	0.2893	0.7108
3	0.1600	0.5600	0.1605	0.5596	0.1609	0.5594
4	0.0998	0.4501	0.1001	0.4500	0.1002	0.4499
5	0.0676	0.3730	0.0678	0.3729	0.0679	0.3729
6	0.0487	0.3171	0.0487	0.3171	0.0488	0.3171

(d) Continued.

Pr	$\beta = 0.5$		$\beta = 1.0$		$\beta = 1.5$	
	$\theta(0)$	$-\theta'(0)$	$\theta(0)$	$-\theta'(0)$	$\theta(0)$	$-\theta'(0)$
7	0.0366	0.2753	0.0367	0.2752	0.0367	0.2752
8	0.0285	0.2429	0.0285	0.2429	0.0286	0.2429
9	0.0228	0.2172	0.0228	0.2172	0.0229	0.2171
10	0.0187	0.1963	0.0187	0.1963	0.0187	0.1963

(e) $s = 1, \lambda = 0.5, \gamma_1 = 1, \varepsilon = -0.5, T_s = 0.5, \gamma = 0.5$

Pr	$\beta = 0.5$		$\beta = 1.0$		$\beta = 1.5$	
	$\theta(0)$	$-\theta'(0)$	$\theta(0)$	$-\theta'(0)$	$\theta(0)$	$-\theta'(0)$
1	0.6884	1.4256	0.6899	1.4204	0.6908	0.6884
2	0.3812	1.1468	0.3824	1.1455	0.3831	0.3812
3	0.2236	0.8741	0.2242	0.8737	0.2246	0.2236
4	0.1431	0.6922	0.1434	0.6921	0.1436	0.1431
5	0.0983	0.5690	0.0985	0.5689	0.0986	0.0983
6	0.0713	0.4815	0.0715	0.4815	0.0715	0.0713
7	0.0540	0.4167	0.0541	0.4167	0.0541	0.0540
8	0.0422	0.3669	0.0422	0.3669	0.0423	0.0422
9	0.0338	0.3276	0.0339	0.3276	0.0339	0.0338
10	0.0277	0.2958	0.0278	0.2958	0.0278	0.0277

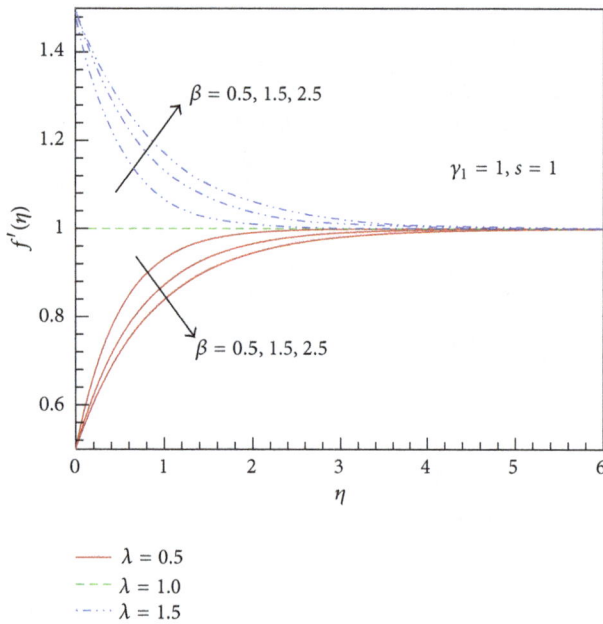

FIGURE 2: Effect of β and λ on the velocity profiles.

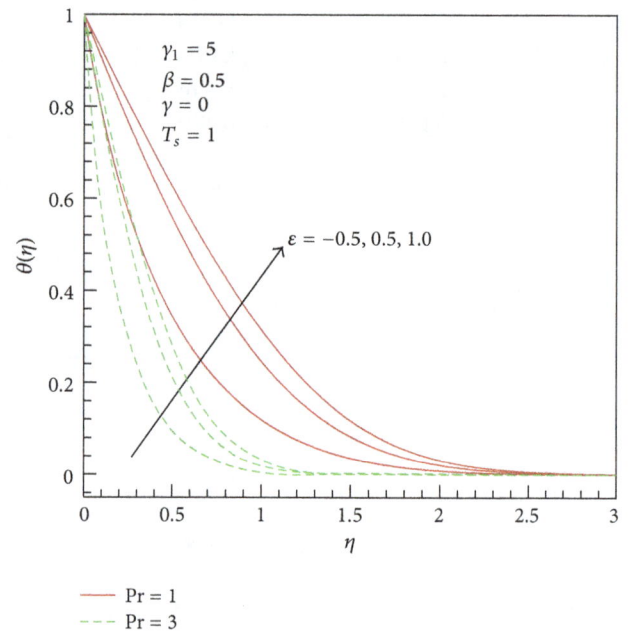

FIGURE 3: Effect of Pr and ε on the temperature profiles.

thickness decreases with an increase in Prandtl number. As the thermal conductivity parameter increases, the temperature in the thermal boundary layer increases. The variaton of skin friction with the ratio of relaxation and retardation times for different parameters is shown in Figures 4(a) and 4(b). When there is no stretching, the skin friction increases with the ratio of relaxation and retardation times and decreases with an increase in Deborah number. As expected, the skin friction reduces with an increase in the suction parameter in both cases. Comparing Figures 4(a) and 4(b), it can be

seen that the skin friction decreases with an increase in the stretching parameter. The variation in heat transfer rates with the ratio of relaxation and retardation times is shown in Figures 5 and 6 for different values of suction and thermal conductivity parameters and Prandtl and Deborah numbers. The other parameters are kept constant. As the ratio of relaxation and retardation times increases, the heat transfer rate increases. For higher values of the suction parameter, the heat transfer rates are found to be higher. It is also observed that the heat transfer rates decrease with an increase in

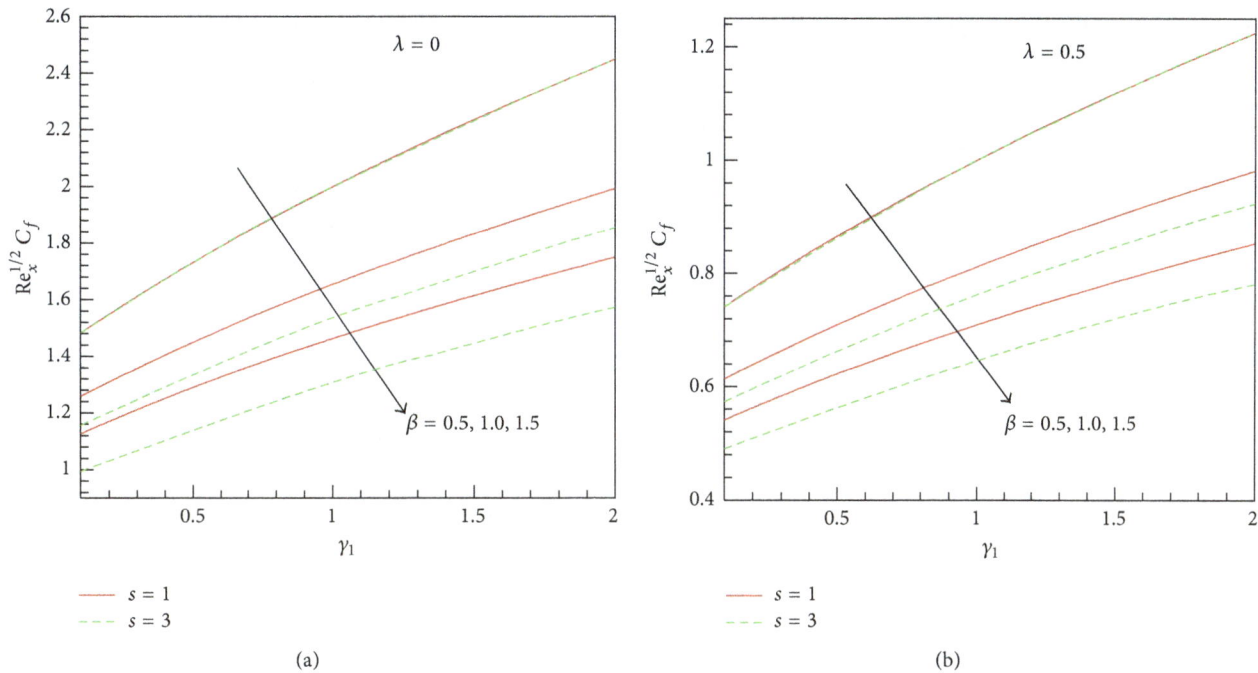

FIGURE 4: Effect of β and λ on the skin friction profiles.

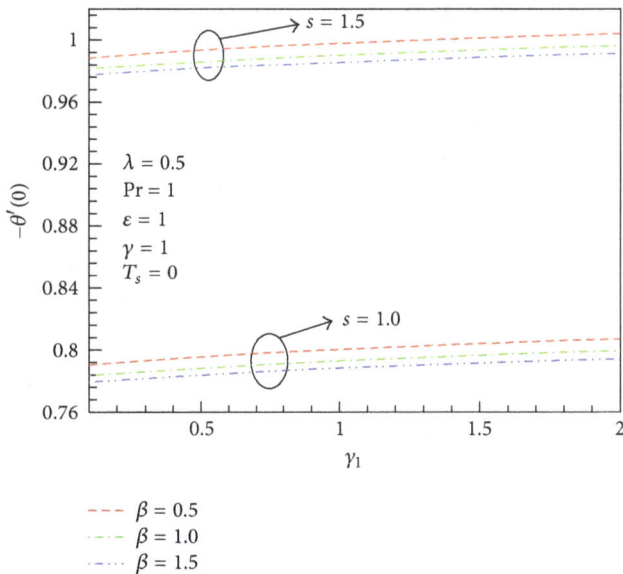

FIGURE 5: Effect of β and s on the heat transfer rate.

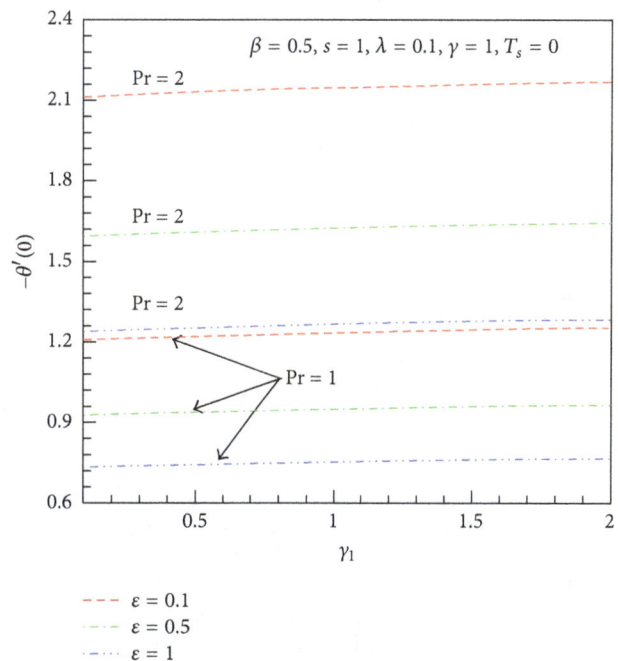

FIGURE 6: Effect of Pr and ε on the heat transfer rate.

Deborah number. This is shown in Figure 5. As evident from Figure 6, the heat transfer rates increase with an increase in the Prandtl number. Figure 6 also shows that the heat transfer rates decrease with an increase in the thermal conductivity parameter.

4. Conclusions

The effects of thermal jump and variable thermal conductivity on flow and heat transfer near the stagnation point on a stretching/shrinking sheet are investigated numerically in a Jeffrey fluid. The effects of governing parameters including ratio of relaxation and retardation times γ_1, Deborah number β, Prandtl number, stretching parameter λ, suction parameter s, and thermal conductivity parameter ε on the dimensionless velocity, temperature, skin friction, and heat transfer rates are investigated and are presented graphically and in tabular form. We conclude the following.

(a) The wall temperature increases slightly while heat transfer values decrease slightly with an increase in Deborah number.

(b) The wall temperature and heat transfer decrease with an increase in the ratio of specific heats.

(c) The decreases of Deborah number and the increases of relaxation and retardation times leads to increases in the velocity.

(d) The skin friction decreases with an increase in the stretching parameter.

(e) As the ratio of relaxation and retardation times increases, the heat transfer rate increases.

(f) For higher values of the suction parameter, the heat transfer rates are found to be higher.

(g) The heat transfer rates decrease with an increase in Deborah number.

References

[1] T. C. Chiam, "Heat transfer with variable conductivity in a stagnation-point flow towards a stretching sheet," *International Communications in Heat and Mass Transfer*, vol. 23, no. 2, pp. 239–248, 1996.

[2] T. C. Chiam, "Heat transfer in a fluid with variable thermal conductivity over a linearly stretching sheet," *Acta Mechanica*, vol. 129, no. 1-2, pp. 63–72, 1998.

[3] P. S. Datti, K. V. Prasad, M. S. Abel, and A. Joshi, "MHD viscoelastic fluid flow over a non-isothermal stretching sheet," *International Journal of Engineering Science*, vol. 42, no. 8-9, pp. 935–946, 2004.

[4] K. V. Prasad, M. S. Abel, and S. K. Khan, "Momentum and heat transfer in visco-elastic fluid flow in a porous medium over a non-isothermal stretching sheet," *International Journal of Numerical Methods for Heat and Fluid Flow*, vol. 10, no. 8, pp. 786–801, 2000.

[5] M. S. Abel, K. V. Prasad, and A. Mahaboob, "Buoyancy force and thermal radiation effects in MHD boundary layer visco-elastic fluid flow over continuously moving stretching surface," *International Journal of Thermal Sciences*, vol. 44, no. 5, pp. 465–476, 2005.

[6] K. V. Prasad and K. Vajravelu, "Heat transfer in the MHD flow of a power law fluid over a non-isothermal stretching sheet," *International Journal of Heat and Mass Transfer*, vol. 52, no. 21-22, pp. 4956–4965, 2009.

[7] M. S. Abel, P. G. Siddheshwar, and N. Mahesha, "Effects of thermal buoyancy and variable thermal conductivity on the MHD flow and heat transfer in a power-law fluid past a vertical stretching sheet in the presence of a non-uniform heat source," *International Journal of Non-Linear Mechanics*, vol. 44, no. 1, pp. 1–12, 2009.

[8] M. A. Seddeek, S. N. Odda, and M. S. Abdelmeguid, "Numerical study for the effects of thermophoresis and variable thermal conductivity on heat and mass transfer over an accelerating surface with heat source," *Computational Materials Science*, vol. 47, no. 1, pp. 93–98, 2009.

[9] K. V. Prasad, D. Pal, V. Umesh, and N. S. P. Rao, "The effect of variable viscosity on MHD viscoelastic fluid flow and heat transfer over a stretching sheet," *Communications in Nonlinear Science and Numerical Simulation*, vol. 15, no. 2, pp. 331–344, 2010.

[10] M. S. Abel, P. G. Siddheshwar, and N. Mahesha, "Numerical solution of the momentum and heat transfer equations for a hydromagnetic flow due to a stretching sheet of a non-uniform property micropolar liquid," *Applied Mathematics and Computation*, vol. 217, no. 12, pp. 5895–5909, 2011.

[11] N. C. Mahanti and P. Gaur, "Effects of varying viscosity and thermal conductivity on steady free convective flow and heat transfer along an isothermal vertical plate in the presence of heat sink," *Journal of Applied Fluid Mechanics*, vol. 2, no. 1, pp. 23–28, 2009.

[12] R. G. Deissler, "An analysis of second-order slip flow and temperature-jump boundary conditions for rarefied gases," *International Journal of Heat and Mass Transfer*, vol. 7, no. 6, pp. 681–694, 1964.

[13] M. M. Rahman and I. A. Eltayeb, "Convective slip flow of rarefied fluids over a wedge with thermal jump and variable transport properties," *International Journal of Thermal Sciences*, vol. 50, no. 4, pp. 468–479, 2011.

[14] J. W. Cipolla Jr., "Heat transfer and temperature jump in a polyatomic gas," *International Journal of Heat and Mass Transfer*, vol. 14, no. 10, pp. 1599–1610, 1971.

[15] T.-T. Kao, "Laminar free convective heat transfer response along a vertical flat plate with step jump in surface temperature," *Letters in Heat and Mass Transfer*, vol. 2, no. 5, pp. 419–428, 1975.

[16] A. V. Latyshev and A. A. Yushkanov, "An analytic solution of the problem of the temperature jumps and vapour density over a surface when there is a temperature gradient," *Journal of Applied Mathematics and Mechanics*, vol. 58, no. 2, pp. 259–265, 1994.

[17] M. Turkyilmazoglu and I. Pop, "Exact analytical solution for the flow and heat transfer near the stagnation point on a stretching/shrinking sheet in a Jeffrey fluid," *International Journal of Heat and Mass Transfer*, vol. 57, no. 1, pp. 82–88, 2013.

[18] S. Akram and S. Nadeem, "Influence of induced magnetic field and heat transfer on the peristaltic motion of Jeffrey fluid in an asymmetric channel: closed form solutions," *Journal of Magnetism and Magnetic Materials*, vol. 328, pp. 11–20, 2013.

[19] C. E. Siewert and D. Valougeorgis, "The temperature-jump problem for a mixture of two gases," *Journal of Quantitative Spectroscopy and Radiative Transfer*, vol. 70, no. 3, pp. 307–319, 2001.

[20] S. Nadeem, A. Hussain, and M. Khan, "Stagnation flow of a Jeffrey fluid over a shrinking sheet," *Zeitschrift fur Naturforschung A*, vol. 65, no. 6-7, pp. 540–548, 2010.

[21] S. K. Pandey and D. Tripathi, "Unsteady model of transportation of Jeffrey-fluid by peristalsis," *International Journal of Biomathematics*, vol. 3, no. 4, pp. 473–491, 2010.

[22] T. Hayat, M. Awais, S. Asghar, and A. A. Hendi, "Analytic solution for the magnetohydrodynamic rotating flow of Jeffrey fluid in a channel," *Journal of Fluids Engineering*, vol. 133, no. 6, Article ID 061201, 2011.

[23] S. K. Pandey and D. Tripathi, "Influence of magnetic field on the peristaltic flow of a viscous fluid through a finite-length cylindrical tube," *Applied Bionics and Biomechanics*, vol. 7, no. 3, pp. 169–176, 2010.

[24] S. K. Pandey and D. Tripathi, "Effects of non-integral number of peristaltic waves transporting couple stress fluids in finite length channels," *Zeitschrift fur Naturforschung A*, vol. 66, no. 3-4, pp. 172–180, 2011.

[25] S. K. Pandey and D. Tripathi, "Unsteady peristaltic flow of micro-polar fluid in a finite channel," *Zeitschrift fur Naturforschung A*, vol. 66, no. 3-4, pp. 181–192, 2011.

[26] D. Tripathi, "A mathematical model for the peristaltic flow of chyme movement in small intestine," *Mathematical Biosciences*, vol. 233, no. 2, pp. 90–97, 2011.

[27] I.-C. Liu and H. I. Andersson, "Heat transfer in a liquid film on an unsteady stretching sheet," *International Journal of Thermal Sciences*, vol. 47, no. 6, pp. 766–772, 2008.

[28] B. S. Dandapat, B. Santra, and K. Vajravelu, "The effects of variable fluid properties and thermocapillarity on the flow of a thin film on an unsteady stretching sheet," *International Journal of Heat and Mass Transfer*, vol. 50, no. 5-6, pp. 991–996, 2007.

[29] T. Hayat, N. Ali, and S. Asghar, "An analysis of peristaltic transport for flow of a Jeffrey fluid," *Acta Mechanica*, vol. 193, no. 1-2, pp. 101–112, 2007.

[30] S. Nadeem, S. Zaheer, and T. Fang, "Effects of thermal radiation on the boundary layer flow of a Jeffrey fluid over an exponentially stretching surface," *Numerical Algorithms*, vol. 57, no. 2, pp. 187–205, 2011.

[31] M. A. A. Hamad, "Analytical solution of natural convection flow of a nanofluid over a linearly stretching sheet in the presence of magnetic field," *International Communications in Heat and Mass Transfer*, vol. 38, no. 4, pp. 487–492, 2011.

[32] M. A. A. Hamad and I. Pop, "Scaling Transformations for Boundary Layer Flow near the Stagnation-Point on a Heated Permeable Stretching Surface in a Porous Medium Saturated with a Nanofluid and Heat Generation/Absorption Effects," *Transport in Porous Media*, vol. 87, no. 1, pp. 25–39, 2011.

[33] R. Nazar, N. Amin, D. Filip, and I. Pop, "Unsteady boundary layer flow in the region of the stagnation point on a stretching sheet," *International Journal of Engineering Science*, vol. 42, no. 11-12, pp. 1241–1253, 2004.

23

Slip-Flow and Heat Transfer in a Porous Microchannel Saturated with Power-Law Fluid

Yazan Taamneh[1] and Reyad Omari[2]

[1] Department of Mechanical Engineering, Tafila Technical University, P.O. Box 179, Tafila 66110, Jordan
[2] Department of Mathematics, Al-Balqa Applied University, Irbid University College, P.O. Box 19117, Irbid 19110, Jordan

Correspondence should be addressed to Yazan Taamneh; taamneh@daad-alumni.de

Academic Editor: Ciprian Iliescu

This study aims to numerically examine the fluid flow and heat transfer in a porous microchannel saturated with power-law fluid. The governing momentum and energy equations are solved by using the finite difference technique. The present study focuses on the slip flow regime, and the flow in porous media is modeled using the modified Darcy-Brinkman-Forchheimer model for power-law fluids. Parametric studies are conducted to examine the effects of Knudsen number, Darcy number, power law index, and inertia parameter. Results are given in terms of skin friction and Nusselt number. It is found that when the Knudsen number and the power law index decrease, the skin friction on the walls decreases. This effect is reduced slowly while the Darcy number decreases until it reaches the Darcy regime. Consequently, with a very low permeability the effect of power law index vanishes. The numerical results indicated also that when the power law index decreases the fully-developed Nusselt number increases considerably especially, in the limit of high permeability, that is, nonDarcy regime. As far as Darcy regime is concerned the effects of the Knudsen number and the power law index of the fully-developed Nusselt number is very little.

1. Introduction

Fluid flow and heat transfer in porous media has been a subject of continuous interest during past decades because of the wide range of engineering applications. In addition to conventional applications including solar receivers, building thermal insulation materials, packed bed heat exchangers, and energy storage units, new applications in the emerging field of microscale heat transfer have existed. However, microchannels are now used in several industries and equipment such as cooling of electronic package, microchannel heat sinks, microchannel heat exchanger, microchannel fabrication, and cooling, and heating of different devices [1–5].

One of the major difficulties in trying to predict the gaseous transport in micron sized devices can be attributed to the fact that the continuum flow assumption implemented in the Navier-Stokes equations breaks down when the mean free path of the molecules (λ) is comparable to the characteristic dimension of the flow domain. Under these conditions, the momentum and heat transfer start to be affected by the discrete molecular composition of the gas and a variety of noncontinuum or rarefaction effects are likely to be exhibited such as velocity slip and temperature jump at the gas-solid interface. Velocity profiles, fluid flow rate, boundary wall shear stresses, temperature profiles, heat transfer rates, and Nusselt number are all influenced by the noncontinuum regime.

However, there is a certain limit of the channel size with which one can still apply Navier-Stokes equations with some modifications on the boundary conditions [6]. This is the case when Knudsen number (Kn) is in the range $0.001 \leq \text{Kn} \leq 0.1$, and the flow under such condition is called slip-flow. Knudsen number is defined as the ratio of the molecular mean free path to the characteristic length of the system. It is also used to measure of the degree of rarefaction of gases encountered in flows through narrow passages, and also to measure the degree of the validity of the continuum model.

The continuum model is valid for very small Knudsen number flows ($\text{Kn} < 10^{-3}$). While the Knudsen number increases, the rarefaction effects become more pronounced, and eventually the continuum assumption breaks down. Therefore, several researchers have suggested that the well-accepted no-slip boundary condition may not be suitable for flows at the micro- and nanoscale [7–10]. Recently, lots of mechanisms have been proposed to explain this phenomenon. In fact, Arkilic et al. [6], Beskok and Karniadaksi [11], and Sparrow and Lin [12], have found that the Navier-Stokes equations, when combined with velocity-slip boundary conditions, yield results for pressure drop and friction factor that are in agreement with experimental data for some microchannel flows.

The appropriate flow and heat transfer models for a given gas flow problem depend on the range of Knudsen number. A classification of the different gas flow regimes is given as follows: $Kn < 10^{-3}$ for the continuum flow, $10^{-3} < Kn < 10^{-1}$ for the slip flow, $10^{-1} < Kn < 10^{+1}$ for the transition flow, and $10^{+1} < Kn$ for the free molecular flow. In this study, the slip flow regime ($10^{-3} < Kn < 10^{-1}$) is considered and the modified extended Darcy-Brinkman-Forchheimer model for power-law fluid is employed to describe the flow behavior in porous medium.

Convection heat transfer in circular and noncircular microchannels has been solved over the years [13–15]. In these studies, the effects of velocity slip and temperature jump at the wall and viscous dissipation were considered. The main consequence was that the velocity slip and temperature jump have opposite effects on heat transfer. Although the velocity slip tends to increase the Nusselt number, the temperature jump tends to reduce it. The inclusion of the viscous heating increases the Nusselt number for the fluid being cooled and decreases it for the fluid being heated.

Laminar forced convection of Newtonian fluid flow in microchannels filled with a porous medium has been solved by numerical and analytical means over the years [16–19]. In these studies, the effect of Knudsen number, Darcy number, Forchheimer number, and Reynolds number on the velocity slip and temperature jump at the wall were considered. The main results were that the skin friction had been increased by (i) decreasing the Knudsen number, (ii) increasing the Darcy number, and (iii) decreasing the Forchheimer number. Heat transfer was found to be (i) decreased as the Knudsen and Forchheimer numbers increase and (ii) increased as the Reynolds and Darcy numbers increase.

A theoretical and numerical analysis of the fully-developed forced convection in a porous channel saturated with a power-law fluid in porous channel has been investigated recently [20–24]. A closed form boundary layer solutions using the integral method was obtained for velocity profiles, temperature fields, and fully-developed Nusselt number. The theoretical solutions can be used to predict primary characteristics of physical phenomena associated with forced convection of nonNewtonian fluids in porous media. These solutions were convenient to serve as a benchmark for more complicated numerical solutions. The results indicated that the nonDarcy regime, the effects of power law index on hydrodynamics and heat transfer behavior in the porous channel are significant, whereas in the Darcy regime the effects of Darcy number were predominant. Researchers have

also found that, in the nonDarcy regime, the Nusselt number increases and the pressure drop decreases as the power law index decreases. Consequently, the combined use of a highly permeable porous matrix with shear thinning fluid appeared to be promising as a heat transfer augmentation technique.

However, little of information on the related literature regarding the flow and heat transfer of power law fluids through porous microchannels. That said, in this study, the forced convection of heat and fluid flow of power law fluids through parallel plate microchannels filled with porous media were considered. The aim of the present study is to investigate the effects of Knudsen number, Darcy number and the inertia parameter on the hydrodynamic and thermal behavior of a power law fluid flow between infinitely long parallel-plates microchannels filled with porous media.

2. Mathematical Formulation

The analysis is carried out for unsteady state, incompressible and laminar forced convection flow between parallel-plates microchannel filled with porous medium and heated with uniform wall temperature at the walls. The flow is assumed to be hydrodynamically fully-developed. The porous medium is saturated with a single phase nonNewtonian fluid described by the power law model and assumed to be in local thermodynamic equilibrium with the fluid.

As a result of the continuity equation, the flow is a unidirectional and is expressed in terms of the axial velocity u alone. That is, the velocity component in the y-direction vanishes and the velocity component in the x-direction (denoted by u) becomes dependent on y. In addition, the flow is assumed to be thermally developing under constant pressure gradient driving force; therefore, the temperature becomes the function of (x, y) only. The present study focuses on the slip flow regime ($10^{-3} \le Kn \le 10^{-1}$), and therefore the Navier-Stokes equations and energy equation combined with slip/jump boundary conditions have been applied. The physical properties of the solid matrix and of the fluid are assumed to be constant except for the viscosity of the power-law fluid which depends on the shear rate. In the present study, a fibrous or foam-metal material is considered such that the porosity and permeability are assumed to be constant even close to the walls. On the other hand, it is assumed that the porous medium is isotropic and homogeneous. Finally, viscous dissipation is neglected in the energy equation. The governing equations can be written as follows [22].

Continuity is

$$\frac{\partial u}{\partial x} = 0. \tag{1}$$

Momentum is

$$\frac{\rho_f}{\varepsilon^n}\frac{\partial u}{\partial t} = -\frac{\partial p}{\partial x} + \frac{\mu^*}{\varepsilon^n}\frac{d}{dy}\left[\left|\frac{du}{dy}\right|^{n-1}\frac{du}{dy}\right] - \frac{\mu^*}{K^*}u^n - \frac{\rho C_F}{\sqrt{K}}u^2. \tag{2}$$

Energy is

$$u\frac{\partial T}{\partial x} = \frac{k_e}{\rho_f c_f}\frac{\partial^2 T}{\partial y^2}. \tag{3}$$

In the momentum equation, a modified Darcy's law for power-law fluids was used where μ^* represents the consistency of the power-law fluid and K^* is the modified permeability which depends on the structure of the porous medium and on the power law index of the fluid [23]. The linear approximation of Darcy's law can directly be derived from the fluid macroscopic, Navier-Stokes, and momentums balance equation [25].

In the energy equation, the thermal dispersion conductivity of the porous media is assumed to be constant and is incorporated into the effective thermal conductivity. The axial heat conduction effects are usually negligible for nearly parallel flows. Momentum and energy transfer between the liquid molecules and the surface requires specification of interactions between the impinging molecules and the surface. From the macroscopic point of view, it is sufficient to know some average parameters in terms of the so-called tangential momentum (σ_v) and thermal (σ_T) accommodation coefficients. These coefficients describe the gas-surface interaction and are functions of the composition and temperature of the gas, the gas velocity over the surface, and the solid surface temperature, chemical state, and roughness. The accommodation coefficients take any value between 0 and 1, where these values represent specular reflection and diffuse reflection, respectively. For most engineering applications, values of the accommodation coefficients are near unity [9].

Under the above assumptions and by using the nondimensional variables listed in the nomenclature, the equations of motion and energy equation are reduced to the following form:

$$\frac{\partial U}{\partial X} = 0,$$

$$\frac{1}{\varepsilon^n}\frac{\partial U}{\partial \tau} = -\frac{dP}{dX} + \frac{1}{\text{Re}\,\varepsilon^n}\frac{d}{dY}\left[\left|\frac{dU}{dY}\right|^{n-1}\frac{dU}{dY}\right]$$
$$- \frac{1}{\text{Re}\,\text{Da}^{(1+n)/2}}U^n - \Gamma U^2, \tag{4}$$

$$U\frac{\partial \theta}{\partial X} = \frac{1}{\text{Re}\,\text{Pr}}\frac{\partial^2 \theta}{\partial Y^2},$$

where $\Gamma = C_F^*/\sqrt{\text{Da}}$ is the inertia parameter and $C_F^* = C_F(K^*)^{1/(1+n)}/K^{1/2}$ is the modified inertia parameter. Also, the boundary conditions, in nondimensional form, are as follows:

$$\frac{\partial U}{\partial Y}\left(\frac{1}{2}\right) = \frac{\partial \theta}{\partial Y}\left(X, \frac{1}{2}\right) = 0,$$

$$U(1) = -\frac{2-\sigma_v}{\sigma_v}\text{Kn}\left(\frac{\partial U}{\partial Y}\right)_w,$$

$$\theta(X, 1) = 1 - \frac{2-\sigma_T}{\sigma_T}\frac{2\gamma}{\gamma+1}\frac{\text{Kn}}{\text{Pr}}\frac{\partial \theta}{\partial Y}(X, 1),$$

$$\theta(0, Y) = 0. \tag{5}$$

The quantities of primary interest in this study are the friction factor and Nusselt number. These are defined as follows:

$$C_f = \frac{\tau_w}{(1/2)\,\rho_f u_m^2} = \frac{2}{\text{Re}}\left|\frac{dU}{dY}\right|^{n-1}\frac{dU}{dY},$$

$$\text{Nu} = \frac{hD_h}{k_f} = \frac{1}{1-\theta_m}\frac{\partial \theta(1, X)}{\partial Y}\bigg|_w, \tag{6}$$

where θ_m is the dimensionless mean temperature, D_h is the hydraulic diameter, h is the local heat transfer coefficient and k_f is thermal conductivity of the fluid.

3. Numerical Method

The governing equations are solved numerically using the finite difference technique. The governing momentum and energy equations are not coupled, consequently the numerical solution proceeds by first solving the velocity distribution from the momentum equation, and then solving the energy equation for the temperature distribution.

The momentum equation (2) is parabolic partial differential equation (unsteady, one-dimensional diffusion equation) if the time term is left in the equation. One way to solve this type of equations is by using the well-known Backward-Time Central-Space (BTCS) method [26]. This particular method is also called the fully implicit method. The finite difference equation (FDE) which approximates the partial differential equation is obtained by replacing the exact partial derivative $\partial U/\partial \tau$ by the backward-time approximation, while the exact partial derivative $\partial^2 U/\partial Y^2$ is replaced by the centered-space approximation. The steady state solution is obtained by marching in time until no further significant change in the solution is obtained with additional marching steps. The present numerical method is advantageous over other available numerical methods in that it is unconditionally stable.

In a similar manner, the energy equation is discretized using the same numerical scheme. In contrast to the momentum equation, it should be mentioned that the steady state form of the energy equation is discretized directly and therefore a time dependent solution was not obtained. When the momentum equation is discretized and applied at every point in the finite difference grid, a system of nonlinear finite difference (algebraic) equations was obtained. To overcome the difficulties in solving such a system, the nonlinear term (last term) in (2) should be linearized. To do so, the well-known time lagging technique is used [26].

Based on the above approach, the resulting systems of algebraic equations obtained by discretizing the momentum and energy equations are tri-diagonal, which are best solved by using Thomas algorithm. The adequacy of the grid is

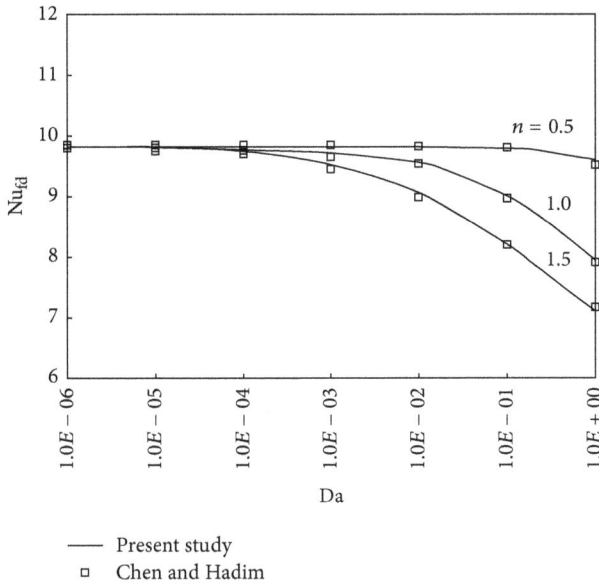

FIGURE 1: Comparison of results for fully developed Nusselt number with analytical results of Chen and Hadim. Kn = 0.001, Γ = 100.

verified by comparing the results of different grid sizes. A mesh refinement study was carried out in order to ensure grid independent solutions. It was found that the obtained numerical solution for the momentum equation is invariant beyond a grid size of 75 points in the y-direction. Therefore, all velocity profiles are obtained using this grid size. Similar refinement study was carried out for the energy equation. It was found that a grid size of (75 × 75) is adequate for the accuracy and any increase in the number of grid points would result in an insignificant effect on the results.

4. Result and Discussion

In order to verify the validity and accuracy of the numerical model, the present numerical results were compared with corresponding integral solution results for the case of fully developed forced convection in porous macrochannel saturated with a power-law fluid [23]. Figure 1 shows a comparison between the two solutions where the agreement is very good at the Darcy and nonDarcy regimes.

The effect of the Knudsen number and the power law index on the axial fully-developed velocity profiles is shown in Figure 2 for a microscopic inertial coefficient equal to 0. From this figure, it is clear that as the Kn number increases, the velocity slip at the wall increases regardless of the power law index value. This is because the increase in Kn number can be due to increase in the mean free path of the molecules, which, in turn, decreases the retarding effect at the wall and thus yields larger flow rates near the channel walls. It is obvious from Figure 2 that as the Kn numberincreases or even when the power law index decreases, the flow velocity near the walls increases while the peak velocity at the centerline decreases to satisfy mass conservation. Figure 2

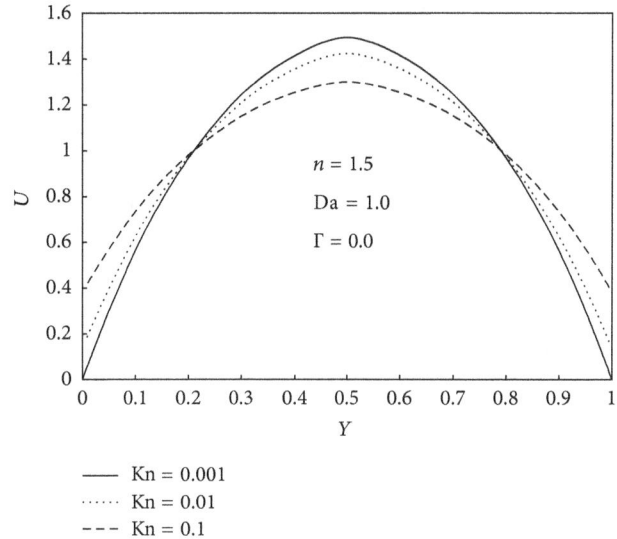

(a)

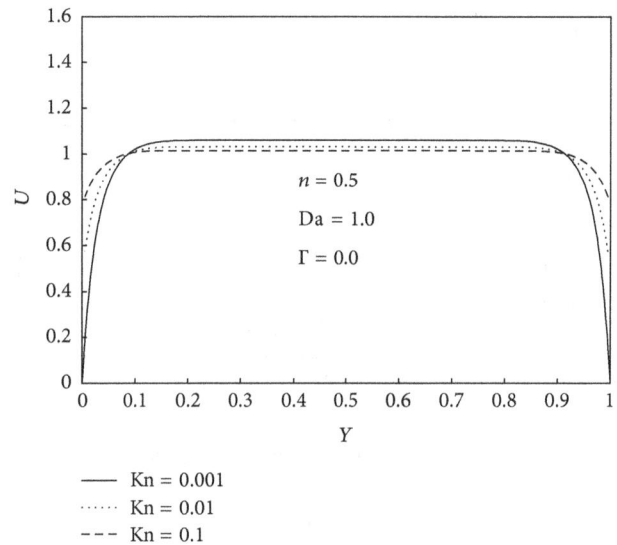

(b)

FIGURE 2: Effect of Nudsen number and power law indexon the axial velocity profile for Γ = 0.0, Da = 1.0. (a) n = 1.5, (b) n = 0.5.

also shows that the effects of Kn number is more evident for shear thinning fluids ($n < 1.0$) and higher values of Kn.

The effect of the Darcy number and the power law index on the axial fully developed velocity profiles is shown in Figure 3 for Kn = 0.001. As the Darcy number decreases, a flat velocity profile occurs in the core region due to bulk damping caused by the presence of the porous matrix and the viscous effects near the walls. In Figure 3, it is shown that the effects of Darcy number are more considerable for shear thickening fluid ($n > 1.0$) than for shear thinning fluids ($n < 1.0$). Thoroughly inspecting Figure 3, it is obvious that as the power law index decreases, the velocity gradient near the wall increases while the peak velocity at the centerline decreases to satisfy mass conservation. It is noted that the

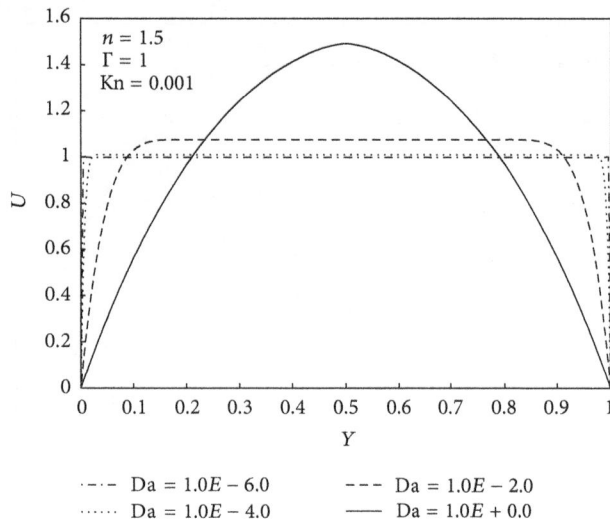

(a)

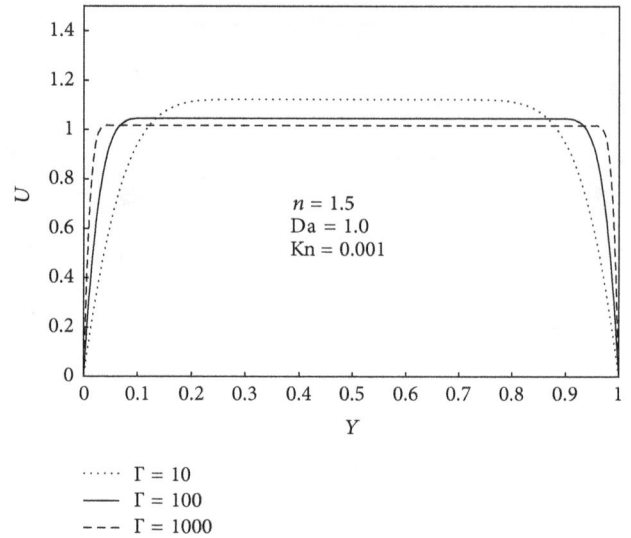

(b)

FIGURE 3: Effect of Darcy number and power law index on the axial velocity profile for $\Gamma = 1.0$, Kn = 0.001. (a) $n = 1.5$, (b) $n = 0.5$.

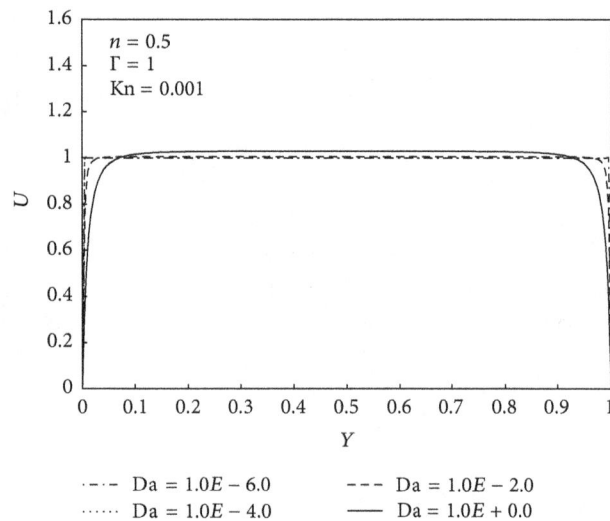

(a)

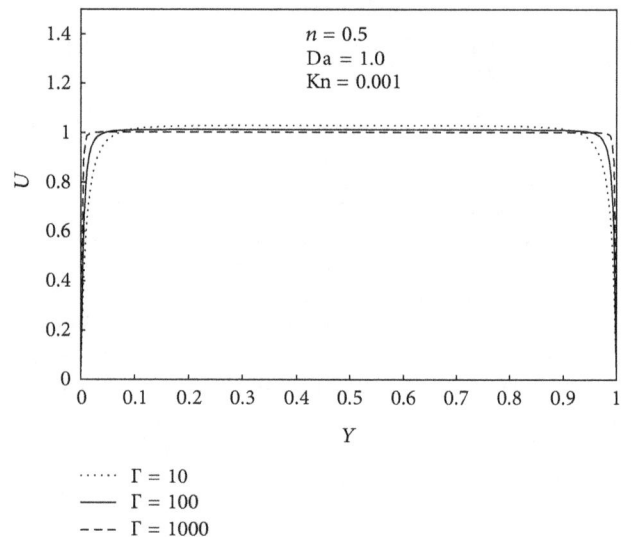

(b)

FIGURE 4: Effects of inertia parameter and power law index on the axial velocity profile for Da = 1.0, Kn = 0.001. (a) $n = 1.5$, (b) $n = 0.5$.

effect of power law index become insignificant at low Darcy number (Da $\leq 10^{-6}$).

In contrast, Figure 4 shows the effect of the inertia parameter and the power law index on the axial fully developed velocity profile for Kn = 0.001. As the inertia parameter increases, the fully developed velocity profile in channel becomes flattened (velocity gradient near the walls increases), and the boundary layer behavior is more pronounced [22]. However, the flattening of the velocity profile is relatively weaker for shear thinning fluids compared with shear thickening fluids.

The combined effects of the Knudsen number, power law index and Darcy number on the skin friction, are clearly presented in Figure 5 for $\Gamma = 1.0$. As the Kn number increases, the skin friction decreases. It is known that any increase in Kn number would increase in Re number due

to the increase in the flow velocity, which is a result of the reduction in the retardation effect of the wall. On the other hand, the skin friction value decreases due to both the decrease in the velocity gradient at the wall and the increase in the flow velocity. The net result of the effect of Kn number is found to decrease C_f Re value. This means that the reduction in skin friction value is more significant so that it overcomes the increase in Re number. The effect of the Knudsen number is more significant at larger values of Darcy number. In Figure 5 also the effects of the Kn number diminish while the Darcy number decreases until they become negligible in the Darcy regime (i.e., very low permeability) due to the obvious effects of the porous matrix. As the Darcy number increases, the skin friction decreases sharply, especially in the Darcy regime, thus approaching the asymptotic value for the flow

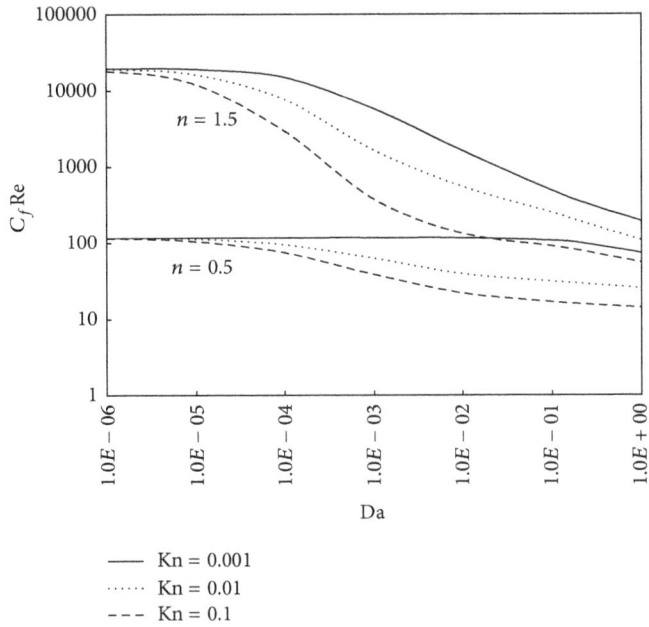

FIGURE 5: Effects of Knudsen number and power law index on the skin friction with Darcy number for $\Gamma = 1.0$.

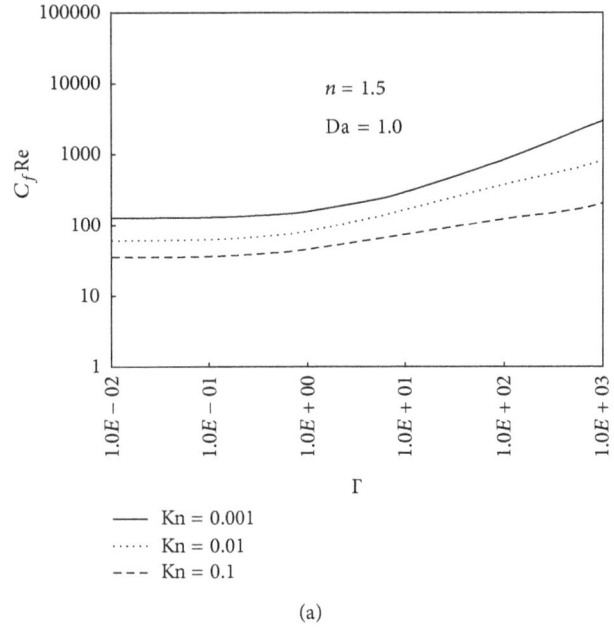

(a)

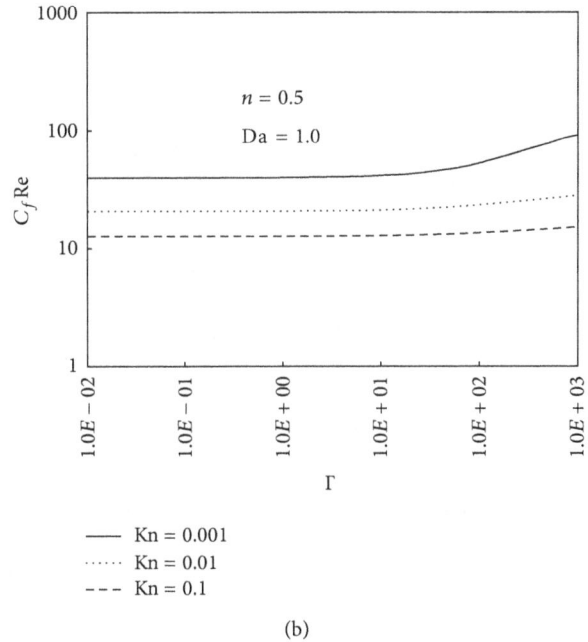

(b)

FIGURE 6: Effect of Knudsen number and power law index on the skin friction with inertia parameter for Da = 1.0. (a) $n = 0.5$ (b) $n = 1.5$.

inside the clear channel. As the power law index decreases, the skin friction also decreases significantly. In the nonDarcy regime, it is noted that the skin friction for shear thickening fluids is about five times the skin friction for shear thinning fluids. In fact, this conclusion supports the numerical results obtained by [22].

The combined effects of the Kn number, power law index, and the inertia parameter on the skin friction are clearly presented in Figure 6 for Da = 1.0. At a fixed value of Kn number, the figure shows that for small values of the inertia parameter, the skin friction remains constant and then increases at a faster rate. When the inertia parameter increases, the velocity gradient near the wall increases therefore increasing the skin friction near the walls. Also, the effect of the Kn value on the skin friction becomes more definite at higher values of inertia parameter because at higher values of inertia parameter, the velocity gradient near the wall gets relatively higher. In the case of very low permeability, as the power law index increases, the skin friction increases.

The effect of the Kn number and power law index on the variation of the fully developed Nusselt number with inertia parameter is shown in Figure 7. Figure 7 shows that at relatively lower Kn number (Kn = 0.001), the inertia effects are very definite while Kn number increases (Kn = 0.1), these effects become very little. Moreover, at relatively small values of inertia parameter, the fully developed Nusselt number remains constant and depends on the power law index and Knudsen number. As the inertia parameter increases, the fully developed Nusselt number increases and then approaches an asymptotic value. It is clear from Figure 7 that for lower Knudsen number case (Kn = 0.001), shear thinning fluids results in higher Nusselt number due to the large velocity gradients near the porous walls. Also, the effects of

the power law index on the fully-developed Nusselt number becomes more evident at lower values of inertia parameter because, at lower values of inertia parameter, the velocity near the walls are relatively lower leading to lower convection heat transfer rate. This implies that any change in the power law index at lower inertia parameter would lead to a significant change in the fully-developed Nusselt number, especially at lower Kn number.

Figure 8 shows the combined effects of the Kn number, power law index, and Darcy number on the fully-developed Nusselt number when $\Gamma = 1.0$. It is noted that the effects of

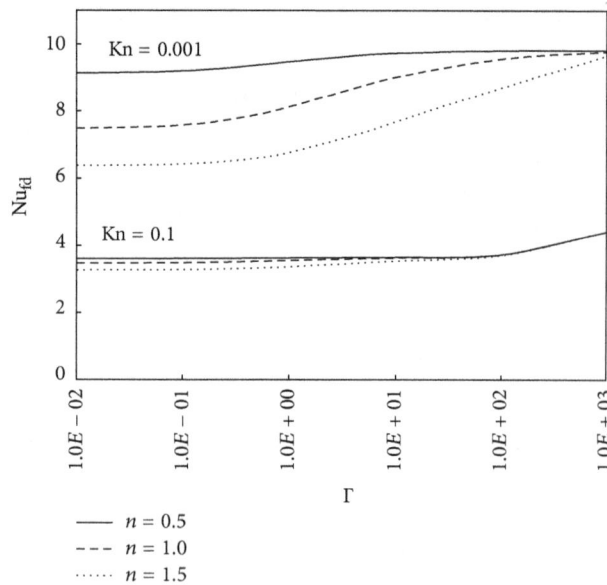

FIGURE 7: Effects of Knudsen number and power law index on the variation of the fully developed Nusselt number with inertia parameter.

Kn number diminishes as the Darcy number decreases until they become negligible in the Darcy regime (i.e., very low permeability) because of the obvious effects of the porous matrix. When the Knvalue increases, the fully developed Nusselt number decrease regardless of the power law index. This is because when normally the Kn number increases, the jump in temperature at the wall increases and a lesser amount of energy is consequently transferred from the wall to the adjacent fluid. As the Darcy number increases, the fully developed Nusslet number for shear thickening ($n = 1.5$) decreases more steeply than at shear thinning ($n = 0.5$) due to the velocity profile attained at the different power law index. When Darcy number increases, the asymptotic regime is identified as the fully developed Nusselt number which approaches the asymptotic value for the flow in clear channel case.

5. Conclusion

The present numerical solutions are conducted for steady laminar forced convection flow between parallel-plate microchannels filled with porous medium and saturated with a power-law fluid. In this study, the slip flow regime ($10^{-3} \leq Kn \leq 10^{-1}$) is considered, and the modified extended Darcy-Brinkman-Forchheimer model is adopted to describe the hydrodynamic and thermal behavior of the power-law fluid in porous microchannels. The present study reports the effects of power law index, Knudsen number, Darcy number, microscopic inertial coefficient on the flow, and heat transfer in the microchannel. The results indicate that in the case of high permeability regime (nonDarcy), the effects of the Knudsen number and the power law index on the flow and heat transfer in the porous microchannel are

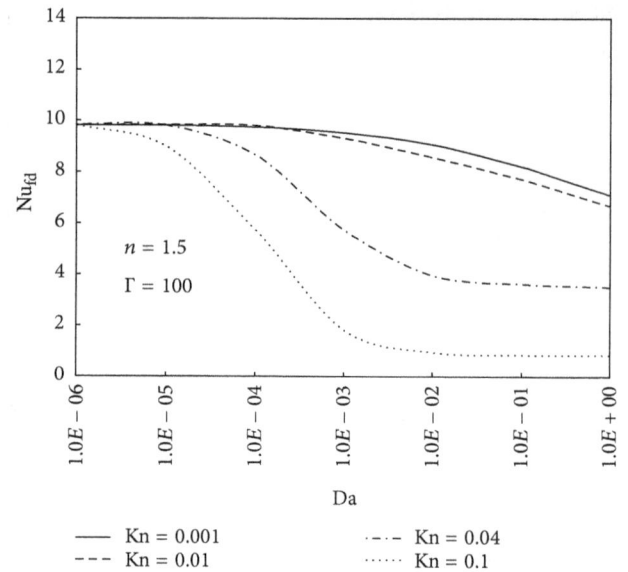

(a)

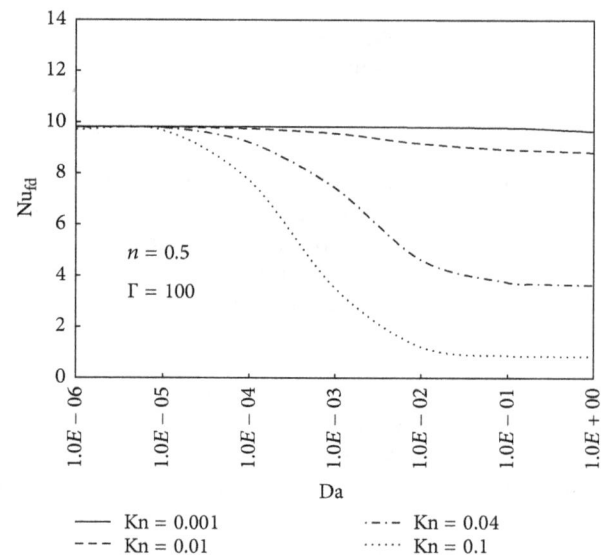

(b)

FIGURE 8: Effects of Knudsen number and power law index on the variation of the fully developed Nusselt number with Darcy number (a) $n = 0.5$ (b) $n = 1.5$.

significant. However, in case of low permeability regime, the effects of Darcy number become more evident. In case of high permeability, the skin friction and the Nusselt number decreases while the Knudsen number and the power law index increase.

Nomenclature

c: Specific heat, J/kg·K
C_F: Inertia coefficient
C_F^*: Modified inertia coefficient
Da: Darcy number, $(K^*)^{2/(n+1)}/D_h^2$

D_h: Hydraulic diameter, H

H: Channel height, m

h: Local heat transfer coefficient

K: Intrinsic permeability of the porous medium, m^2

K^*:   Modified permeability of the porous medium, m^{n+1}

Kn: Knudsen number, (λ/L)

k_f: Thermal conductivity of the fluid, W/m-K

k_e: Effective thermal conductivity of the fluid saturated porous medium, W/m-K

n: Power law index, m

Nu: Local Nusselt number, (hD_h/k_f)

Nu_{fd}:   Fully developed Nusselt number

p: Pressure, Pa

P: Dimensionless pressure, $p/\rho u_m^2$

Pr: Prandtl number, $(\mu^*/\rho\alpha)_f (u_m/D_h)^{n-1}$

Re: Reynolds number, $(\rho u_m^{2-n} D_h^n/\mu^*)$

T: Temperature, K

t: Time, s

t_o: Reference time, (D_h/u_m)

U: Nondimensional axial velocity, (u/u_m)

u: Axial velocity component, m/s

X: Dimensionless axial coordinate, (x/D_h)

x: Axial coordinate, m

Y: Dimensionless transverse coordinate, (y/D_h)

y: Transverse coordinate, m.

Greek Symbols

α: Thermal diffusivity $(k_f/(\rho c)_f)$

λ: Mean free path of the gas molecules

σ_v: Tangential momentum accommodation coefficient

σ_T: Thermal accommodation coefficient

γ: Specific heat ratio, (C_p/C_v)

Γ: Microinertia parameter, $(C_F^*/\sqrt{Da})$

ε: Porosity of the porous media

μ: Dynamic viscosity

μ^*:   Consistency index of the power law fluid

ρ_f: Fluid density

θ: Nondimensional temperature, $((T-T_\infty)/(T_w-T_\infty))$

θ_m: Dimensionless mean temperature, $((T_m-T_\infty)/(T_w-T_\infty))$

τ: Dimensionless time, (t/t_o)

τ_w: Shear stress at the wall $((-\mu(\partial u/\partial y)|_w)$.

Subscripts

e: Effective (i.e., fluid-saturated porous medium)

f: Fluid

m: Mean

p: Pressure

s: Solid

v: Volume

w: Wall.

References

[1] G. Karniadakis, A. Beskok, and N. Aluru, *Micro Flows and Nano Flows: Fundamentals and Simulation*, Springer, New York, NY, USA, 2005.

[2] G. Karniadakis and A. Beskok, *Micro Flows*, Springer, New York, NY, USA, 2002.

[3] D. Nield and A. Bejan, *Convection in Porous Media*, Springer, New York, NY, USA, 1999.

[4] K. Vafai and C. L. Tien, "Boundary and inertia effects on flow and heat transfer in porous media," *International Journal of Heat and Mass Transfer*, vol. 24, no. 2, pp. 195–203, 1981.

[5] K. Watanabe, Y. Yanuar, and H. Mizunuma, "Slip of Newtonian fluids at slid boundary," *JSME International Journal B*, vol. 41, no. 3, pp. 525–529, 1998.

[6] E. B. Arkilic, K. S. Breuer, and M. A. Schmidt, "Gaseous flow in microchannels," in *Proceedings of the Applications of Micro-Fabrication to Fluid Mechanics (ASME '94)*, vol. 197, pp. 57–66, November 1994.

[7] F. Ezquerra Larrodé, C. Housiadas, and Y. Drossinos, "Slip-flow heat transfer in circular tubes," *International Journal of Heat and Mass Transfer*, vol. 43, no. 15, pp. 2669–2680, 2000.

[8] R. F. Barron, X. Wang, T. A. Ameel, and R. O. Warrington, "The Graetz problem extended to slip-flow," *International Journal of Heat and Mass Transfer*, vol. 40, no. 8, pp. 1817–1823, 1997.

[9] S. Yu and T. A. Ameel, "Slip-flow heat transfer in rectangular microchannels," *International Journal of Heat and Mass Transfer*, vol. 44, no. 22, pp. 4225–4234, 2001.

[10] J. Liu, Y.-C. Tai, and C.-M. Ho, "MEMS for pressure distribution studies of gaseous flows in microchannels," in *Proceedings of the IEEE Micro Electro Mechanical Systems Conference*, pp. 209–215, February 1995.

[11] A. Beskok and G. E. Karniadakis, "Simulation of heat and momentum transfer in complex microgeometries," *Journal of Thermophysics and Heat Transfer*, vol. 8, no. 4, pp. 647–655, 1994.

[12] E. M. Sparrow and S. H. Lin, "Laminar heat transfer in tubes under slip-flow conditions," *Journal of Heat Transfer*, vol. 84, pp. 363–369, 1962.

[13] T. A. Ameel, R. F. Barron, X. Wang, and R. O. Warrington Jr., "Laminar forced convection in a circular tube with constant heat flux and slip flow," *Microscale Thermophysical Engineering*, vol. 1, no. 4, pp. 303–320, 1997.

[14] W. Qu, G. M. Mala, and D. Li, "Heat transfer for water flow in trapezoidal silicon microchannels," *International Journal of Heat and Mass Transfer*, vol. 43, no. 21, pp. 3925–3936, 2000.

[15] Y. Zhu and S. Granick, "Rate-dependent slip of Newtonian liquid at smooth surfaces," *Physical Review Letters*, vol. 87, no. 9, Article ID 096105, pp. 961051–961054, 2001.

[16] O. M. Haddad, M. A. Al-Nimr, and Y. Taamneh, "Hydrodynamic and thermal behavior of gas flow in microchannels filled with porous media," *Journal of Porous Media*, vol. 9, no. 5, pp. 403–414, 2006.

[17] O. M. Haddad, M. A. Al-Nimr, and M. S. Sari, "Forced convection gaseous slip flow in circular porous micro-channels," *Transport in Porous Media*, vol. 70, no. 2, pp. 167–179, 2007.

[18] C.-H. Chen, "Thermal transport characteristics of mixed pressure and electro-osmotically driven flow in micro- and nanochannels with Joule heating," *Journal of Heat Transfer*, vol. 131, no. 2, pp. 1–10, 2009.

[19] M. Kaviany, "Laminar flow through a porous channel bounded by isothermal parallel plates," *International Journal of Heat and Mass Transfer*, vol. 28, no. 4, pp. 851–858, 1985.

[20] A. Nakayama and A. V. Shenoy, "Non-Darcy forced convective heat transfer in a channel embedded in a non-Newtonian inelastic fluid-saturated porous medium," *Canadian Journal of Chemical Engineering*, vol. 71, no. 1, pp. 168–173, 1993.

[21] A. V. Shenoy, "Non-Newtonian fluid heat transfer in porous media," *Advances in Heat Transfer*, vol. 24, pp. 101–190, 1994.

[22] G. Chen and H. A. Hadim, "Forced convection of a power-law fluid in a porous channel—numerical solutions," *Heat and Mass Transfer*, vol. 34, no. 2-3, pp. 221–228, 1998.

[23] G. Chen and A. Hadim, "Forced convection of power-law fluid in porous channel-Integral solution," *Journal of Porous Media*, vol. 2, pp. 59–70, 1999.

[24] A. Hadim, "Forced convection in a porous channel with localized heat sources," *ASME Journal of Heat Transfer*, vol. 116, no. 2, pp. 465–472, 1994.

[25] S. Sorek, D. Levi-Hevroni, A. Levy, and G. Ben-Dor, "Extensions to the macroscopic Navier-Stokes equation," *Transport in Porous Media*, vol. 61, no. 2, pp. 215–233, 2005.

[26] J. D. Hoffman, *Numerical Methods for Engineers and Scientists*, McGraw Hill, New York, NY, USA, 1992.

24

Analysis of Heat and Mass Transfer on MHD Peristaltic Flow through a Tapered Asymmetric Channel

M. Kothandapani,[1] J. Prakash,[2] and V. Pushparaj[3]

[1]*Department of Mathematics, University College of Engineering Arni (A Constituent College of Anna University, Chennai), Arni, Tamil Nadu 632 326, India*
[2]*Department of Mathematics, Arulmigu Meenakshi Amman College of Engineering, Vadamavandal, Tamil Nadu 604 410, India*
[3]*Department of Mathematics, C. Abdul Hakeem College of Engineering & Technology, Melvisharam, Tamil Nadu 632 509, India*

Correspondence should be addressed to M. Kothandapani; mkothandapani@gmail.com

Academic Editor: Miguel Onorato

This paper describes the peristaltic flow of an incompressible viscous fluid in a tapered asymmetric channel with heat and mass transfer. The fluid is electrically conducting fluid in the presence of a uniform magnetic field. The propagation of waves on the nonuniform channel walls to have different amplitudes and phase but with the same speed is generated the tapered asymmetric channel. The assumptions of low Reynolds number and long wavelength approximations have been used to simplify the complicated problem into a relatively simple problem. Analytical expressions for velocity, temperature, and concentration have been obtained. Graphically results of the flow characteristics are also sketched for various embedded parameters of interest entering the problem and interpreted.

1. Introduction

The study of peristaltic transport has enjoyed increased interest from investigators in several engineering disciplines. From a mechanical point of view, peristalsis offers the opportunity of constructing pumps in which the transported medium does not come in direct contact with any moving parts such as valves, plungers, and rotors. This could be of great benefit in cases where the medium is either highly abrasive or decomposable under stress. This has led to the development of fingers and roller pumps which work according to the principle of peristalsis. Applications include dialysis machines, open-heart bypass pump machines, and infusion pumps. After the first investigation reported by Latham [1], several theoretical and experimental investigations [2–5] about the peristaltic flow of Newtonian and non-Newtonian fluids have been made under different conditions with reference to physiological and mechanical situations.

In view of the processes like hemodialysis and oxygenation, some progress is shown in the theory of peristalsis

with heat transfer [6–8]. Such analysis of heat transfer is of great value in biological tissues, dilution technique in examining blood flow, destruction of undesirable cancer tissues, metabolic heat generation, and so forth. In addition, the mass transfer effect on the peristaltic flow of viscous fluid has been examined in the studies [9–11]. The effect of magnetic field on a Newtonian fluid has been reported for treatment of gastronomic pathologies, constipation, and hypertension.

Recently, few attempts have been made in the peristaltic literature to study the combined effects of heat and mass transfer. Eldabe et al. [10] analyzed the mixed convective heat and mass transfer in a non-Newtonian fluid at a peristaltic surface with temperature-dependent viscosity. The influence of heat and mass transfer on MHD peristaltic flow through a porous space with compliant walls was studied by Srinivas and Kothandapani [11]. The effects of elasticity of the flexible walls on the peristaltic transport of viscous fluid with heat transfer in a two-dimensional uniform channel have been investigated by Srinivas and Kothandapani [12]. Ogulu [13]

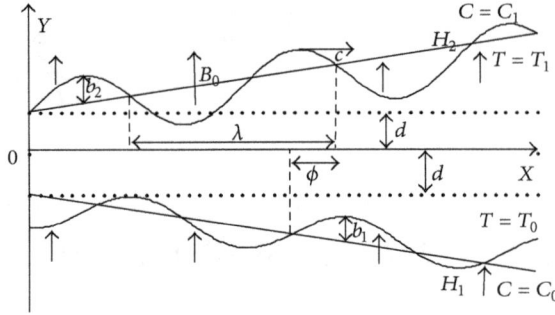

FIGURE 1: Schematic diagram of the tapered asymmetric channel.

examined heat and mass transfer of blood under the influence of a uniform magnetic field.

The problem of intrauterine fluid motion in a non-pregnant uterus caused by myometrial contractions is a peristaltic-type fluid motion and the myometrial contractions may occur in both symmetric and asymmetric directions. Further it is observed that the intrauterine fluid flow in a sagittal cross section of the uterus discloses a narrow channel enclosed by two fairly parallel walls with wave trains having different amplitudes and phase difference [14–16]. Keeping in view of the abovementioned reasons, a new theoretical investigation of peristaltic motion of a Newtonian fluid in the presence of heat and mass transfer in the most generalized form of the channel, namely, the tapered asymmetric channel, is carried out. The governing equations of motion, energy, and concentration are simplified by using the assumptions of long wavelength and low Reynolds number approximations. The exact solutions of velocity, temperature, and concentration of the fluid are generated. Also interesting flow quantities are analyzed by plotting various graphs.

2. Mathematical Formulation

Consider the unsteady, combined convective heat and mass transfer, MHD flow of an electrically conducting viscous fluid in a two-dimensional tapered asymmetric channel. Let $\overline{Y} = \overline{H}_1$ and $\overline{Y} = \overline{H}_2$, respectively, the lower and upper wall boundaries of the tapered asymmetric channel. The flow is generated by sinusoidal wave trains propagating with constant speed c along the tapered asymmetric channel (Figure 1). The geometry of the wall surface [15, 16] is defined as

$$\overline{H}_2\left(\overline{X}, \overline{t}\right) = d + m'\overline{X} + b_2 \sin\left[\frac{2\pi}{\lambda}\left(\overline{X} - c\overline{t}\right)\right].$$
(1a)

upper wall,

$$\overline{H}_1\left(\overline{X}, \overline{t}\right) = -d - m'\overline{X} - b_1 \sin\left[\frac{2\pi}{\lambda}\left(\overline{X} - c\overline{t}\right) + \phi\right].$$
(1b)

lower wall,

where d is the half width of the channel at the inlet, b_1 and b_2 are the amplitudes of lower and upper walls, respectively, c is the phase speed of the wave, m' ($m' \ll 1$) is the nonuniform

parameter, λ is the wavelength, the phase difference ϕ varies in the range $0 \leq \phi \leq \pi$, $\phi = 0$ corresponds to symmetric channel with waves out of phase (i.e., both walls move towards outward or inward simultaneously), and further b_1, b_2, and ϕ satisfy the following conditions for the divergent channel at the inlet:

$$b_1^2 + b_2^2 + 2b_1 b_2 \cos\left(\phi\right) \leq \left(2d\right)^2.$$
(2)

It is assumed that the temperature and concentration at lower wall are T_0 and C_0, respectively, while the temperature and concentration at the upper wall are T_1 and C_1, respectively ($T_0 < T_1$ & $C_0 < C_1$). The equations of continuity, momentum, energy, and concentration are described as follows [11, 12]:

$$\frac{\partial \overline{U}}{\partial \overline{X}} + \frac{\partial \overline{V}}{\partial \overline{Y}} = 0,$$

$$\rho\left[\frac{\partial \overline{U}}{\partial \overline{t}} + \overline{U}\frac{\partial \overline{U}}{\partial \overline{X}} + \overline{V}\frac{\partial \overline{U}}{\partial \overline{Y}}\right]$$

$$= -\frac{\partial \overline{P}}{\partial \overline{X}} + \mu\left[\frac{\partial^2 \overline{U}}{\partial \overline{X}^2} + \frac{\partial^2 \overline{U}}{\partial \overline{Y}^2}\right] - \sigma B_0^2 \overline{U},$$

$$\rho\left[\frac{\partial \overline{V}}{\partial \overline{t}} + \overline{U}\frac{\partial \overline{V}}{\partial \overline{X}} + \overline{V}\frac{\partial \overline{V}}{\partial \overline{Y}}\right] = -\frac{\partial \overline{P}}{\partial \overline{Y}} + \mu\left[\frac{\partial^2 \overline{V}}{\partial \overline{X}^2} + \frac{\partial^2 \overline{V}}{\partial \overline{Y}^2}\right],$$

$$\rho\zeta\left[\frac{\partial \overline{T}}{\partial \overline{t}} + \overline{U}\frac{\partial \overline{T}}{\partial \overline{X}} + \overline{V}\frac{\partial \overline{T}}{\partial \overline{Y}}\right]$$

$$= \kappa\left[\frac{\partial^2 \overline{T}}{\partial \overline{X}^2} + \frac{\partial^2 \overline{T}}{\partial \overline{Y}^2}\right]$$

$$+ \mu\left\{2\left[\left(\frac{\partial \overline{U}}{\partial \overline{X}}\right)^2 + \left(\frac{\partial \overline{V}}{\partial \overline{Y}}\right)^2\right] + \left(\frac{\partial \overline{U}}{\partial \overline{Y}} + \frac{\partial \overline{V}}{\partial \overline{X}}\right)^2\right\},$$

$$\left[\frac{\partial \overline{C}}{\partial \overline{t}} + \overline{U}\frac{\partial \overline{C}}{\partial \overline{X}} + \overline{V}\frac{\partial \overline{C}}{\partial \overline{Y}}\right]$$

$$= D_m\left[\frac{\partial^2 \overline{C}}{\partial \overline{X}^2} + \frac{\partial^2 \overline{C}}{\partial \overline{Y}^2}\right] + \frac{D_m K_T}{T_m}\left[\frac{\partial^2 \overline{T}}{\partial \overline{X}^2} + \frac{\partial^2 \overline{T}}{\partial \overline{Y}^2}\right],$$
(3)

where $\overline{U}$, $\overline{V}$ are the components of velocity along $\overline{X}$ and $\overline{Y}$ directions, respectively, $\overline{t}$ is the dimensional time, μ is the coefficient of viscosity, σ is the electrical conductivity of the fluid, B_0 is the uniform applied magnetic field, ρ is the fluid density, ζ is the specific heat at constant volume, $\overline{P}$ is the pressure, $\overline{T}$ is the temperature, $\overline{C}$ is the concentration of the fluid, T_m is the mean temperature, κ is the thermal conductivity, D_m is the coefficient of mass diffusivity, and K_T is the thermal diffusion ratio.

The corresponding boundary conditions are

$$\overline{U} = 0, \quad \overline{T} = T_0, \quad \overline{C} = C_0 \quad \text{at } \overline{Y} = \overline{H}_1, \qquad (4a)$$

$$\overline{U} = 0, \quad \overline{T} = T_1, \quad \overline{C} = \overline{C}_1, \quad \text{at } \overline{Y} = \overline{H}_2. \qquad (4b)$$

By introducing the following set of nondimensional variables in (3),

$$x' = \frac{\overline{X}}{\lambda}, \quad y' = \frac{\overline{Y}}{d}, \quad t' = \frac{ct}{\lambda},$$

$$u' = \frac{\overline{U}}{c}, \quad v' = \frac{\overline{V}}{c}, \quad \delta = \frac{d}{\lambda},$$

$$h_1 = \frac{\overline{H}_1}{d}, \quad h_2 = \frac{\overline{H}_2}{d}, \quad p' = \frac{d^2 \overline{P}}{c\lambda\mu}, \qquad (5)$$

$$\theta = \frac{\overline{T} - T_0}{T_1 - T_0}, \quad \Theta = \frac{\overline{C} - C_0}{C_1 - C_0},$$

we obtain

$$R\delta \left[\frac{\partial u'}{\partial t'} + u' \frac{\partial u'}{\partial x'} + v' \frac{\partial u'}{\partial y'} \right]$$

$$= -\frac{\partial p'}{\partial x'} + \delta^2 \frac{\partial^2 u'}{\partial x'^2} + \frac{\partial^2 u'}{\partial y'^2} - M^2 u',$$

$$R\delta^3 \left[\frac{\partial v'}{\partial t'} + u' \frac{\partial v'}{\partial x'} + v' \frac{\partial v'}{\partial y'} \right]$$

$$= -\frac{\partial p'}{\partial y'} + \delta^2 \left[\frac{\partial^2 v'}{\partial x'^2} + \frac{\partial^2 v'}{\partial y'^2} \right],$$

$$R\delta \left[\frac{\partial \theta}{\partial t'} + u' \frac{\partial \theta}{\partial x'} + v' \frac{\partial \theta}{\partial y'} \right]$$

$$= \frac{1}{\text{Pr}} \left[\delta^2 \frac{\partial^2 \theta}{\partial x'^2} + \frac{\partial^2 \theta}{\partial y'^2} \right] \qquad (6)$$

$$+ E \left[2 \left[\delta^2 \left(\left(\frac{\partial u'}{\partial x'} \right)^2 + \left(\frac{\partial v'}{\partial y'} \right)^2 \right) \right] \right.$$

$$\left. + \left[\frac{\partial u'}{\partial y'} + \delta \frac{\partial v'}{\partial x'} \right]^2 \right],$$

$$R\delta \left[\frac{\partial \Theta}{\partial t'} + u' \frac{\partial \Theta}{\partial x'} + v' \frac{\partial \Theta}{\partial y'} \right]$$

$$= \frac{1}{\text{Sc}} \left[\delta^2 \frac{\partial^2 \Theta}{\partial x'^2} + \frac{\partial^2 \Theta}{\partial y'^2} \right] + \text{Sr} \left[\delta^2 \frac{\partial^2 \theta}{\partial x'^2} + \frac{\partial^2 \theta}{\partial y'^2} \right],$$

where a ($= b_1/d$) and b ($= b_2/d$) are nondimensional amplitudes of lower and upper walls, respectively, $k_1 = (\lambda m'/d)$ is nonuniform parameter, R ($= cd\rho/\mu$) is the Reynolds number, $\text{Sc} = \mu/D_m\rho$ is the Schmidt number, $\text{Sr} = \rho D_m K_T (T_1 - T_0)/\mu(C_1 - C_0)T_m$ is the Soret number, $M = \sqrt{\sigma/\mu}B_0 d$ is the Hartmann number, $\text{Pr} = \rho\nu\zeta/\kappa$ is the Prandtl number, and $E = c^2/\zeta(T_1 - T_0)$ is the Eckert number.

In order to discuss the results quantitatively, we assume the instantaneous volume rate of the flow $F(x, t)$, periodic in $(x - t)$ [17–19], as

$$F(x, t) = Q + a \sin [2\pi (x - t) + \phi] + b \sin 2\pi (x - t), \quad (7)$$

where Q is the time-average of the flow over one period of the wave and

$$F = \int_{h_1}^{h_2} u \, dy. \qquad (8)$$

3. Exact Analytical Solution

Using the long wavelength approximation and neglecting the wave number along with low Reynolds number and omitting prime, one can find from (6) that

$$0 = -\frac{\partial p}{\partial x} + \frac{\partial^2 u}{\partial y^2} - M^2 u, \qquad (9)$$

$$0 = -\frac{\partial p}{\partial y}, \qquad (10)$$

$$0 = \frac{1}{\text{Pr}} \frac{\partial^2 \theta}{\partial y^2} + E \left(\frac{\partial u}{\partial y} \right)^2, \qquad (11)$$

$$0 = \frac{\partial^2 \Theta}{\partial y^2} + \text{ScSr} \frac{\partial^2 \theta}{\partial y^2}. \qquad (12)$$

Equation (10) shows that p is not function of y.
The corresponding boundary conditions are

$$u = 0, \quad \theta = 1, \text{ and } \Theta = 1$$
$$\text{at } y = h_2 = 1 + k_1 x + b \sin (2\pi (x - t)), \qquad (13a)$$

$$u = 0, \quad \theta = 0, \text{ and } \Theta = 0$$
$$\text{at } y = h_1 = -1 - k_1 x - a \sin (2\pi (x - t) + \phi) \qquad (13b)$$

which satisfy, at the inlet of channel,

$$a^2 + b^2 + 2ab \cos (\phi) \le 4. \qquad (14)$$

The set of (9)–(12), subject to the conditions (13a) and (13b), are solved exactly for u, θ, and Θ, and we have

$$u = \frac{(\partial p/\partial x)}{M^2}$$

$$\times \left(1 - \frac{(\cosh (Mh_2) - \cosh (Mh_1)) \sinh (Mh_1)}{M^2 \sinh (M (h_1 - h_2))} \right)$$

$$\times \frac{\cosh (My)}{\cosh (Mh_1)} + \frac{(\partial p/\partial x)}{M^2}$$

$$\times \left(\frac{(\cosh (Mh_2) - \cosh (Mh_1)) \sinh (My)}{M^2 \sinh (M (h_1 - h_2))} - 1 \right),$$

$$\theta = A_1 + B_1 y - \frac{\Pr EA^2}{8}\left[\cosh\left(2My\right) - 2y^2 M^2\right]$$

$$- \frac{\Pr EB^2}{8}\left[\cosh\left(2My\right) + 2y^2 M^2\right]$$

$$- \frac{\Pr EAB}{8}\sinh\left(2My\right),$$

$$\Theta = 4D_3 \text{ScSr}M^2 h_1^2 - D_4 h_1 - D_1 \text{ScSr}e^{2Mh_1}$$

$$- D_2 \text{ScSr}e^{-2Mh_1} + D_4 y + D_1 \text{ScSr}e^{2My}$$

$$+ D_2 \text{ScSr}e^{-2My} - 4D_3 \text{ScSr}M^2 y^2,$$

$$\frac{\partial p}{\partial x} = FM^3 \cosh\left(Mh_1\right)\sinh\left(M\left(h_1 - h_2\right)\right)$$

$$\times \left(\left(\sinh\left(Mh_2\right) - \sinh\left(Mh_1\right)\right)\right.$$

$$\times \left[\sinh\left(M\left(h_1 - h_2\right)\right)\right.$$

$$\left.\left. - \left[\cosh\left(Mh_2\right) - \cosh\left(Mh_1\right)\right]\sinh\left(Mh_1\right)\right]\right)^{-1}$$

$$+ \frac{FM^3 \sinh\left(M\left(h_1 - h_2\right)\right)}{\left[\cosh\left(Mh_2\right) - \cosh\left(Mh_1\right)\right]^2} - \frac{FM^2}{h_2 - h_1},$$

$$(15)$$

where

$$A = \frac{\left(\left(\partial p/\partial x\right) - BM^2 \sinh\left(Mh_1\right)\right)}{M^2 \cosh\left(Mh_1\right)},$$

$$B = \frac{\left(\partial p/\partial x\right)\left(\cosh\left(Mh_2\right) - \cosh\left(Mh_1\right)\right)}{M^2 \sinh\left(M\left(h_1 - h_2\right)\right)},$$

$$A_1 = \frac{\Pr EA^2}{8}\left[\cosh\left(2Mh_1\right) - 2h_1^2 M^2\right]$$

$$+ \frac{\Pr EB^2}{8}\left[\cosh\left(2Mh_1\right) + 2h_1^2 M^2\right]$$

$$+ \frac{\Pr EAB}{8}\sinh\left[2Mh_1\right] - Dh_1,$$

$$B_1 = \frac{1}{h_2 - h_1}\left[1 + \frac{\Pr EA^2}{8}\left[\cosh\left(2Mh_2\right) - 2h_2^2 M^2\right.\right.$$

$$\left. - \cosh\left(2Mh_1\right) + 2h_1^2 M^2\right]$$

$$+ \frac{\Pr EB^2}{8}\left[\cosh\left(2Mh_2\right) + 2h_2^2 M^2\right.$$

$$\left. - \cosh\left(2Mh_1\right) - 2h_1^2 M^2\right]$$

$$\left. + \frac{\Pr EAB}{8}\left[\sinh\left(2Mh_2\right) - \sinh\left(2Mh_1\right)\right]\right],$$

$$D_1 = \frac{\Pr EA^2}{16} + \frac{\Pr EB^2}{16} + \frac{\Pr EAB}{8},$$

$$D_2 = \frac{\Pr EA^2}{16} + \frac{\Pr EB^2}{16} - \frac{\Pr EAB}{8},$$

$$D_3 = \frac{\Pr EA^2}{16} - \frac{\Pr EB^2}{16},$$

$$D_4 = \frac{1}{h_1 - h_2}\left(4D_3 \text{ScSr}M^2\left(h_1^2 - h_2^2\right)\right.$$

$$- D_1 \text{ScSr}\left(e^{2Mh_1} - e^{2Mh_2}\right)$$

$$\left. - D_2 \text{ScSr}\left(e^{-2Mh_1} - e^{-2Mh_2}\right) - 1\right).$$

$$(16)$$

The coefficient of heat transfer at the lower wall is given by

$$Z = h_{1x}\theta_y. \tag{17}$$

4. Numerical Results and Discussion

The effect of various flow parameters on temperature is plotted in Figure 2 for the fixed values of $x = 0.6$ and $t = 0.4$. The influence of the nonuniform parameter (k_1) on θ is depicted in Figure 2(a). It is noticed that the temperature (θ) increases nearer to the lower wall of the tapered channel while the situation is reversed as nonuniform parameter (k_1) increases. Figure 2(b) reveals that the temperature profile increases with increase of Eckert number. It is considered from Figure 2(c) that the temperature profile increases as the amplitude of lower tapered channel increases. The variation of the Prandtl number (Pr) on θ is shown in Figure 2(d). This figure indicates that an increase in Prandtl number results in increase in the temperature of the fluid. Figure 2(e) depicts the effect of temperature for the various values of Q. It is observed that the temperature θ increases with an increase in the time-average flow rate Q in the entire tapered channel. The variations of the Hartmann number M against θ are shown in Figure 2(f). This figure indicates that by increasing the Hartmann number the temperature decreases. The result presented in Figure 3 indicates the behavior of E, a, Q, M, ϕ, and Pr on the heat transfer coefficient (Z). These figures display the typical oscillatory behavior of heat transfer which may be due to the phenomenon of peristalsis. Figures 3(a)–3(f) reveal that the absolute value of the heat transfer coefficient increases by increasing E, a, Q, M, ϕ, and Pr.

The effect of various physical parameters on the concentration of the fluid (Θ) is shown in Figure 4. Figure 4(a) exposes that the fluid concentration decreases with an increase of E. It is noticed from Figure 4(b) that the concentration of the fluid increases as a increases. Figure 4(c) shows that the absolute value of concentration distribution increases at the central part of channel when k_1 is increased. Figure 4(d) is plotted to see the influence of M on the concentration. We notice that an increase in M increases Θ. The concentration for the phase difference is shown in Figure 4(e). It is observed that an increase in ϕ causes increase in Θ. In Figure 4(f), the cause of Prandtl number Pr on Θ is captured. It is

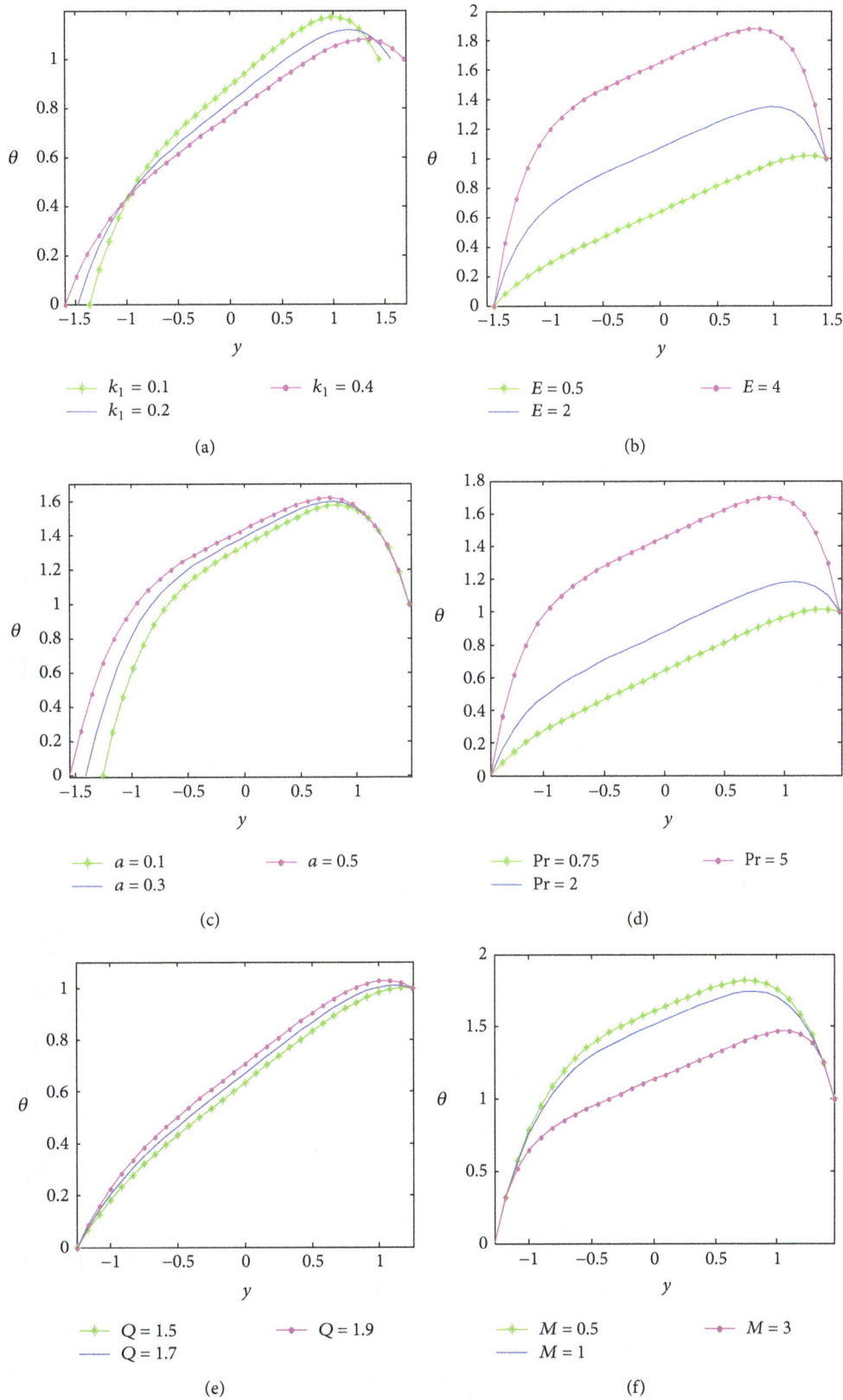

FIGURE 2: Temperature distribution (for fixed value of $t = 0.4$ and $x = 0.6$): (a) $Q = 1.7$, $\phi = \pi/3$, $M = 1$, $E = 0.5$, $\Pr = 1.7$, $a = 0.3$, and $b = 0.4$; (b) $E = 0.5$, $k_1 = 0.2$, $Q = 1.85$, $M = 2$, $\phi = \pi/4$, $a = 0.35$, and $b = 0.3$; (c) $E = 1$, $\phi = \pi/3$, $\Pr = 2$, $M = 0.5$, $k_1 = 0.3$, $Q = 1.75$, and $b = 0.3$; (d) $E = 0.5$, $k_1 = 0.2$, $Q = 1.85$, $M = 2$, $\phi = \pi/4$, $a = 0.35$, and $b = 0.3$; (e) $E = 0.7$, $\phi = \pi/6$, $\Pr = 0.6$, $k_1 = 0.25$, $M = 0.8$, $a = 0.1$, and $b = 0.2$; (f) $\phi = \pi/2$, $E = 2$, $\Pr = 1.5$, $b = 0.3$, $a = 0.2$, $k_1 = 0.3$, and $Q = 1.6$.

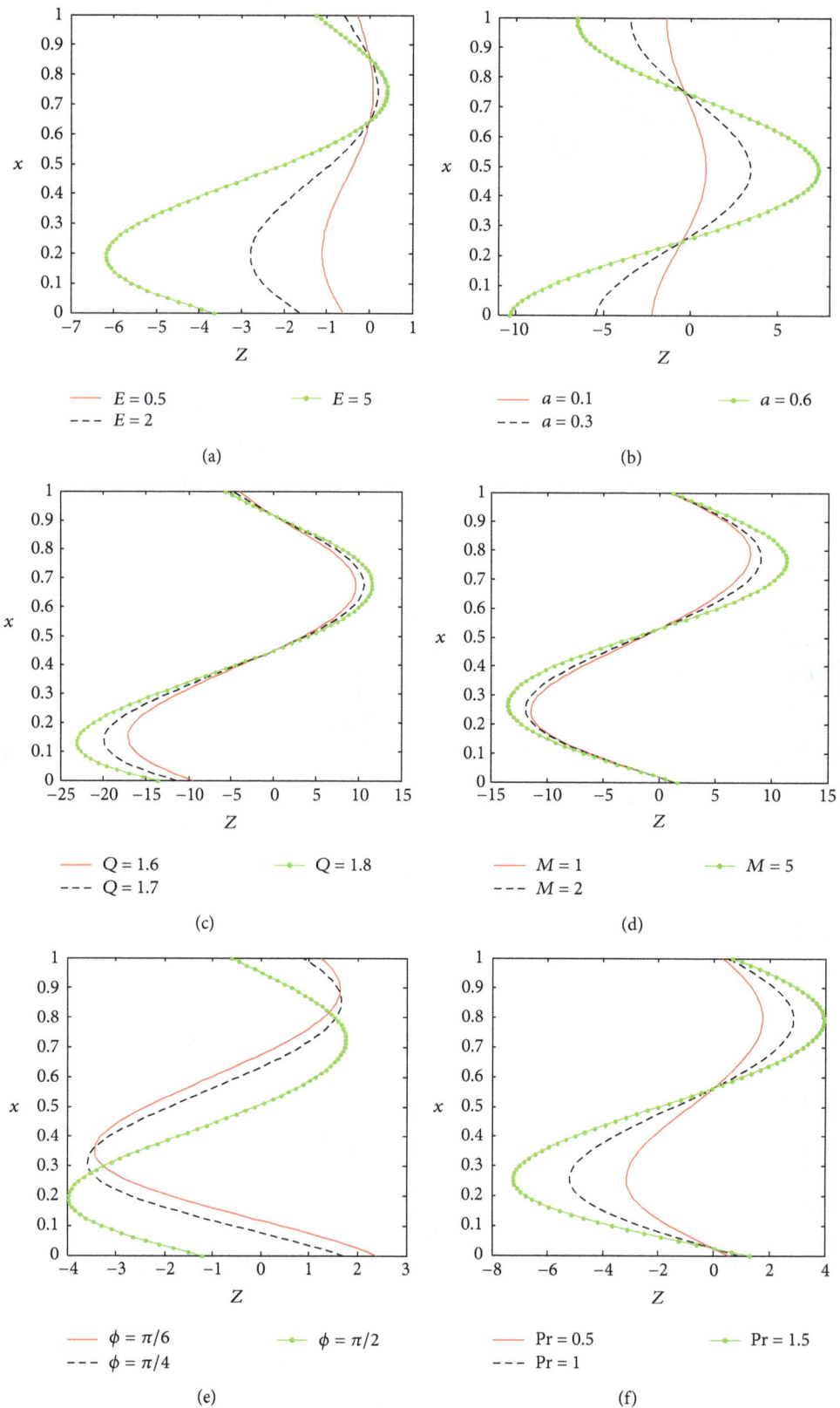

FIGURE 3: Coefficient of heat transfer distribution: (a) $Q = 1.8$, $Pr = 0.5$, $k_1 = 0.5$, $M = 3$, $t = 0.5$, $\phi = \pi/2$, $b = 0.2$, and $a = 0.1$; (b) $Q = 1.6$, $M = 0.8$, $t = 0.5$, $\phi = \pi$, $k_1 = 0.2$, $E = 1$, $b = 0.2$, and $Pr = 1.1$; (c) $t = 0.35$, $E = 2$, $a = 0.4$, $k_1 = 0.25$, $M = 2.5$, $Pr = 1.5$, $\phi = \pi/3$, $k_1 = 0.1$, and $b = 0.3$; (d) $Q = 1.6$, $t = 0.4$, $\phi = \pi/4$, $E = 2$, $b = 0.35$, $Pr = 1$, and $a = 0.4$; (e) $Q = 1.78$, $x = 0.48$, $E = 1$, $Pr = 0.7$, $a = 0.26$, $b = 0.3$, $k_1 = 0.29$, and $M = 1.6$; (f) $Q = 1.72$, $x = 0.42$, $E = 1$, $a = 0.2$, $k_1 = 0.23$, $M = 1.2$, $\phi = \pi/4$, and $b = 0.3$.

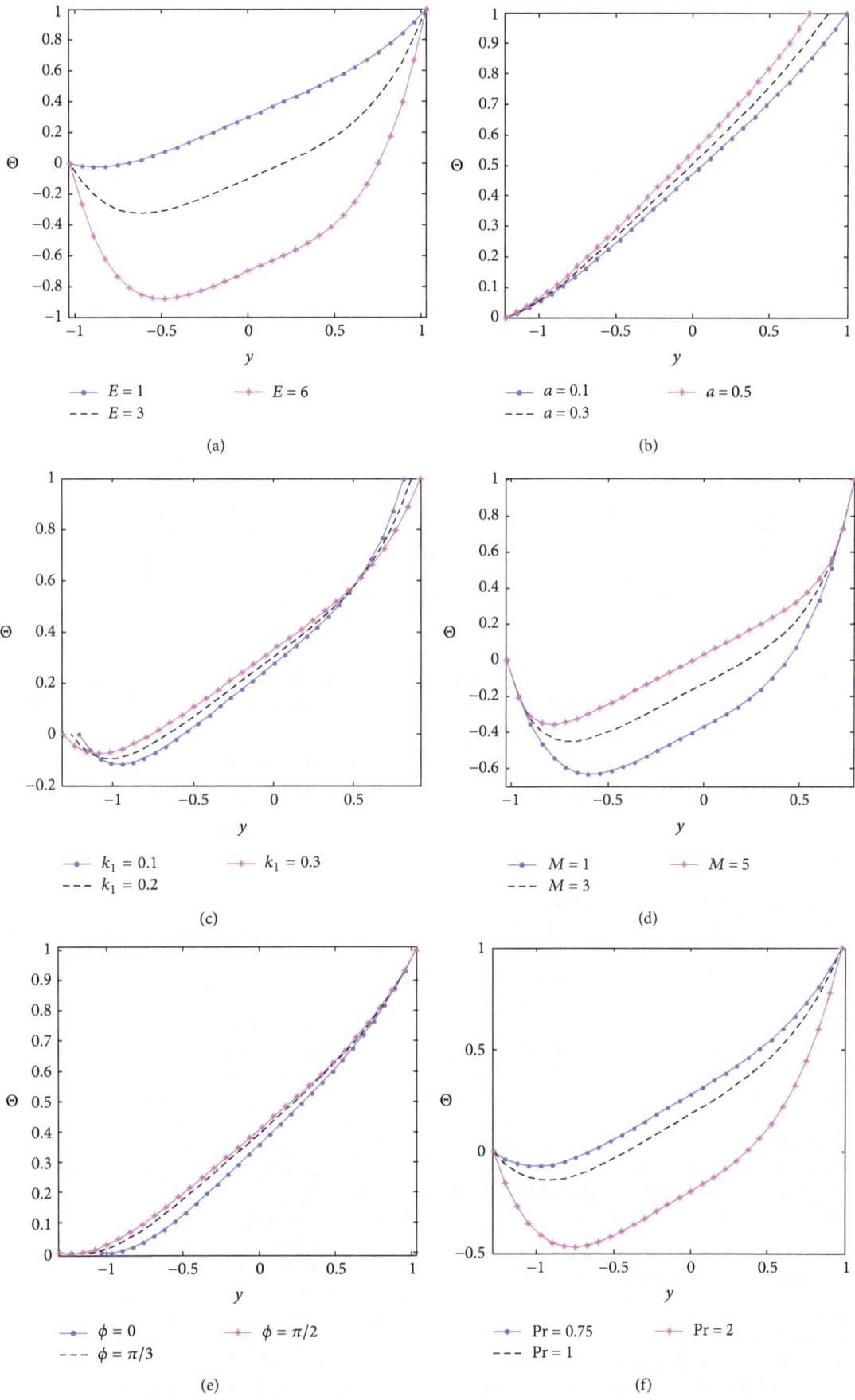

(a)

(b)

(c)

(d)

(e)

(f)

FIGURE 4: Continued.

(g)

(h)

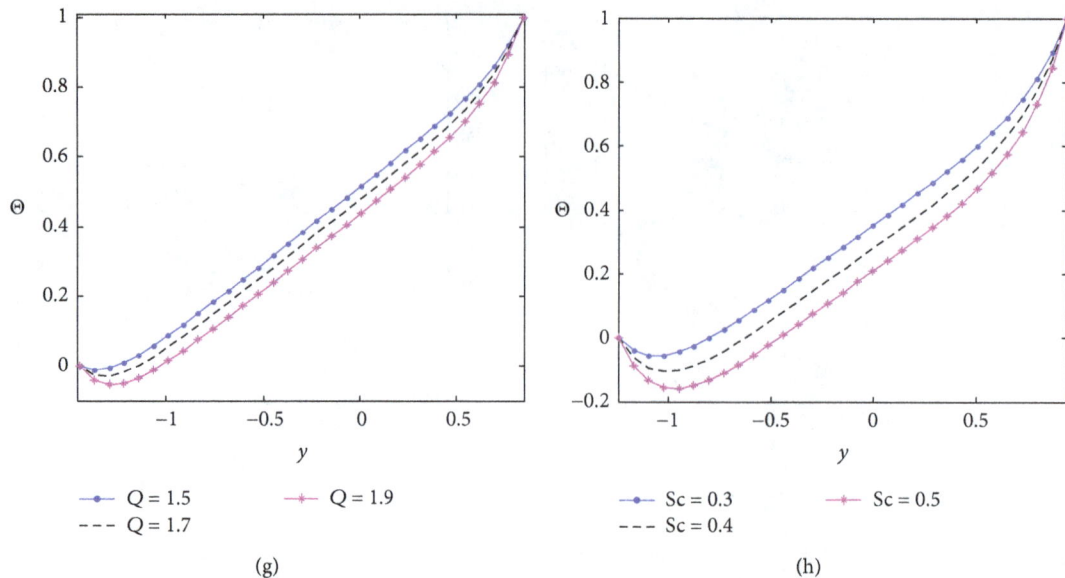

FIGURE 4: Mass transfer distribution (fixed value of $x = 0.5$ and $t = 0.6$): (a) $Q = 1.7$, $a = 0.2$, $b = 0.1$, $\phi = 0$, $Pr = 0.7$, $k_1 = 0.5$, $M = 0.8$, $Sc = 0.6$, and $Sr = 1.6$; (b) $E = 0.5$, $a = 0.2$, $Q = 1.6$, $\phi = \pi/2$, $Pr = 0.5$, $k_1 = 0.1$, $M = 0.5$, $Sc = 0.5$, and $Sr = 1.4$; (c) $a = 0.4$, $b = 0.3$, $E = 1.25$, $Q = 1.65$, $\phi = \pi/3$, $Pr = 1.5$, $M = 0.8$, $Sc = 0.38$, and $Sr = 1.25$; (d) $b = 0.5$, $a = 0.4$, $E = 2$, $Q = 1.55$, $\phi = \pi/6$, $Pr = 2$, $k_1 = 0.15$, $Sc = 0.52$, and $Sr = 1.42$; (e) $b = 0.2$, $a = 0.1$, $E = 0.9$, $Q = 1.8$, $Pr = 0.75$, $k_1 = 0.3$, $M = 0.2$, $Sc = 0.4$, and $Sr = 1.22$; (f) $b = 0.25$, $a = 0.2$, $E = 1.5$, $Q = 1.75$, $\phi = \pi$, $k_1 = 0.25$, $M = 0.5$, $Sc = 0.44$, and $Sr = 1.3$; (g) $a = 0.6$, $b = 0.5$, $E = 0.75$, $\phi = \pi/3$, $Pr = 2$, $k_1 = 0.4$, $M = 3$, $Sc = 0.3$, and $Sr = 1.5$; (h) $b = 0.4$, $a = 0.3$, $E = 1.2$, $Q = 1.65$, $\phi = \pi/4$, $Pr = 1.6$, $M = 2.5$, and $Sr = 1.53$.

detected that with an increase in Pr the concentration of the fluid decreases. Figure 4(g) depicts the concentration profile corresponding to various values of Q. It is seen that the concentration distribution decreases with an increase in the mean flow rate Q. From Figure 4(h), one can view that the concentration profile decreases with increase of Sc.

5. Conclusion

The present analysis can serve as a model which may help in understanding the mechanism of physiological flow of a Newtonian fluid in tapered asymmetric channel. The analytical solutions for the temperature, concentration, and coefficient of heat transfer have been obtained under long wavelength and low Reynolds number approximations. The features of the flow characteristics are analyzed by plotting graphs and discussed in detail. The main observations found from the present study are given as follows.

(1) There is an increase in the temperature (θ) when the Prandtl number Pr, time-average flow rate Q, occlusion parameter a, and Eckert number are increased while it decreases when M is increased.

(2) Mass transfer (Θ) increases with the increase of a, M, and ϕ while it decreases with the increase of E, Pr, Q, and Sc.

(3) The temperature distribution of the fluid increases at the core part of channel when nonuniform parameter is increased and the reverse situation is observed in respect of the concentration of the fluid.

(4) The absolute value of the heat transfer coefficient increases when the E, a, Q, M, ϕ, and Pr are increased.

Conflict of Interests

The authors declare that there is no conflict of interests regarding the publication of this paper.

References

[1] T. W. Latham, *Fluid Motion in a Peristaltic Pump*, MIT Press, Cambridge, Mass, USA, 1966.

[2] A. H. Shapiro, M. Y. Jaffrin, and S. L. Weinberg, "Peristaltic pumping with long wave lengths at low Reynolds number," *Journal of Fluid Mechanics*, vol. 37, no. 4, pp. 799–825, 1969.

[3] G. Radhakrishnamacharya and V. Radhakrishna Murty, "Heat transfer to peristaltic transport in a non-uniform channel," *Defence Science Journal*, vol. 43, no. 3, pp. 275–280, 1993.

[4] T. Hayat, Y. Wang, K. Hutter, S. Asghar, and A. M. Siddiqui, "Peristaltic transport of an Oldroyd-B fluid in a planar channel," *Mathematical Problems in Engineering*, vol. 2004, no. 4, pp. 347–376, 2004.

[5] K. Vajravelu, S. Sreenadh, and V. R. Babu, "Peristaltic transport of a Herschel-Bulkley fluid in an inclined tube," *International Journal of Non-Linear Mechanics*, vol. 40, no. 1, pp. 83–90, 2005.

[6] S. Nadeem, T. Hayat, N. S. Akbar, and M. Y. Malik, "On the influence of heat transfer in peristalsis with variable viscosity," *International Journal of Heat and Mass Transfer*, vol. 52, no. 21-22, pp. 4722–4730, 2009.

[7] T. Hayat, M. U. Qureshi, and Q. Hussain, "Effect of heat transfer on the peristaltic flow of an electrically conducting fluid in

a porous space," *Applied Mathematical Modelling*, vol. 33, no. 4, pp. 1862–1873, 2009.

[8] K. S. Mekheimer, S. Z. A. Husseny, and Y. A. Elmaboud, "Effects of heat transfer and space porosity on peristaltic flow in a vertical asymmetric channel," *Numerical Methods for Partial Differential Equations*, vol. 26, no. 4, pp. 747–770, 2010.

[9] A. Ogulu, "Effect of heat generation on low Reynolds number fluid and mass transport in a single lymphatic blood vessel with uniform magnetic field," *International Communications in Heat and Mass Transfer*, vol. 33, no. 6, pp. 790–799, 2006.

[10] N. T. M. Eldabe, M. F. El-Sayed, A. Y. Ghaly, and H. M. Sayed, "Mixed convective heat and mass transfer in a non-Newtonian fluid at a peristaltic surface with temperature-dependent viscosity," *Archive of Applied Mechanics*, vol. 78, no. 8, pp. 599–624, 2008.

[11] S. Srinivas and M. Kothandapani, "The influence of heat and mass transfer on MHD peristaltic flow through a porous space with compliant walls," *Applied Mathematics and Computation*, vol. 213, no. 1, pp. 197–208, 2009.

[12] S. Srinivas and M. Kothandapani, "Peristaltic transport in an asymmetric channel with heat transfer—a note," *International Communications in Heat and Mass Transfer*, vol. 35, no. 4, pp. 514–522, 2008.

[13] T. A. Ogulu, "Effect of heat generation on low Reynolds number fluid and mass transport in a single lymphatic blood vessel with uniform magnetic field," *International Communications in Heat and Mass Transfer*, vol. 33, no. 6, pp. 790–799, 2006.

[14] O. Eytan, A. J. Jaffa, and D. Elad, "Peristaltic flow in a tapered channel: application to embryo transport within the uterine cavity," *Medical Engineering and Physics*, vol. 23, no. 7, pp. 473–482, 2001.

[15] M. Kothandapani and J. Prakash, "The peristaltic transport of Carreau nanofluids under effect of a magnetic field in a tapered asymmetric channel: application of the cancer therapy," *Journal of Mechanics in Medicine and Biology*, vol. 15, Article ID 1550030, 2014.

[16] M. Kothandapani and J. Prakash, "Effect of radiation and magnetic field on peristaltic transport of nanofluids through a porous space in a tapered asymmetric channel," *Journal of Magnetism and Magnetic Materials*, vol. 378, pp. 152–163, 2015.

[17] L. M. Srivastava, V. P. Srivastava, and S. N. Sinha, "Peristaltic transport of a physiological fluid. Part-I. Flow in non-uniform geometry," *Biorheology*, vol. 20, no. 2, pp. 153–166, 1983.

[18] M. Kothandapani and J. Prakash, "Influence of heat source, thermal radiation and inclined magnetic field on peristaltic flow of a Hyperbolic tangent nanofluid in a tapered asymmetric channel," *IEEE Transactions on NanoBioscience*, 2014.

[19] M. Kothandapani and J. Prakash, "Effects of thermal radiation parameter and magnetic field on the peristaltic motion of Williamson nanofluids in a tapered asymmetric channel," *International Journal of Heat and Mass Transfer*, vol. 81, pp. 234–245, 2015.

25

On the Stability of a Compressible Axial Flow with an Axial Magnetic Field

M. Subbiah[1] and M. S. Anil Iype[2]

[1] *Department of Mathematics, Pondicherry University, Kalapet, Pondicherry 605014, India*
[2] *Geomagnetic Observatory, Indian Institute of Geomagnetism, Pondicherry University Campus, Kalapet, Pondicherry 605014, India*

Correspondence should be addressed to M. S. Anil Iype; aniliype@gmail.com

Academic Editor: Miguel Onorato

We consider the stability problem of inviscid compressible axial flows with axial magnetic fields following the work of Dandapat and Gupta (Quarterly of Applied Mathematics, 1975). A numerical study of the stability of some basic flows has been carried out and it is found that an increase in the magnetic field strength has a stabilizing effect on subsonic flows and a destabilizing effect on supersonic flows. An analytical study of the stability problem has also been done in the present paper, but this analytical study is restricted by the approximation $M \ll 1$ and $c_i \ll 1$, where M is the Mach number and c_i is the imaginary part of the complex phase velocity c. A semicircular region depending on the magnetic field parameter and the Mach number is found for subsonic disturbances and as a consequence it is found that sufficiently strong magnetic field stabilizes all subsonic disturbances. Under a weak magnetic field, it is shown that short subsonic disturbances are stable.

1. Introduction

The stability of inviscid shear flows of a compressible fluid was studied in Blumen [1]. In order to simplify the stability problem, Blumen [1] focused attention on the stability of basic parallel shear flow of a perfect gas whose thermodynamic state is constant. From the equations of motion, the pressure of the shear flow is a constant only and if the basic flow temperature is also a constant, it follows that the basic flow density is also a constant. In the context of magnetogasdynamics, Dandapat and Gupta [2] studied the stability of a parallel flow of an inviscid perfectly conducting gas in the presence of a uniform magnetic field. Following Blumen [1], Dandapat and Gupta [2] also considered the case of a constant basic thermodynamic state. In Dandapat and Gupta [3] the stability of a nondissipative axial flow of a compressible conducting fluid between two concentric cylinders in the presence of a uniform axial field was studied. In cylindrical polar coordinates (r, θ, z) they considered the basic flow with velocity $(0, 0, W(r))$, magnetic field $(0, 0, H_0)$, where H_0 is a constant, $P_0(r)$ is the pressure, and $\rho_0(r)$ is the density. In the absence of the swirl component of the velocity in the basic flow it follows from the equations of magnetohydrodynamics that P_0 is a constant. In their stability analysis, Dandapat and Gupta [3] noticed that the equation involving the density perturbation ρ_0 is not needed in deriving the stability equation. Dandapat and Gupta [3] explained this by observing that there is no basic swirl velocity and therefore the mechanism of the centrifugal acceleration playing the role of a radial effective gravity is absent in this stability analysis. Consequently we consider the stability of a basic flow of a compressible fluid, whose thermodynamic state is constant.

The instability of an incompressible MHD basic flow with an axial velocity profile with a constant shear has been studied numerically in Shumlak and Roderick [4]. Four simulations were performed with varying values of peak velocity. Since Shumlak and Roderick [4] have considered basic flows with linear basic velocity profiles and variable density basic profiles, their numerical results show that increasing the value of the peak velocity reduces the Rayleigh-Taylor instability due to density variations. It may be noted here that Shumlak and Roderick [4] have also done an analytical study of the stability problem but in slab geometry and not in cylindrical geometry as in the case with their numerical study. However it has been observed in Zhang and Ding [5] that the incompressibility condition makes the effects of the plasma compressibility and magnetic field undetectable. But in the early stage of implosion, due to the relatively low plasma temperature, the effect of the compressibility becomes important. Consequently, the impact of compressibility on the stability in Z-pinch implosions in which equilibrium axial flows are included has been studied in Zhang and Ding [5]. However Zhang and Ding [5] have considered only the slab geometry and not cylindrical annular geometry. The basic axial velocity profile is a linear profile as in Shumlak and Roderick [4]. It is found from the study of Zhang and Ding [5] that the effect of plasma compressibility is to reduce the growth rate of unstable modes. It is observed in Zhang and Ding [5] that the incompressible approximation neglects the effects of compressibility and magnetic field, which will become important when the plasma temperature is relatively low. Therefore in the early stage of implosion, the compressibility model is much more suitable than the incompressible one. With the cooperation of sheared axial flow and magnetic field, it is found that plasma compressibility improves the stability of the basic flow significantly.

In their work, Dandapat and Gupta [3] have discussed the axisymmetric stability of a pure axial flow with velocity $(0, 0, W(r))$ of a compressible perfectly conducting fluid between two concentric cylinders permeated by a uniform axial magnetic field. They have shown that the complex wave speed c ($= c_r + ic_i$, $c_i > 0$) for any unstable wave lies in a semicircle in the upper half plane, having the range of the axial velocity as the diameter. In the incompressible case the same problem has been studied earlier in Howard and Gupta [6] and it is shown that the instability region is given by a semicircle of radius $((1/2)(b-a)^2 - V_A^2)$, where $a = \min W(r)$, $b = \max W(r)$, and V_A is the Alfven velocity. It is seen that the instability region depends on the magnetic field and one can conclude from this that a sufficient condition for stability to axisymmetric perturbations of axial flow of an incompressible fluid with uniform axial magnetic field is that the Alfven speed of the axial field should exceed half the maximum velocity difference.

A natural question that arises here is whether we can get a semicircle theorem for compressible flows where the radius depends on the magnetic field and consequently we can get a sufficient condition for stability as the magnetic field to be sufficiently strong. In this paper we find an answer to this question by obtaining such a semicircle for a special class of basic flows and for a special class of disturbances. We consider a basic flow with velocity field $(0, 0, W(r))$ and we denote the Mach number by M. The special class of basic flows considered in the analysis satisfies the conditions $M \ll 1$ and $c_i \ll 1$ so that their product can be ignored in comparison to unity. The special class of disturbances considered in the analysis is the subsonic ones which satisfy the condition $1 - M^2(W - c_r)^2 > 0$. Since a sufficiently strong magnetic field stabilizes the axial flow we study the problem under the weak magnetic field approximation and show that short waves are stable.

Then we make a numerical study of the same problem for a family of symmetric velocity profiles but without making any approximations and without any restrictions on the disturbances. The growth rate αc_i versus the wave number α and the neutral curves in the $\alpha - M$ plane are plotted for various values of the Mach number M and the magnetic field parameter S. It is found that an increase in the magnetic field strength has a stabilizing effect on subsonic flows and a destabilizing effect on supersonic flows. Furthermore it is found that all unstable modes are Holmboe modes, that is, propagating modes, and that they are long wave modes.

2. Formulation of the Problem

Consider nondissipative axial flow of a compressible conducting fluid between two concentric cylinders at $r = R_1, R_2$ in the presence of a uniform axial magnetic field. Thus, in cylindrical coordinates (r, θ, z) the basic velocity and magnetic fields are given by $(0, 0, W(r))$ and $(0, 0, H_0)$, respectively, and the pressure and density are constants P_0 and ρ_0, respectively.

We take the perturbed state as $\mathbf{q} = (u, v, W + w)$, $p^* = p_0 + p$, $\mathbf{H} = (h_r, h_\theta, H_0 + h_z)$, and $\rho^* = \rho_0 + \rho$. All perturbed quantities are assumed to be axisymmetric; that is, $u(r, \theta, z, t) = u(r)e^{i\alpha(z - ct)}$, where α is the real wave number and $c = c_r + ic_i$ is the complex wave velocity. Then the stability problem is governed by the equation (cf. Dandapat and Gupta [3]):

$$\left[\rho_0 \left\{ \frac{a_0^2(W - c)^2}{(W - c)^2 - a_0^2} + V_A^2 \right\} \frac{(rF)'}{r} \right]'$$
$$+ \rho_0 \alpha^2 \left[(W - c)^2 - V_A^2 \right] F = 0, \tag{1}$$

with the boundary conditions

$$F = 0 \quad \text{at } r = R_1, R_2, \tag{2}$$

where the variable $F(r)$ is obtained from $u(r)$ by the transformation $u(r) = (W - c)F(r)$.

Here $a_0 = (\gamma p_0/\rho_0)^{1/2}$ is the adiabatic sound speed, $V_A = (\mu_e H_0^2/4\pi\rho_0)^{1/2}$ is the Alfven velocity, and a prime denotes differentiation with respect to r. Let V be a characteristic velocity and let L be a characteristic length; for example, we can take $V = \max W(r)$ and $L = R_2 - R_1$. We nondimensionalize the variables by $r' = r/L$, $W' = W/V$, $c' = c/V$, and $V_A' = V_A/V$. Then (1) becomes, after dropping the prime,

$$D \left[\left\{ \frac{(W - c)^2}{1 - M^2(W - c)^2} - S \right\} D_* F \right]$$
$$+ \alpha^2 \left[S - (W - c)^2 \right] F = 0, \tag{3}$$

and the boundary conditions are $F(R_1) = 0 = F(R_2)$.

Here $M = V/a_0$ is the Mach number which is a measure of the compressibility of the fluid and $S = V_A^2/V^2$ is the magnetic field parameter which is a measure of the strength of the basic magnetic field. We have chosen to denote the magnetic field parameter by S following the works of S. C. Agrawal and G. S. Agrawal [7] and Rathy and Kishan [8]. Here $D = d/dr$ and $D_* = D + 1/r$.

3. Analytical Results

We consider the stability problem under an approximation introduced in Shivamoggi [9]. We take $M \ll 1$ and $c_i \ll 1$ so that their product can be neglected in comparison to unity. The condition $M \ll 1$ means that the basic flow is subsonic, while the condition $c_i \ll 1$ means that the unstable modes under consideration are very close to neutral modes. Consequently the stability equation becomes

$$D\left[\left\{\frac{(W-c)^2}{1-M^2(W-c_r)^2} - S\right\} D_* F\right]$$

$$+ \alpha^2 \left[S - (W-c)^2\right] F = 0, \tag{4}$$

and the boundary conditions are unchanged.

The disturbances are classified (under Shivamoggi's approximation) as subsonic, sonic, or supersonic according to $1 - M^2(W - c_r)^2 >, =, < 0$. Sonic disturbances satisfy the condition $1 - M^2(W - c_r)^2 = 0$ and their existence is possible only when W is a constant. From the semicircle theorem of Dandapat and Gupta [3] it follows that such a basic flow is stable and the neutral modes move with speed $c_r = W \pm 1/M$. For subsonic modes the wave velocity c_r of any disturbance satisfies the inequalities $W_{min} - 1/M < c_r < W_{max} + 1/M$ and this means that the wave velocity of subsonic modes is bounded. For supersonic modes the wave velocity c_r satisfies the inequalities $c_r \leq W_{min} - 1/M$ or $c_r \geq W_{max} + 1/M$ and this means that the wave velocity of supersonic modes needs not to be bounded. Thus the classification of the disturbances is based on the speed of propagation of the disturbances. Here it may be recalled that the basic flow (i) is a subsonic flow if $M < 1$, (ii) a sonic flow if $M = 1$, and (iii) a supersonic flow if $M > 1$. Hence our classification of the disturbances as subsonic, sonic, and supersonic is different from the classification of the basic flows.

Multiplying (4) by (rF^*) (where F^* is the complex conjugate of F), integrating the resultant equation over (R_1, R_2), and using the boundary conditions (2) we get

$$\int \left[\frac{(W-c)^2}{1-M^2(W-c_r)^2} - S\right] |D_* F|^2 r \, dr$$

$$+ \alpha^2 \int \left[(W-c)^2 - S\right] |F|^2 r \, dr = 0. \tag{5}$$

Taking the imaginary part of this equation gives

$$\int \frac{WQ \, dr}{1-M^2(W-c_r)^2}$$

$$= c_r \int \frac{Q \, dr}{1-M^2(W-c_r)^2}, \tag{6}$$

where $Q = (|D_* F|^2 + \alpha^2 |F|^2) r \geq 0$, while the real part gives

$$\int \frac{W^2 Q \, dr}{1-M^2(W-c_r)^2}$$

$$= (c_r^2 + c_i^2) \int \frac{Q \, dr}{1-M^2(W-c_r)^2} \tag{7}$$

$$+ S \int Q \, dr.$$

Following the standard procedure for proving the semicircle theorem we can get

$$\left[\left(c_r - \frac{a+b}{2}\right)^2 + c_i - \left(\frac{b-a}{2}\right)^2\right] \int \frac{Q \, dr}{1-M^2(W-c_r)^2}$$

$$+ S \int Q \, dr \leq 0. \tag{8}$$

From this we get the instability region as

$$\left(c_r - \frac{a+b}{2}\right)^2 + c_i \leq \left(\frac{b-a}{2}\right)^2. \tag{9}$$

This semicircle theorem has been already obtained in Dandapat and Gupta [3] without the restrictions imposed here. However we can improve upon this semicircle theorem. It follows from (9) that

$$\frac{1}{1-M^2(W-c_r)^2} \leq \frac{1}{1-M^2(b-a)^2}. \tag{10}$$

Using this in (8) we get

$$\left[\left(c_r - \frac{a+b}{2}\right)^2 + c_i - \left(\frac{b-a}{2}\right)^2\right] \int \frac{Q \, dr}{1-M^2(b-a)^2}$$

$$+ S \int Q \, dr \leq 0, \tag{11}$$

and from this we get the improved instability region as

$$\left(c_r - \frac{a+b}{2}\right)^2 + c_i^2$$

$$\leq \left(\frac{b-a}{2}\right)^2 - S\left(1 - M^2(b-a)^2\right). \tag{12}$$

For an incompressible flow $M = 0$, this reduces to the semicircle theorem of Howard and Gupta [6].

From (12) it follows that a sufficient condition for stability is that

$$S \geq \frac{(b-a)^2}{4\left(1 - M^2(b-a)^2\right)}. \tag{13}$$

Thus we see that a sufficiently strong axial magnetic field stabilizes an axial flow (under the restrictions mentioned earlier). This result reduces to the sufficient conditions of Howard and Gupta [6] for incompressible flows.

Since the flow is stable under a strong magnetic field let us study the problem under the weak magnetic field approximation. In particular we take $S \ll 1$ so that the term SD_*F in (4) can be neglected. However the other magnetic field term $\alpha^2 S$ is not neglected as it contains the wave number α which may be large. Under this weak magnetic field approximation the stability equation (4) becomes

$$D\left[\frac{(W-c)^2 D_* F}{1 - M^2(W - c_r)^2}\right] + \alpha^2\left[S - (W-c)^2\right]F = 0, \quad (14)$$

and the boundary conditions remain unchanged.

For an unstable mode $c_i > 0$ and the transformation $G = (W-c)^{1/2}F$ is well defined.

Substituting for F in (14) gives the equation

$$D\left[\frac{(W-c)\,D_* G}{1 - M^2(W - c_r)^2}\right]$$

$$+ \frac{(DW)\,G}{2r\left(1 - M^2(W - c_r)^2\right)}$$

$$- \frac{1}{2}D\left[\frac{(DW)}{1 - M^2(W - c_r)^2}\right]G \quad (15)$$

$$- \frac{1}{4}\frac{(DW)^2 G}{(W-c)\left(1 - M^2(W - c_r)^2\right)}$$

$$+ \alpha^2\left[S - (W-c)^2\right]\frac{G}{(W-c)} = 0,$$

and the corresponding boundary conditions are

$$G(R_1) = 0 = G(R_2). \quad (16)$$

Multiplying (15) by rG^* (where G^* is the complex conjugate of G), integrating the resultant equation over (R_1, R_2), and using the boundary conditions (16) we get

$$\int \frac{(W-c)\left|D_* G\right|^2 r\,dr}{1 - M^2(W - c_r)^2}$$

$$- \frac{1}{2}\int \frac{(DW)\left|G\right|^2 dr}{1 - M^2(W - c_r)^2}$$

$$+ \frac{1}{2}\int D\left[\frac{(DW)}{1 - M^2(W - c_r)^2}\right]|G|^2 r\,dr \quad (17)$$

$$+ \frac{1}{4}\frac{(DW)^2 |G|^2 r\,dr}{(W-c)\left(1 - M^2(W - c_r)^2\right)}$$

$$- \alpha^2 \int \left[S - (W-c)^2\right]\frac{|G|^2 r\,dr}{(W-c)} = 0.$$

Taking the imaginary part in (17) gives

$$\int\left(\frac{|D_* G|^2}{1 - M^2(W - c_r)^2} + \alpha^2|G|^2\right)r\,dr$$

$$+ \int\left[\alpha^2 S - \frac{(DW)^2}{4\left(1 - M^2(W - c_r)^2\right)}\right]\frac{|G|^2 r\,dr}{|W - c|^2} \quad (18)$$

$$= 0.$$

Since the first term is nonnegative it is necessary for instability that

$$\alpha^2 S - \frac{(DW)^2}{4\left(1 - M^2(W - c_r)^2\right)} < 0, \quad (19)$$

at least once in (R_1, R_2). It follows that the normal mode with wave number α is stable if

$$\alpha^2 \geq \frac{(DW)^2_{\max}}{4S\left(1 - M^2(b - a)^2\right)}; \quad (20)$$

that is, short waves are stable under the weak magnetic field approximation.

4. Numerical Results

We make a numerical study of the influence of a constant axial magnetic field on the stability of an axial flow of a compressible fluid. The numerical study carried out in this paper consists of finding the real eigenvalues corresponding to neutral modes and the complex eigenvalues corresponding to unstable modes. The numerical computations are made here with the use of two computer programs given in Hazel [10]. These programs of Hazel [10] are implemented by us in Mathematica 7 using the inbuilt NDSolve for solving the stability equation with homogeneous boundary conditions. Since we do not make the approximations of Section 3 that were used in getting the analytical results the equation to be used is (3) and the boundary conditions are (4). We consider the flow in the region between the two cylinders at $R_1 = 2$ and $R_2 = 4$. The axial velocity considered is

$$W(r) = \lambda_1\left[\text{Sech}\left\{\left(r - \frac{R_1 + R_2}{2}\right)\frac{\lambda_2}{R_2 - R_1}\right\} - \lambda_3\right]. \quad (21)$$

Here the value of the parameter λ_3 does not affect the stability of the flow, but we choose a suitable value of λ_3 for making $W(r)$ zero on the boundaries.

In Figure 1 we have plotted a family of velocity profiles corresponding to different values of λ_1 in (21) and a fixed value of λ_2. All velocity profiles are symmetric profiles about $r = (R_1 + R_2)/2$. An increase in λ_1 corresponds to an increase in $(W_{\max} - W_{\min})$.

From Figure 2 it is seen that that an increase in λ_1 corresponds to an increase in the range of unstable wave numbers and consequently we can conclude that an increase in the value of $(W_{\max} - W_{\min})$ corresponds to an increase in

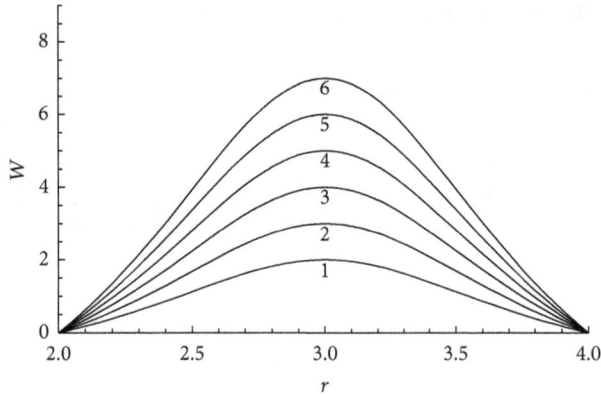

FIGURE 1: Family of Sech velocity profiles (1 to 6) for different values of the parameters given in Table 1.

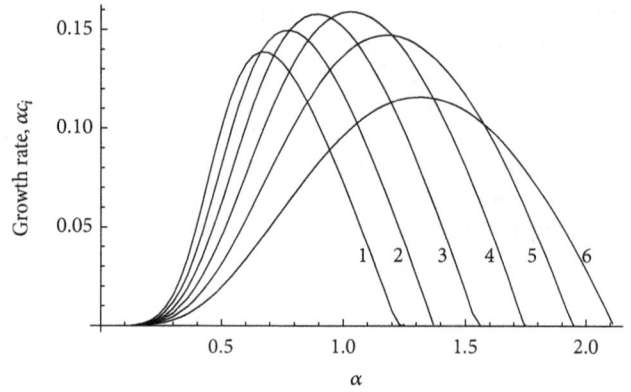

FIGURE 2: Growth rate versus wave number plotted for the velocity profiles (1 to 6) shown in Figure 1 for $S = 0$ and $M = 0.5$.

TABLE 1: Axial velocity profiles.

Profile number	λ_1	λ_2	λ_3
1	3.0183	3.50	0.33736
2	4.5274	3.50	0.33736
3	6.0365	3.50	0.33736
4	7.5456	3.50	0.33736
5	9.0547	3.50	0.33736
6	10.5638	3.50	0.33736
7	3.193	7.00	0.0603
8	3.331	6.00	0.0993
9	3.584	5.00	0.1631
10	3.924	4.25	0.2355
11	4.528	3.50	0.3374

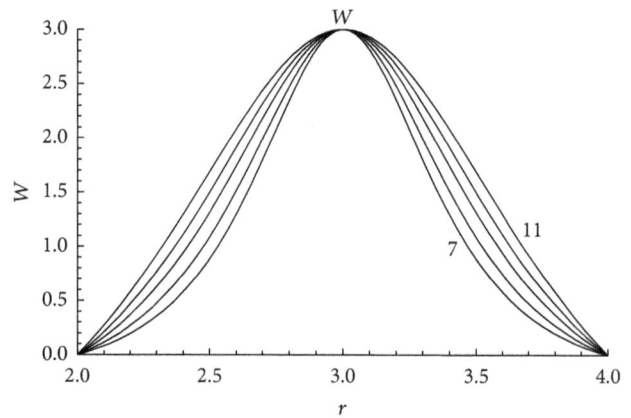

FIGURE 3: Family of Sech velocity profiles (7 to 11). The values of the parameters given in Table 1.

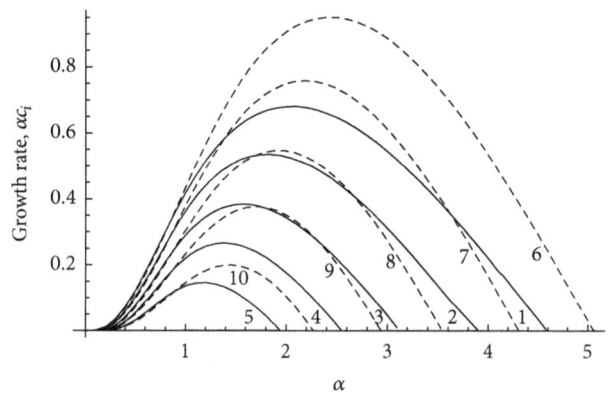

FIGURE 4: Growth rate versus wave number plotted for the velocity profiles (7 to 11) shown in Figure 3. The dashed lines correspond to $S = 0$ and $M = 0$ and the continuous lines correspond to $S = 0$ and $M = 0.5$.

instability. However we should observe from Figure 2 that the maximum growth rate of unstable modes does not always increase with an increase in the values of λ_1. It is seen that the maximum growth rate increases with an increase in λ_1 up to a certain value of λ_1 and when λ_1 increase beyond that value the maximum growth rate is found to be decreasing. Hence the conclusion we can make from Figure 2 is that the enhancement of instability due to an increase in λ_1 is by the increase in the spectral range of instability and not by an increase in the maximum growth rate of unstable modes.

In Figure 3 we have plotted some velocity profiles corresponding to different values of λ_1 and λ_2. All these velocity profiles have the same value at the end points and the same value of $W_{\max}$. From Figure 4 it is seen that when $S = 0$ and $M = 0.5$, the spectral range is small for the outer profile and the spectral range is increased for the inner profiles. This may be because $|W'|_{\max}$ increases as we move from the outer profile to the inner profile. It is also seen that the growth rate curves for $S = 0$, $M = 0.5$ case are always below the growth rate curves for the purely hydrodynamic case of $S = 0$ and $M = 0$. Thus compressibility is seen to have a stabilizing effect on all the velocity profiles.

In Figure 5 we have plotted the classifier expression $1 - M^2(W - c_r)^2$ between the radial region from $r = 2$ to 4 for the eigenvalue c_r obtained in each case for the axial velocity

profile number 2. Mach number M is taken as 1.0 and magnetic field strength parameter S is increased from 0 through 0.2, 0.4, 0.6, and 0.8 and the corresponding five plots are shown above. Here we observe that the disturbance mode satisfies the condition of being subsonic, sonic, and supersonic modes. The supersonic condition is satisfied in

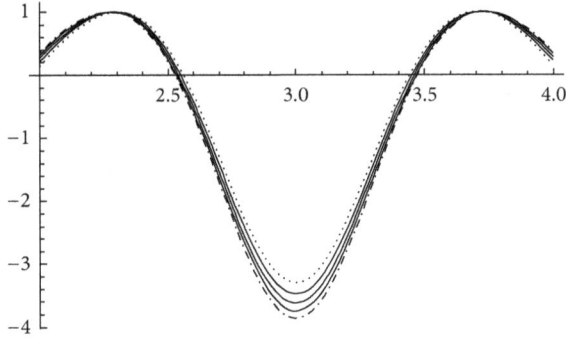

FIGURE 5: Plot of the expression $1 - M^2(W - c_r)^2$ as a function of r.

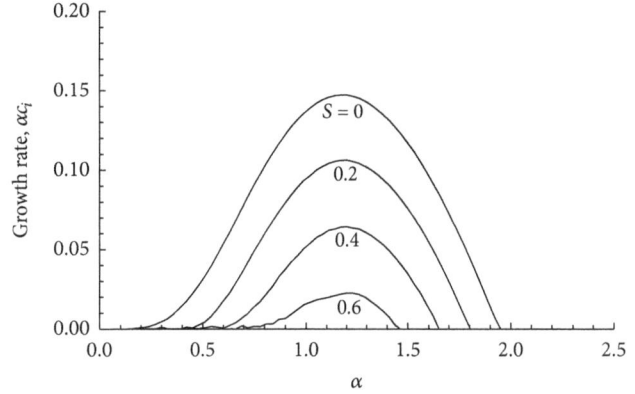

FIGURE 7: $M = 0.5$.

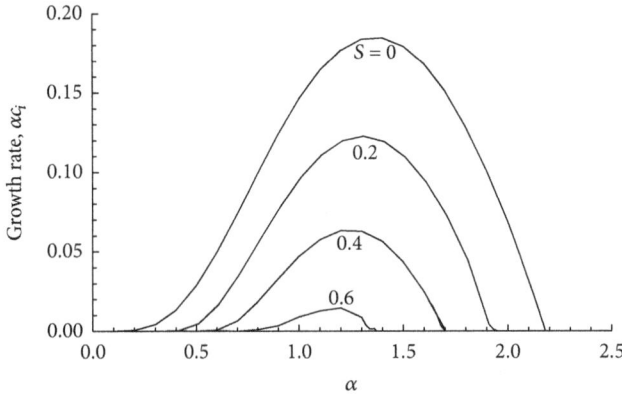

FIGURE 6: $M = 0.25$.

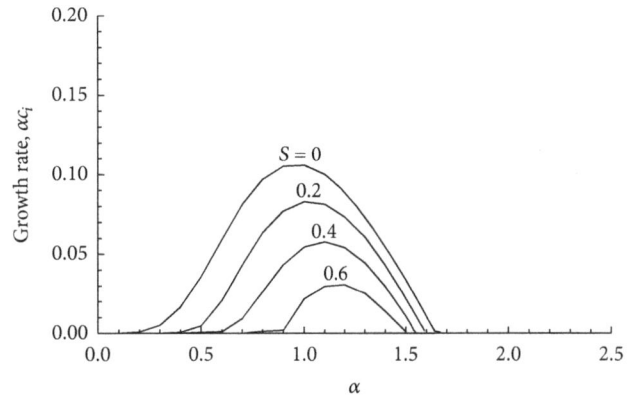

FIGURE 8: $M = 0.75$.

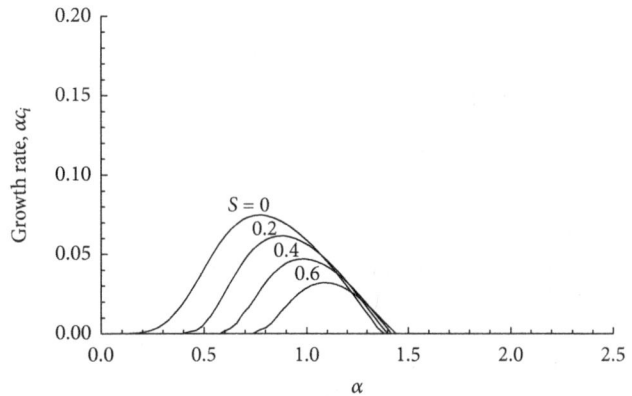

FIGURE 9: $M = 1$.

the central region of the flow profile. The dot-dashed curve corresponds to $S = 0.8$ and the dotted one corresponds to $S = 0$. From the above plots we can infer that an increase in axial magnetic field strength enhances the speed of supersonic disturbances slightly. The axial velocity profile and other parameters used here are the same which we have used in Figure 5. There we have seen that, beyond the magnetic field strength of 0.9, the flow becomes stabilized. In the above figure, the value of $(b - a)^2/4(1 - M^2(b - a)^2)$ is -0.333333 for all the five curves and is less than S in each case. The sufficient condition for stabilization of subsonic flows with strong magnetic field, which we have obtained in the analytical Section 3, that is $S \geq (b - a)^2/4(1 - M^2(b - a)^2)$, the subsonic disturbances are stabilized as per the above condition and since the flow is still unstable as seen from the growth rate plots of Figure 9, it is the supersonic disturbances that cause the flow instability. But further increase in the value of S (beyond 0.9) ensures complete stability. So we may infer from the above discussions that the subsonic disturbances get stabilized earlier than the supersonic disturbances as we increase the magnetic field strength parameter S. In other words, the magnetic field strength required for stabilization of supersonic disturbances has to be more than that needed for the stabilization of subsonic disturbances.

In Figures 6, 7, 8, 9, 10, and 11 we have plotted the growth rate αc_i versus the wave number α for various values of S and by choosing particular value for the Mach number M. We

have considered three subsonic basic flows with $M = 0.25$, $M = 0.5$, and $M = 0.75$, a sonic basic flow for which $M = 1$, and two supersonic basic flows with $M = 1.25$ and $M = 1.5$. It is seen that the spectral range (the range of α for which the instability occurs) in the magnetic case (i.e., $S \neq 0$) lies within the spectral range for the nonmagnetic range ($S = 0$) only for the subsonic flows, while the spectral range gets shifted to the right for the sonic and supersonic flows with an increase in the magnetic field parameter S. This means that the stabilizing role of compressibility is reduced by an

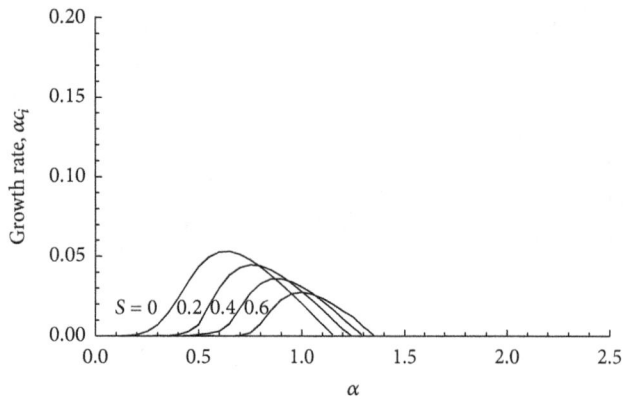

FIGURE 10: $M = 1.25$.

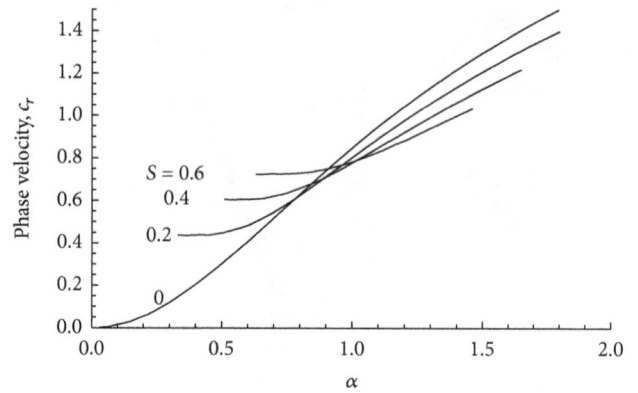

FIGURE 11: $M = 1.5$.

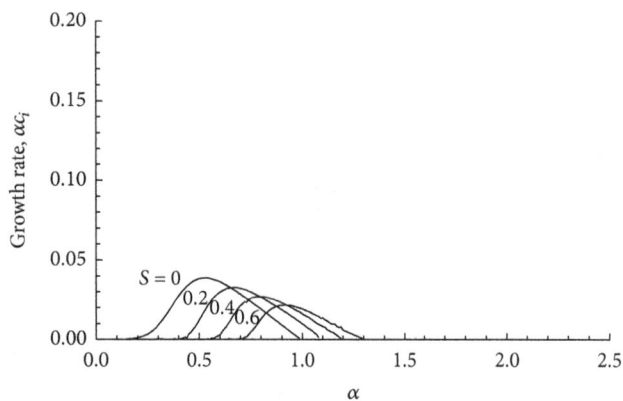

FIGURE 12: Neutral curves in in the $\alpha - M$ plane for values of $S = 0$, 0.1, 0.2, 0.3, and 0.4.

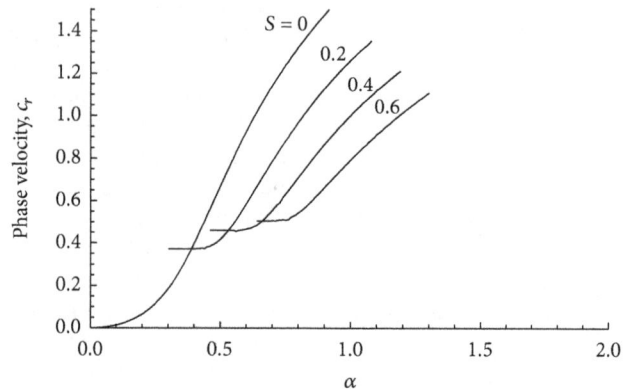

FIGURE 13: Phase velocity versus wave number for $M = 0.5$.

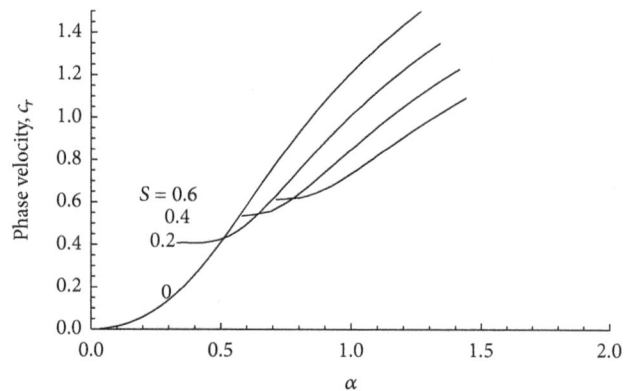

FIGURE 14: Phase velocity versus wave number for $M = 1$.

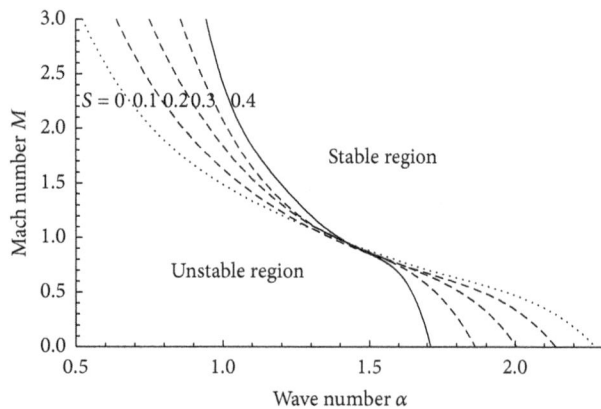

FIGURE 15: Phase velocity versus wave number for $M = 1.5$.

increase in the magnetic field strength. This is also seen from the neutral curves.

In Figure 12 we have plotted the neutral curves, that is, curves in which $c_i = 0$, in the $\alpha - M$ plane for various values of S. All these neutral curves intersect at a particular point. In comparison to the neutral curve for $S = 0$ (i.e., the nonmagnetic case) the other neutral curves lie below the $S = 0$ curve in the right side of the intersection point, while the neutral curves for $S \neq 0$ lie above the neutral curves for $S = 0$ to the left of the intersection point. It is seen from Figure 12 that the magnetic field has a stabilizing effect on the subsonic flows ($M < 1$) and a destabilizing effect on supersonic ones ($M > 1$). In Figures 13, 14, and 15 we have plotted the phase velocity c_r versus the wave number for different values of the Mach number M. In all the cases it is found that $c_r \neq 0$. This means that the unstable modes are Holmboe mode (i.e., propagating modes).

5. Discussion and Concluding Remarks

Dandapat and Gupta [2] have discussed the physical interpretation of the stabilizing effect of the compressibility of the fluid and that of the magnetic field. Compressibility is stabilizing because certain amount of the basic flow energy must be used to do work against the force due to the elasticity of the medium before it becomes available to cause instability, while the stabilizing effect of the magnetic field is due to the restoring force offered by the tension along the magnetic lines. In the context of the stability of axial flows of a compressible fluid in the presence of an uniform axial magnetic field, Dandapat and Gupta [3] have obtained a semicircle theorem for the axisymmetric unstable modes. In the present paper we have considered this problem under the approximation $M \ll 1$ and $c_i \ll 1$ (cf. Shivamoggi [9]) and have found an improved instability region which includes the magnetic field parameter S and the compressibility parameter M in the radius of the semicircle. The stabilizing effect of a strong magnetic field follows from this result. Consequently we have studied the problem under the weak magnetic field approximation and we have found that short waves are stable. From this it follows that a weak magnetic field is also stabilizing the disturbances.

Because of the complexity of the mathematical problem, analytical study of the stability of axial flows of a compressible fluid in the presence of a uniform magnetic field is limited. Consequently we have made a numerical study of this stability problem in the present work. Here it may be emphasized that in the numerical study of the problem, the approximation made in the analytical study presented in Section 3 of the present paper is not made and also that the disturbances are not restricted to be of any particular type. We have chosen a family of symmetric velocity profiles and the numerical study of their stability consists of finding the growth rate of the unstable modes and also their propagation speed. The growth rate αc_i versus the wave number α and the neutral curves in the $\alpha - M$ plane are plotted for various values of the Mach number M and the magnetic field parameter S. It is found that an increase in the magnetic field strength has a stabilizing effect on subsonic flows and a destabilizing effect on supersonic flows. Another conclusion of our numerical study is that the unstable disturbances are all Holmboe modes, that is, propagating modes, and that they are all long wave modes.

Acknowledgments

The authors are thankful to the referee for his constructive comments that helped them to improve their paper. The second author is grateful to Professor B. Veenadhari, Area Chairperson, Observatory and Data Analysis Division, and the Director, Indian Institute of Geomagnetism, Panvel, Mumbai, India for giving encouragement to do this research work.

References

[1] W. Blumen, "Shear layer instability of an inviscid compressible fluid," *Journal of Fluid Mechanics*, vol. 40, no. 4, pp. 769–781, 1970.

[2] B. S. Dandapat and A. S. Gupta, "Stability of magnetogasdynamic shear flow," *Acta Mechanica*, vol. 28, no. 1–4, pp. 77–83, 1977.

[3] B. S. Dandapat and A. S. Gupta, "On the stability of swirling flow in magnetogasdynamics," *Quarterly of Applied Mathematics*, vol. 33, no. 2, pp. 182–186, 1975.

[4] U. Shumlak and N. F. Roderick, "Mitigation of the Rayleigh-Taylor instability by sheared axial flows," *Physics of Plasmas*, vol. 5, no. 6, pp. 2384–2389, 1998.

[5] Y. Zhang and N. Ding, "Effects of compressibility on the magneto-Rayleigh-Taylor instability in Z-pinch implosions with sheared axial flows," *Physics of Plasmas*, vol. 13, no. 2, Article ID 022701, 2006.

[6] L. N. Howard and A. S. Gupta, "On the hydrodynamic and hydromagnetic stability of swirling flows," *Journal of Fluid Mechanics*, vol. 14, no. 3, pp. 463–476, 1962.

[7] S. C. Agrawal and G. S. Agrawal, "Hydromagnetic stability of heterogeneous shear flow," *Journal of the Physical Society of Japan*, vol. 27, no. 1, pp. 218–223, 1969.

[8] R. K. Rathy and H. Kishan, "Stability of hydromagnetc stratified shear flow," *Indian Journal of Pure and Applied Mathematics*, vol. 12, no. 6, pp. 764–768, 1981.

[9] B. K. Shivamoggi, "Inviscid theory of stability of parallel compressible flows," *Journal de Mecanique*, vol. 16, no. 2, pp. 227–255, 1977.

[10] P. Hazel, "Numerical studies of the stability of inviscid stratified shear flows," *Journal of Fluid Mechanics*, vol. 51, no. 1, pp. 39–61, 1972.

26

Pulsatile Non-Newtonian Laminar Blood Flows through Arterial Double Stenoses

Mir Golam Rabby, Sumaia Parveen Shupti, and Md. Mamun Molla

School of Engineering & Applied Science, Department of Electrical & Computer Engineering, North South University, Dhaka 1229, Bangladesh

Correspondence should be addressed to Md. Mamun Molla; mmamun@northsouth.edu

Academic Editor: Kuo-Kang Liu

The paper presents a numerical investigation of non-Newtonian modeling effects on unsteady periodic flows in a two-dimensional (2D) pipe with two idealized stenoses of 75% and 50% degrees, respectively. The governing Navier-Stokes equations have been modified using the Cartesian curvilinear coordinates to handle complex geometries. The investigation has been carried out to characterize four different non-Newtonian constitutive equations of blood, namely, the (i) Carreau, (ii) Cross, (iii) Modified Casson, and (iv) Quemada models. The Newtonian model has also been analyzed to study the physics of fluid and the results are compared with the non-Newtonian viscosity models. The numerical results are represented in terms of streamwise velocity, pressure distribution, and wall shear stress (WSS) as well as the vorticity, streamlines, and vector plots indicating recirculation zones at the poststenotic region. The results of this study demonstrate a lower risk of thrombogenesis at the downstream of stenoses and inadequate blood supply to different organs of human body in the Newtonian model compared to the non-Newtonian ones.

1. Introduction

Stenosis is characterized by localized arterial narrowing that is initiated due to deposition of lipid, cholesterol, and some other substances on the endothelium and is of major concern to most of the Western world. Atherosclerotic lesions preferentially occur in arteries and arterioles in regions of high curvature or bifurcations and junctions causing major changes in flow structure and consequently large changes in fluid loading on vessel walls [1]. Such plaques or arterial constrictions usually disturb normal blood flow through the artery and there is considerable evidence that hydrodynamic factors can play a significant role in the development and progression of these lesions. It has been established that once a mild stenosis is developed inside the arterial lumen, the resulting flow disorder further influences the development of the disease and the arterial deformability to some extent, which eventually changes the regional blood rheology as well [2].

The rheological behavior of blood can be identified by non-Newtonian viscosity. Halder [3] demonstrated that the rheology of blood and the fluid dynamical properties of blood flow can play an important role in the basic understanding, diagnosis, and treatment of many cardiovascular and arterial diseases. Now, stenosis not only develops in one position of artery but also it may develop at more than one location of the cardiovascular system. However, in many medical situations, the patient is found to have multiple stenoses in the same arterial segment. So, several studies were conducted by Misra et al. [4], Minagar et al. [5], Johnston and Kilpatrick [6], and Mustapha and Amin [7] to understand the effects of double stenoses on blood flow in arteries. A numerical investigation has been conducted for generalized Newtonian blood flows past a couple of irregular arterial stenoses [8].

A few studies have been carried out on multiple stenoses using the momentum integral method and finite element method. There are also many papers devoted to studying these phenomena experimentally. Kilpatrick et al. [9] performed one of the most extensive works on double stenoses where they worked on the vascular resistance of arterial stenoses in series. However, Ang and Mazumdar [10] worked on triplet stenoses and their research presented that multiple stenoses have more significant effects on blood flow compared to the sum of the consequences of the individual stenoses. Blood flow through irregular multistenoses has been investigated with and without magnetohydrodynamic effect by Mustapha et al. [11, 12].

Using computed outcomes founded on Galerkin finite element method Tu et al. [13] executed numerical simulations of models for steady and pulsatile blood flow for distinct constriction levels and Reynolds numbers inside the artery with rigid wall. Talukder et al. [14] experimented the consequences of multiple stenoses on pressure. Young and Tsai [15] experimentally investigated the steady and pulsatile flow aspects through stenotic arteries and found significant pressure decrease across the stenosis. On the contrary, Tu and Deville [16] presented a theoretical analysis of pulsatile flow of blood in stenosed arteries.

Blood is a complex mixture of cells, proteins, lipoproteins, and ions by which nutrients and wastes are transported. Red blood cells typically comprise approximately 40% of blood by volume. As red blood cells are small semisolid particles, they increase the viscosity of blood and affect the behavior of the fluid. Blood is approximately four times more viscous than water. Moreover, blood does not exhibit a constant viscosity at all flow rates and is especially non-Newtonian in the microcirculatory system [17]. Most of the authors have paid more attention in the investigation of the blood flow by assuming that the blood is Newtonian and homogeneous fluid. However, the non-Newtonian behavior is most evident at very low shear rates when the red blood cells clump together into larger particles. According to Berger and Jou [1] and Huang et al. [18], the shear rates fall below that asymptotic level when the viscosity of blood increases and the non-Newtonian properties of blood are exhibited, especially when the shear rates drop below $10 \, \text{s}^{-1}$. Blood also exhibits non-Newtonian behavior in small branches and capillaries, where the cells squeeze through microvasculature and a cell-free skimming layer reduces the effective viscosity through the tube.

The presence of moderate or severe stenoses in the artery can cause the flow to transit from laminar to transition in the downstream region. Moreover, wall pressure and wall shear stress initiated by stenoses play important roles in hemodynamics. Fry [19] revealed that high wall shear stress caused by atherosclerosis is a strong factor for endothelial or inner side damage in an artery. It can again overstimulate platelet thrombosis causing blockage [20]. Therefore, it is important to study the hemodynamic factors to understand the fundamental scenario behind the physiology of arterial diseases.

The aim of the present study is to investigate the non-Newtonian modeling effects on the unsteady periodic flow through an arterial segment with two stenoses of different degree using the most well-documented blood constitutive equations, namely, the Carreau [21], Cross [22], Modified González and Moraga [23], and Quemada [24] models. Newtonian and non-Newtonian flow computations have been carried out elaborately in order to examine the modeling

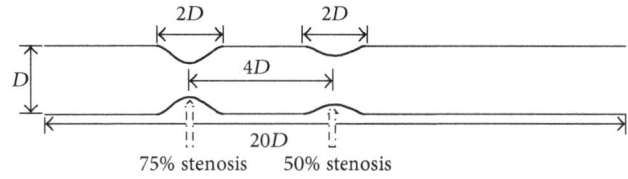

FIGURE 1: Schematic diagram for the double arterial stenoses.

effects with respect to the vortex formation, wall pressure, wall shear stress, and streamlines to achieve very good flow insight into a stenotic artery.

2. Governing Equation

Incompressible, homogeneous, and sinusoidal pulsatile flow is simulated for Reynolds number 300. The arterial wall is assumed rigid and blood is modeled as both Newtonian and non-Newtonian fluids for the flow field computation.

The geometry of the two-dimensional (2D) pipe with two cosine-shaped stenoses constricted symmetrically on both sides of the central axis is shown in Figure 1. The artery upstream and downstream to the stenoses have been considered as a straight rigid pipe. Due to the presence of the stenoses, the height of the pipe, δ, is a variable in the streamwise direction (i.e., $\delta = \delta(x)$). Away from the stenosis, the height of the pipe is constant and is represented here using D (i.e., $\delta = D$ in the region either upstream or downstream of the stenosis). First stenosis is centered $5D$ downstream of the pipe inlet (i.e., the inlet location is $x/D = -5$) and second stenosis is centered $4D$ downstream of the first stenosis. The stenoses are centered at $x/D = 0.0$ and $x/D = 4.0$ axial locations and each stenosis length is $2D$. The length of the prestenotic region has been considered to be smaller than the poststenotic region for detailed flow insight into the downstream region. The cosine-shaped symmetrically constricted regions are modeled using the following formula:

$$\frac{y}{D} = 1 - \frac{f_c}{2}\left(1 + \cos\frac{x\pi}{D}\right); \quad -D \leq x \leq D, \quad (1)$$

where $f_c = (1/2)[1 - ((100 - \text{percentage})/100)^2]$ is a parameter that controls the reduction of the cross-sectional area of stenosis. In the present study, a 75% and 50% reduction of the cross-sectional area has been considered at the center of the first and second stenosis, correspondingly. Generally, the 75% constricted stenosis is referred to as critical stenosis whereas the 50% constricted one is referred to as severe stenosis.

Here x and y are used to represent the streamwise and radial coordinates, respectively. Also the tensor notation is used in these two directions and it is represented by indices 1 and 2, respectively:

$$\frac{\partial u}{\partial x} + \frac{\partial v}{\partial y} = 0,$$

$$\rho \left(\frac{\partial u}{\partial t} + u\frac{\partial u}{\partial x} + v\frac{\partial u}{\partial y} \right)$$

$$= -\frac{\partial p}{\partial x} + \frac{\partial}{\partial x}\left(2\mu\frac{\partial u}{\partial x} \right) + \frac{\partial}{\partial y}\left(\mu\frac{\partial u}{\partial y} \right) + \frac{\partial}{\partial y}\left(\mu\frac{\partial u}{\partial x} \right), \quad (2)$$

$$\rho \left(\frac{\partial v}{\partial t} + u\frac{\partial v}{\partial x} + v\frac{\partial v}{\partial y} \right)$$

$$= -\frac{\partial p}{\partial y} + \frac{\partial}{\partial x}\left(\mu\frac{\partial v}{\partial x} \right) + \frac{\partial}{\partial y}\left(2\mu\frac{\partial u}{\partial y} \right) + \frac{\partial}{\partial x}\left(\mu\frac{\partial u}{\partial y} \right).$$

The blood viscosity, $\mu = \mu(|\dot\gamma|)$, depends on the shear rate $\dot\gamma = (1/2)((\partial u_i/\partial x_j)+(\partial u_j/\partial x_i))$, and its magnitude is defined as $|\dot\gamma| = \sqrt{2\dot\gamma_{ij}\dot\gamma_{ji}}$. When blood is treated as a Newtonian fluid, its viscosity tends to a constant value which is denoted by $\mu_\infty = 3.45 \times 10^{-3}$ Pa · s, while for a non-Newtonian model, constitutive relations are used for the apparent viscosity of the blood that are presented in Section 3.

To compute the blood flow though double constricted artery, the governing equations are transformed into curvilinear coordinates. Thompson et al. [25] introduced an approach where the finite difference equations are formulated in a transformed curvilinear coordinate system that coincides with the boundaries of the fluid domain. In this approach flow domain in physical space is mapped onto a rectangular domain in computational space, as shown in Figure 3. For mapping $x_j \rightarrow \xi_j$, if J_{ij} represents the elements of the Jacobian matrix, $\mathbf{J}$, of the transformation then

$$J_{ij} = \frac{\partial x_i}{\partial \xi_j}. \quad (3)$$

The determinate of the Jacobian matrix, $\mathbf{J}$, is denoted by $|\mathbf{J}|$ and given by

$$|\mathbf{J}| = \frac{\partial x_i}{\partial \xi_j} A_{ij}, \quad (4)$$

where A_{ij} are the elements of the cofactor matrix, $\mathbf{A}$, of the Jacobian, defined as

$$|\mathbf{A}| = |\mathbf{J}|\mathbf{J}^{-1}. \quad (5)$$

By applying the chain rule, the derivatives can now be expressed in the transformed variables in the following way:

$$\frac{\partial \phi}{\partial x_i} = \frac{\partial \phi}{\partial \xi_j}\frac{\partial \xi_j}{\partial x_i} = \frac{A_{ij}}{|\mathbf{J}|}\frac{\partial \phi}{\partial \xi_j}, \quad (6)$$

where ϕ is a generic variable.

The governing equations for an incompressible flow take the following forms in the general Cartesian curvilinear coordinate system:

$$\frac{A_{11}}{|\mathbf{J}|}\frac{\partial u}{\partial \xi_1} + \frac{A_{12}}{|\mathbf{J}|}\frac{\partial u}{\partial \xi_2} + \frac{A_{21}}{|\mathbf{J}|}\frac{\partial v}{\partial \xi_1} + \frac{A_{22}}{|\mathbf{J}|}\frac{\partial v}{\partial \xi_2} = 0,$$

$$\rho\frac{\partial u}{\partial t} + \frac{\rho}{|\mathbf{J}|}\left[uA_{11} + vA_{21}\right]\frac{\partial u}{\partial \xi_1} + \frac{\rho}{|\mathbf{J}|}\left[uA_{12} + vA_{22}\right]\frac{\partial u}{\partial \xi_2}$$

$$= -\left[\frac{A_{11}}{|\mathbf{J}|}\frac{\partial p}{\partial \xi_1} + \frac{A_{12}}{|\mathbf{J}|}\frac{\partial p}{\partial \xi_2}\right] + \frac{1}{|\mathbf{J}|^2}\frac{\partial}{\partial \xi_1}\left(\mu\frac{\partial u}{\partial \xi_1}\right)$$

$$\times \left(2A_{11}^2 + A_{21}^2\right)$$

$$+ \frac{1}{|\mathbf{J}|^2}\frac{\partial}{\partial \xi_1}\left(\mu\frac{\partial u}{\partial \xi_2}\right)\left(2A_{11}A_{12} + A_{21}A_{22}\right)$$

$$+ \frac{1}{|\mathbf{J}|^2}\frac{\partial}{\partial \xi_2}\left(\mu\frac{\partial u}{\partial \xi_1}\right)\left(2A_{11}A_{12} + A_{21}A_{22}\right)$$

$$+ \frac{1}{|\mathbf{J}|^2}\frac{\partial}{\partial \xi_2}\left(\mu\frac{\partial u}{\partial \xi_2}\right)\left(2A_{12}^2 + A_{22}^2\right)$$

$$+ \frac{A_{21}}{|\mathbf{J}|}\frac{\partial}{\partial \xi_1}\left(\mu\left\{\frac{A_{11}}{|\mathbf{J}|}\frac{\partial v}{\partial \xi_1} + \frac{A_{12}}{|\mathbf{J}|}\frac{\partial v}{\partial \xi_2}\right\}\right)$$

$$+ \frac{A_{22}}{|\mathbf{J}|}\frac{\partial}{\partial \xi_2}\left(\mu\left\{\frac{A_{11}}{|\mathbf{J}|}\frac{\partial v}{\partial \xi_1} + \frac{A_{12}}{|\mathbf{J}|}\frac{\partial v}{\partial \xi_2}\right\}\right),$$

$$\rho\frac{\partial v}{\partial t} + \frac{\rho}{|\mathbf{J}|}\left[uA_{11} + vA_{21}\right]\frac{\partial v}{\partial \xi_1} + \frac{\rho}{|\mathbf{J}|}\left[uA_{12} + vA_{22}\right]\frac{\partial u}{\partial \xi_2}$$

$$= -\left[\frac{A_{21}}{|\mathbf{J}|}\frac{\partial p}{\partial \xi_1} + \frac{A_{22}}{|\mathbf{J}|}\frac{\partial p}{\partial \xi_2}\right] + \frac{1}{|\mathbf{J}|^2}\frac{\partial}{\partial \xi_1}\left(\mu\frac{\partial v}{\partial \xi_1}\right)$$

$$\times \left(A_{11}^2 + 2A_{21}^2\right)$$

$$+ \frac{1}{|\mathbf{J}|^2}\frac{\partial}{\partial \xi_1}\left(\mu\frac{\partial v}{\partial \xi_2}\right)\left(A_{11}A_{12} + 2A_{21}A_{22}\right)$$

$$+ \frac{1}{|\mathbf{J}|^2}\frac{\partial}{\partial \xi_2}\left(\mu\frac{\partial v}{\partial \xi_1}\right)\left(A_{11}A_{12} + 2A_{21}A_{22}\right)$$

$$+ \frac{1}{|\mathbf{J}|^2}\frac{\partial}{\partial \xi_2}\left(\mu\frac{\partial v}{\partial \xi_2}\right)\left(A_{12}^2 + 2A_{22}^2\right)$$

$$+ \frac{A_{21}}{|\mathbf{J}|}\frac{\partial}{\partial \xi_1}\left(\mu\left\{\frac{A_{11}}{|\mathbf{J}|}\frac{\partial u}{\partial \xi_1} + \frac{A_{12}}{|\mathbf{J}|}\frac{\partial u}{\partial \xi_2}\right\}\right)$$

$$+ \frac{A_{22}}{|\mathbf{J}|}\frac{\partial}{\partial \xi_2}\left(\mu\left\{\frac{A_{11}}{|\mathbf{J}|}\frac{\partial u}{\partial \xi_1} + \frac{A_{12}}{|\mathbf{J}|}\frac{\partial u}{\partial \xi_2}\right\}\right),$$

$$(7)$$

where A_{ij} are the elements of the cofactor matrix, $\mathbf{A}$, of the Jacobian $|\mathbf{J}|$.

2.1. Boundary Conditions. In the present study, no slip condition has been used at the arterial wall, where velocity u and v are zero. The sinusoidal pulsatile laminar velocity profile $u(r) = 2U[1 - (y/R)^2][1 + 0.3\sin t]$ is used to generate

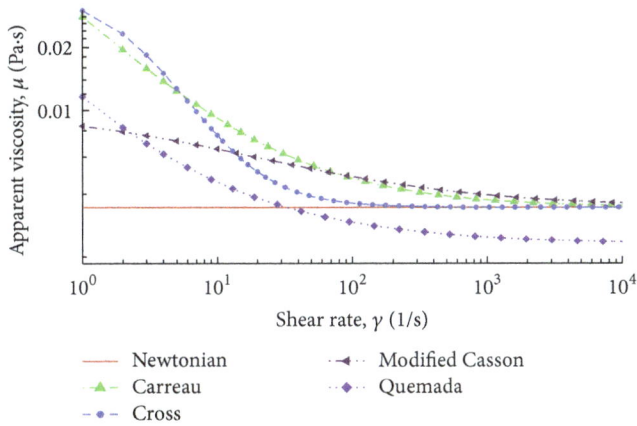

FIGURE 2: Relations between the shear rate and the apparent blood viscosity for the different models.

the time-dependent pulsatile boundary condition at the inlet of pipe, where bulk velocity, U, depends on the Reynolds number and y is the radial length of artery. The zero gradient condition is applied at the outlet of artery, where velocity gradient u and v are zero along the streamwise direction.

3. Non-Newtonian Viscosity Models

Newtonian fluid viscosity is always constant against the shear rate. On the other hand, non-Newtonian fluid viscosity changes depending on the shear rate. Figure 2 represents the relationship between blood viscosity and the shear rate for Newtonian and four different non-Newtonian models. The Newtonian and these four non-Newtonian models are summarized in Table 1.

The relationship between the shear rates and viscosity for the non-Newtonian blood viscosity models, that is, Carreau, Cross, Modified Casson, and Quemada models, along with the Newtonian viscosity model is presented in Figure 2. Blood viscosity is constant in the Newtonian model shown by the solid line. On the other hand, the viscosity of blood produced by non-Newtonian models for low shear rates (less than $100\,\mathrm{s}^{-1}$) is higher than that of the Newtonian model. Viscosity in the Carreau and Modified Casson models tends to asymptotic constant viscosity, μ_∞, at the shear rate, $\dot{\gamma}$, greater than $10^4\,\mathrm{s}^{-1}$. The Quemada and the Cross models exhibit the non-Newtonian properties of blood at shear rates rate, $\dot{\gamma}$, less than $10^4\,\mathrm{s}^{-1}$. Particularly, viscosity asymptotically matches the constant viscosity at the shear rates, $\dot{\gamma}$, greater than $10^2\,\mathrm{s}^{-1}$ in the Cross model. Quemada model shows the asymptotic nature below the constant viscosity, μ_∞.

4. Numerical Procedures

A three-point backward difference formula is used for time derivation of the velocity where the central difference is used for the convective and diffusion terms. Here a pressure correction algorithm is used and pressure as well as the velocity components are stored at the center of a control volume according to the collocated grid arrangement. The

Poisson like pressure correction equation is discretised using the pressure smoothing approach, which prevents the even-odd node uncoupling in the pressure and velocity fields. A BI-CGSTAB [26] solver is used for solving the matrix of velocity vectors, while for the Poisson like pressure correction equation a ICCG [27] solver is applied due to its symmetric and positive definite nature. Overall, the code is second-order accurate in both time and space.

The three-dimensional (3D) version of the present code has been successfully used for various numerical simulation involving LES and DNS techniques. The code is named as BOFFIN (Body Fitted Flow Integrator) which was developed in Imperial College, London, and the details of the program can be found in [28]. It has extensively been used in different 3D pulsatile flow simulations [29–35].

5. Results and Discussion

The geometry of arteries has a vital control on blood flow pattern and a local luminal constriction like stenosis greatly disturbs the velocity field. In stenotic flows, of particular interest are the phenomena of the vortex generation and propagation as well as the distribution of the wall shear stress (WSS). According to Neofytou and Drikakis [36], these are considered the most prominent attributes for blood flows because of their relation to atheroma formation in arteries. In line with these two, a detailed description about a set of hemodynamic factors like wall pressure and centerline velocity in streamwise direction with vector plot and vorticity can give a better understanding about the relationship between the fluid dynamics in pulsatile blood flow and arterial disease like stenosis. In this section, plots are demonstrated to show the results of the numerical investigation of blood flow for both Newtonian and non-Newtonian cases through a double stenoses model.

The grid independence test has been carried out to establish a suitable combination of the grid configuration to adequately resolve the flow for different viscous fluid in the stenoses. Fixing the Reynolds number at 300, three computations have been performed for three different grid systems with 150×70 (Case 1), 180×80 (Case 2), and 210×90 (Case 3) control volumes (in the x and y directions, resp.).

The number of streamwise grid points upstream of the first stenosis is always fixed at 30 while the rest of the grid points are distributed nonuniformly within and downstream of the first stenosis. Figure 4 shows the $x-y$ view of a portion of grid system and here it is clearly seen that the grid is significantly refined in order to accurately resolve the wall shear stress in the near-wall region.

The results of Cases 1, 2, and 3 are compared in Figure 5 in terms of the nondimensionalized streamwise velocity at different streamwise locations. It is observed from the figure that results for all three cases collapse to almost the same solution throughout the artery having a bit variation in frames 5(i) and (j) which are the poststenotic arterial positions. So, the agreement found in particular for the streamwise velocity is quite good and the flow is well resolved by the grids used in simulation. Therefore, it can be concluded that the present grid is capable of providing convergent solution independent

TABLE 1: Non-Newtonian models with given molecular viscosity of blood.

Model	Effective viscosity[*]
Newtonian	$\mu = 3.45 \times 10^{-3}$ Pa·s
Carreau $\longrightarrow$ Carreau [21]	$\mu\left(\lvert\dot{\gamma}\rvert\right) = \mu_\infty + \left(\mu_0 - \mu_\infty\right)\left[1 + (\lambda\dot{\gamma})^2\right]^{(n-1)/2}$ $\mu_0 = 0.056$ Pa·s viscosity at zero shear rate $\lambda = 3.131$ time constant $n = 0.3568$
Cross $\longrightarrow$ Cross [22]	$\mu\left(\lvert\dot{\gamma}\rvert\right) = \mu_\infty + \dfrac{(\mu_0 - \mu_\infty)}{\left[1 + (\dot{\gamma}/\gamma_c)^n\right]}$ $\mu_0 = 0.0364$ Pa·s, at a very low shear rate $\gamma_c = 2.63\,\text{s}^{-1}$ $n = 1.45$
Modified Casson $\longrightarrow$ Gonzalez and Moraga [23]	$\mu\left(\lvert\dot{\gamma}\rvert\right) = \left(\sqrt{\eta_c} + \dfrac{\sqrt{\tau_0}}{\sqrt{\lambda} + \sqrt{\dot{\gamma}}}\right)$ $\eta_c = 3.45 \times 10^{-3}$ $\tau_0 = 2.1 \times 10^{-2}\,\text{s}^{-1}$ $\lambda = 11.5\,\text{s}^{-1}$ $n = 1.45$
Quemada $\longrightarrow$ Quemada [24]	$\mu\left(\lvert\dot{\gamma}\rvert\right) = \mu_p\left(1 - \dfrac{1}{2}\dfrac{k_0 + k_\infty\sqrt{\lvert\dot{\gamma}\rvert/\gamma_c}}{1 + \sqrt{\lvert\dot{\gamma}\rvert/\gamma_c}}\phi\right)^{-2}$ $\phi = 0.45$ for haematocrit $\mu_p = 1.2 \times 10^{-3}$ $\gamma_c = 1.88\,\text{s}^{-1}$ $k_\infty = 2.07$ and $k_0 = 4.33$

[*]Viscosity is a function of Global shear rate, $\lvert\dot{\gamma}\rvert$ in non-Newtonian models.

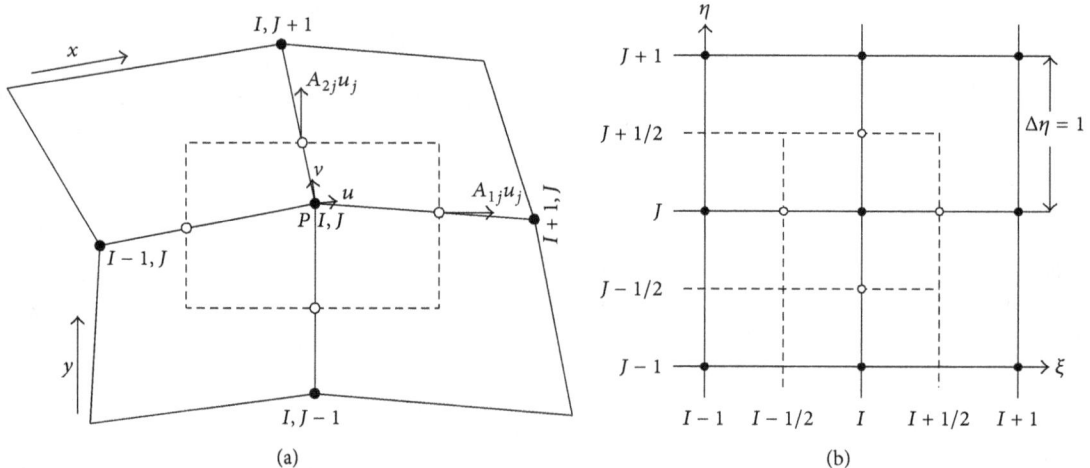

FIGURE 3: Grid arrangement and notation in two-dimensional case in both physical space (a) and in computational space (b). Solid lines indicated the grid lines and dashed lines indicated the faces of the control volume.

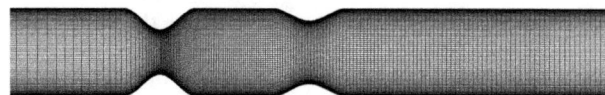

FIGURE 4: A portion of grid system ($x - y$ view).

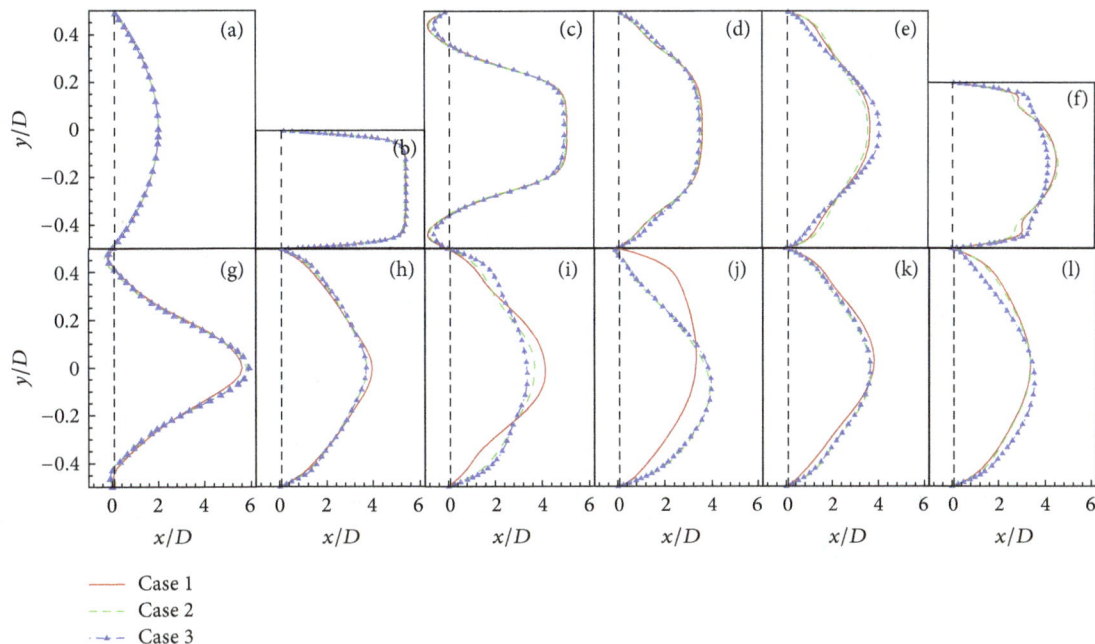

FIGURE 5: Grid independence test with respect to streamwise velocity, u/U at (a) $x/D = -5.0$ (inlet), (b) $x/D = 0.0$, (c) $x/D = 1.0$, (d) $x/D = 2.0$, (e) $x/D = 3.0$, (f) $x/D = 4.0$, (g) $x/D = 5.0$, (h) $x/D = 7.0$, (i) $x/D = 9.0$, (j) $x/D = 11.0$, (k) $x/D = 13.0$, and (l) $x/D = 15.0$ (outlet). Based on three-grid arrangements, Case 1: solid line for 150×70 control volumes, Case 2: dashed line for 180×80 control volumes, and Case 3: line with symbol for 210×90 control volumes.

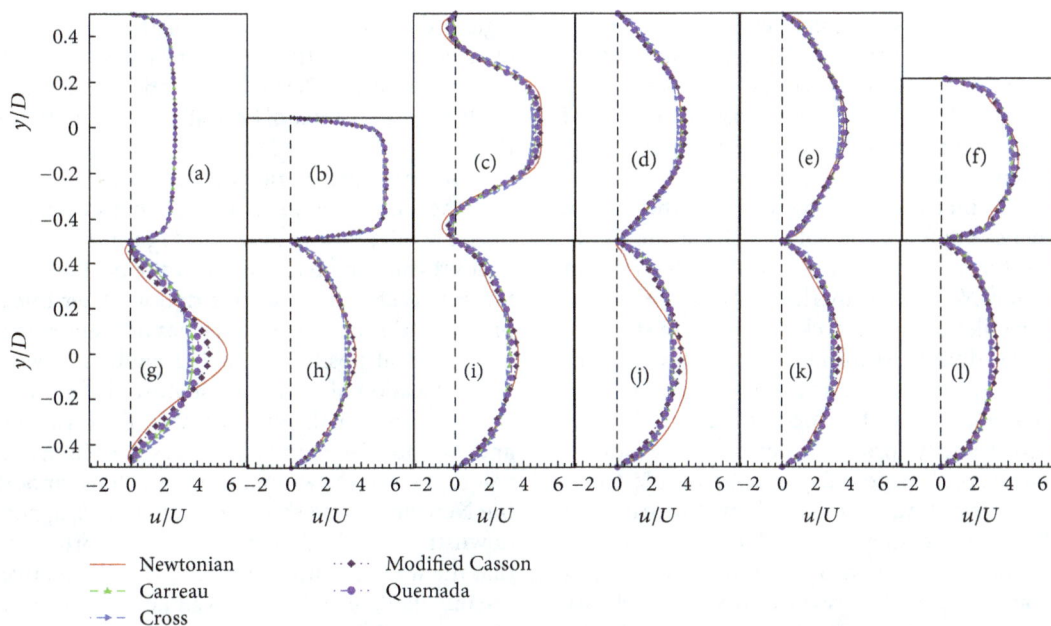

FIGURE 6: Streamwise velocity u/U, for different oscillation at the different axial position (a) $x/D = -5.0$ (inlet), (b) $x/D = 0.0$, (c) $x/D = 1.0$, (d) $x/D = 2.0$, (e) $x/D = 3.0$, (f) $x/D = 4.0$, (g) $x/D = 5.0$, (h) $x/D = 7.0$, (i) $x/D = 9.0$, (j) $x/D = 11.0$, (k) $x/D = 13.0$, and (l) $x/D = 15.0$ (outlet).

of different grid sizes. Based on the satisfactory agreement above the grid arrangement of 180×80 (Case 2) has been used for all other simulation.

As we have discussed earlier, the presence of stenosis causes a very disturbed flow inside an arterial segment.

Hence, it is important to resolve the magnitude of velocity at every point inside the artery in order to study the pulsatile flow behavior.

The nondimensionalized streamwise velocity (u/U) recorded at different axial locations is presented in

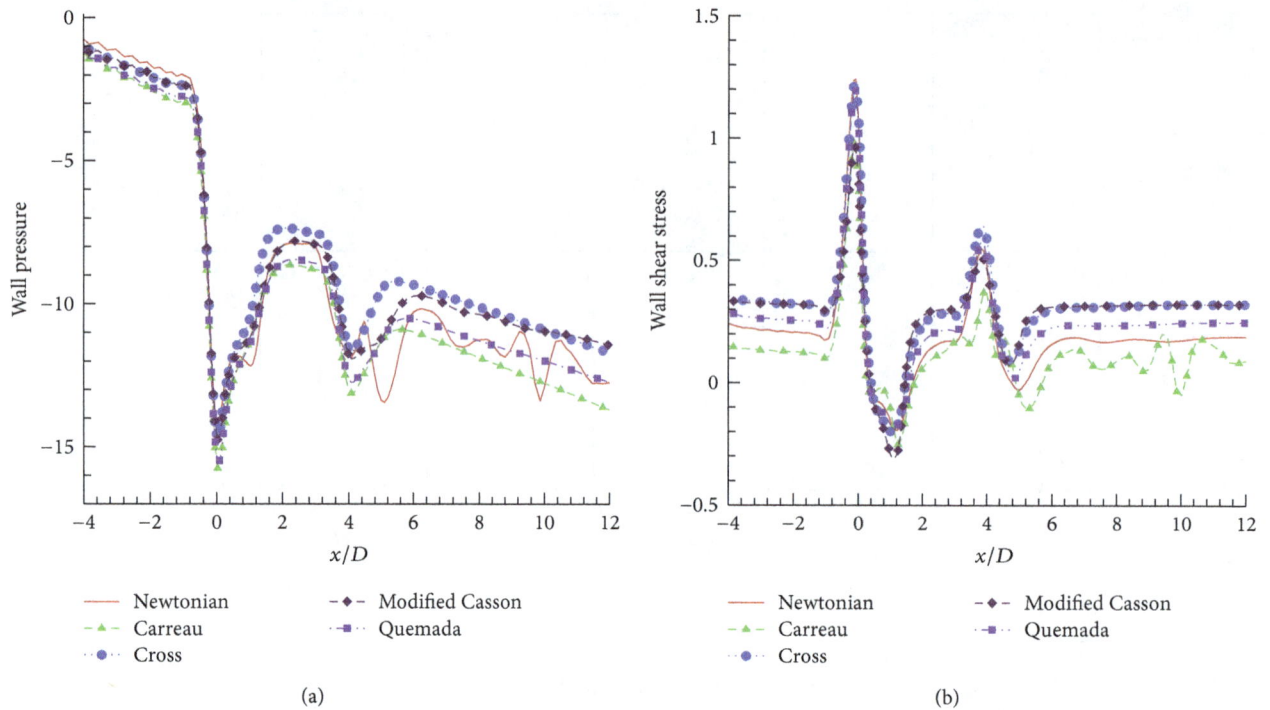

FIGURE 7: (a) Wall pressure, $p/\rho U^2$ and (b) wall shear stress, $\tau_w/\rho U^2$ for different viscosity models, while Re = 300.

Figures 6(a)–6(l) for the Newtonian and different non-Newtonian models considering Re = 300. The streamwise velocity, whose pattern in the inlet corresponds to sinusoidal laminar pulsatile profile, increases the most with a value of 5.5 at the neck of the critical stenosis shown in frame Figure 6(b). It then decreases in the downstream locations. The velocity again increases in upstream of the severe stenosis and reaches another peak value of 4.5 at the throat location which is shown in frame Figure 6(f). The Newtonian and four other non-Newtonian models show more or less similar velocity profiles up to the neck of the severe stenosis. But their patterns diverse erotically in the downstream location of the severe stenosis shown in frame Figure 6(g). The Newtonian model shows the highest magnitude with Gaussian shaped velocity profile followed by Modified Casson model while with distorted parabolic profile Cross model shows the lowest value followed by Carreau and Quemada models (in increasing order). All five models show almost similar velocity profiles with the some exceptions in the Newtonian one in frames Figures 6(h)–6(k). The velocity decreases and does not change significantly towards the further downstream of the second stenosis since the flow settles down.

Newtonian fluid always maintains constant viscosity and the rate of this viscosity is always lower than the non-Newtonian fluid viscosity. It happens because, after the poststenotic region the velocity of Newtonian fluid is comparatively higher than the non-Newtonian fluid. Negative values of velocity in the downstream location of both stenoses near the upper and lower wall correspond to the presence of permanent recirculation zones shown in frames 6(c) and (g).

Again, due to the narrowing of the artery segment caused by stenosis formation, flow gets a slender region inside. According to the Bernoulli equation, the velocity increases the most at the center of both stenoses since the area is smaller therein.

Wall pressure, p, normalized by ρU^2 for the Newtonian and the non-Newtonian models considering physiological pulsatile inlet flow is shown in Figure 7(a). It is seen that the wall pressure suddenly drops at the neck of the first stenosis having ($x/D = 0$) 75% constriction. After this position wall pressure tries to recover but does not attain as higher values as the initial positions. Having another lowest peak at the throat location of the severe stenosis, it follows an oscillating pattern through the downstream regions. The Newtonian and four non-Newtonian models show almost similar pattern where the non-Newtonian models show smooth curves but the Newtonian one shows a bit more zigzag pattern near the downstream of both stenoses. An important observation is that the wall pressure always maintains negative value for all viscous fluids in this rigid wall and double stenosed model artery segment. The diverging-converging shape of the artery seems to have an effect on this pressure distribution.

The blood pressure in most arteries is unsteady and more specifically it is pulsatile in nature which causes variation in pattern of blood pressure during different cardiac phases. The aorta serves as a compliance chamber that provides a reservoir of high pressure during diastole as well as systole and thus blood pressure does not go to zero during diastole even [17]. However, the lowest wall pressure values are found at the throat locations of the stenoses in the present study which are again negative. As a result, the artery might collapse

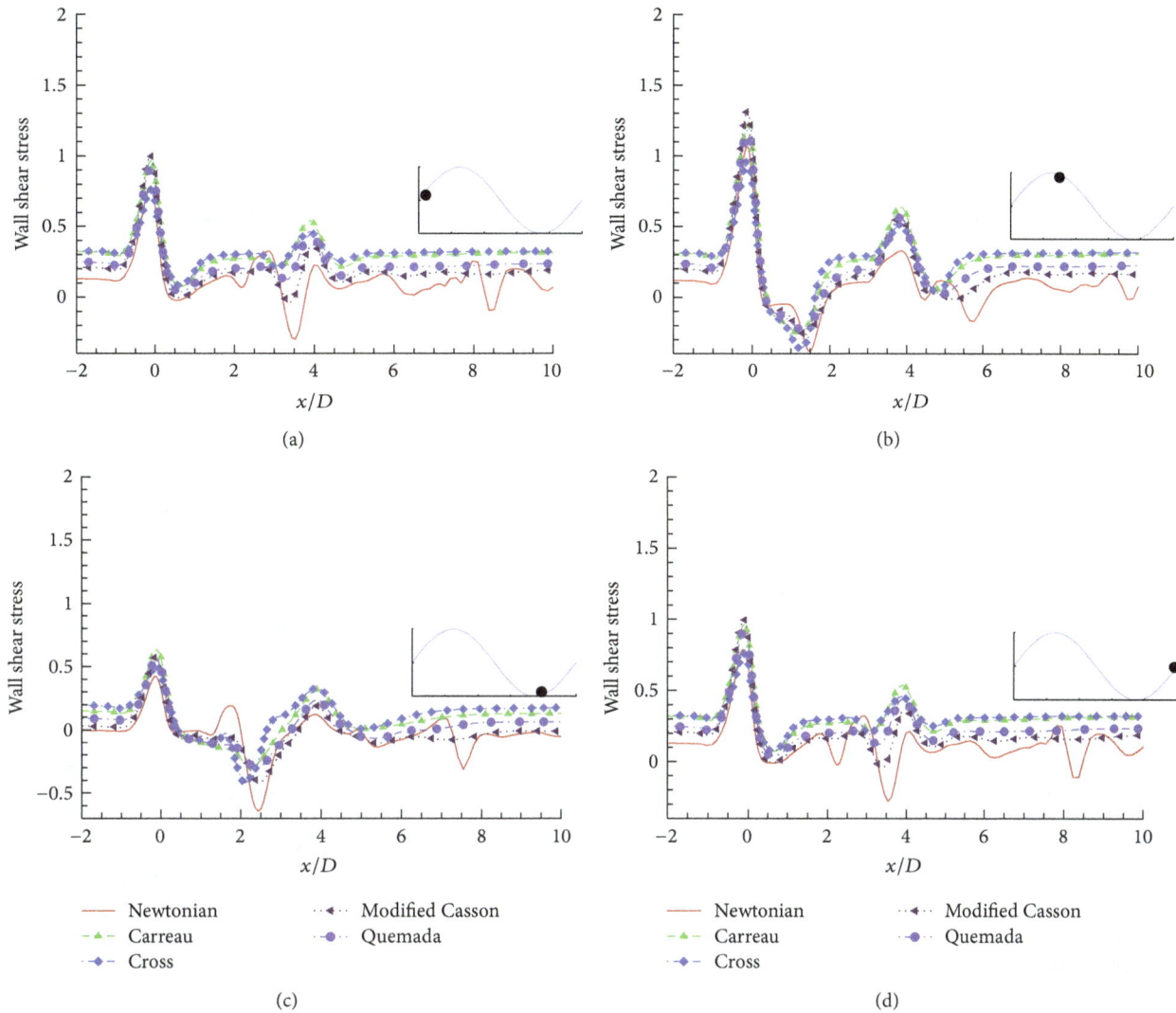

FIGURE 8: Wall shear stress, $\tau_w/\rho U^2$, at different pulsatile phase (a) $t/T = 9.0$, (b) $t/T = 9.3$, (c) $t/T = 9.7$, and (d) $t/T = 9.9$, while Re = 300.

at the neck due to insufficient pressure to maintain the opening of the lumen. Geometric influences on pressure losses have also been studied here and it is found that the pressure drop is so significant in the critical stenosis that the effect of the 50% stenosis is negligible since the two stenoses are close to one another which matches the results of Seeley and Young [37].

Distribution of the wall shear stress, τ_w, normalized by ρU^2 caused by the Newtonian and non-Newtonian fluids is depicted in Figure 7(b). Both the Newtonian and non-Newtonian fluids follow the same pattern for wall shear stress in this stenosed artery where the highest peak occurs at the center of the critical stenosis ($x/D = 0$ position) and another peak is observed at the center of the severe stenosis ($x/D = 4$ position). A similar result was found in case of asymmetric shaped single stenosis model where peak WSS occurred at the throat location by Neofytou and Drikakis [36]. Very high shear stresses near the throat of the stenosis can activate platelets and thereby induce thrombosis, which can totally

block blood flow to the heart or brain [17]. The lowest peaks are observed to occur at the distal ends of both stenoses at $x/D = 1$ and $x/D = 5$ positions, respectively. In the pre- and poststenotic regions, the wall shear stress remains almost constant for all models whereas the Carreau model maintains a very irregular pattern with zigzags and also the minimum magnitudes among all viscosity models including the Newtonian one. Additionally, the Cross and Modified Casson models show comparatively higher wall shear stress throughout the arterial segment than the other three viscosity variant fluids.

According to Ku [17], the cyclic nature of the heart pump creates pulsatile conditions in all arteries. Moreover, heart always maintains two cyclic phases. It ejects and fills with blood in alternating cycles called systole and diastole. Blood is pumped out of the heart during systole. The heart rests during diastole and no blood is ejected.

The WSS through an arterial segment with critical and severe stenoses for Re = 300 considering different phases

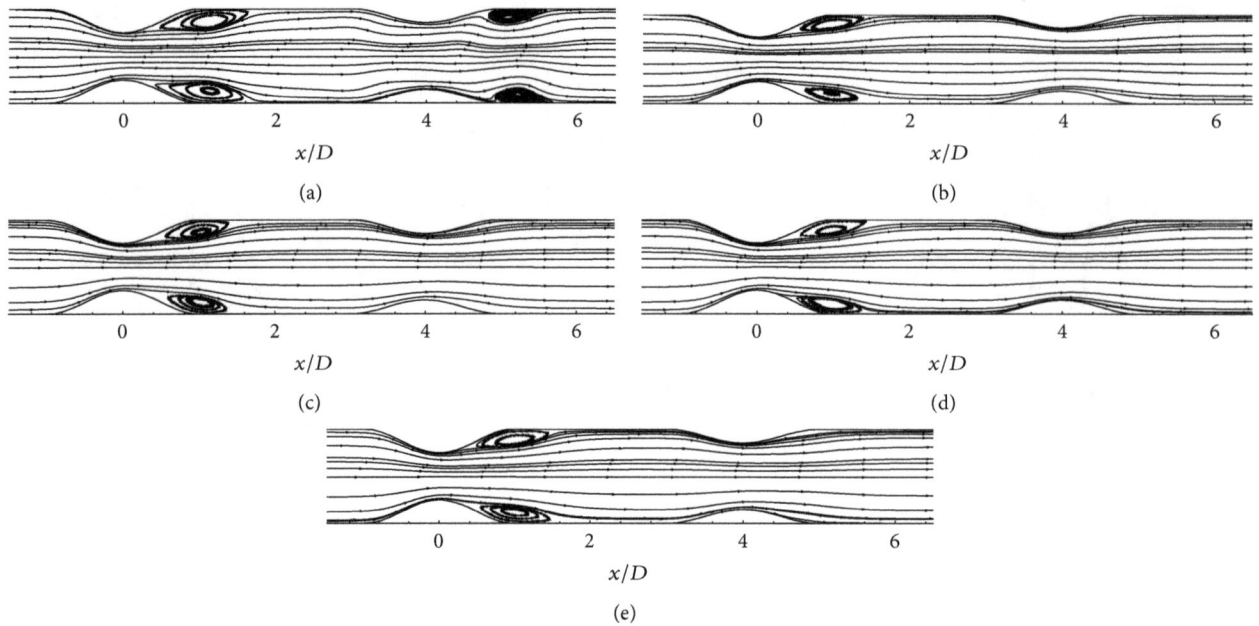

FIGURE 9: Streamlines for five different viscosity models: (a) Newtonian, (b) Carreau, (c) Cross, (d) Modified Casson, and (e) Quemada models for Re = 300.

of a cardiac cycle has been evaluated in this work which is depicted in Figure 8. These WSS diagrams are illustrated based on a sinusoidal cycle which is equivalent to a cardiac cycle. Four distinct phases of a sinusoidal cycle, that is, early systole, peak systole, peak diastole, and late diastole, are plotted here for WSS which is also shown inset. At a glance over the diagram, it is clearly seen that all five viscous fluids follow similar patterns and two peak values of WSS occur at the throats of two stenoses. Variation in values is mostly prominent in the Newtonian model which again maintains the lowest values in all phases among all five models. On the contrary, the Carreau and Cross models show comparatively higher values in different cardiac phases.

An important observation is that the maximum WSS occurs at the peak systole (t/T = 9.3) shown in frame 8(b) while the minimal WSS is observed at the peak diastole (t/T = 9.7) shown in frame 8(c) for both Newtonian and non-Newtonian models. Moreover, it is clearly seen that wall shear stress is approximately the same at the beginning and the end of the cycle. This phenomenon proves that our results become steady eventually.

Figure 9 holds the streamlines caused by pulsatile flow in a double stenosed arterial segment for Re = 300. Frame of Figure 9(a) represents the streamline for Newtonian case and frames of Figures 9(b) to 9(e) represent the streamlines for four different non-Newtonian viscosity models (Carreau, Cross, Modified Casson, and Quemada, resp.). In the case of Newtonian fluid, the flow is fully developed inside the constriction having higher magnitudes of velocity field in the contour plot which is significant in the throats and downstream regions of both stenoses. The flow is partially developed in the second constriction with a lower magnitude of velocity field. The high velocity region is comparatively

smaller in the non-Newtonian viscous models around the critical stenosis and it is almost absent in the severe stenosis. Only Cross model shows a presence of velocity field inside the severe constriction with a magnitude of 4. However, the extent of the breadth of the recirculation region from the first stenosis and its effects on the flow field downstream of the second stenosis depend on the stenosis spacing ratio, constriction ratio, and the Reynolds number [38].

Downstream of the critical stenosis two-stationary eddies are formed near the upper and lower wall, the sizes of which are different for every model. For the Newtonian case these eddies are the largest in size, while a second pair of smaller eddies in the downstream of severe stenosis is also generated. In the flow field of the non-Newtonian models, eddies downstream of the constriction are smaller than the Newtonian case and the second pair of eddies is absent. The eddies for the Cross model are the largest while the eddy for the Carreau model is the smallest one. The difference in size of the eddy for each of the models can be explained from the fact that—the behavior of the Carreau model is the most viscous followed by the Modified Casson, the Quemada, Cross, and finally the Newtonian model.

In Table 2, comparisons of the point of separation of the shear layer from the nose of the stenoses and its position of reattachment on the wall at the poststenosis region for the different viscous models are given. The separation of the Newtonian model starts early, as point of separation (PS) is recorded at about 0.02671254 which is an upstream location of the nose of the critical stenosis followed by another PS due to the severe stenosis at 4.990938, while comparatively late separation is predicted by all the non-Newtonian models. The PS of the Cross and the Quemada models is exactly the same which is at 0.03667778; separation occurs a bit later by

TABLE 2: Nonlinear model results.

Model	Point of separation (PS)	Reattachment point (RA)
Newtonian	0.267125, 4.990938	1.789380, 5.774552
Carreau	0.401262	1.504084
Cross	0.366777	1.504084
Modified Casson	0.332934, 4.779434	1.730357, 5.207934
Quemada	0.366777	1.615282

the Carreau model. Only the Newtonian and the Modified Casson models cause flow separation in more than one places.

Comparing all the reattachment (RA) points, it is clear that the Carreau model underpredicts the regime of the post-stenosis recirculation of blood, while the Newtonian model has an overall maximum prediction of the recirculation regime in case of critical stenosis. For the severe stenosis, the Newtonian model predicts larger recirculation region than the Modified Casson model.

The viscous effects of blood on the development of the flow along the streamwise direction are presented in Figure 10 for Re = 300. In this figure, the streamwise velocity vectors are appended on the contours of the streamwise velocity u/U. We find that for all viscous models, the primary recirculation region develops near the postlip of the critical stenosis due to the separation of the shear layer from nose of the stenosis. A small recirculation region is observed at the downstream of the severe stenosis in the Newtonian model which is completely absent in case of the non-Newtonian models. Due to these recirculation, Newtonian fluid provides some negative contour values where the contour level ranges from −0.5 to 5.5. It is also found that non-Newtonian fluid contour values are always positive.

Vorticity describes the local spinning motion of fluid and it can also be said that vorticity highly depends on viscosity. More insight into the flow separation seen in Figure 10 is given through the streamwise vorticity contours, $\omega = (\partial v/\partial x) - (\partial u/\partial y)$, in Figure 11. Figures 11(a) to 11(e) represent five different viscosity models, that is, the Newtonian, Carreau, Cross, Modified Casson, and Quemada models, respectively. It is noted that a total of 15 unequal contour levels are plotted between their maximum and minimum values, which can be viewed easily through the legend color bar. The vortex units rotated in the clockwise and anticlockwise direction that gives positive and negative values of ω, respectively. The clockwise rotations are represented by the solid lines where the anticlockwise rotations are represented by dashed lines.

Vorticity is very high in low viscosity models. As a result, more vortices are generated from the nose of the stenosis where the flow separation begins in the Newtonian fluid shown in frame of Figure 11(a) where two vortical structures form in the downstream region of the critical stenosis at $x/D = 1.0$; one acts in the anticlockwise direction at the upper wall and the other acts in the clockwise direction near lower wall. Both of them interact with each other and then roll up to downstream region. Another pair of vortical structures

of clockwise and anticlockwise direction is also present at the downstream of the second stenosis at $x/D = 5.0$. It is also evident from the color legend that the maximum magnitude of clockwise vortices lies in the region of $0 < x/D < 2.0$, $4.0 < x/D < 6.0$, and $9.0 < x/D < 10$. An important observation is that the anticlockwise vortex is absent at the location $9.0 < x/D < 10$ and its strength is very negligible compared to the clockwise ones.

Berger and Jou [1] found that whereas the individual vortex moves at a speed proportional to that of the flow, the front of a train of vorticity, the vorticity wave, propagates at a much higher speed. The reason for the faster propagation speed of the vorticity wave is vortex multiplication, in which a corotating and a contrarotating vortices are generated out of the original vortex. As a result of this vortex multiplication, the extent of the vortex wave in Newtonian fluid grows on an order of magnitude faster than an individual vortex moves.

The vortex structures at the location of $9.0 < x/D < 10$ are absent in case of all non-Newtonian models. Moreover, the strength of the vortices is also weaker in these models than that of the Newtonian fluid due to more viscosity. Among all of the viscous models, Modified Casson model shows the largest and the Cross model shows the smallest vortex structures.

6. Conclusion

Finite volume numerical simulation of unsteady, incompressible, and homogeneous blood flow in two-dimensional rigid models with double constriction has been presented in this paper. Flow field, flow induced wall pressure, and wall shear stress have been compared for Newtonian and non-Newtonian models (Carreau, Cross, Modified Casson, and Quemada) in a rigid pipe with two axisymmetric shaped stenoses of different degree under pulsatile condition for Re = 300. The maximum shear stresses are observed to occur at the throat locations of the 75% and 50% constricted regions which follow a very oscillating manner. The WSS is also characterized by means of a sequence of different flow stages in one period of the cardiac pulse. The highest value of WSS is found at the peak systole while the lowest one occurs during peak diastole.

Pressure loss can be an important reason of stroke or heart attack since it causes inadequate blood supply to the brain, heart, and other organs which is found at the neck of both stenoses. Moreover, Newtonian fluid causes a very disturbed pattern where pressure drop frequently fluctuates between higher and lower values in the downstream of severe stenosis causing more risk of potential heart attack than the non-Newtonian fluids. The wall pressure maintains a negative value throughout the artery segment which might be explained by the geometric influences. Study of the velocity patterns along the streamwise direction shows the peak velocity at the center of each constriction which can be explained by the Bernoulli equation and the one-dimensional continuity equation.

Streamlines demonstrate the presence of recirculation zones in the flow field which is found in the largest scale

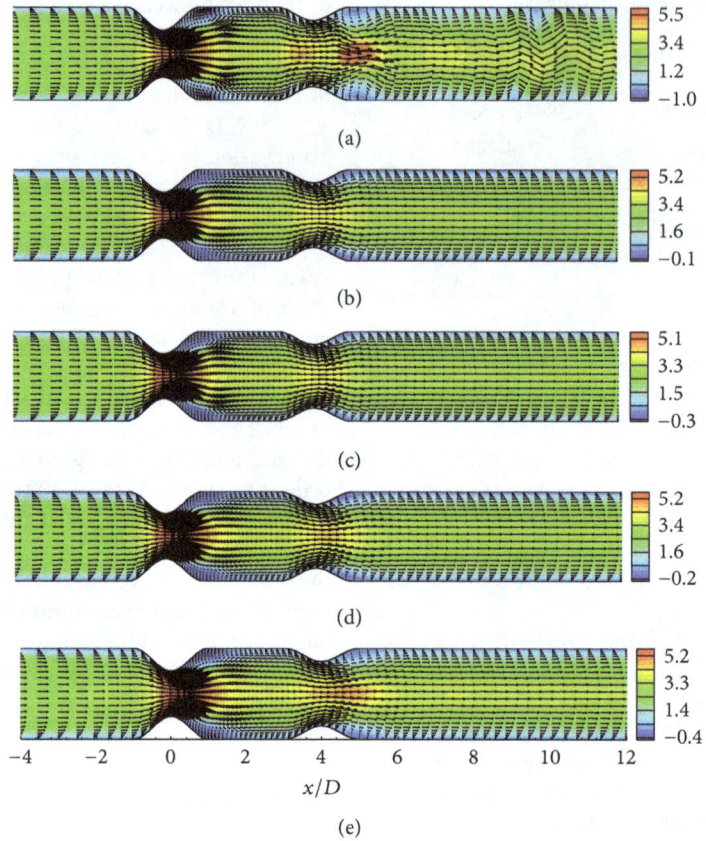

FIGURE 10: Vector plot for the different viscosity model appended on the streamwise velocity contour (a) Newtonian, (b) Carreau, (c) Cross, (d) Modified Casson, and (e) Quemada models for Re = 300.

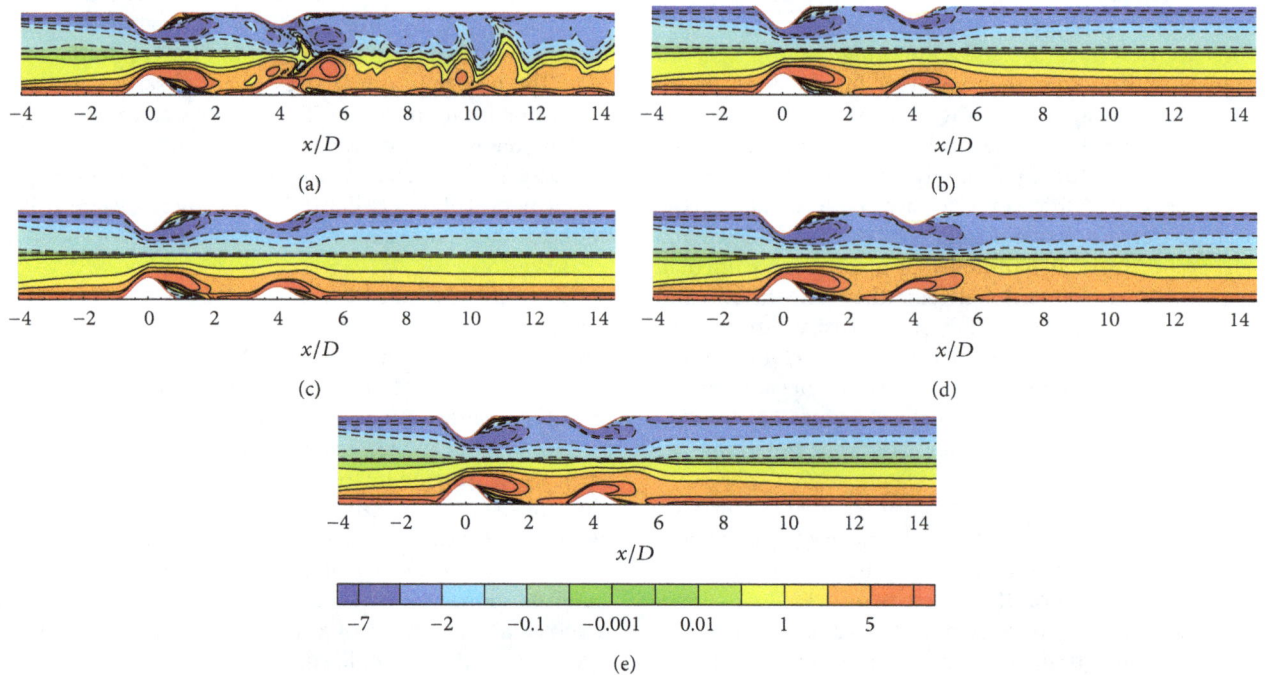

FIGURE 11: Vorticity for different viscosity models: (a) Newtonian, (b) Carreau, (c) Cross, (d) Modified Casson, and (e) Quemada models for Re = 300.

by the Newtonian fluid where the flow is fully developed inside both constricted regions. This again increases the possibility of thrombosis. However, the smallest recirculation region is caused by the Carreau model. A very interesting phenomenon of vorticity is observed while investigating its characteristics; the strength of clockwise vortices is higher than the anticlockwise ones. Moreover, the speed or propagation of a vortex train is higher than an individual vortex which is observed in case of Newtonian fluids due to low viscosity.

In conclusion, it can be stated that the Newtonian fluid is more likely to cause heart attack or blockage due to its characteristics of high wall shear stress, pressure loss, and the largest recirculation region at the throat locations of the stenoses than the non-Newtonian models. Limitations of this investigation include the consideration of rigid wall and simple sinusoidal pulsatile inlet profile instead of compliant arterial wall and physiological realistic inlet profile.

Nomenclature

English Symbols

A: Amplitude of the wall oscillation (m)
A_{ij}:   Elements of the cofactor matrix
D: Diameter of artery (m)
J: Jacobian
p: Pressure (Pa)
r: Radius of the pipe (m)
Re: Reynolds number (UD/ν)
t: Time (s)
U: Bulk velocity (m $\cdot$ s^{-1})
u: Velocity along the streamwise direction (m $\cdot$ s^{-1})
v: Velocity along the radial direction (m $\cdot$ s^{-1}).

Greek Symbols

μ: Viscosity of blood (kg $\cdot$ m^{-1} $\cdot$ s^{-1})
δ: Height of stenosis (m)
ν: Kinematic viscosity (m^2 $\cdot$ s^{-1})
ω: Vorticity (s^{-1})
ρ: Density of blood (kg $\cdot$ m^{-3})
τ_w: Wall shear stress (kg $\cdot$ m^{-1} $\cdot$ s^{-3})
ξ_1:   Coordinate along the streamwise direction (m)
ξ_2:   Coordinate along the radial direction (m).

Conflict of Interests

The authors declare that there is no conflict of interests regarding the publication of this paper.

Acknowledgment

Mir Golam Rabby and Sumaia Parveen Shupti wish to acknowledge gratefully the funding from the North South University, Bangladesh, during the period of this research.

References

[1] S. A. Berger and L.-D. Jou, "Flows in stenotic vessels," *Annual Review of Fluid Mechanics*, vol. 32, pp. 347–382, 2000.

[2] N. Mustapha, S. Chakravarty, P. K. Mandal, and N. Amin, "Unsteady response of blood flow through a couple of irregular arterial constrictions to body acceleration," *Journal of Mechanics in Medicine and Biology*, vol. 8, no. 3, pp. 395–420, 2008.

[3] K. Haldar, "Effects of the shape of stenosis on the resistance of blood flow through an artery," *Bulletin of Mathematical Biology*, vol. 47, no. 4, pp. 545–550, 1985.

[4] J. C. Misra, A. Sinha, and G. C. Shit, "Mathematical modeling of blood flow in a porous vessel having double stenoses in the presence of an external magnetic field," *International Journal of Biomathematics*, vol. 4, no. 2, pp. 207–225, 2011.

[5] A. Minagar, W. Jy, J. J. Jimenez, and J. S. Alexander, "Multiple sclerosis as a vascular disease," *Neurological Research*, vol. 28, no. 3, pp. 230–235, 2006.

[6] P. R. Johnston and D. Kilpatrick, "Mathematical modelling of paired arterial stenoses," in *Proceedings of the Computers in Cardiology*, pp. 229–232, September 1990.

[7] N. Mustapha and N. Amin, "The unsteady power law blood flow through a multiirregular stenosed artery," *Matematika*, vol. 24, no. 2, pp. 189–200, 2008.

[8] N. Mustapha, P. K. Mandal, I. Abdullah, N. Namin, and T. Hayat, "Numerical simulation of generalized newtonian blood flow past a couple of irregular arterial stenoses," *Numerical Methods for Partial Differential Equations*, vol. 27, no. 4, pp. 960–981, 2011.

[9] D. Kilpatrick, S. D. Webber, and J.-P. Colle, "The vascular resistance of arterial stenoses in series," *Angiology*, vol. 41, no. 4, pp. 278–285, 1990.

[10] K. C. Ang and J. Mazumdar, "Mathematical modelling of triple arterial stenoses," *Australasian Physical and Engineering Sciences in Medicine*, vol. 18, no. 2, pp. 89–94, 1995.

[11] N. Mustapha, N. Amin, S. Chakravarty, and P. K. Mandal, "Unsteady magnetohydrodynamic blood flow through irregular multi-stenosed arteries," *Computers in Biology and Medicine*, vol. 39, no. 10, pp. 896–906, 2009.

[12] N. Mustapha, P. K. Mandal, P. R. Johnston, and N. Amin, "A numerical simulation of unsteady blood flow through multi-irregular arterial stenoses," *Applied Mathematical Modelling*, vol. 34, no. 6, pp. 1559–1573, 2010.

[13] C. Tu, M. Deville, L. Dheur, and L. Vanderschuren, "Finite element simulation of pulse tile flow through arterial stenosis," *Journal of Biomechanics*, vol. 25, no. 10, pp. 1141–1152, 1992.

[14] N. Talukder, P. E. Karayannacos, R. M. Nerem, and J. S. Vasko, "An experimental study of fluid mechanics of arterial stenosis," *Journal of Biomechanical Engineering*, vol. 99, no. 2, pp. 74–82, 1977.

[15] D. F. Young and F. Y. Tsai, "Flow characteristics in models of arterial stenoses: I. Steady flow," *Journal of Biomechanics*, vol. 6, no. 4, pp. 395–410, 1973.

[16] C. Tu and M. Deville, "Pulsatile flow of non-newtonian fluids through arterial stenoses," *Journal of Biomechanics*, vol. 29, no. 7, pp. 899–908, 1996.

[17] D. N. Ku, "Blood flow in arteries," *Annual Review of Fluid Mechanics*, vol. 29, no. 1, pp. 399–434, 1997.

[18] C. R. Huang, W. D. Pan, H. Q. Chen, and A. L. Copley, "Thixotropic properties of whole blood from healthy human subjects," *Biorheology*, vol. 24, no. 6, pp. 795–801, 1987.

[19] D. L. Fry, "Acute vascular endothelial changes associated with increased blood velocity gradients," *Circulation Research*, vol. 22, no. 2, pp. 165–197, 1968.

[20] J. D. Folts, E. B. Crowell Jr., and G. G. Rowe, "Platelet aggregation in partially obstructed vessels and its elimination with aspirin," *Circulation*, vol. 54, no. 3, pp. 365–370, 1976.

[21] P. J. Carreau, "Rheological equations from molecular network theories," *Journal of Rheology*, vol. 16, no. 1, pp. 99–127, 1972.

[22] M. M. Cross, "Rheology of non-Newtonian fluids: a new flow equation for pseudoplastic systems," *Journal of Colloid Science*, vol. 20, no. 5, pp. 417–437, 1965.

[23] H. A. González and N. O. Moraga, "On predicting unsteady non-newtonian blood flow," *Applied Mathematics and Computation*, vol. 170, no. 2, pp. 909–923, 2005.

[24] D. Quemada, "Rheology of concentrated disperse systems III. General features of the proposed non-newtonian model. Comparison with experimental data," *Rheologica Acta*, vol. 17, no. 6, pp. 643–653, 1978.

[25] J. F. Thompson, F. C. Thames, and C. W. Mastin, "Automatic numerical generation of body-fitted curvilinear coordinate system for field containing any number of arbitrary two-dimensional bodies," *Journal of Computational Physics*, vol. 15, no. 3, pp. 299–319, 1974.

[26] D. S. Kershaw, "The incomplete Cholesky-conjugate gradient method for the iterative solution of systems of linear equations," *Journal of Computational Physics*, vol. 26, no. 1, pp. 43–65, 1978.

[27] H. A. D. Vorst, "Bi-cgstab: a first and smoothly converging variant of bi-cg for the solution of the non-symmetric linear systems," *SIAM Journal on Scientific and Statistical Computing*, vol. 155, pp. 631–644, 1992.

[28] W. P. Jones, F. di Mare, and A. J. Marquis, *LES-BOFFIN: Users Guide*, Mechanical Engineering Department, Imperial College London, London, UK, 2002.

[29] M. C. Paul, M. Mamun Molla, and G. Roditi, "Large-Eddy simulation of pulsatile blood flow," *Medical Engineering and Physics*, vol. 31, no. 1, pp. 153–159, 2009.

[30] M. C. Paul and M. Mamun Molla, "Investigation of physiological pulsatile flow in a model arterial stenosis using large-eddy and direct numerical simulations," *Applied Mathematical Modelling*, vol. 36, no. 9, pp. 4393–4413, 2012.

[31] M. Mamun Molla, M. C. Paul, and G. Roditi, "Physiological flow in a model of arterial stenosis," *Journal of Biomechanics*, vol. 41, supplement 1, p. S243, 2008.

[32] M. Mamun Molla and M. C. Paul, "LES of non-newtonian physiological blood flow in a model of arterial stenosis," *Medical Engineering & Physics*, vol. 34, no. 8, pp. 1079–1087, 2012.

[33] M. Mamun Molla, M. C. Paul, and G. Roditi, "LES of additive and non-additive pulsatile flows in a model arterial stenosis," *Computer Methods in Biomechanics and Biomedical Engineering*, vol. 13, no. 1, pp. 105–120, 2010.

[34] M. Mamun Molla, B. C. Wang, and D. C. Kuhn, "Numerical study of pulsatile channel flows undergoing transition triggered by a modelled stenosis," *Physics of Fluids*, vol. 24, no. 12, Article ID 121901, 2012.

[35] M. Mamun Molla, A. Hossain, B. C. Wang, and D. C. Kuhn, "Large-eddy simulation of pulsatile non-newtonian flow in a constricted channel," *Progress in Computational Fluid Dynamics*, vol. 12, no. 4, pp. 231–242, 2012.

[36] P. Neofytou and D. Drikakis, "Effects of blood models on flows through a stenosis," *International Journal for Numerical Methods in Fluids*, vol. 43, no. 6-7, pp. 597–635, 2003.

[37] B. D. Seeley and D. F. Young, "Effect of geometry on pressure losses across models of arterial stenoses," *Journal of Biomechanics*, vol. 9, no. 7, pp. 439–448, 1976.

[38] T. S. Lee, W. Liao, and H. T. Low, "Numerical simulation of turbulent flow through series stenoses," *International Journal for Numerical Methods in Fluids*, vol. 42, no. 7, pp. 717–740, 2003.

27

Peristaltic Motion of Non-Newtonian Fluid with Heat and Mass Transfer through a Porous Medium in Channel under Uniform Magnetic Field

Nabil T. M. Eldabe,[1] **Bothaina M. Agoor,**[2] **and Heba Alame**[2]

[1] *Department of Mathematics, Faculty of Education, Ain Shams University, Cairo 11566, Egypt*
[2] *Department of Mathematics, Faculty of Science, Fayoum University, P.O. Box 63514, Fayoum, Egypt*

Correspondence should be addressed to Bothaina M. Agoor; bma00@fayoum.edu.eg

Academic Editor: Kuo-Kang Liu

This paper is devoted to the study of the peristaltic motion of non-Newtonian fluid with heat and mass transfer through a porous medium in the channel under the effect of magnetic field. A modified Casson non-Newtonian constitutive model is employed for the transport fluid. A perturbation series' method of solution of the stream function is discussed. The effects of various parameters of interest such as the magnetic parameter, Casson parameter, and permeability parameter on the velocity, pressure rise, temperature, and concentration are discussed and illustrated graphically through a set of figures.

1. Introduction

Peristaltic motion is a phenomenon that occurs when expansion and contraction of an extensible tube in a fluid generate progressive waves which propagate along the length of the tube, mixing and transporting the fluid in the direction of wave propagation. In some biomedical instruments, such as heart-lung machines, peristaltic motion is used to pump blood and other biological fluids [1]. Peristaltic pumping is a form of fluid transport generally from a region of lower to higher pressure, by means of a progressive wave of area contraction or expansion, which propagates along the length of a tube like structure. Some electrochemical reactions are held responsible for this phenomenon. This mechanism occurs in swallowing of food through oesophagus, in the ureter, the gastro intestinal tract, the bile duct, and even in small blood vessels. It has now been accepted that most of the physiological fluids behave like a non-Newtonian fluids. The peristaltic flows have attracted a number of researchers because of wide applications in physiology and industry. The theoretical work of peristaltic transport primarily with the inertia free Newtonian flow driven by a sinusoidal transverse wave of small amplitude is investigated by Fung et al. [2]. Burns and Parkes [3] studied the peristaltic motion of a viscous fluid through a pipe and channel by considering sinusoidal variations at the walls. A mathematical study of the peristaltic transport of Casson fluid is given by Mernone and Mazumdar [4, 5]; they used the perturbation method to solve the problem. Mekheimer [6, 7] studied the peristaltic transport of MHD flow. Peristaltic transport of Casson fluid in a channel is discussed by Nagarani and Sarojamma [8, 9]. El Shehawy et al. [10] Studied the peristaltic transport in a symmetric channel through a porous medium. Finite element solutions for non-Newtonian pulsatile flow in a non-Darcian porous medium are given by Bharagava et al. [11]. Mekheirmer and Abd elmaboud [12] discussed the influence of heat transfer and magnetic field on peristaltic transport. Nadeem et al. [13] have discussed the influence of heat and mass transfer on peristaltic flow of third order fluid in a diverging tube. Abdelmaboud and Mekheimer [14] analyzed the transport of second order fluid through a porous medium. Abd Elmaboud [15] studied the heat transfer characteristics of micropolar fluid through an isotropic porous medium in a two-dimensional channel with rhythmically contracting walls. El-dabe et al. [16] studied the effects of radiation on the unsteady flow of an incompressible non-Newtonian (Jeffrey) fluid through porous medium. Mustafa et al. [17] studied the peristaltic transport of nanofluid in a channel with complaint walls. Anwr Beg and Tripathi [18] introduced a theoretical study to examine the peristaltic pumping with double-diffusive convection in nanofluids through a deformable channel. El-dabe et al. [19] have discussed the effects of heat and mass transfer on the MHD flow of an incompressible, electrically conducting couple stress fluid through a porous medium in an asymmetric flexible channel over which a traveling wave of contraction and expansion is produced, resulting in a peristaltic motion. El-dabe et al. [20] studied the peristaltic motion of incompressible micropolar fluid through a porous medium in a two-dimensional channel under the effects of heat absorption and chemical reaction in the presence of magnetic field. Ebaid and Emad Aly [21] showed the mathematical model describing the slip peristaltic flow of nanofluid application to the cancer treatment. Emad and Ebaid [22] applied two different analytical and numerical methods to solve the system describing the mixed convection boundary layer nanofluids flow along an inclined plate embedded in a porous medium. Abd Elmaboud [23] investigated the magneto thermodynamic aspects micropolar fluid (blood model) through an isotropic porous medium in a nonuniform channel with rhythmically contracting walls. Noreen et al. [24] studied the mathematical model to investigate the mixed convective heat and mass transfer effects on peristaltic flow of magnetohydrodynamic pseudoplastic fluid in a symmetric channel. Hayat et al. [25] discussed the effects of heat and mass transfer on the peristaltic flow in the presence of an induced magnetic field. Noreen [26] consider the peristaltic flow of third order nanofluid in an asymmetric channel with an induced magnetic field.

The main aim of this work is to study the peristaltic motion of non-Newtonian fluid with heat and mass transfer through a porous medium in the channel under the effect of magnetic field. A modified Casson non-Newtonian constitutive model is employed for the transport fluid. A perturbation series' method of solution of the stream function is discussed. The effects of various parameters of interest such as the magnetic parameter, Casson parameter, and permeability parameter on the velocity, pressure rise, and temperature are discussed and illustrated graphically through a set of figures.

2. Mathematical Analysis

Consider the peristaltic motion of non-Newtonian fluid through a porous medium in two-dimensional channel, having width d. A rectangular coordinate system x, y is chosen such that x-axis lies along the direction of wave progression and y-axis normal to it. The fluid is subjected to a constant magnetic field $\underline{B} = (0, B_0, 0)$. Let u and v be the velocity components. The vertical displacements for the upper and lower walls are ζ and $-\zeta$, see Figure 1, where ζ is defined by

$$\zeta(x, t) = a \cos \frac{2\pi}{\lambda} (x - ct).$$

(1)

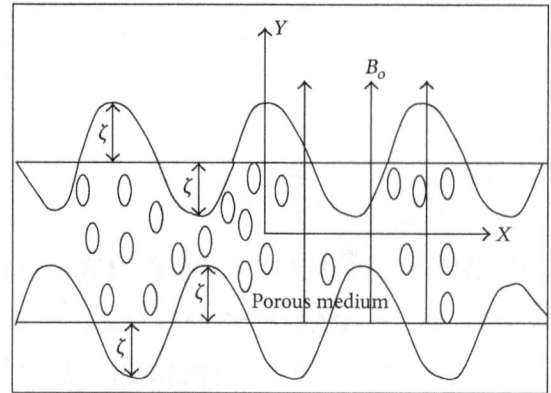

FIGURE 1: Sketch of the problem.

λ is the wavelength, t is the time and a is the amplitude of the sinusoidal waves travelling along the channel at velocity c. The constitutive equation for the non-Newtonian Casson fluid can be written as in [27].

Consider

$$\tau_{ij} = \begin{cases} 2\left(\mu_B + \dfrac{p_y}{\sqrt{2\pi}}\right) e_{ij}, & \pi > \pi_c, \\ 2\left(\mu_B + \dfrac{p_y}{\sqrt{2\pi_c}}\right) e_{ij}, & \pi < \pi_c, \end{cases}$$

(2)

where τ_{ij} is the components of the stress tensor, $\pi = e_{ij}e_{ji}$ and e_{ij} are the (i, j)th components of the deformation rate, π is the product of the component of deformation rate by itself, π_c is a critical value of this product based on the Nakamura-Sawada model [27], μ_B is the plastic dynamic viscosity of the non-Newtonian fluid, and p_y is yield stress of slurry fluid.

The equations governing the fluid motion can be written as follows.

The continuity equation is

$$\frac{\partial u}{\partial x} + \frac{\partial v}{\partial y} = 0.$$

(3)

The momentum equations are

$$\rho\left(\frac{\partial u}{\partial t} + u\frac{\partial u}{\partial x} + v\frac{\partial u}{\partial y}\right)$$
$$= -\frac{\partial P}{\partial x} + (\mu + \mu_B)\left(\frac{\partial^2 u}{\partial x^2} + \frac{\partial^2 u}{\partial y^2}\right) - \left(\sigma B_0^2 + \frac{\mu}{k}\right) u,$$

$$\rho\left(\frac{\partial v}{\partial t} + u\frac{\partial v}{\partial x} + v\frac{\partial v}{\partial y}\right)$$
$$= -\frac{\partial P}{\partial y} + (\mu + \mu_B)\left(\frac{\partial^2 v}{\partial x^2} + \frac{\partial^2 v}{\partial y^2}\right) - \left(\sigma B_0^2 + \frac{\mu}{k}\right) v.$$

(4)

The energy equation is

$$\left(\frac{\partial T}{\partial t} + u\frac{\partial T}{\partial x} + v\frac{\partial T}{\partial y}\right) = \frac{k_1}{\rho c_p}\left(\frac{\partial^2 T}{\partial x^2} + \frac{\partial^2 T}{\partial y^2}\right).$$

(5)

The concentration equation is

$$\left(\frac{\partial C}{\partial t} + u \frac{\partial C}{\partial x} + v \frac{\partial C}{\partial y} \right) = D_m \left(\frac{\partial^2 C}{\partial x^2} + \frac{\partial^2 C}{\partial y^2} \right). \tag{6}$$

Lorentz force:

$$\underline{F} = \underline{J} \wedge \underline{B}, \qquad \underline{J} = \sigma \underline{V} \wedge \underline{B}, \tag{7}$$

where k is the permeability of the medium, J is the current density, c_p is the specific heat of the fluid, k_1 is the coefficient of heat conduction, T is the temperature of the fluid, D_m is the coefficient of mass diffusivity, C is the concentration of the fluid, μ is the coefficient of viscosity, σ is the electrical conductivity, P is pressure, and B_o is the strength of the applied magnetic field.

The appropriate boundary conditions are

$$u = 0, \qquad v = \frac{\partial \zeta}{\partial t}, \qquad T = T_1, \qquad C = C_1$$
$$\text{at } y = d + \zeta,$$
$$u = 0, \qquad v = -\frac{\partial \zeta}{\partial t}, \qquad T = T_0, \qquad C = C_0$$
$$\text{at } y = -d - \zeta. \tag{8}$$

Using the following nondimensional variables:

$$x' = \frac{x}{d}, \qquad y' = \frac{y}{d}, \qquad u' = \frac{u}{c}, \qquad v' = \frac{v}{c}, \qquad t' = \frac{ct}{d},$$

$$p' = \frac{p}{\rho c^2}, \qquad \zeta' = \frac{\zeta}{d}, \qquad \varepsilon = \frac{a}{d},$$

$$T = (T_1 - T_0)\theta + T_0, \qquad C = (C_1 - C_0)\phi + C_0, \tag{9}$$

equations (4)–(8) after dropping the stars mark reduce to

$$\left(\frac{\partial u}{\partial t} + u \frac{\partial u}{\partial x} + v \frac{\partial u}{\partial y} \right)$$
$$= -\frac{\partial P}{\partial x} + \left(\tau_0 + \frac{1}{R_e} \right) \left(\frac{\partial^2 u}{\partial x^2} + \frac{\partial^2 u}{\partial y^2} \right) - \left(M + \frac{1}{K} \right) u,$$

$$\left(\frac{\partial v}{\partial t} + u \frac{\partial v}{\partial x} + v \frac{\partial v}{\partial y} \right)$$
$$= -\frac{\partial P}{\partial y} + \left(\tau_0 + \frac{1}{R_e} \right) \left(\frac{\partial^2 v}{\partial x^2} + \frac{\partial^2 v}{\partial y^2} \right) - \left(M + \frac{1}{K} \right) v,$$

$$\left(\frac{\partial \theta}{\partial t} + u \frac{\partial \theta}{\partial x} + v \frac{\partial \theta}{\partial y} \right) = \frac{k_1}{\rho c_p} \left(\frac{\partial^2 \theta}{\partial x^2} + \frac{\partial^2 \theta}{\partial y^2} \right),$$

$$\left(\frac{\partial \phi}{\partial t} + u \frac{\partial \phi}{\partial x} + v \frac{\partial \phi}{\partial y} \right) = \frac{k_1}{\rho c_p} \left(\frac{\partial^2 \phi}{\partial x^2} + \frac{\partial^2 \phi}{\partial y^2} \right), \tag{10}$$

with the boundary conditions

$$u = 0, \qquad v = \alpha \varepsilon \sin \alpha (x - t), \qquad \theta = 1, \qquad \phi = 1$$
$$\text{at } y = 1 + \zeta,$$
$$u = 0, \qquad v = -\alpha \varepsilon \sin \alpha (x - t), \qquad \theta = 0, \qquad \phi = 0$$
$$\text{at } y = -1 - \zeta, \tag{11}$$

where $M = \sigma B_0^2 d / \rho c$ is the magnetic parameter, $K = ck/\vartheta d$ is the permeability parameter, $\alpha = 2d\pi/\mu$ is the wave number, $\tau_0 = \mu_B/\rho cd$ is the Casson parameter, and $R_e = \rho cd/\mu$ is the Reynolds number.

Now, we shall define a stream function ψ as $u = \psi_y$ and $v = -\psi_x$ then (10) can be written as

$$\psi_{yt} + \psi_y \psi_{yx} - \psi_x \psi_{yy}$$
$$= -\frac{\partial p}{\partial x} + \left(\tau_0 + \frac{1}{R_e} \right) (\psi_{yyy} + \psi_{yxx}) - \left(M + \frac{1}{K} \right) \psi_y,$$

$$- \psi_{xt} - \psi_y \psi_{xx} + \psi_x \psi_{xy}$$
$$= -\frac{\partial p}{\partial y} - \left(\tau_0 + \frac{1}{R_e} \right) (\psi_{xxx} + \psi_{yyx}) + \left(M + \frac{1}{K} \right) \psi_x,$$

$$\left(\frac{\partial \theta}{\partial t} + \psi_Y \frac{\partial \theta}{\partial x} - \psi_X \frac{\partial \theta}{\partial y} \right) = \frac{k_1}{\rho c_p} \left(\frac{\partial^2 \theta}{\partial x^2} + \frac{\partial^2 \theta}{\partial y^2} \right),$$

$$\left(\frac{\partial \phi}{\partial t} + \psi_Y \frac{\partial \phi}{\partial x} - \psi_X \frac{\partial \phi}{\partial y} \right) = D_m \left(\frac{\partial^2 \phi}{\partial x^2} + \frac{\partial^2 \phi}{\partial y^2} \right), \tag{12}$$

with conditions

$$\psi_y = 0, \qquad \psi_x = -\alpha \varepsilon \sin \alpha (x - t), \qquad \theta = 1, \qquad \phi = 1$$
$$\text{at } y = 1 + \zeta,$$
$$\psi_y = 0, \qquad \psi_x = \alpha \varepsilon \sin \alpha (x - t), \qquad \theta = 0, \qquad \phi = 0$$
$$\text{at } y = -1 - \zeta. \tag{13}$$

Express a stream function ψ, p, θ, and ϕ as a series in terms of small amplitude ratio ε, we have

$$\psi(x, y, t) = \psi_0 + \varepsilon \psi_1 + \cdots,$$
$$p(x, y) = p_0 + \varepsilon p_1 + \cdots,$$
$$\theta(x, y, t) = \theta_0 + \varepsilon \theta_1 + \cdots,$$
$$\phi(x, y, t) = \phi_0 + \varepsilon \phi_1 + \cdots, \tag{14}$$

where ψ_0 is a function of y only. Substituting (14) in (12) and collecting the terms in ε, we get the following system of equations.

Coefficient of ε^0:

$$\frac{\partial p_0}{\partial y} = 0, \text{ this leads to } P_o \text{ is a function of } x, \ p_0 = c_1(x),$$

(15)

$$\psi_{oyyy} - \lambda_1^2 \psi_{0y} + L = 0,$$

(16)

$$\frac{\partial^2 \theta_0}{\partial y^2} = 0,$$

(17)

$$\frac{\partial^2 \phi_0}{\partial y^2} = 0,$$

(18)

where

$$\lambda_1 = \sqrt{\frac{(M + (1/K))}{(\tau_0 + (1/R_e))}}, \qquad L = \left(\frac{\partial p_0/\partial x}{(\tau_0 + (1/R_e))}\right).$$

(19)

and coefficient of ε:

$$\psi_{1yt} + \psi_{0y}\psi_{1yx} - \psi_{1x}\psi_{0yy}$$

$$= -\frac{\partial p_1}{\partial x} + \left(\tau_0 + \frac{1}{R_e}\right)\left(\psi_{1yyy} + \psi_{1yxx}\right) - \left(M + \frac{1}{K}\right)\psi_{1y},$$

(20)

$$-\psi_{1xt} - \psi_{0y}\psi_{1xx}$$

$$= -\frac{\partial p_1}{\partial y} - \left(\tau_0 + \frac{1}{R_e}\right)\left(\psi_{1xxx} + \psi_{1yyx}\right) + \left(M + \frac{1}{K}\right)\psi_{1x},$$

(21)

$$\left(\frac{\partial \theta_1}{\partial t} + \psi_{0y}\frac{\partial \theta_1}{\partial x} - \psi_{1x}\frac{\partial \theta_1}{\partial y}\right) = \frac{k_1}{\rho c_p}\left(\frac{\partial^2 \theta_1}{\partial x^2} + \frac{\partial^2 \theta_1}{\partial y^2}\right),$$

(22)

$$\left(\frac{\partial \phi_1}{\partial t} + \psi_{0y}\frac{\partial \phi_1}{\partial x} - \psi_{1x}\frac{\partial \phi_1}{\partial y}\right) = D_m\left(\frac{\partial^2 \phi_1}{\partial x^2} + \frac{\partial^2 \phi_1}{\partial y^2}\right).$$

(23)

Also, boundary conditions (13) can be written after using the Taylor series expansions about $y = \pm 1 \pm \zeta$ as follows:

$$\psi_{0y}(1) = 0, \qquad \theta_0(1) = 1, \qquad \phi_0(1) = 1,$$

$$\psi_{1y}(1) = -\psi_{0yy}(1)\cos\alpha(x - t),$$

$$\theta_1(1) = -\theta_{0y}(1)\cos\alpha(x - t),$$

$$\psi_0(-1) = 0, \qquad \theta_0(-1) = 0, \qquad \phi_0(-1) = 0,$$

$$\psi_{1x}(1) = -\alpha\sin\alpha(x - t),$$

$$\psi_{0y}(-1) = 0, \qquad \phi_1(1) = -\phi_{0y}(1)\cos\alpha(x - t).$$

(24)

Using conditions (24) with (15), the solution of (16) can be written as

$$\psi_0(y) = \frac{-L}{\lambda_1}\left(y - \frac{\sinh\lambda_1 y}{\lambda_1 \cosh\lambda_1}\right).$$

(25)

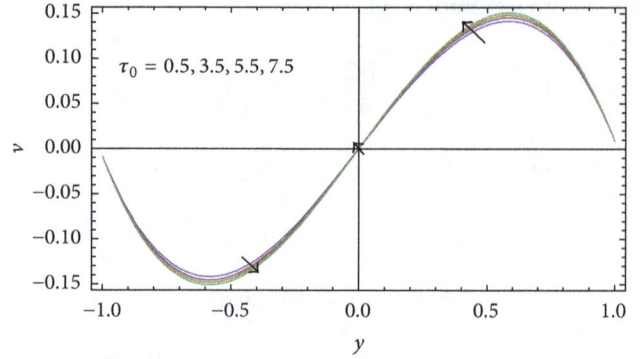

FIGURE 2: The velocity v is plotted against the distance for different values of τ_o at $\alpha = 0.1$, $\varepsilon = 0.1$, $M = 5$, $k = 0.9$.

The flow rate, q, is given by

$$q = \int_0^1 u\,dy = \int_0^1 \frac{\partial \psi}{\partial y}\,dy = \psi(1) - \psi(0)$$

$$= \frac{-1}{\lambda_1(\tau_0 + (1/R_e))}\left(1 - \frac{\sinh\lambda_1}{\lambda_1\cosh\lambda_1}\right)\frac{\partial p_0}{\partial x}.$$

(26)

The pressure rise is given by

$$\Delta p_0 = \int_0^1 \frac{\partial p_0}{\partial x}\,dx = \frac{-(\tau_0 + (1/R_e))\lambda_1 q}{(1 - (\sinh\lambda_1/\lambda_1\cosh\lambda_1))}.$$

(27)

Eliminating the pressure terms in (20) and (21), we have

$$\nabla^2\psi_{1t} + \psi_{0y}\nabla^2\psi_{1x} - \psi_{1x}\nabla^2\psi_{0y}$$

$$= \left(\tau_0 + \frac{1}{R_e}\right)\nabla^4\psi_1 - \left(M + \frac{1}{K}\right)\nabla^2\psi_1.$$

(28)

From conditions (24) and (25) we can write ψ_1, θ_1, and ϕ_1 in the form

$$\psi_1(x, y, t) = f(y)\cos\alpha(x - t) + g(y)\sin\alpha(x - t).$$

$$\theta_1(x, y, t) = h_1(y)\cos\alpha(x - t),$$

(29)

$$\phi_1(x, y, t) = h_2(y)\cos\alpha(x - t).$$

Equation (28) can be simplified by using (29) and assuming that the wave number $\alpha = 2\pi d/\lambda$ is small, so the terms of α^2 and higher can be neglected.

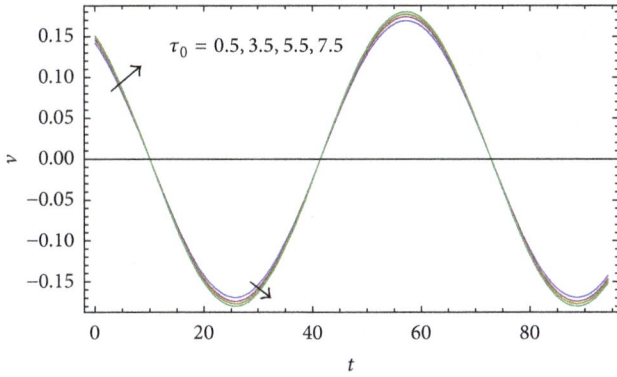

FIGURE 3: The velocity is plotted against the time for different values of τ_o at $\alpha = 0.1$, $\varepsilon = 0.1$, $M = 5$, $k = 0.9$.

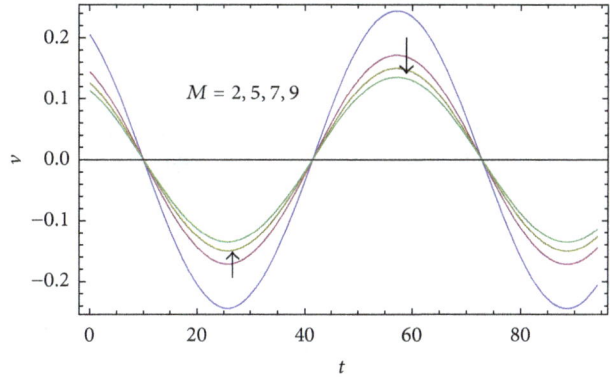

FIGURE 5: The velocity v is plotted against the time t for different values of M at $\alpha = 0.1$, $\varepsilon = 0.1$, $\tau_o = 2$, $k = 0.9$.

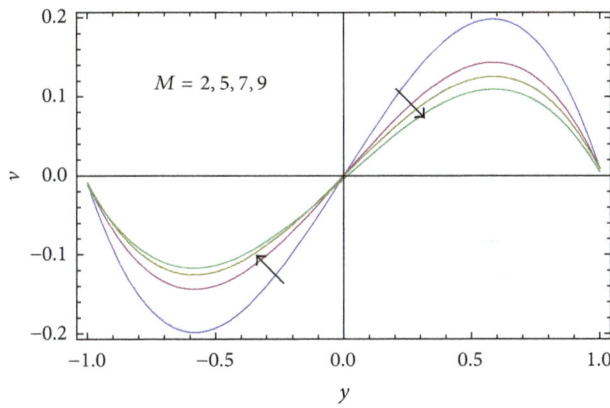

FIGURE 4: The velocity v is plotted against the distance for different values of M at $\alpha = 0.1$, $\varepsilon = 0.1$, $\tau_o = 2$, $k = 0.9$.

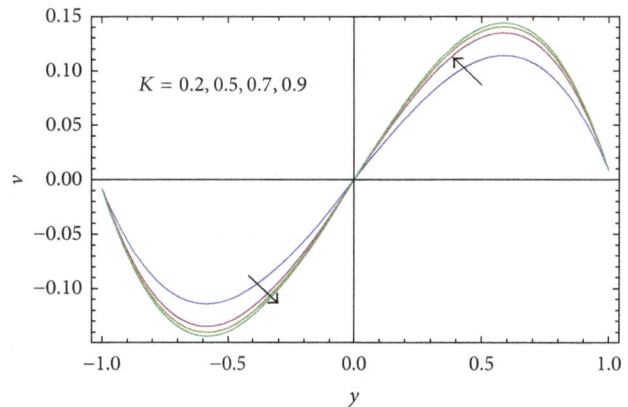

FIGURE 6: The velocity v is plotted against the distance for different values of k at $\alpha = 0.1$, $\varepsilon = 0.1$, $M = 5$, $\tau_o = 2$.

We get

$$
\alpha \left[\left(\left(1 - \frac{L}{\lambda_1} \right) + \frac{\cosh \lambda_1 y}{\cosh \lambda_1} \right) f'' \right.
$$
$$
\left. + L \frac{\cosh \lambda_1 y}{\cosh \lambda_1} f \right] \sin \alpha (x - t)
$$
$$
- \alpha \left[\left(\left(1 - \frac{L}{\lambda_1} \right) + \frac{\cosh \lambda_1 y}{\cosh \lambda_1} \right) g'' \right.
$$
$$
\left. - L \frac{\cosh \lambda_1 y}{\cosh \lambda_1} g \right] \cos \alpha (x - t)
$$
$$
= \left[\left(\tau_0 + \frac{1}{R_e} \right) f'''' - \left(M + \frac{1}{R_e} \right) f'' \right] \cos \alpha (x - t)
$$
$$
+ \left[\left(\tau_0 + \frac{1}{R_e} \right) g'''' - \left(M + \frac{1}{K} \right) g'' \right] \sin \alpha (x - t).
$$

$$(30)$$

Collecting coefficients of $\cos \alpha(x-t)$ and $\sin \alpha(x-t)$ on either side of (30), two differential equations for $f(y)$ and $g(y)$ are obtained as follows:

$$
- \alpha \left[\left(\left(1 - \frac{L}{\lambda_1} \right) + \frac{\cosh \lambda_1 y}{\cosh \lambda_1} \right) g'' + L \frac{\cosh \lambda_1 y}{\cosh \lambda_1} g \right]
$$
$$
= \left[\left(\tau_0 + \frac{1}{R_e} \right) f'''' - \left(M + \frac{1}{K} \right) f'' \right],
$$
$$
\alpha \left[\left(\left(1 - \frac{L}{\lambda_1} \right) + \frac{\cosh \lambda_1 y}{\cosh \lambda_1} \right) f'' + L \frac{\cosh \lambda_1 y}{\cosh \lambda_1} f \right]
$$
$$
= \left[\left(\tau_0 + \frac{1}{R_e} \right) g'''' - \left(M + \frac{1}{K} \right) g'' \right].
$$

$$(31)$$

Equation (31) can be simplified by assuming that

$$
f(y) = f_0 + \alpha f_1 + \cdots ,
$$
$$
g(y) = g_0 + \alpha g_1 + \cdots .
$$

$$(32)$$

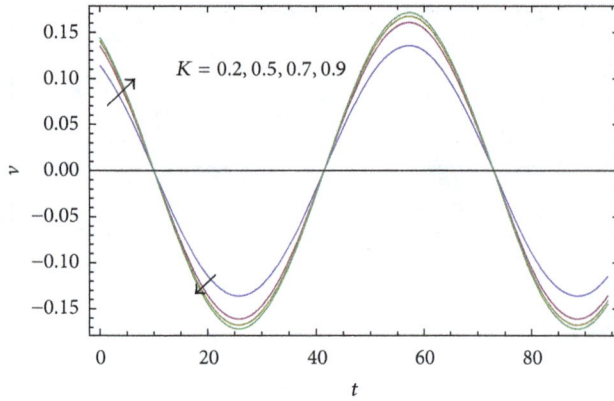

FIGURE 7: The velocity v is plotted against the time t for different values of k at $\alpha = 0.1$, $\varepsilon = 0.1$, $M = 5$, $\tau_o = 2$.

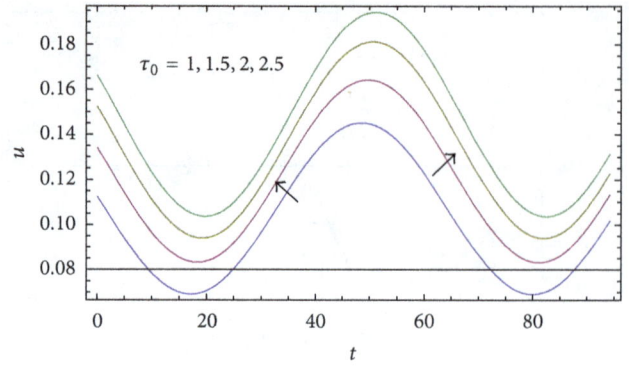

FIGURE 9: The velocity u is plotted against the time t for different values of τ_o at $\alpha = 0.1$, $\varepsilon = 0.1$, $M = 5$, $k = 0.9$.

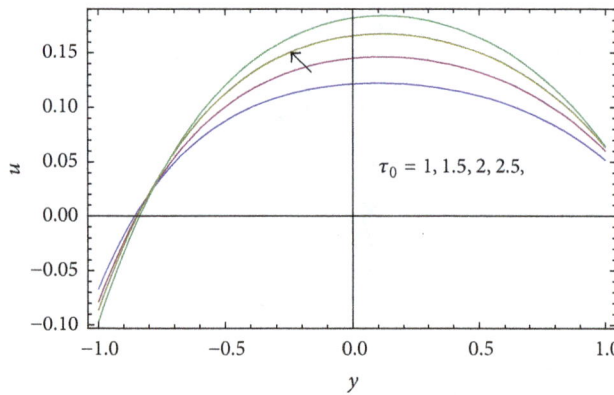

FIGURE 8: The velocity u is plotted against the distance for different values of τ_o at $\alpha = 0.1$, $\varepsilon = 0.1$, $M = 5$, $k = 0.9$.

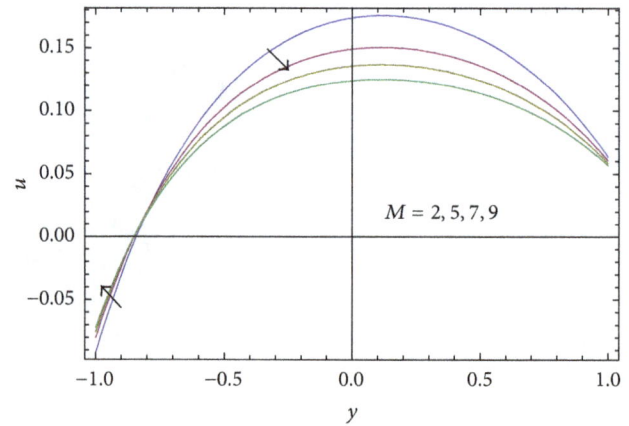

FIGURE 10: The velocity u is plotted against the distance for different values of M at $\alpha = 0.1$, $\varepsilon = 0.1$, $\tau_o = 2$, $k = 0.9$.

Substituting (32) into (31), and equating terms in α, the following ordinary differential equations are obtained for f_o, f_1, g_0, and g_1, respectively:

$$f_0'''' - \lambda_1^2 f_0'' = 0,$$

$$g_0'''' - \lambda_1^2 g_0'' = 0,$$

$$-\left[\left(\left(1 - \frac{L}{\lambda_1}\right) + \frac{\cosh \lambda_1 y}{\cosh \lambda_1}\right) g_0'' - L\frac{\cosh \lambda_1 y}{\cosh \lambda_1} g_0\right]$$

$$= \left[\left(\tau_0 + \frac{1}{R_e}\right) f_1'''' - \left(M + \frac{1}{K}\right) f_1'\right], \qquad (33)$$

$$\left[\left(\left(1 - \frac{L}{\lambda_1}\right) + \frac{\cosh \lambda_1 y}{\cosh \lambda_1}\right) f_0'' - L\frac{\cosh \lambda_1 y}{\cosh \lambda_1} f_0\right]$$

$$= \left[\left(\tau_0 + \frac{1}{R_e}\right) g_1'''' - \left(M + \frac{1}{K}\right) g_1''\right]$$

with boundary conditions:

$$f_0(1) = 1, \qquad f_0'(1) = -A,$$

$$f_0(-1) = -1, \qquad f_0'(-1) = -A,$$

$$g_0(1) = 0, \qquad g_0'(1) = 0,$$

$$g_0(-1) = 0, \qquad g_0'(-1) = 0,$$

$$f_1(1) = 0, \qquad f_1'(1) = 0, \qquad (34)$$

$$f_1(-1) = 0, \qquad f_1'(-1) = 0,$$

$$g_1(1) = 0, \qquad g_1'(1) = 0,$$

$$g_1'(-1) = 0, \qquad g_1(-1) = 0.$$

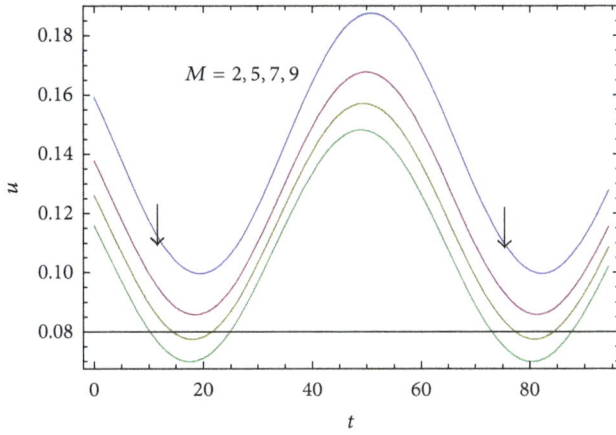

FIGURE 11: The velocity u is plotted against the time t for different values of M at $\alpha = 0.1,\ \varepsilon = 0.1,\ \tau_o = 2,\ k = 0.9$.

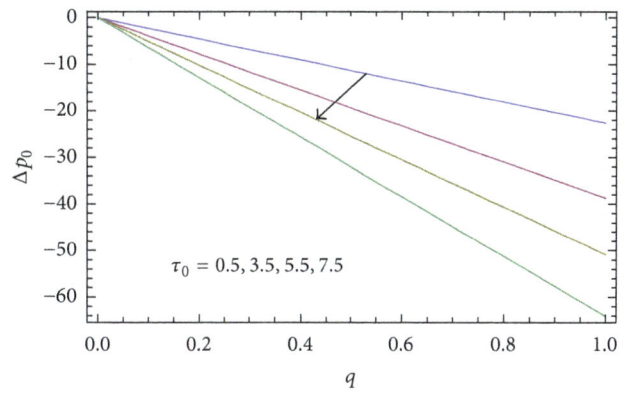

FIGURE 14: Δp_o is plotted against q for different values of τ_o at $\alpha = 0.1,\ \varepsilon = 0.1,\ M = 5,\ k = 0.2$.

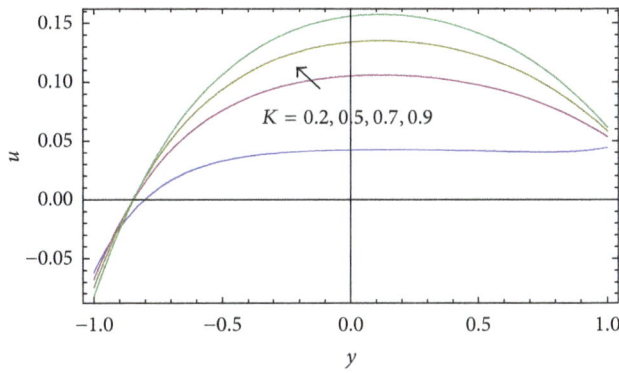

FIGURE 12: The velocity u is plotted against the distance for different values of k at $\alpha = 0.1,\ \varepsilon = 0.1,\ M = 5,\ \tau_o = 2$.

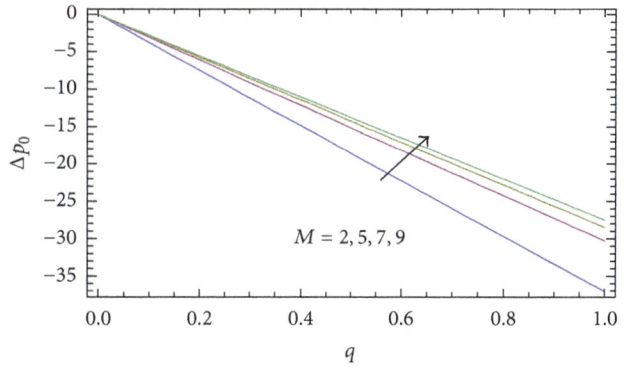

FIGURE 15: Δp_o is plotted against q for different values of τ_o at $\alpha = 0.1,\ \varepsilon = 0.1, M = 5, k = 0.2$.

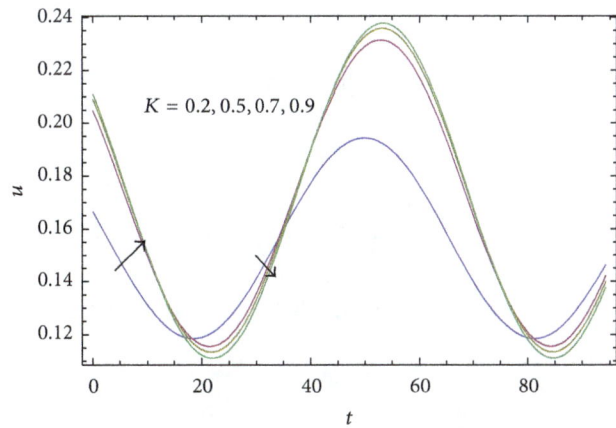

FIGURE 13: The velocity u is plotted against the time t for different values of k at $\alpha = 0.1,\ \varepsilon = 0.1,\ M = 5,\ \tau_o = 2$.

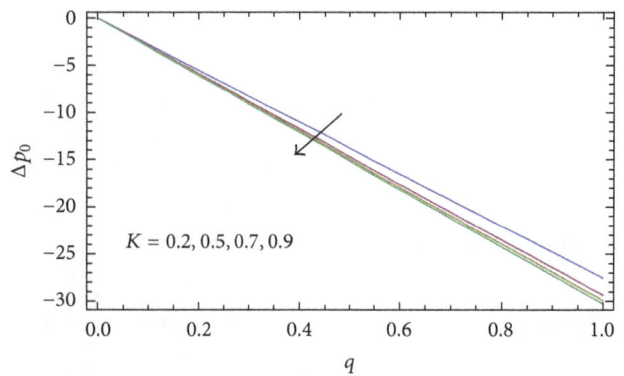

FIGURE 16: Δp_o is plotted against q for different values of k at $\alpha = 0.1,\ \varepsilon = 0.1, M = 5, \tau_o = 2$.

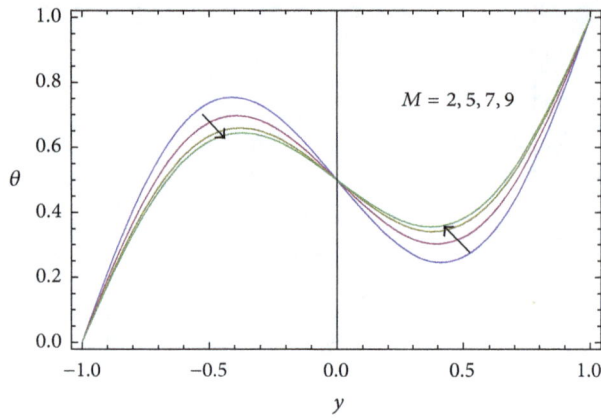

FIGURE 17: The temperature θ is plotted against the distance for different values of M at $\alpha = 0.1$, $\varepsilon = 0.1$, $\tau_o = 2$, $k = 0.9$.

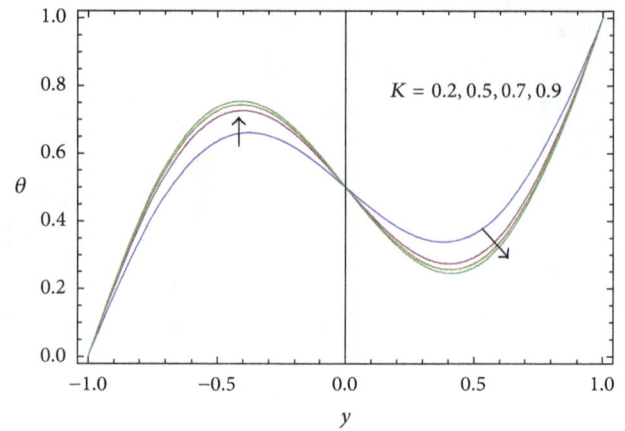

FIGURE 19: The temperature θ is plotted against the distance for different values of k at $\alpha = 0.1$, $\varepsilon = 0.1$, $\tau_o = 2$, $M = 5$.

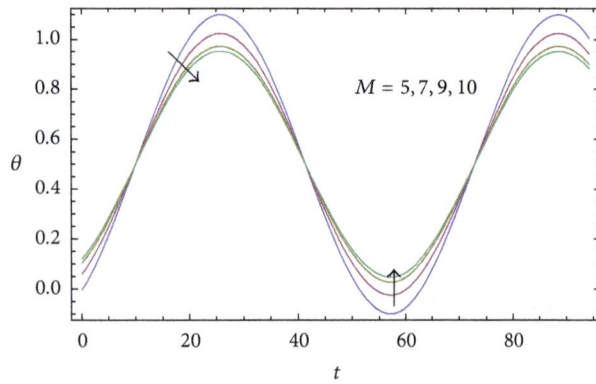

FIGURE 18: The temperature θ is plotted against the time t for different values of M at $\alpha = 0.1$, $\varepsilon = 0.1$, $\tau_o = 2$, $k = 0.9$.

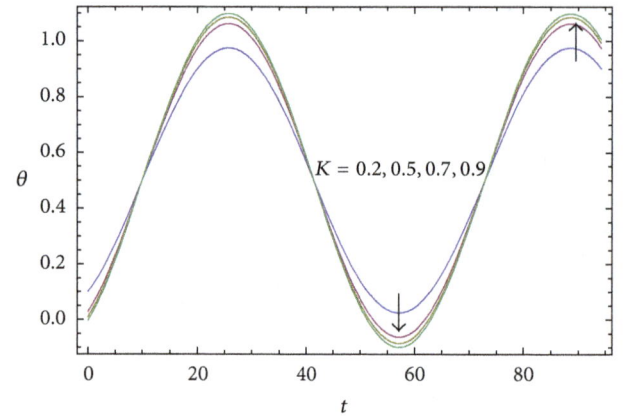

FIGURE 20: The temperature θ is plotted against the time t for different values of k at $\alpha = 0.1$, $\varepsilon = 0.1$, $\tau_o = 2$, $M = 5$.

Solving (33) by using (34), we get

$$f_0(y) = \lambda_2 y - \lambda_3 \sinh \lambda_1 y, \tag{35}$$

$$f_1(y) = 0, \qquad g_0(y) = 0, \tag{36}$$

$$g_1(y) = \lambda_7 y \cosh \lambda_1 y + \left(\lambda_8 + \lambda_9 y^2\right) \sinh \lambda_1 y$$
$$+ \lambda_{10} \sinh 2\lambda_1 y + \lambda_{11} y. \tag{37}$$

Substituting (35) in (29), we get

$$\psi_1(x, y, t) = \left(\lambda_2 y - \lambda_3 \sinh \lambda_1 y\right) \cos \alpha (x - t)$$
$$+ \alpha \sin \alpha (x - t)$$
$$\times \left(\lambda_7 y \cosh \lambda_1 y + \left(\lambda_8 + \lambda_9 y^2\right) \sinh \lambda_1 y\right.$$
$$\left. + \lambda_{10} \sinh 2\lambda_1 y + \lambda_{11} y\right). \tag{38}$$

Substituting (25) and (38) in (14), we get

$$\psi(x, y, t) = \frac{-L}{\lambda_1} \left(y - \frac{\sinh \lambda_1 y}{\lambda_1 \cosh \lambda_1}\right)$$
$$+ \varepsilon \left(\left(\lambda_2 y - \lambda_3 \sinh \lambda_1 y\right) \cos \alpha (x - t)\right.$$
$$+ \alpha \sin \alpha (x - t)$$
$$\times \left(\lambda_7 y \cosh \lambda_1 y + \left(\lambda_8 + \lambda_9 y^2\right) \sinh \lambda_1 y\right.$$
$$\left.\left. + \lambda_{10} \sinh 2\lambda_1 y + \lambda_{11} y\right)\right). \tag{39}$$

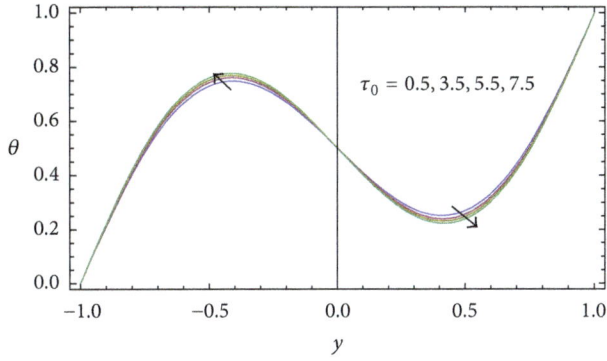

FIGURE 21: The temperature θ is plotted against the distance for different values of τ_o at $\alpha = 0.1$, $\varepsilon = 0.1$, $k = 0.9$, $M = 5$.

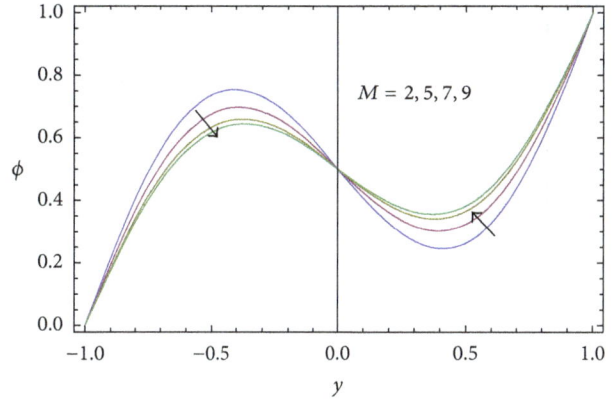

FIGURE 23: The concentration distribution ϕ is plotted against the distance for different values of M at $\alpha = 0.1$, $\varepsilon = 0.1$, $\tau_o = 2$, $k = 0.9$.

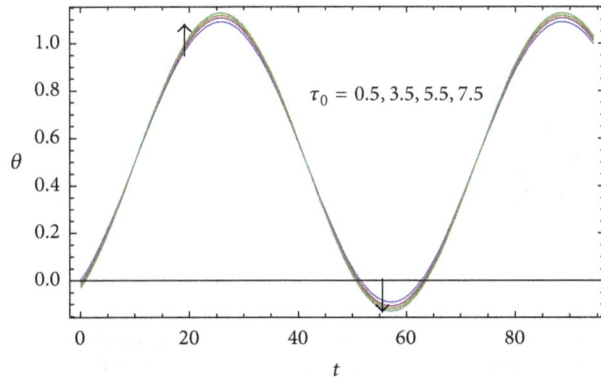

FIGURE 22: The temperature θ is plotted against the time t for different values of τ_o at $\alpha = 0.1$, $\varepsilon = 0.1$, $k = 0.9$, $M = 5$.

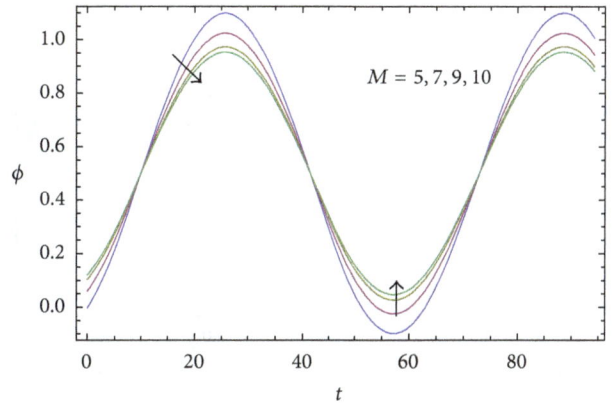

FIGURE 24: The concentration distribution ϕ is plotted against the time t for different values of M at $\alpha = 0.1$, $\varepsilon = 0.1$, $\tau_o = 2$, $K = 0.9$.

The velocity components can be written as

$$u(x, y, t) = \frac{-L}{\lambda_1} \left(1 - \frac{\cosh \lambda_1 y}{\cosh \lambda_1} \right)$$

$$+ \varepsilon \big((\lambda_2 - \lambda_3 \lambda_1 \cosh \lambda_1 y) \cos \alpha (x - t)$$

$$+ \alpha \sin \alpha (x - t)$$

$$\times (\lambda_7 \cosh \lambda_1 y + \lambda_1 \lambda_7 y \sinh \lambda_1 y$$

$$+ \lambda_1 (\lambda_8 + \lambda_9 y^2) \cosh \lambda_1 y$$

$$+ 2\lambda_1 \lambda_{10} \cosh 2\lambda_1 y$$

$$+ \lambda_{11} + 2\lambda_9 y \sinh \lambda_1 y) \big).$$

(40)

$$v(x, y, t) = \alpha \varepsilon \left[(\lambda_2 y - \lambda_3 \sinh \lambda_1 y) \sin \alpha (x - t) \right]$$

Substituting (39) in (17), (18), (22), and (23) and using (24) we get

$$\theta_o (y) = \frac{y + 1}{2}, \qquad \varphi_o (y) = \frac{y + 1}{2},$$

$$\theta_1 (x, y, t) = \Big(\big(-\lambda_1^2 \lambda_{12} y + \lambda_1^2 \lambda_{12} y^3 + 6\lambda_{13} y \sinh \lambda_1$$

$$-6\lambda_{13} \sinh \lambda_1 y \big) \big(6\lambda_1^2 \big)^{-1} \Big)$$

$$\times \cos \alpha (x - t),$$

$$\varphi_1 (x, y, t) = \Big(\big(-\lambda_1^2 \lambda_{14} y + \lambda_1^2 \lambda_{12} y^3 + 6\lambda_{15} y \sinh \lambda_1$$

$$-6\lambda_{15} \sinh \lambda_1 y \big) \big(6\lambda_1^2 \big)^{-1} \Big) \cos \alpha (x - t).$$

(41)

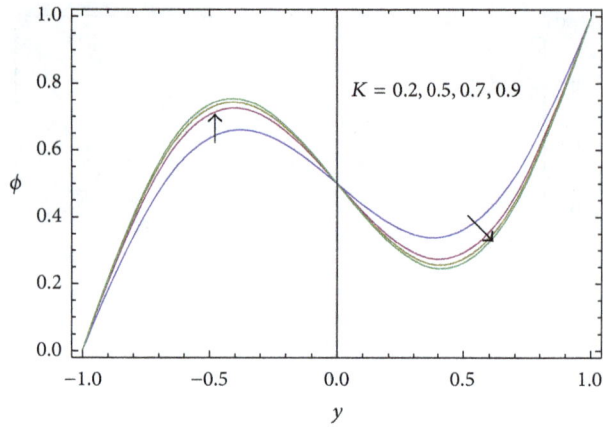

FIGURE 25: The concentration distribution ϕ is plotted against the distance for different values of k at $\alpha = 0.1$, $\varepsilon = 0.1$, $\tau_o = 2$, $M = 5$.

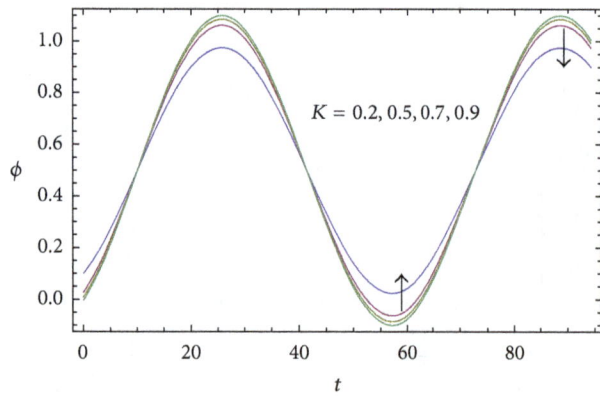

FIGURE 27: The concentration distribution ϕ is plotted against the distance for different values of τ_o at $\alpha = 0.1$, $\varepsilon = 0.1$, $M = 5$, $k = 0.9$.

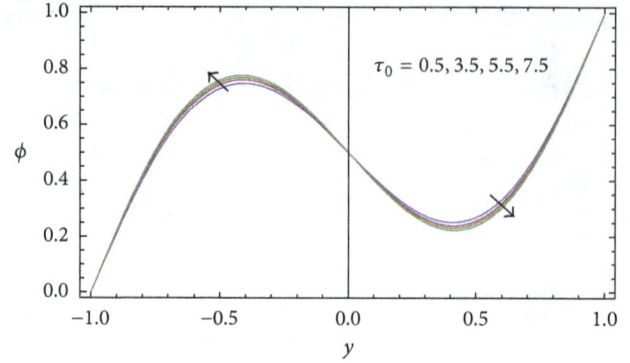

FIGURE 26: The concentration distribution ϕ is plotted against the time t for different values of k at $\alpha = 0.1$, $\varepsilon = 0.1$, $\tau_o = 2$, $M = 5$.

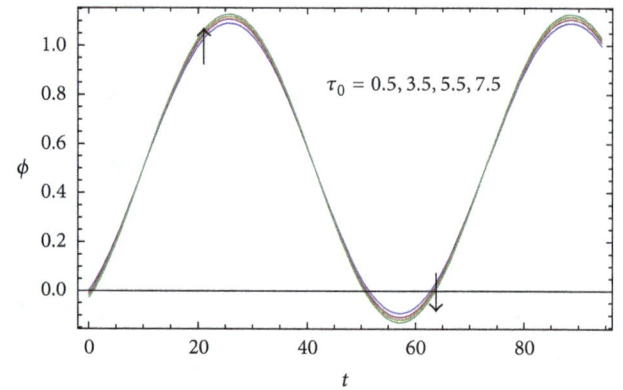

FIGURE 28: The concentration distribution ϕ is plotted against the time t for different values of τ_0 at $\alpha = 0.1$, $\varepsilon = 0.1$, $K = 0.9$, $M = 5$.

Substituting (41) in (14), we get

$$
\theta(x, y, t)
$$
$$
= \left(\frac{y+1}{2} \right)
$$
$$
+ \varepsilon\alpha\left(\left(-\lambda_1^2\lambda_{12}y + \lambda_1^2\lambda_{12}y^3 \right.\right.
$$
$$
\left.\left. +6\lambda_{13}y\sinh\lambda_1 - 6\lambda_{13}\sinh\lambda_1 y \right)\left(6\lambda_1^2\right)^{-1} \right)
$$
$$
\times \cos\alpha(x-t),
$$

$$(42)$$

$$
\phi(x, y, t)
$$
$$
= \frac{y+1}{2} + \varepsilon\alpha\left(\left(-\lambda_1^2\lambda_{14}y + \lambda_1^2\lambda_{14}y^3 \right.\right.
$$

$$
+6\lambda_{15}y\sinh\lambda_1 - 6\lambda_{15}\sinh\lambda_1 y)
$$
$$
\times\left(6\lambda_1^2\right)^{-1} \cos\alpha(x-t),
$$

$$(43)$$

where

$$
A = \psi_{0yy}(1) = \frac{L\sinh\lambda_1}{\lambda_1\cosh\lambda_1},
$$
$$
\lambda_2 = \frac{A\sinh\lambda_1 + \lambda_1\cosh\lambda_1}{-\sinh\lambda_1 + \lambda_1\cosh\lambda_1},
$$
$$
\lambda_3 = \frac{1+A}{-\sinh\lambda_1 + \lambda_1\cosh\lambda_1},
$$
$$
\lambda_4 = \frac{\lambda_1^2\lambda_3(1-(L/\lambda_1))}{(\tau_0 + (1/R_e))},
$$
$$
\lambda_5 = \frac{(L+\lambda_1^2)\lambda_2}{(\tau_0 + (1/R_e))\cosh\lambda_1}, \qquad \lambda_6 = \frac{L\lambda_2}{\cosh\lambda_1},
$$

$$\lambda_7 = \Big((24\lambda_1\lambda_4 + 60\lambda_6)\sinh\lambda_1$$

$$- \left(24\lambda_1^2\lambda_4 + 60\lambda_1\lambda_6\right)\cosh\lambda_1\Big)$$

$$\times \left(48\lambda_1^4\left(\lambda_1\cosh\lambda_1 - \sinh\lambda_1\right)\right)^{-1},$$

$$\lambda_8 = \left(\left(24\lambda_1^2\lambda_4 + 48\lambda_1\lambda_6\right)\sinh\lambda_1\right.$$

$$- 12\lambda_1^2\lambda_6\cosh\lambda_1 - 2\lambda_5\sinh2\lambda_1$$

$$\left.+4\lambda_1\lambda_5\cosh2\lambda_1\right)$$

$$\times \left(48\lambda_1^4\left(\lambda_1\cosh\lambda_1 - \sinh\lambda_1\right)\right)^{-1},$$

$$\lambda_9 = \frac{\left(12\lambda_1^2\lambda_6\right)\cosh\lambda_1 - 12\lambda_1\lambda_6\sinh\lambda_1}{48\lambda_1^4\left(\lambda_1\cosh\lambda_1 - \sinh\lambda_1\right)},$$

$$\lambda_{10} = \frac{\left(2\lambda_5\right)\sinh\lambda_1 - 2\lambda_1\lambda_5\cosh\lambda_1}{48\lambda_1^4\left(\lambda_1\cosh\lambda_1 - \sinh\lambda_1\right)},$$

$$\lambda_{11} = \left(\left(24\lambda_1^2\lambda_4\right) - 12\lambda_1\lambda_5\sinh2\lambda_1 + 3\lambda_1\lambda_5\sinh\lambda_1\right.$$

$$- \lambda_1\lambda_5\sinh3\lambda_1 + 48\lambda_1\lambda_6$$

$$\left.+ \left(12\lambda_1\lambda_6 - 30\lambda_6\right)\cosh2\lambda_1\right)$$

$$\times \left(48\lambda_1^4\left(\lambda_1\cosh\lambda_1 - \sinh\lambda_1\right)\right)^{-1},$$

$$\lambda_{12} = \frac{\lambda_2\tan\alpha\,(x-t)}{C_p}, \qquad \lambda_{13} = \frac{\lambda_3\tan\alpha\,(x-t)}{C_p},$$

$$\lambda_{14} = \frac{\lambda_2\sin\alpha\,(x-t)}{D_m\cos\alpha\,(x-t)}, \qquad \lambda_{15} = \frac{\lambda_3\sin\alpha\,(x-t)}{D_m\cos\alpha\,(x-t)}.$$

$$(44)$$

3. Results and Discussion

In this work, we have studied the effect of different parameters of the considered problem on the solutions of the momentum, heat, and mass equations. This discussion is illustrated graphically through a set of Figures 2–28. Since Figures 2 and 3 illustrated the influence of the Casson parameter τ_0 on the velocity component v, hence we noticed that the velocity component v increases with the increase of the Casson parameter τ_0 for $0 \le y \le 1$ and decreases for $-1 \le y \le 0$. The effect of the magnetic parameter M on the velocity component v is shown in Figures 4 and 5. These figures reveal that the velocity component v decreases with the increase of M at $0 \le y \le 1$ and increases at $-1 \le y \le 0$. Figures 6 and 7 depicted the behavior of permeability parameter K on the velocity component v. It is noticed that the velocity component v decreases with K in the region $-1 \le y \le 0$, and it increases in the region $0 \le y \le 1$. Figures 8 and 9 showed the effect of τ_0 on the velocity component u; it is clear that the velocity component u increases with increasing of τ_0. Also, the velocity component u decreases when the magnetic parameter M increases, then shown through Figures 10 and 11. From Figures 12 and 13, since the motion is sinusoidal,

we have seen that the longitudinal velocity u increases or decreases as the permeability parameter K increases. Figures 14 and 16 illustrated the influence of τ_0 and K on the pressure rise Δp_0. These figures show that the pressure rise Δp_0 decreases with the increase of both τ_0 and K. Figure 15 displayed the effected of the magnetic parameter M on the pressure rise Δp_0; it is noticed that the magnitude of Δp_0 increases with M. We can see from Figures 17 and 18 that the temperature θ decreases when M increases in the interval $-1 \le y \le 0$, while it increases in the interval $0 \le y \le 1$. We observed from Figures 19, 20, 21, and 22 that the temperature θ increases when K and τ_0 increase in the interval $-1 \le y \le 0$, and it decreases in the interval $0 \le y \le 1$. In Figures 23 and 24, it is seen that the concentration distribution ϕ decreases with M in the region $-1 \le y \le 0$, but it increases in the region $0 \le y \le 1$. The concentration distribution ϕ increases when K and τ_0 increase in the interval $-1 \le y \le 0$, and it decreases in the interval $0 \le y \le 1$, this is shown in Figures 25, 26, 27, and 28.

Conflict of Interests

The authors declare that there is no conflict of interests regarding the publication of this paper.

References

[1] P. Muthu, B. V. R. Kumar, and P. Chandra, "On the influence of wall properties in the peristaltic motion of micropolar fluid," *ANZIAM Journal*, vol. 45, no. 2, pp. 245–260, 2003.

[2] Y. C. Fung, C. S. Yih, and J. Asme, "Peristaltic transport," *Journal of Applied Mechanics*, vol. 35, no. 4, pp. 669–675, 1968.

[3] J. C. Burns and T. Parkes, "Peristaltic motion," *Journal of Fluid Mechanics*, vol. 29, pp. 731–743, 1970.

[4] A. V. Mernone and J. N. Mazumdar, "Biomathematical modelling of physiological fluids using a Casson fluid with emphasis to peristalsis," *Australasian Physical and Engineering Sciences in Medicine*, vol. 23, no. 3, pp. 94–100, 2000.

[5] A. V. Mernone, J. N. Mazumdar, and S. K. Lucas, "A mathematical study of peristaltic transport of a Casson fluid," *Mathematical and Computer Modelling*, vol. 35, no. 7-8, pp. 895–912, 2002.

[6] Kh. S. Mekheimer and T. H. Al-Arabi, "Nonlinear peristaltic transport of MHD flow through a porous medium," *International Journal of Mathematics and Mathematical Sciences*, vol. 2003, no. 26, pp. 1663–1682, 2003.

[7] Kh. S. Mekheimer, "Non-linear peristaltic transport of magnetohydrodynamic flow in an inclined planar channel," *Arabian Journal for Science and Engineering A*, vol. 28, no. 2, pp. 183–201, 2003.

[8] P. Nagarani and G. Sarojamma, "Peristaltic transport of a Casson fluid in an asymmetric channel," *Australasian Physical and Engineering Sciences in Medicine*, vol. 27, no. 2, pp. 49–59, 2004.

[9] P. Nagarani, "Peristaltic transport of a casson fluid in an inclined channel," *Korea Australia Rheology Journal*, vol. 22, no. 2, pp. 105–111, 2010.

[10] E. F. Elshehawey, N. T. Eldabe, E. M. Elghazy, and A. Ebaid, "Peristaltic transport in an asymmetric channel through a

porous medium," *Applied Mathematics and Computation*, vol. 182, no. 1, pp. 140–150, 2006.

[11] R. Bharagava, H. S. Takhar, S. Rawat, A. Tasveer Beg, and O. Anwar Beg, "finite element solutions for non-Newtonian pulsatile flow in a non Darican porous medium conduit," *Nonlinear Analysis Modeling and Control*, vol. 12, no. 3, pp. 317–327, 2007.

[12] Kh. S. Mekheimer and Y. Abd elmaboud, "The influence of heat transfer and magnetic field on peristaltic transport of a Newtonian fluid in a vertical annulus: application of an endoscope," *Physics Letters A: General, Atomic and Solid State Physics*, vol. 372, no. 10, pp. 1657–1665, 2008.

[13] S. Nadeem, N. S. Akbar, N. Bibi, and S. Ashiq, "Influence of heat and mass transfer on peristaltic flow of a third order fluid in a diverging tube," *Communications in Nonlinear Science and Numerical Simulation*, vol. 15, no. 10, pp. 2916–2931, 2010.

[14] Y. Abd elmaboud and K. S. Mekheimer, "Non-linear peristaltic transport of a second-order fluid through a porous medium," *Applied Mathematical Modelling*, vol. 35, no. 6, pp. 2695–2710, 2011.

[15] Y. Abd Elmaboud, "Thermomicropolar fluid flowin a porous channel with peristalsis," *Journal of Porous Media*, vol. 14, no. 11, pp. 1033–1045, 2011.

[16] N. T. El-dabe, S. N. Sallam, M. A. Mohamed, M. Y. Abo Zaid, and A. Abd-Emonem, "Magnetohydrodynamic peristaltic motion with heat and mass transfer of a Jeffery fluid in a tube through porous medium," *Innovative System Design and Engineering*, vol. 2, no. 4, 2011.

[17] M. Mustafa, S. Hina, T. Hayat, and A. Alsaedi, "Influence of wall properties on the peristaltic flow of a nanofluid: Analytic and numerical solutions," *International Journal of Heat and Mass Transfer*, vol. 55, no. 17-18, pp. 4871–4877, 2012.

[18] O. Anwr Beg and D. Tripathi, "Mathematica simulation of peristaltic pumoing with double-diffusive convection in nanofluids: a bio-nano-engineering model," *The Journal of Nanoengineering and Nanosystems*, 2012.

[19] N. T. El-dabe, S. M. Elshaboury, A. A. Hasan, and M. A. Elogail, "MHD Peristaltic flow of a couple stress fluids with heat and mass transfer through a porous medium," *Innovative System Design and Engineering*, vol. 3, no. 5, 2012.

[20] N. T. El-dabe, K. A. Kamel, M. Galila abd-Allah, and S. F. Ramadan, "Heat absorption and chemical reaction effects on Peristaltic motion of micropolar fluid through a porous medium in the presence of magnetic field," *The African Journal of Mathematics and Computer Science Research*, vol. 6, no. 5, pp. 94–101, 2013.

[21] A. Ebaid and H. Emad Aly, "Exact analytical solution of the peristaltic nanofluids flow in an asymmetric channel with flexible walls and slip condition: application to cancer treatment," *Computational and Mathematical Methods in Medicine*, vol. 2013, Article ID 825376, 8 pages, 2013.

[22] H. Emad and A. Ebaid, "New analytical and numerical solutions for mixed convection boundary- layer nanofluid flow along an inclined plate embedded in porous medium," *Journal of Applied Mathematics*, vol. 2013, Article ID 219486, 7 pages, 2013.

[23] Y. Abd Elmaboud, "Unsteady flow of magneto thermomicropolar fluid in a porous channel with peristalsis: unsteady separation," *Journal of Heat Transfer-Transactions of the ASME*, vol. 135, no. 7, Article ID 072602, 2013.

[24] S. Noreen, T. Hayat, A. Alsaedi, and Qasim, "Mixed convection heat and mass transfer in perisltic flow," *Indian Journal of Physics*, vol. 87, no. 9, pp. 889–896, 2013.

[25] T. Hayat, S. Noreen, and M. Qasim, "Influence of heat and mass transfer on the peristaltic transport of a phan-thien-tanner fluid," *Zeitschrift fur Naturforschung A*, vol. 68, pp. 751–758, 2013.

[26] S. Noreen, "Correction: mixed convection peristaltic flow of third order nanofluid with an induced magnetic field," *PLoS ONE*, vol. 8, no. 12, 2013.

[27] M. Nakamura and T. Sawada, "Numerical study on the flow of a non-Newtonian fluid through an axisymmetric stenosis," *Journal of Biomechanical Engineering*, vol. 110, no. 2, pp. 137–143, 1988.

Permissions

List of Contributors

K. Ramesh and M. Devakar
Department of Mathematics, Visvesvaraya National Institute of Technology, Nagpur 440010, India

Stephen Wan, Jason Leong, Te Ba, Arthur Lim and Chang Wei Kang
Institute of High Performance Computing, 1 Fusionopolis Way, No. 16-16 Connexis, Singapore 138632

Mary D. Saroka
United Technologies Research Center, 411 Silver Lane, MS 129-19, East Hartford, CT 06108, USA

Nasser Ashgriz
Department of Mechanical and Industrial Engineering, University of Toronto, Toronto, ON, Canada M5S 3G8

Samir Kumar Nandy
Department of Mathematics, A.K.P.C Mahavidyalaya, Bengai, Hooghly 712 611, India

Swati Mukhopadhyay
Department of Mathematics, The University of Burdwan, West Bengal 713104, India

Haihua Yuan, Yang Liu, Wanqian Wei
Department of Biology and Guangdong Provincial Key Laboratory of Marine Biotechnology, Shantou University, Shantou, Guangdong 515063, China

Yongjie Zhao
Department of Mechanical Engineering, College of Engineering, Shantou University, Shantou, Guangdong 515063, China

Macha Madhu and Naikoti Kishan
Department of Mathematics, Osmania University, Hyderabad, Telangana 500007, India

Ramesh Chand
Department of Mathematics, Government College Dhaliara, Himachal Pradesh 177103, India

G. C. Rana
Department of Mathematics, Government College Nadaun, Himachal Pradesh 177033, India

J. Prakash
Department of Mathematics, University of Botswana, Private Bag 0022, Gaborone, Botswana

S. GouseMohiddin
Department of Mathematics, Madanapalle Institute of Technology & Science, Madanapalle 517325, Andhra Pradesh, India

S. Vijaya Kumar Varma
Department of Mathematics, Sri Venkateswara University, Tirupati 517502, Andhra Pradesh, India

Mahinder Singh
Department of Mathematics, Govt. Post Graduate College, Seema (Rohru), Shimla, India

Rajesh Kumar Gupta
Department of Mathematics, Lovely School of Engineering and Technology, Lovely Professional University, Phagwara, India

Takashi Kitaura and Toshio Tagawa
Tokyo Metropolitan University, 6-6, Asahigaoka, Hinoshi 191-0065, Japan

M. M. Hamza
Department of Mathematics, Usmanu Danfodiyo University, PMB 2346, Sokoto, Nigeria

I. G. Usman and A. Sule
Department of Mathematics, Zamfara State College of Education, PMB 1002, Maru, Nigeria

Naser Moosavian
Civil Engineering Department, University of Torbat-e-Heydarieh, Torbat-e-Heydarieh, Iran

Mohammad Reza Jaefarzadeh
Civil Engineering Department, Ferdowsi University of Mashhad, Mashhad, Iran

Mahinder Singh
Department of Mathematics, Government Post Graduate College Seema (Rohru), Shimla District, Himachal Pradesh 171207, India

Chander Bhan Mehta
Department of Mathematics, Centre of Excellence, Government Degree College Sanjauli, Shimla District, Himachal Pradesh 171006, India

Yoshiaki Haneda, Akiko Souma and Hideo Kurasawa
Nagano National College of Technology, Nagano 381-8550, Japan

Shouichiro Iio and Toshihiko Ikeda
Shinshu University, Nagano 380-8553, Japan

D. R. V. S. R. K. Sastry
Aditya Engineering College, Surampalem, Andhra Pradesh 533437, India

A. S. N. Murti
GITAM University, Visakhapatnam, Andhra Pradesh 530045, India

Zhi Shang, Jing Lou and Hongying Li
Institute of High Performance Computing (IHPC), Agency for Science, Technology and Research (A*STAR), 1 Fusionopolis Way, No. 16-16 Connexis, Singapore 138632

Bikash Sahoo
Department of Mathematics, National Institute of Technology Rourkela, Rourkela 769008, India

Sébastien Poncet
Faculté de Génie, Université de Sherbrooke, Sherbrooke, QC, Canada J1K 2R1
Aix-Marseille Université, CNRS, École Centrale, Laboratoire M2P2 UMR 7340, 13451 Marseille, France

Fotini Labropulu
Department of Mathematics, Luther College, University of Regina, Regina, SK, Canada S4S 0A2

Jafar M. Hassan, Wahid S.Mohammed and Wissam H. Alawee
Department of Mechanical Engineering, University of Technology, Baghdad, Iraq

Thamer A. Mohamed
Department of Civil Engineering, Faculty of Engineering, Universiti Putra Malaysia (UPM), 43400 Serdang, Selangor, Malaysia

Snehamoy Majumder, Debajit Saha and Partha Mishra
Department of Mechanical Engineering, Jadavpur University, Kolkata 700 032, West Bengal, India

R. N. Barik
Department of Mathematics, Trident Academy of Technology, Infocity, Bhubaneswar, Odisha 751024, India

G. C. Dash
Department of Mathematics, S.O.A. University, Bhubaneswar, Odisha 751030, India

M. Kar
Department of Mathematics, Christ College, Cuttack, Odisha, India

S. D. Ram
Department of Mathematics, Mata Sundri College, University of Delhi, Delhi 110002, India

R. Singh and L. P. Singh
Department of Applied Mathematics, Indian Institute of Technology (BHU), Varanasi 221005, India

M. A. A. Hamad, S. M. Abd El-Gaied
Mathematics Department, Faculty of Science, Assiut University, Assiut 71516, Egypt

W. A. Khan
Department of Engineering Sciences, PN Engineering College, National University of Science, Pakistan

Yazan Taamneh
Department of Mechanical Engineering, Tafila Technical University, P.O. Box 179, Tafila 66110, Jordan

Reyad Omari
Department of Mathematics, Al-Balqa Applied University, Irbid University College, P.O. Box 19117, Irbid 19110, Jordan

M. Kothandapani
Department of Mathematics, University College of Engineering Arni (A Constituent College of Anna University, Chennai), Arni, Tamil Nadu 632 326, India

J. Prakash
Department of Mathematics, Arulmigu Meenakshi Amman College of Engineering, Vadamavandal, Tamil Nadu 604 410, India

V. Pushparaj
Department of Mathematics, C. Abdul Hakeem College of Engineering & Technology, Melvisharam, Tamil Nadu 632 509, India

M. Subbiah
Department of Mathematics, Pondicherry University, Kalapet, Pondicherry 605014, India

M. S. Anil Iype
Geomagnetic Observatory, Indian Institute of Geomagnetism, Pondicherry University Campus, Kalapet, Pondicherry 605014, India

Mir Golam Rabby, Sumaia Parveen Shupti and Md. MamunMolla
School of Engineering & Applied Science, Department of Electrical & Computer Engineering, North South University, Dhaka 1229, Bangladesh

Nabil T.M. Eldabe
Department of Mathematics, Faculty of Education, Ain Shams University, Cairo 11566, Egypt

Bothaina M. Agoor and Heba Alame
Department of Mathematics, Faculty of Science, Fayoum University, P.O. Box 63514, Fayoum, Egypt

www.ingramcontent.com/pod-product-compliance
Lightning Source LLC
Chambersburg PA
CBHW080459200326
41458CB00012B/4033